Learn Audio Electronics with Arduino

Learn Audio Electronics with Arduino: Practical Audio Circuits with Arduino Control teaches the reader how to use Arduino to control analogue audio circuits and introduces electronic circuit theory through a series of practical projects, including a MIDI drum controller and an Arduino-controlled two-band audio equalizer amplifier.

Learn Audio Electronics with Arduino provides all the theoretical knowledge needed to design, analyse, and build audio circuits for amplification and filtering, with additional topics like C programming being introduced in a practical context for Arduino control. The reader will learn how these circuits work and also how to build them, allowing them to progress to more advanced audio circuits in the future. Beginning with electrical fundamentals and control systems, DC circuit theory is then combined with an introduction to C programming to build Arduino-based systems for audio (tone sequencer) and MIDI (drum controller) output. The second half of the book begins with AC circuit theory to allow analogue audio circuits for amplification and filtering to be analysed, simulated, and built. These circuits are then combined with Arduino control in the final project – an Arduino-controlled two-band equalizer amplifier.

Building on high-school physics and mathematics in an accessible way, *Learn Audio Electronics with Arduino* is suitable for readers of all levels. An ideal tool for those studying audio electronics, including as a component within other fields of study, such as computer science, human-computer interaction, acoustics, music technology, and electronics engineering.

Charlie Cullen is Assistant Head of the School of Electrical & Electronic Engineering in the Technological University Dublin, Ireland.

Learn Audio Electronics with Arduino

Practical Audio Circuits with Arduino Control

Charlie Cullen

LONDON AND NEW YORK

First published 2020
by Routledge
2 Park Square, Milton Park, Abingdon, Oxon OX14 4RN

and by Routledge
605 Third Avenue, New York, NY 10017

Routledge is an imprint of the Taylor & Francis Group, an informa business

British Library Cataloguing-in-Publication Data
A catalogue record for this book is available from the British Library

Library of Congress Cataloging-in-Publication Data
A catalog record has been requested for this book

ISBN 13: 978-0-367-18664-7 (hbk)
ISBN 13: 978-0-367-18665-4 (pbk)
ISBN 13: 978-0-429-19749-9 (ebk)

DOI: 10.4324/9780429197499

Typeset in Times New Roman
by Newgen Publishing UK

Visit the eResources: www.routledge.com/9780367186654

'If you think of something clever or funny, tell them your father said it – it works better that way.'

– Charles J. Cullen Snr

You've done so much for all of us Dad – thank you.

Contents

Figures

Tables

Preface

The idea for this book first occurred while teaching audio electronics, where basic concepts were proving difficult to explain quickly enough to allow audio amplifier and filter projects to be built. Thus, many students would struggle with the fundamentals and then lose interest before the more interesting circuits could be discussed. There are many good books on audio electronics that require a prior knowledge of electronic circuit theory and conversely the books that discuss these theories do not cover audio in detail. Similarly, there are many good audio project tutorials available, but they do not introduce fundamentals like programming and system control. After trying to teach theory before project, then project before theory, it was becoming clear that a text combining both in a series of smaller steps would be more useful – the chapters in this book are those resulting steps.

The Arduino is a great learning resource, and this book tries to use it to get circuits built quickly whilst introducing audio electronics fundamentals. Part of the reason for this is the large user base for Arduino – there are many great resources available that significantly extend the concepts introduced in these chapters. In addition, DC circuit theory is much easier to learn when building circuits, and the Arduino can quickly extend these circuits into digital control systems. In this book, building MIDI circuits is a great way to put DC circuit theory into practice and there is significant scope to progress beyond the MIDI drum trigger that represents the first milestone project.

The second half of the book uses AC circuit theory to introduce capacitors, which are fundamental to time-varying signals and can often be the 'extra' components in an audio circuit. Although this is only an introductory text, amplification and filtering circuits are arguably much easier to learn when the role of capacitors in DC blocking, AC decoupling and load balancing is also understood. Many students spend significant amounts of time starting at schematics for well-known pedals and amplifiers, thinking that those circuits are unbelievably complex. In reality, these circuits often take simpler amplification and filtering principles and combine them with practical techniques for stability and noise reduction (mostly capacitors) – the reality is much less complex than the circuit may initially suggest. It is partly for this reason that some transistor theory is included in the text – though it is provided for information rather than application. Many classic effects pedals use transistors, but operational amplifiers are a much better option for modern circuits.

In combining audio circuits with Arduino control, the main aim of this book is to show how digital control of analogue signals is a rich area of potential investigation. It is argued that this is not well covered by existing texts, even though most commercial audio equipment incorporates these techniques. The final project in chapter 9 is much more complex (and challenging) than any of the previous chapters, but it helps to underline how quickly the Arduino can be used for control in an audio circuit. If you work through the book it is hoped that you will finish chapter 9 with lots of ideas for potential circuits of your own – this is where electronics becomes really interesting!

Acknowledgements

I would like to thank Shannon Neill, Hannah Rowe and Claire Margerison from Taylor & Francis for all their patience and assistance during this process – it has not been a linear production! I would also like to thank my research collaborators at Xperi (Ming, Ton, James and Ted) who as audio engineering experts have been invaluable in their support of this work. Within my academic institution, particular thanks go to Professor Michael Conlon for giving me the time and space needed to complete the book. On a personal level, I would like to thank my brother John for all his input and discussions during the writing of this book – it really helped keep me focussed! The rest of my family always deserve thanks for their help, and I would particularly like to thank my wife Clara for her invaluable support and guidance – I hope that Ella will be our best project yet.

Introduction

Welcome to *Learn Audio Electronics with Arduino*! In this book, we will learn about the fundamentals of audio electronic circuits and how to use the Arduino to control them. This introduction section will be brief, aiming only to provide you with the relevant information you need to begin learning about audio electronics, so please take some time to read through the sections on software tools and equipment requirements. This book assumes no prior knowledge of electronics and will provide all the information you need to progress your learning through a combination of theory and practice. The book is broadly divided into three areas – electronics concepts, Arduino control and audio circuits (Figure 1).

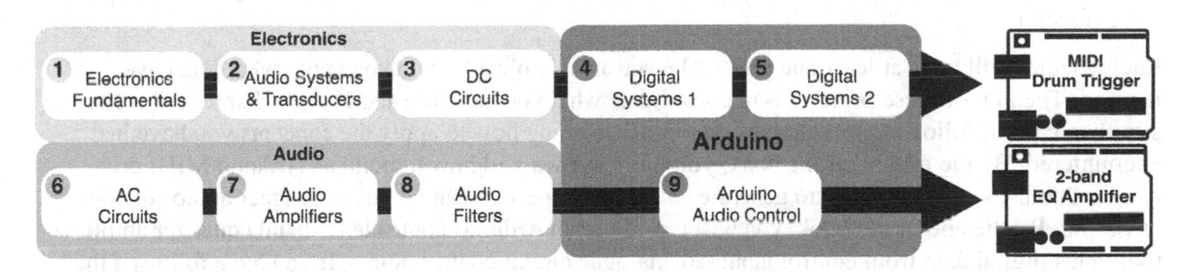

Figure 1 Textbook organizational structure. ***This diagram shows the three main areas of this textbook: electronics concepts, Arduino control and audio circuits. The first three chapters will introduce systems and DC circuit theory, whilst chapters 4 and 5 cover programming the Arduino to build the milestone project of a MIDI drum trigger. Chapters 6–8 introduce audio-related topics of AC circuit theory, amplification and filtering, with chapter 9 combining all of these areas with Arduino control to build the final milestone project of an Arduino controlled 2-band equalizer amplifier.***

The diagram shows how the book begins with an introduction to electronics concepts that then leads into systems analysis, where each project in the book will be built around a system that consists of inputs, processes and outputs. Chapter 3 then introduces the direct current (DC) circuit concepts of Ohm's Law and Kirchoff's Laws that will be used throughout the book. Serial and parallel combination of components can be difficult to understand at first as the scales and symbols used can take a little time to become familiar with, so don't be too concerned if this section of the book appears somewhat abstract – simple LED circuits can help to demonstrate how these fundamental concepts are used throughout electronics, so be sure not to skip any of the chapter projects.

The second section of the book looks at using the Arduino as a controller for electronics systems. The Arduino is used simply as a power source for the first three chapters, but in chapters 4 and 5 it forms the basis of all project circuits. These chapters aim to provide a short introduction to Arduino programming with the C language, and whilst not a full text on programming concepts the book provides practical examples of the three main instruction types (sequence, selection, iteration) and how they can be executed with larger groups of instructions (functions) that manipulate data (variables and arrays) within the system. Chapter 5 ends with the first of two milestone projects – an Arduino-controlled MIDI drum trigger.

The third section of the book begins in chapter 6 with alternating current (AC) theory, which is needed to work with the sinusoidal waveforms that are created by audio sensors that detect the compression of air (heard by humans as sound). AC theory can become mathematically complex, so the chapter aims to provide an overview of the essential core elements needed to work with audio circuits. This chapter should not be considered as a substitute for more dedicated study, particularly of concepts related to the phase of a signal (which are not covered in detail in this book). Chapter 7 looks at audio amplification, where transistor circuits (which can be found in many equipment schematics) have been largely superseded by operational amplifiers that combine semiconductor components to provide more stable and linear amplification. The LM386 audio amplifier is used in this book as it is fairly straightforward to work with, allowing audio filter concepts to be introduced in chapter 8 with a 2-band filter amplifier being built as the chapter project. In chapter 9, the project is then combined with the Arduino as a controller as the final milestone project of the book, where the Arduino responds to digital control inputs (a push-button and potentiometer) to change audio outputs (filter levels).

Each chapter will have at least one practical electronics project to help you apply what you have learned. The aim of these projects is to consolidate what you have learned in each chapter – it is not simply a case of following instructions, but rather learning how to apply the concepts you have just encountered. By the middle of the book, you will have learned how to build an Arduino MIDI drum trigger that uses a piezo sensor to generate MIDI messages that can be linked to other audio software as output. By the end of the book, you will have built an Arduino-controlled 2-band equalizer amplifier that maps digital data from control inputs to analogue circuit control points. If you have followed the material (including the self-study questions), you should be able to understand and potentially extend on these milestone projects in many different ways – they are designed to be building block systems for your own ideas.

What you will learn

Chapter 1: The importance of scales and symbols in electronics
- What current, potential difference (voltage) and resistance are (covered again in chapter 3)
- How to use the Arduino with a resistor to light an LED

Chapter 2: How to understand electronic systems in terms of input/process/output
- How transducers convert sensor inputs to electrical signals
- How transducers convert electrical signals to actuator outputs

Chapter 3: Ohm's Law to relate voltage, current and resistance
- Kirchoff's Voltage Law (KVL) for series circuits and Kirchoff's Current Law (KCL) for parallel circuits
- How to use voltage dividers to create output voltage ratios

Chapter 4: The three basic programming instruction types (there is no fourth instruction!)
- How variables (and constants) are used to give named access to data in memory locations
- How functions encapsulate code instructions for reuse

Chapter 5: How to use selection instructions to process Arduino digital input
- How to use the Arduino to sample (and analyse) analogue input
- How to analyse analogue piezo input to build a MIDI Drum Trigger

Chapter 6: The fundamental elements of a time-varying (AC) signal
    How capacitors store charge over time
    How to use LTspice as a circuit analysis tool
Chapter 7: How operational amplifiers use differential inputs to reduce noise
    How negative feedback works
    How to build a minimal audio amplifier
Chapter 8: How to design first-order low- and high-pass filters
    How to use LTspice to analyse possible circuit designs using Bode plots
    How to build an audio amplifier with a 2-band equalizer
Chapter 9: How to build state and data storage into a digital control system
    How to use digital potentiometer chips to control audio signals
    How to build an Arduino-controlled 2-band equalizer amplifier

Software tools

All software tools used in this book are free to use either online (Tinkercad) or by download. Although there are many other electronics software tools available (a notable example being Autocad by Autodesk), one of the aims of this book is to keep costs down for the new learner and so free applications have been chosen (though Autodesk do provide extensive free options for those registered in education).

Tinkercad

Tinkercad by Autodesk Inc. is a free online simulator for electronic circuits based around the Arduino Uno (Figure 2).

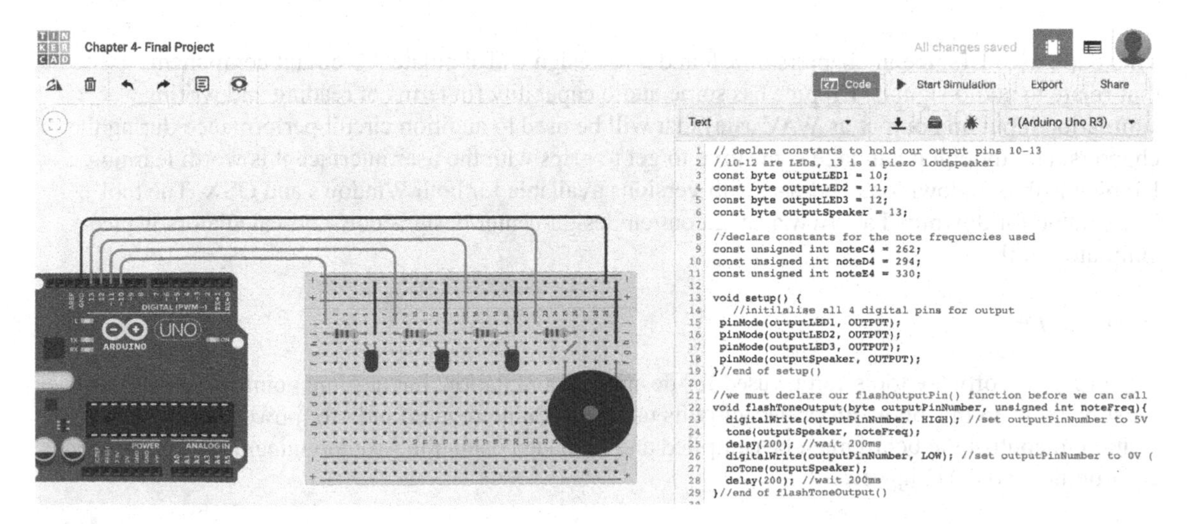

Figure 2 Tinkercad (from Tinkercad, by Autodesk Inc.). ***The image shows the component layout area and code editor for the Tinkercad online application. Arduino code and electronic components can be simulated in real time using this application, which makes it very useful for both learning and prototyping.***

Tinkercad is a really great way to learn the basics of Arduino control programming, as it can simulate component layouts, code and system operation in real time. This book uses Tinkercad extensively for the first five chapters, when the more bespoke elements of MIDI control must be prototyped using real components. All circuit layouts in this book are made using Tinkercad, which is free, runs in most browsers and is fairly light on processing requirements so can (feasibly) be run by most machines. If you are just beginning with electronics (audio or otherwise!) then this will really speed up your understanding of how circuits actually work. A free Tinkercad account can be created by signing up at www.tinkercad.com/.

LTspice

LTspice by Analog Devices is a powerful simulation tool that builds upon the Simulation Program with Integrated Circuit Emphasis (SPICE) simulator originally developed by Laurence Nagel (Figure 3).

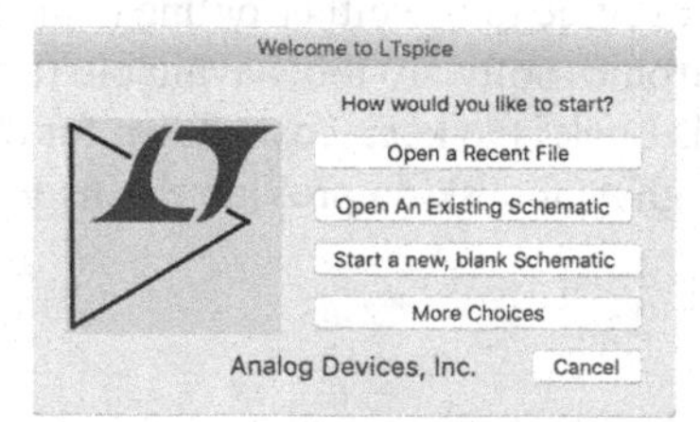

Figure 3 LTspice (from LTspice by Analog Devices). ***The image shows the LTspice front screen on the OSX platform. The LTspice program has powerful simulation and analysis capabilities, which include allowing audio signals to be simulated (as both sine waves and WAV data) for prototyping circuits.***

This book uses LTspice in chapters 6, 7, 8 and 9 to design and simulate AC circuit components that relate to audio signals. LTspice has some audio capability (in terms of reading and writing simulation input and output as WAV data) that will be used to audition circuit performance during the chapters, and though it can take some time to get to grips with the user interface it is worth learning. LTspice is free to download and use, with versions available for both Windows and OSX. The tool is available for download at www.analog.com/en/design-center/design-tools-and-calculators/ltspice-simulator.html.

Arduino IDE

The previous software tools can be used for design and simulation, but at some point an actual circuit must be built. In this book, all project circuits use the Arduino (even if only for power supply) and code written to control the board must be compiled and uploaded using the Arduino integrated development environment (IDE) (Figure 4).

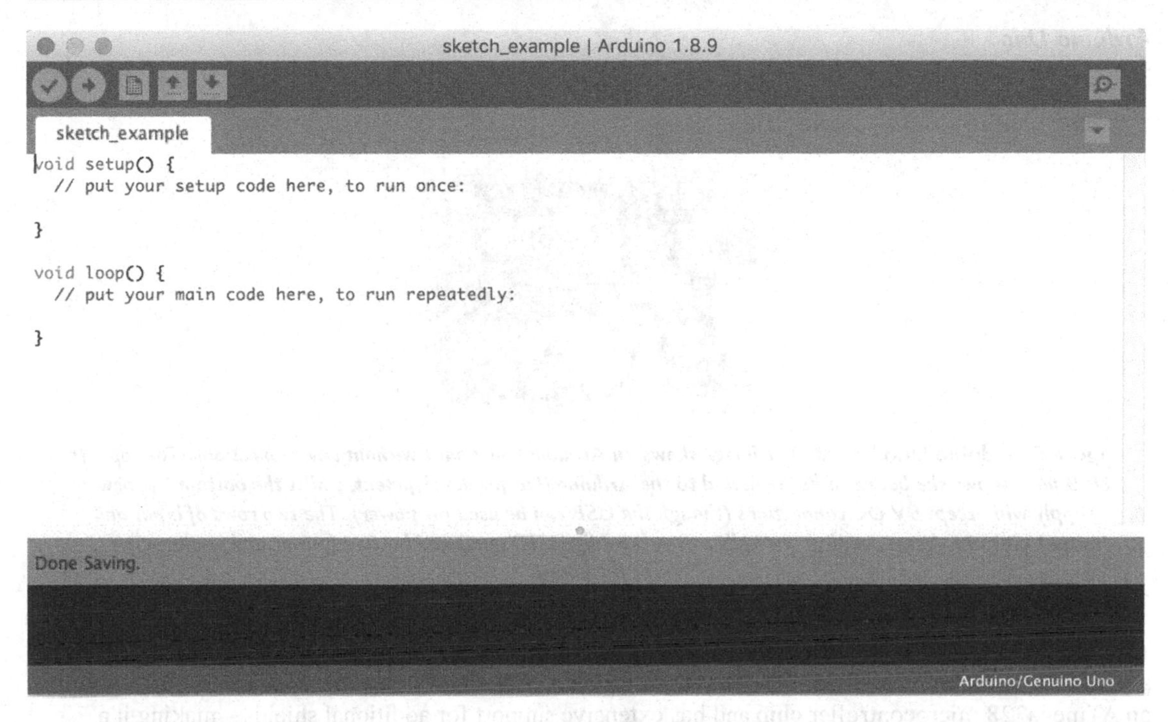

Figure 4 The Arduino IDE. ***The image shows the Arduino IDE editor, where code can be written and compiled. The setup() and loop() functions are shown in the listing above, which are fundamental to all Arduino sketches. The IDE can then upload this code to an Arduino board linked over USB, which can also provide power to the Arduino. Once code is compiled and uploaded, it can be executed by the Arduino at runtime.***

This book uses the Arduino Uno board – in line with the recommendations of the Arduino group for future development. This board can be set up and configured by the Arduino IDE, which can then upload code sketches to the Arduino for execution. The Arduino IDE has versions available for Windows, OSX and Linux and can be downloaded from www.arduino.cc/en/main/software.

Equipment

Electronics can be an expensive discipline to begin learning, as the components, tools and supporting items required can quickly mount up costs. This book aims to work with the bare minimum of components and tools wherever possible, to allow you to get an introduction to working with circuits as quickly and cheaply as possible. Specific suppliers are not provided as they can change, and with online competition may also not provide the cheapest option for a given component – an online search for any of the items listed below will provide a variety of options for purchase.

Arduino Uno

The most important piece of equipment in this book is the Arduino Uno board (Figure 5).

Figure 5 Arduino Uno board. ***The image shows an Arduino Uno board without any connections. The top left USB port allows the board to be connected to the Arduino IDE for development, whilst the bottom left power supply will accept 9V DC connections (though the USB can be used for power). The two rows of input and output pins are shown running vertically at the top (digital IO pins) and bottom (power and analogue pins) of the board.***

The Uno has been chosen both on the Arduino group's recommendations regarding future development and also because an Arduino Uno is provided within the Tinkercad simulator. The Uno is built around an ATmega328 microcontroller chip and has extensive support for additional shields – making it a very powerful tool for learning. There are a huge number of Arduino resources available online, but all information needed to prototype the project circuits used in this book is provided to avoid confusion when beginning with Arduino.

Breadboard

All circuits in this book are designed and prototyped using a 30-column breadboard (Figure 6).

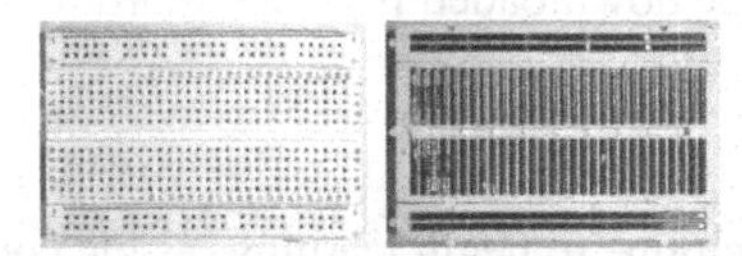

Figure 6 A 30-column breadboard. ***The left-hand image shows the layout for a standard 30-column breadboard, where vertical columns are numbered (1–30) and horizontal rows are alphabetized (a–j). All pins on each column are connected together up to the centre gap (the right-hand image shows the bottom of the board), which forms the basis of how breadboard circuits are built. The centre of the board includes a gap that can be used to connect integrated circuits (ICs). The paired top and bottom rows (known as bus strips) are designated for power rails (labelled +/–), which can be used to route power to different components placed on the board.***

Breadboards are fairly cheap to purchase as they are used solely for prototyping circuits, and so do not provide a high level of signal-to-noise performance. This will become apparent later in the book when encountering noise with audio input and output signals. It is beneficial (though not essential) to have several breadboards available if possible, to allow multiple circuits (particularly the milestone projects) to be built without having to then remove the work to build the next project.

The final project uses all the available space on a 30-column breadboard, and though this could have been set out more easily on a bigger board the aim of this book is to introduce concepts – ideally for the lowest study costs possible. In addition, space is always at a premium in electronics as smaller designs are not only cheaper but also more energy efficient (and arguably more usable as a result). One of the primary goals in circuit design is to produce a layout that is as small and efficient as possible, and so the final project retains the smaller 30-column breadboard to illustrate some of the techniques and workarounds that can be used to reduce space. Circuit board layout can be a very creative discipline and it is important to be aware of the need to minimise components and space wherever possible in all your projects.

Switches and LEDs

In chapter 2, we will learn how systems process inputs (from sensors like switches and potentiometers) to generate outputs (to actuators like LEDs and loudspeakers). The simplest examples of these are switches and LEDs, which are used in most of the projects in this book, as inputs to and outputs from the Arduino. Push-button (also called momentary action) switches are simple to work with and lend themselves well to digital control circuits as they have only two states – on or off (Figure 7):

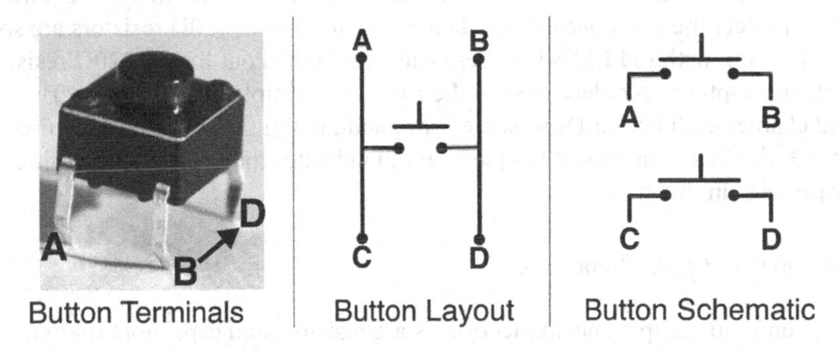

Figure 7 Push-button switch example. ***The left-hand image shows a push-button switch, where the connected terminals are bent inwards towards each other in pairs (A-C, B-D). The middle image shows the internal layout of the switch and the right-hand image shows its schematic representation. In practice, the connection layout means that only one input/output configuration is possible – effectively from one side (A/C) to the other (B/D).***

Push-button switches have two pairs of linked terminals (A-C, B-D) which are indicated by the direction of curvature of the pins (see top-left image in Figure 7). It is important to orient the switch correctly, as current will always flow between these terminals regardless of whether the switch has been pressed. When the switch is pressed, the contact completes the circuit path between all four terminals (middle image in the diagram), which allows current to flow between them. This means a push-button switch has effectively one path between input (either A or B) and output (either C or D), as all terminals are now connected – the schematic representation of this is shown in the right-hand image in the diagram.

Light-emitting diodes (LEDs) are semiconductors that convert the flow of electrons (current) into photons of light (Figure 8).

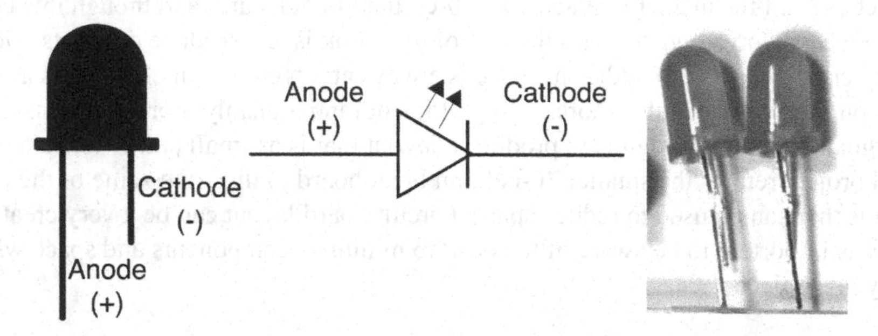

Figure 8 Light-emitting diode (LED) example. ***The left-hand diagram shows the anode (+) and cathode (−) of an LED, where the positive cathode has a longer terminal. The middle diagram shows the schematic symbol for an LED, which is similar to that of a diode (the upward arrows indicate light emission). The right-hand image shows examples of LEDs used in the project circuits in this book.***

LEDs are widely used as output actuators in electronic circuits, as they can quickly indicate a change in the state of the system (chapter 9). There are two things to remember about LEDs: as diodes they only allow current to flow (and hence emit light) in one direction, and they always require a current-limiting resistor to protect the component from damage. In this book, 150Ω resistors are specified for optimum current-limiting with red LEDs but these can be swapped out for the 220Ω resistors used in the MIDI projects in chapter 5 to reduce costs – the LEDs will simply be less bright. The projects in chapters 1–5 and chapter 9 all use LEDs in some form, and though six are ideally required for the first project in chapter 3, the final component list (see below) indicates that a minimum of three can be used for all the other projects in this book.

Resistors, capacitors and potentiometers

Perhaps the most common components in electronics are resistors and capacitors (particularly for audio), and as you progress beyond this book it will become common to maintain a stock of various values, tolerances and types of both components for use when designing and prototyping circuits. Resistors limit the flow of electrical current (see chapter 1) and are used throughout electronics to control the path that electrons take through a circuit (Figure 9).

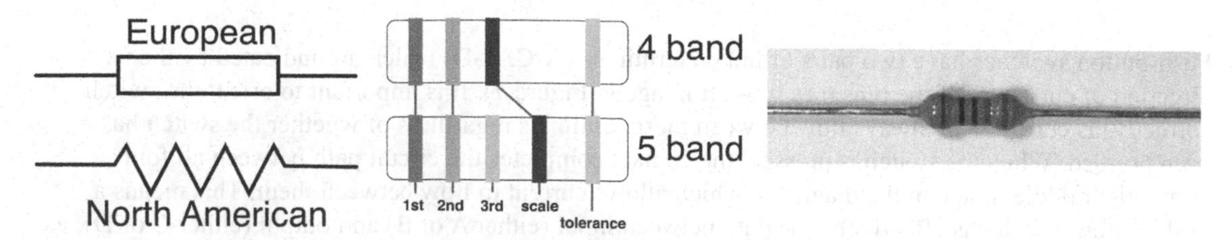

Figure 9 Resistor example. ***The left-hand diagram shows the standard European and North American schematic symbols for a resistor. The middle diagram shows how resistor bands indicate the value (and tolerance) of the component (see chapter 1), whilst the right-hand image shows an example of a typical resistor.***

The left-hand diagram shows the standard European and North American schematic symbols for a resistor, where the European symbol will be used in the schematics in this book. The middle diagram shows how resistor values are indicated by coloured bands, and the right-hand image shows how this is achieved in practice. Resistors are fairly small and thus the bands can often be difficult to see, particularly if you have difficulty distinguishing between certain colours. For this reason, it is recommended to clearly label and store resistors separately (see below) to avoid confusion – it is not uncommon to find yourself working on a project until late at night when fatigue and lower light conditions make it easier to make mistakes when selecting components.

The higher the resistance value the greater the restriction on current, which is also known as impedance when working with alternating current (AC) signals. AC signals are crucial to audio electronics as they vary over time (see chapter 6), but resistors do not vary and so capacitors are often used as their response is time-related (Figure 10).

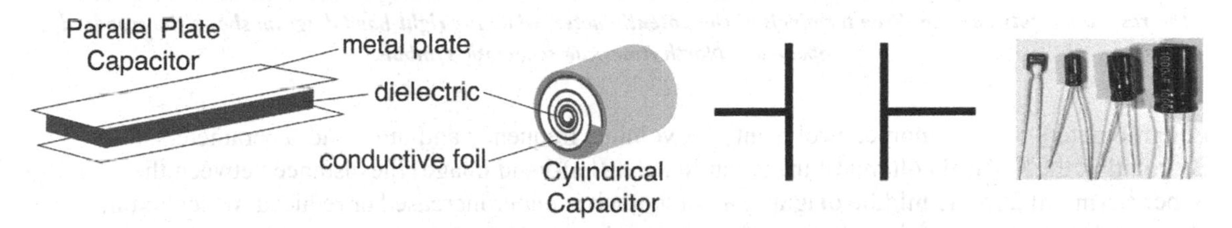

Figure 10 Capacitor examples. ***The left-hand diagram shows the internal structure of parallel plate and cylindrical capacitors. The middle diagram shows the standard schematic symbol for a capacitor, whilst the right-hand images shows various capacitor examples.***

Capacitors store charge between their plates, and as the charge in a capacitor increases it impedes the flow of any more current (not unlike a traffic jam slowing down cars further back on the road). Charging takes time (known as the time constant, τ) and so a capacitor varies its impedance based on the time taken for it to charge. For this reason, capacitors are essential in audio electronics, where they can be used to vary the response of the circuit based on the input frequency of the signal – this is known as a filter. Capacitors have many other uses too, and later in this book (chapter 7) we will learn how they are commonly used in noise reduction – this may help to explain some of the 'extra' components that are often found on an audio schematic. The resistors and capacitors chosen for the projects in this book are based on standard component values (see Appendix 3), where chapter 8 builds audio filters using certain values to achieve a specific frequency response. For your learning, it is possible to replace these values with the closest ones you have available (this is also true for the blocking and decoupling capacitors used in all the audio project circuits). Although the response will differ when components are changed, the main aim of this book is to provide an introduction to audio electronics and so costs can be minimised by using whatever is available.

Potentiometers are resistors that can be manually controlled, where the resistance of the component is varied by changing the position of a wiper contact (Figure 11).

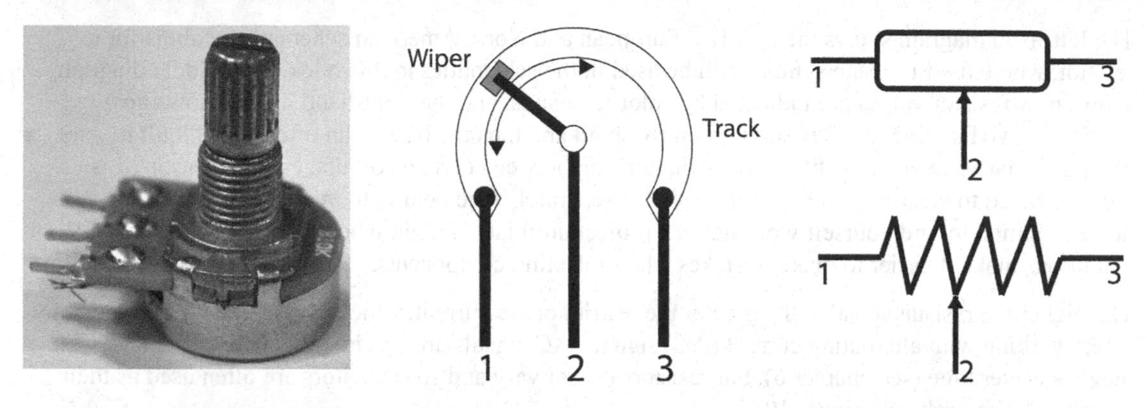

Figure 11 Potentiometer example. ***The left-hand image shows a potentiometer component, where the rotating wiper is connected by the vertical column to the top of the track. The middle diagram shows how this wiper varies the resistance between the three terminals of the potentiometer, while the right-hand diagram shows the standard European and North American schematic symbols.***

Potentiometers are common control points for volume, frequency and other audio parameters. By rotating the vertical column of the potentiometer (left-hand image) the distance between the wiper (terminal 2 in the middle diagram) and the track is either increased or reduced, which in turn changes the resistance of the component (resistance is partly based on the length of a conductor). Potentiometers are used in chapter 8 to control audio filter levels (also optionally in chapter 7 as an amplifier volume control), and in chapter 9 a potentiometer is used as an analogue control input to the Arduino. In addition, digital potentiometers are also used in chapter 9 to allow the Arduino to directly control filter levels (Figure 12).

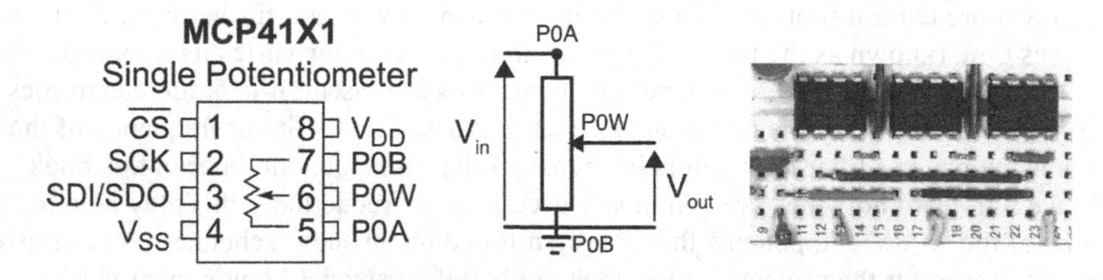

Figure 12 The MCP413-103 10kΩ digital potentiometer (from Microchip Technology Inc.). ***The left-hand diagram shows the pin layout for the digital potentiometer chip, whilst the middle diagram shows how the wiper pins (P0A, P0W, P0B) act in the same manner to an analogue potentiometer. The right-hand image shows the chips mounted on a breadboard in the chapter 9 project to filter audio signals.***

Digital potentiometers effectively switch between resistor levels based on control signals from the Arduino. In so doing, they are pivotal in this book to the combination of analogue audio circuits with digital Arduino control. The MCP413-103 chip can be controlled using the serial peripheral interface (SPI), which allows it to be connected to the Arduino for digital communication (see chapter 9).

Audio and MIDI components

This book focusses on audio, and this means using specific components that are not as commonly required in other areas of electronics. Chapters 7–9 work directly with audio signals, which requires an audio amplifier (Figure 13).

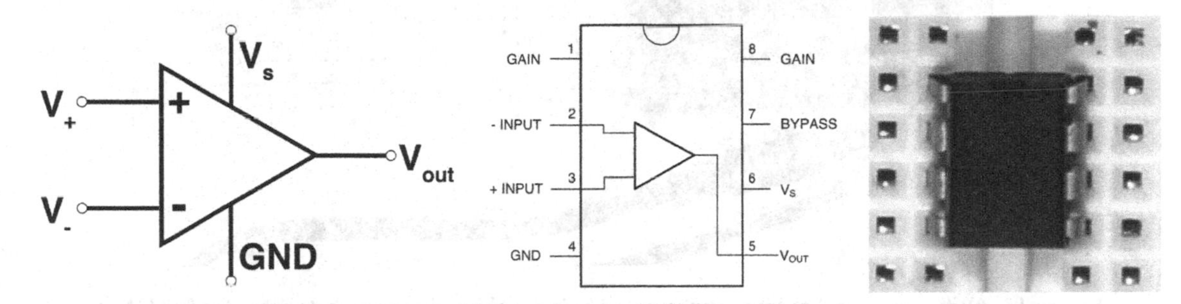

Figure 13 The LM386 audio amplifier (from LM386 datasheet, courtesy of Texas Instruments Inc.). ***The left-hand diagram shows the standard schematic symbol for an operational amplifier. The middle diagram shows the pin layout for the LM386 audio amplifier, where the operation amplifier symbol is embedded in the layout. The right-hand image shows an LM386 mounted on a breadboard, using the middle gap between columns to separate the terminal connections.***

Audio amplification is a highly complex (and often subjective) topic, so to reduce the scope of this book the aim is to introduce core concepts using a simple (and cheap) amplifier that is widely used in hobby electronics circuits. Although the LM386 has arguably been superseded in commercial designs by other operational amplifiers, it is easy to use and can produce very usable results for learning purposes. This is noted to avoid direct comparisons with commercial audio equipment you may own, which inevitably performs at a far higher level. It is important not to expect commercial circuits as outputs from this book, as the aim of an introductory text is to build prototypes that demonstrate techniques and methods rather than industrial standard products.

Extending this idea of prototyping, the audio input and output components used in this book are kept as open as possible to allow you to work with whatever equipment you have available. It is not recommended to use expensive connectors when prototyping for learning, where the requirement of a 3.5mm audio input jack is stipulated to allow any audio source (such as a mobile phone) to be used as an audio input to your circuits (Figure 14).

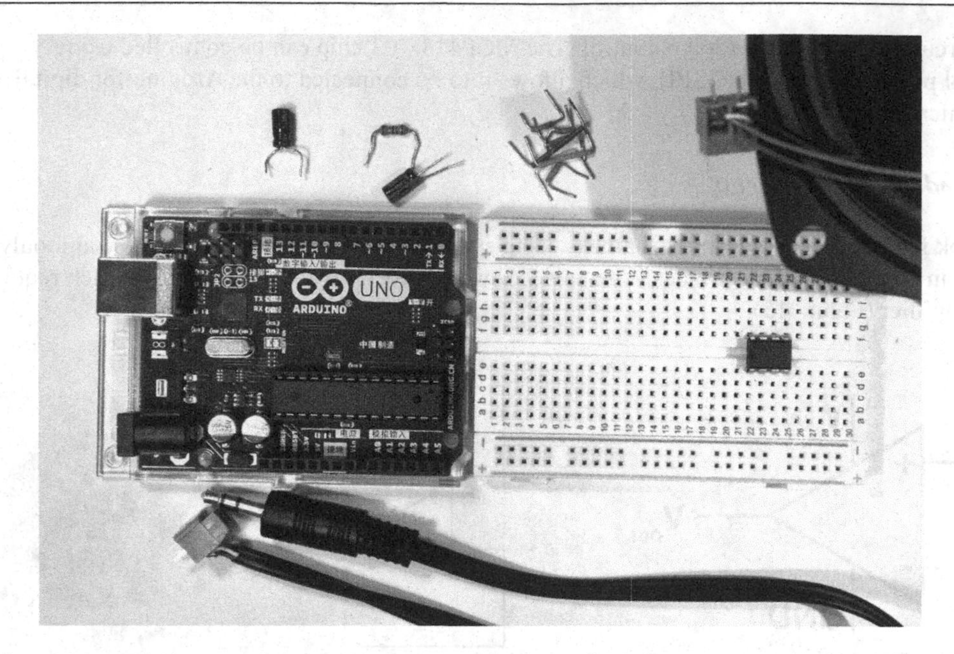

Figure 14 Audio input and output connectors. ***The image shows the components needed to build the audio amplifier circuit in chapter 7. The LM386 operational amplifier is already placed on the breadboard, with screw terminal block connectors being used to connect the audio input jack (bottom left) and loudspeaker (top right) to the board.***

In this book, all audio connections to breadboard are made using 2-pin screw terminal connector blocks (bottom left and top right of Figure 14), which allow quick and stable connections to be made. Although it is possible to work without these connector blocks, it is not recommended to do so as the possibility of loose audio connections is significantly increased. Such poor connections can easily add significant time to circuit prototyping, where other components on the board may be checked (and rechecked) when the actual problem lies with a poor input or output connection.

In addition to audio connectors, the chapter 5 project uses a 5-pin DIN connector to implement a MIDI interface (Figure 15).

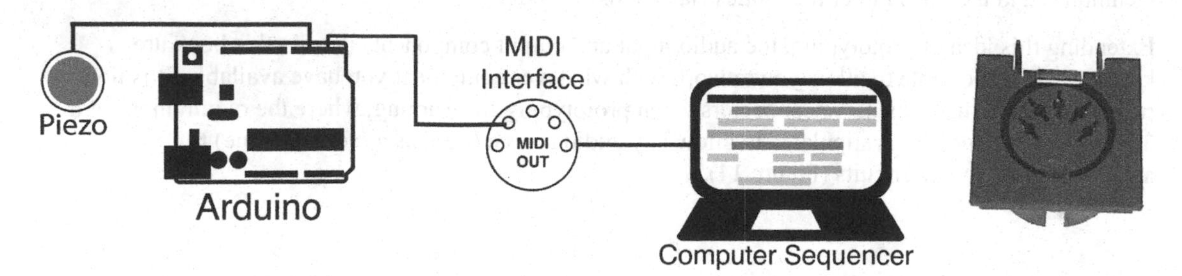

Figure 15 MIDI interface connector. ***The left-hand diagram shows a possible output connection for the midi drum trigger project in chapter 5. The piezo sensor acts as a trigger which is processed by the Arduino to generate MIDI Note messages that are sent through a 5-pin DIN MIDI OUT connector to a computer sequencer that has MIDI DIN connector. The right-hand image shows a 5-pin DIN connector, which can be connected and used to send and receive MIDI data.***

The chapter 5 MIDI drum trigger project uses a piezoelectric sensor that detects vibrations (from drum strokes) as an input to the Arduino, which uses selection instructions to determine how to process these input signals. The selection instructions use the serial bus that is built in to the ATMega328P microcontroller that is mounted on the Arduino to send MIDI Note messages as outputs to a 5-pin DIN connector, which can be built to function as a MIDI interface. A suitable 5-pin DIN connector can be mounted directly onto breadboard, with additional 220Ω resistors being used to limit the current through the MIDI IN and OUT pins. As noted in section 1.2.3, these 220Ω resistors can also be used in place of the 150Ω resistors used in chapters 1–4 to keep costs down.

Other components

Although not strictly components like those listed in the sections above, the common elements in all project circuits are connecting wires and cables (Figure 16).

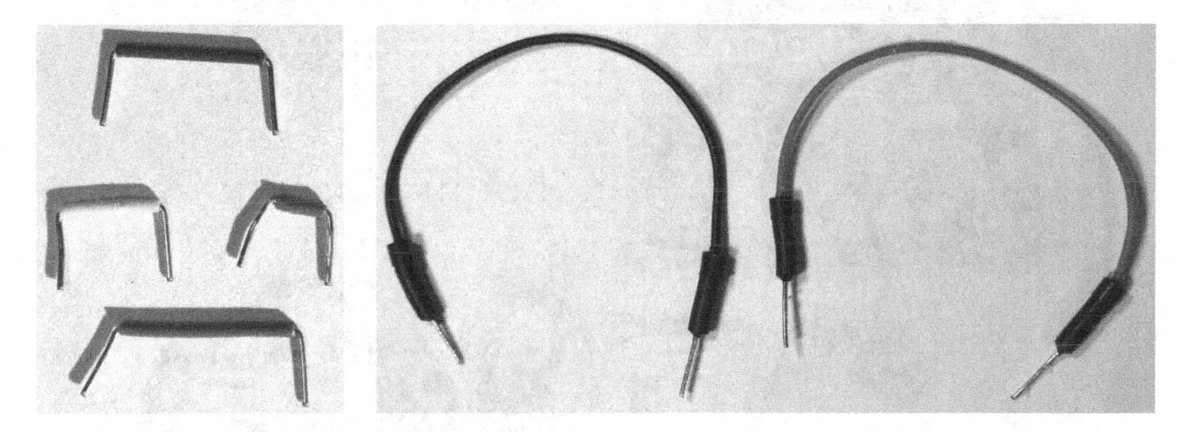

Figure 16 Connecting wires and cables. ***The left-hand image shows connecting wires, while the right-hand image shows connecting cables. Connecting wires are much easier to see in more complex circuit layouts (such as chapters 7–9) as they take up far less physical space. They are also much less likely to become accidently displaced if the board is moved, but connecting cables (if correctly seated) will also function effectively.***

The image shows connecting wires and cables, which are both used in the project circuits throughout this book. In the project layouts, connecting wires (left-hand image) are used in instances where pin to pin connections on the breadboard are required as they take up much less space and are far easier to see in the accompanying photographs. In addition, connecting wires are much less likely to become displaced or have loose connections than connecting cables (right-hand image).

Having said this, connecting cables can provide perfectly serviceable connections if correctly seated and are often included in Arduino starter kits – you may have bought one prior to reading this book. For the early chapters, connecting cables can be used as the DC and digital signals involved are much less impacted by noise than an AC or audio signal connection – an LED will light (even if it flickers slightly) and a switch will either be +5V (HIGH) or 0V (LOW). In the audio projects, it is recommended to use connecting wires if at all possible to reduce noise and keep the circuit layouts manageable – particularly in the final chapter 9 project.

Although there are not a particularly large number of resistors, capacitors and connecting wires/cables used in the projects in this book, the first layout image of the final chapter 9 project build can help to illustrate how messy a workspace can quickly become if not properly organised (Figure 17).

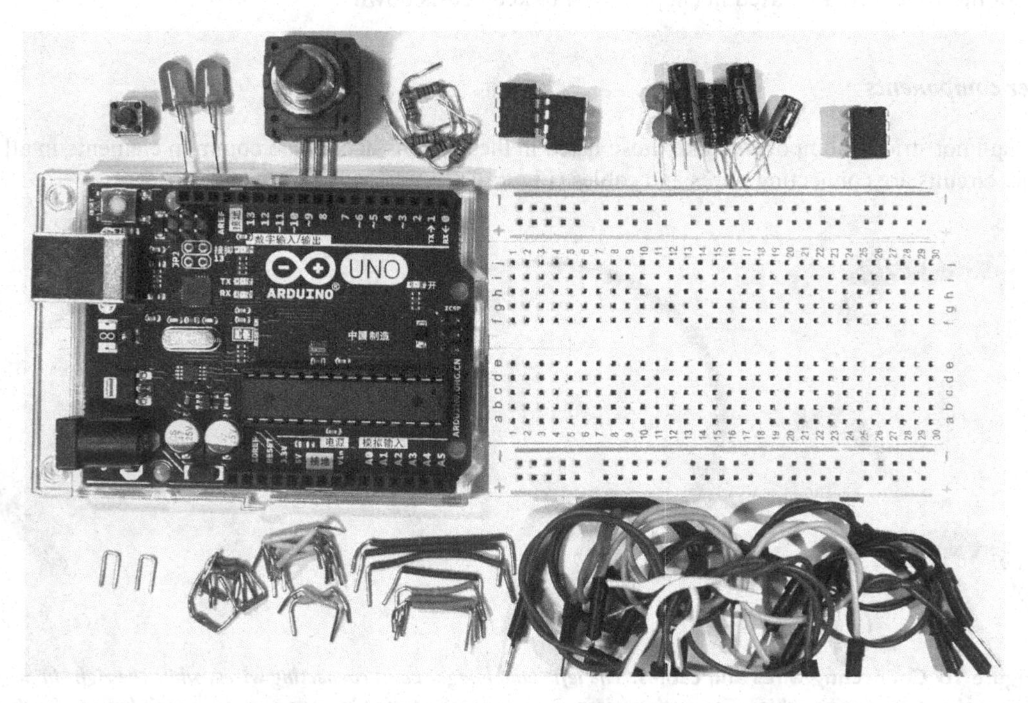

Figure 17 Component requirements. ***The image shows the components needed for the final chapter 9 project, as they are laid out at the beginning of the build instructions. Whilst it is important to have these components close at hand before starting the build (to save time and reduce errors), if they are simply laid out on a surface as shown above then displacement and loss of components becomes highly likely.***

The project build instructions in each chapter begin by showing all the components required for that specific build, arranged around an Arduino and 30-row breadboard for consistency. The image above shows the components required for the final chapter 9 project, which helps to illustrate the increasing number of components and wires required as the projects in the book progress. While this layout is intended to provide clarity in relation to each build, it is **not recommended** to work with components set out in this way as they can quickly become lost or confused – resistor colour codes can quickly tire the eyes when working on a complex build.

To avoid this, it is recommended to use some form of storage boxes/containers to hold each component type (Figure 18).

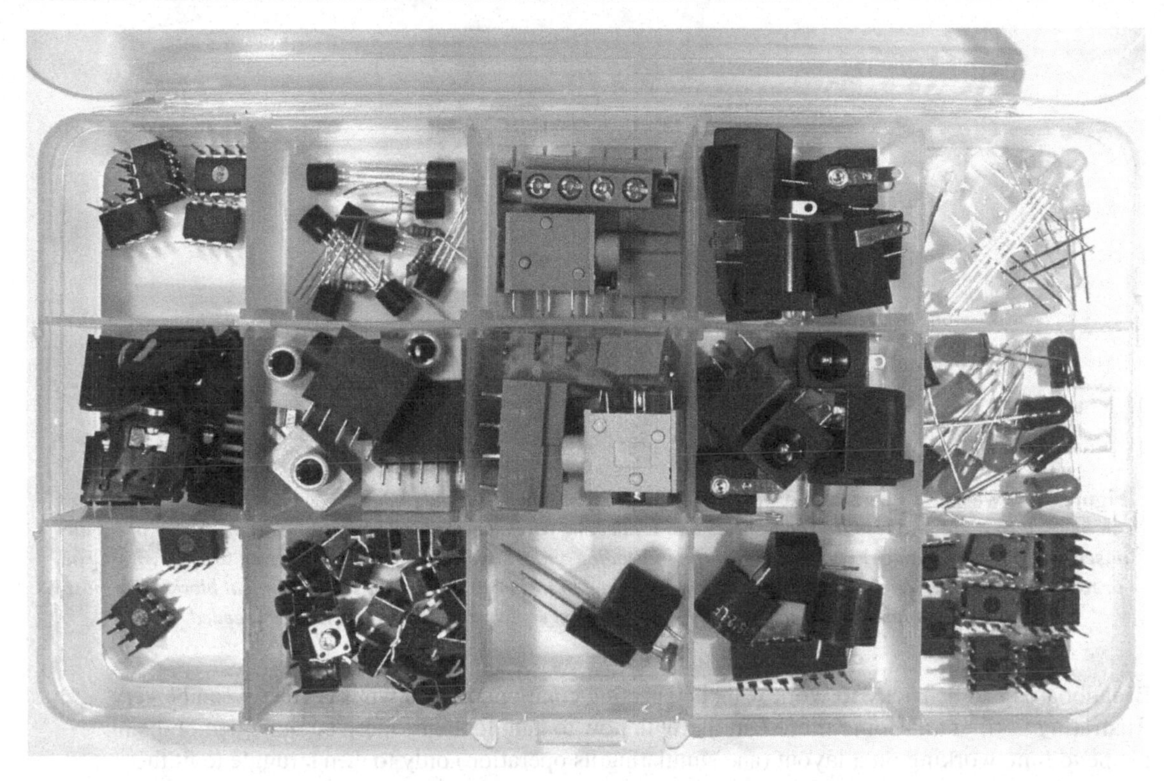

Figure 18 Storage box example. ***The image shows various components stored in a segmented box, allowing each to be quickly accessed during a build. As an example, the top-left compartment contains 555 timer chips, the bottom left compartment contains MCP413 digital potentiometers and the bottom right compartment contains LM386 audio amplifier chips – a fairly disparate set of integrated circuit (IC) components that are easily stored within a single sealed box (lid partially shown).***

The image shows an example of a segmented storage box, where each compartment can hold different components as required. This allows components to be securely stored (if the lid closes tightly!) and quickly accessed as required. An addition step would be to **label the lid of the box with the contents** of each compartment, which becomes more useful over time as you begin working with different component types (e.g. this book does not use 555 timer chips, but many simple audio synth projects can be built with them). An additional benefit of using storage boxes is increased mobility, particularly if you do not have a dedicated permanent working area in which to build, test and store your projects. The ability to take a partially completed project and move it to another location to continue working on it can often be very useful, allowing work to continue without significant interruption or loss of components. There are many cheap storage options available, and the key is to find a segmented solution that has a securely fastening lid to avoid spills of large numbers of components.

This book aims to introduce audio electronics concepts and methods as quickly and cheaply as possible, and purposely avoids using any more equipment than is absolutely necessary in all chapter projects. For this reason, the only tools shown in any project build images are pliers and tweezers, which could be considered as most of the tools needed to begin breadboard electronics project work (Figure 19).

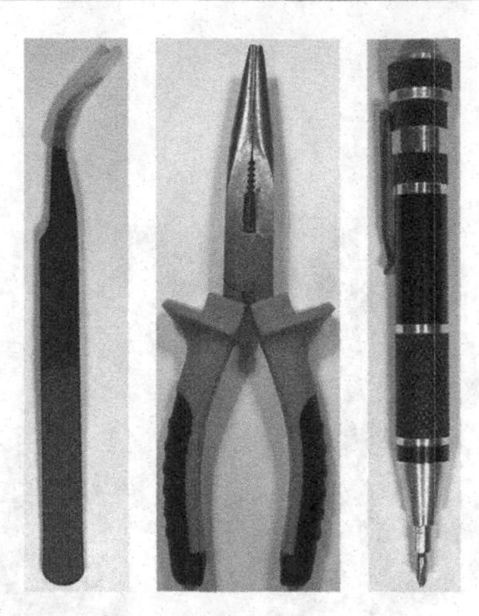

Figure 19 Tweezers, pliers and a precision screwdriver. ***The left-hand image shows the type of tweezers that can be used to place and replace components and connecting wires quickly and easily. In addition, a pair of needle-nosed pliers (middle image) can be very helpful in bending and trimming components and even to strip wires. The precision screwdriver (right-hand image) is only required in this book to connect screw terminal blocks, but most audio equipment will use some type of precision screws and so this tool is often needed.***

Breadboards can be tricky to work with, and seating components and wires correctly can be very frustrating until you become more familiar with how to handle them. It can be quite demoralising to spend time working on a layout (and simulating its operation) only to then struggle to fit the components and connecting wires onto the breadboard correctly – it is not uncommon to place a terminal or wire on the wrong row/column and then spend a long time breaking down the layout to discover this. Tweezers (left-hand image) can take a little time to become comfortable with lifting components like resistors, but are far more accurate than fingers and thumbs – particularly when reseating a component that is incorrectly placed. Needle-nosed pliers (middle image) can be slightly cumbersome (and can damage components if used as tweezers!) but are very useful for bending and shaping terminals. In addition, pliers can be used to strip wires if used carefully – saving on the expense of purchasing a wire stripper when beginning learning. A precision screwdriver (right-hand image) uses driver bits for screwheads much smaller than construction screws, and is a very useful addition to any audio engineer's toolbox as most audio equipment will contain precision screws of some description.

Soldering

This book does not use any soldered joints other than the potentiometer terminal cables used in the chapter 8 filter project – for prototyping these can connected by simply wrapping the wire around the potentiometer terminals without solder. The decision not to include soldered connections was made partly to avoid the cost of a dedicated soldering station, but primarily to keep the projects in this book as simple and safe as possible. Soldering is an essential technique that all electronics engineers must learn, but at an introductory level the equipment and working area needed are arguably a barrier to designing and building simple circuits quickly and cheaply. As all the projects in this book use breadboard, it was decided to work with non-soldered components and connections as much as possible to keep things simple – the potentiometers in chapter 8 are the only exception to this rule (and a temporary workaround is available).

Final component list

At this point, a full component listing for all projects in this book is provided for reference – you do not need to purchase everything at once, and the breakdown notes indicate where and when different groups of components can be purchased to complete different projects in the book (Table 1).

Table 1 Final component list. ***The table lists the components required for the project builds in each chapter of the book. The total column indicates the number of each component required to complete all projects, with the notes indicating where reduced numbers or workarounds are possible.***

Chapter:	1	2	3	4	5	6	7	8	9	Total	Notes
Arduino Uno	1	1	1	1	1		1	1	1	1	Highlighted chapters only use the Arduino for power – a 9V battery can also be used for these projects.
30-row breadboard	1	1	1	1	1		1	1	1	1	
LED	1	2	6	3	1				2	6	A minimum of three LED/resistors/switches are needed for chapter 3, where series branches can be built sequentially. The 220Ω resistors used in the MIDI projects (chapters 4 and 5) can replace the 150Ω resistors used in chapters 1–3 if necessary. Extra LEDs and resistors are needed for stage 2 of chapter 9.
Push-button switch		3	3						1	3	
Resistor 150Ω	1	2	6	3	1				1	6	
Resistor 220Ω					3					3	
Resistor 1MΩ					1					1	Only needed for chapter 5 MIDI drum trigger project (alongside 220Ω resistors).
Piezo sensor					1					1	
MIDI Out (DIN)					1					1	
Piezo speaker				1						1	Only needed for chapter 4 project.
2-pin screw terminal block connector					1		2	2	2	2	Used in both MIDI and audio projects (chapter 4 onwards).
Resistor 10kΩ									1	1	These components are only needed for the second half of the book. They can be purchased over time, to reduce initial costs. A single potentiometer can be used with filters being swapped for testing. Any small loudspeaker should be usable (<1W) with the LM386 – though audible output from filters will be impacted. A headphone jack can be spliced to make the audio input connector, allowing the cheapest-possible headphones to be used!
Resistor 16kΩ								1	1	1	
Resistor 160kΩ								1	1	1	
Resistor 10Ω							1	1	1	1	
Analogue potentiometer 10kΩ							1	2	1	2	
MCP413-103 10kΩ digital potentiometer									2	2	
Capacitor 10nF								2	2	2	
Capacitor 0.05μF							1	1	1	1	
Capacitor 0.1μF							1	1	1	1	
Capacitor 1μF							1	1	1	1	
Capacitor 250μF							1	1	1	1	
LM386 amplifier							1	1	1	1	
8Ω loudspeaker							1	1	1	1	
Audio input connector (3.5mm jack)							1	1	1	1	
Connector cable	2	2	4	6	4		2	4	6	6	Connecting wires are used for visual clarity (and preferred), but cables can be used for chapters 1–5 (though chapter 9 is too complex).
Connector wire	2	5	6	4	10		6	9	12	12	

The list provided in the table aims to help you get started quickly with as few components as possible. Chapters 1–3 can be completed with a battery, breadboard, push-button switches, 150Ω resistors (220Ω resistors could be used as replacements and also for chapters 5 and 9), LEDs and connecting wires and/or cables. The MIDI drum trigger project requires a 1MΩ resistor, piezo sensor and 5-pin DIN connector, and the piezo speaker is only used for the chapter 4 project (and so could be simulated in Tinkercad if not available). The rest of the components are used by the audio projects in chapters 7–9, and thus do not need to be purchased immediately to begin learning about electronics or the Arduino. By taking the book in stages (see Figure 1), a set of components can be built up over time that can be extended as needed – this is the best way to approach learning this new subject.

Conclusions

Each chapter in the book will end with a summary of what has been covered in that chapter, alongside a brief link to the material in the following chapter. This chapter introduced the structure of the book, which has three areas (electronics concepts, Arduino control and audio circuits) that are combined in the final chapter project to build an Arduino-controlled 2-band equalizer amplifier. The software tools needed for this book were discussed, noting that Tinkercad, LTspice and the Arduino IDE are all free applications that run on multiple platforms. The last section of the chapter provided a brief overview of the various components and equipment required to complete the chapter projects, noting that costs and simplicity have been considered when designing the projects in this book to make learning electronics as accessible as possible to the new learner.

The next chapter will introduce some of the fundamentals of electronics, beginning with the scales, quantities and equations required to work with DC and AC circuits. Engineering mathematics can become complex and is often seen as challenging by students, so this first section aims to show how simple rules and principles are all that is actually required to learn the basics. The following sections of the chapter will explain the three basic elements of electricity and electronics – current (the flow of electrons), resistance (the reduction in the flow of electrons) and voltage (the potential for electrons to flow to a given point).

At the end of most chapters (exceptions will provide another activity), self-study questions are provided to assist in your learning. Please do not skip them as they are deliberately designed to reinforce the concepts involved – this book does not include large numbers of worked examples and does not ask trick questions to confuse the learner. After the self-study questions, a brief summary of the main points covered in each chapter is provided as a memory aid for revision – it can help to look at this chapter summary to ensure that you are comfortable with the material you have just learned.

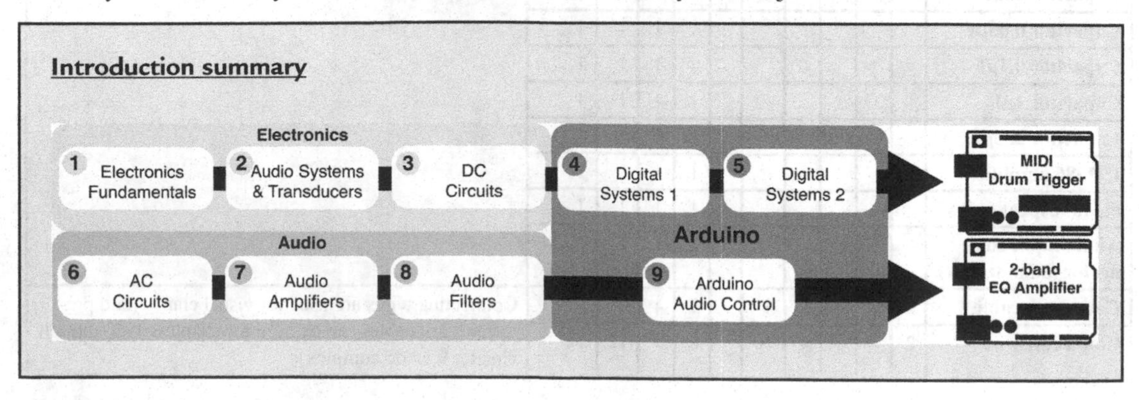

CHAPTER 1

Electronics fundamentals

This chapter will introduce some of the fundamental concepts that you will need to learn about electronics. Electronics can be a little overwhelming when you first start to learn it, so the chapter begins with a brief introduction to scales, symbols and equations. A lot of the mistakes that new electronics learners make are caused by the many different scales, quantities and symbols that they have to contend with – not their own lack of knowledge or aptitude. This can quickly become very demoralizing, so from the outset try to remember that learning the language of electronics will take **time** as well as effort.

The chapter continues by addressing the fundamental concepts of electronics – current, voltage and resistance. These have been covered in great detail by many other sources, but they can still be confusing at first – particularly when presented as 'easy' concepts that you are expected to understand straight away. The idea of current being the movement of electrons and resistance being the means of slowing this movement should hopefully become clearer, but the concept of voltage being the ***potential*** for electrons to move can often take time to fully understand – we will return to all of these concepts again in later chapters so don't be too concerned if certain aspects seem a little confusing at first!

The tutorial in this chapter introduces the Autodesk Tinkercad application, which is an online circuit simulator for simple Arduino projects. We will use this application in the first half of the book to test our Arduino code and simulate control systems for our circuits, so this chapter takes you through the basics of circuit simulation. After testing and simulation, we can then progress to prototyping our circuit using an actual Arduino, basic electronic components (resistor with LED) and breadboard. The chapter ends with our first circuit of an LED being lit by power from the Arduino – this is not a complex (nor impressive!) project, but it helps to introduce the project structure employed throughout the book. The chapter ends with some self-study questions, which are always popular with learners! It is hoped that you do not skip them to get to the more 'interesting' material – electronics is a language, and all languages require repetition for effective acquisition.

What you will learn

The importance of scales and symbols in electronics
What current, potential difference (voltage) and resistance are (covered again in chapter 3)
How a battery uses oxidization and reduction to create cations and anions
What open and closed circuits are
What a simple electronic circuit schematic looks like
How to use the Arduino with a resistor to light an LED

1.1 Scales, symbols and equations

When beginning to learn electronics, many of the terms used will be new to you. Just like mathematics and physics, electronics has its own language and so this chapter begins by discussing some of the fundamental elements involved. Before we do this, it is important to know a little about the mathematical scales used and how they relate to each other – they can often be a major source of confusion when learning some of the concepts and equations used in later chapters. Table 1.1 lists the common mathematical powers of ten and the symbols used to represent them.

Table 1.1 Powers of ten and their symbolic equivalent. ***The table lists the mathematical powers of ten, their relative scale and the symbols used to abbreviate them in electronics equations. The example quantities used are commonly found in computing (e.g. B for bytes) and engineering (e.g. V for volts, A for amps) – we will cover these quantities in more detail as the book progresses. Note that for each prefix the powers scale in three decimal places each time, with the exception of centi, which is commonly used for distance (cm) and liquid volume (cl).***

Prefix	Power of ten	Multiplier	Symbol	Example
Terra	10^{12}	1,000,000,000,000	**T**	4,000,000,000,000B = **4TB**
Giga	10^{9}	1,000,000,000	**G**	3,000,000,000B = **3GB**
Mega	10^{6}	1,000,000	**M**	3,000,000 = **3M**
Kilo	10^{3}	1,000	**k**	25,000W = **25kW**
None	10^{0}	1	none	
Centi	10^{-2}	0.01	**c**	0.1m = **10cm**
Milli	10^{-3}	0.001	**m**	0.005A = **5mA**
Micro	10^{-6}	0.000001	**µ**	0.000002V = **2µV**
Nano	10^{-9}	0.000000001	**n**	0.000000003F = **3nF**
Pico	10^{-12}	0.000000000001	**p**	0.000000000047F = **47pF**

The table shows how large and small quantities can be scaled to make them easier to work with – it is much easier to refer to a 3MΩ resistor than a 3,000,000Ω resistor when listing components for a circuit. The key to working with powers of ten is to carry out the following steps:

1. Check whether the **power** is positive or negative – positive is larger than 1, negative is smaller than 1.
2. Count the number of **decimal places** from the lowest digit – the number of decimal places should be equal to the power of the number.
3. Convert the power to the **nearest factor** of 3 – we will be working with scales like M, k, m, µ, n, p so always scale the value to the nearest 3/6/9/12 power.

Often the most confusing thing about powers of ten is when a scaling prefix is applied to a number that is already scaled – you may sometimes encounter a capacitance value like 0.22µF which is equivalent to 220nF (0.000000022F)! Another downside of scaling in this way occurs when performing calculations with the actual quantities involved – particularly smaller values like µ, n and p. Let's say we are working with two voltages that have values of 10mV and 8µV (we'll learn about voltage later in the chapter), and that we want to add them together. The following example illustrates where problems can occur due to scaling if each step of the calculation is not performed correctly.

1.1.1 Worked example – adding voltages

We will learn more about adding voltages in chapter 3, but for now let's look at how scaling affects the mathematical process involved.

***Q1**: What is the total voltage V_{tot}, as a result of adding 10mV and 8µV?*

$$V_1 = 10mV = 10\times10^{-3}V \qquad V_2 = 8\mu V = 8\times10^{-6}V$$

$$\textbf{Total Voltage, } V_{tot} = V_1 + V_2$$

$$V_{tot} = 10\times10^{-3} + 8\times10^{-6} = 0.01 + 0.000008$$

$$= 0.010008V \approx \mathbf{0.01V}$$

Answer: The total voltage V_{tot} is *approximately* 0.01V (rounded to 2 decimal places).

Notes:

The total value is approximately 0.01V (to 2 decimal places) – this is accurate enough for the calculations we will perform in this book

The symbol ≈ is used in mathematics to denote **approximately equal to**

The voltage value of 10mV has a scaling factor of 10^{-3} which gives 0.01V (10×0.001)

The voltage value of 8µV has a scaling factor of 10^{-6} which gives 0.000008V (8×0.000001)

Follow these steps when working with scaled numbers:

1. Write out all the quantities in the equation – *don't skip this step, it will only cost you time if an error is made*
2. Convert all units to include scaling factors – *this avoids making errors when combining different 10^x values*
3. Count the number of digits from the decimal point – *it should be the same number as the power used*
4. State the answer to 2 decimal places – *with certain values (like frequencies), you can often round up to the nearest whole number*

The example above seems simple enough, but what if the wrong scaling factor was used for either the 10mV or the 8µV values – the result would be very different! This can be a problem when learning electronics, where the quantities used are often related by completely different orders of magnitude. The reason for this is partly due to the evolution of knowledge over a long period of time, where the definition of a quantity and its scale were made prior to their current usage – for example, an audio signal capacitor will often be in the n or ρ range, whilst the resistors we will use will mainly be in the kΩ range. We will see another example of this later in the chapter when we encounter conventional current flow – it is conventional because of its historical precedent, even though electrical charge actually moves in the opposite direction! Although you may have learned other ways of doing arithmetic with powers of ten, in this book we will always write out the full quantity to avoid arithmetic errors – it may seem a little pedantic, but you can always double check the full value by counting the number of decimal places involved.

How to progress your learning

As we will learn during this book, many of the mistakes you can make will not necessarily be due to a lack of understanding but rather a lack of experience of the processes involved. This is a very important point, as learning can be significantly impeded by an increasing belief that *'this is too difficult'* based on errors or mistakes in your work.

If you take the above example, using the wrong scaling factor for one of the voltages would give you a completely different result, which could immediately lead you to believe that you do not understand how to add voltages! In actual fact, using the wrong scaling factor is an arithmetic error not a learning problem. Thus, the example above highlights two things:

1. **Scaling** factors are important because they can significantly **change** your results
2. Making a **procedural** mistake does **not** mean you are **struggling** to learn electronics

Text boxes like this are used throughout the book to reinforce important concepts or provide further information on common pitfalls and problems that you may encounter. In all instances, if you follow the steps for each example then you should get the correct result – this is where **experience can only be built over time**.

Another difficult aspect of electronics can be the symbols used for scales and quantities involved – you have already seen scaling symbols like k, c, m, μ, n, ρ in the table above, but symbols like ohms (Ω), watts (W), amperes (A), volts (V) and farads (F) were also used to define some of the different quantities used during the book. There is no easy way to learn symbols other than by practice, and as with scaling factors it is important to remember that **learning a new set of concepts takes time** – learning the language that represents them is a separate (though related) task that will also take a similar amount of time and effort. We will encounter this problem again when we learn how to program the Arduino in chapters 4 and 5 – programming concepts are what an Arduino can do, the syntax of the C language is how you tell the Arduino to do it! In all instances, if you take your time to follow all the steps in each task you will begin to see the **method** behind them.

In comparison to scales and symbols, the electronics equations we will use are relatively straightforward and do not employ complex mathematical operations like those found in other audio areas such as acoustics or digital signal processing. The main problem with learning electronics equations is that the scales and symbols involved introduce a new layer of complexity to adding, subtracting and multiplying quantities. For example, one of the most widely used equations in electronics is Ohm's Law:

$$\text{Voltage, } V = IR \tag{1.1}$$

V is the potential difference in volts (V)

***I** is the current in amperes (A)*

***R** is the resistance in ohms (Ω)*

The equation above is arranged to allow you to learn the relationship first, then consider the quantities involved afterwards – this convention is used for all equations in this book. We will learn more about Ohm's Law in chapter 3, but for now consider the mathematical operation involved, where one quantity (voltage) is equal to another two quantities multiplied together. This is not a difficult mathematical relationship to grasp, but when you combine it with scales, symbols and other circuit analysis methods that we will learn (like Kirchoff's Laws) it can often feel like it has become quite complicated. Thus, before you learn how to analyse a circuit, try to remember that you already know the mathematics involved – it's the other elements that are new and sometimes confusing to learn!

An important thing to note about electronics equations is that many of them deal with **proportions** that are based on **fractions.** In the modern digital world computers are used for speed and accuracy, but when circuit theory was first proposed this means of calculation did not yet exist. Thus, when working with series and parallel circuits in chapter 3 you will learn about adding and multiplying fractions and also about reciprocals – 1 divided by the quantity. This will perhaps be unfamiliar to you (even if you studied it at school) as fractions are not used very much in modern computation – the fraction is usually converted to a percentage or simply to a decimal number, which in itself can create a significant level of scaling to quantify the value accurately (e.g. the fraction $\frac{1}{3}$ can only be *approximated* using decimal numbers). The important point about fractions in electronics is that they are often used to represent **ratios** between quantities, as in the next example:

1.1.2 Worked example – working with fractions

Although we will often convert fractions to decimal values, it is important in electronics to understand the ***proportions*** *of the quantity that they represent.*

Q1: Add $\frac{1}{2}+\frac{1}{4}$ *and* $\frac{1}{4}+\frac{1}{4}$ *– how do the answers compare as fractions of a 5V power supply?*

Answer 1: $\frac{1}{2}+\frac{1}{4}=\frac{2}{4}+\frac{1}{4}=\mathbf{\frac{3}{4}}$ *so* $\frac{3}{4}$ ***of a 5V*** *supply* $=\frac{3}{4}\times 5=\frac{3}{4}\times\frac{5}{1}=\frac{15}{4}=\mathbf{3.75V}$

Answer 2: $\frac{1}{4}+\frac{1}{4}=\frac{2}{4}=\mathbf{\frac{1}{2}}$ *so* $\frac{1}{2}$ ***of a 5V*** *supply* $=\frac{1}{2}\times 5=\frac{1}{2}\times\frac{5}{1}=\frac{5}{2}=\mathbf{2.5V}$

$\therefore$ *Difference between voltages* $= 3.75 - 2.5V = \mathbf{1.25V}$

Answer: $\frac{3}{4}$ *of a 5V supply (3.75V) is bigger than* $\frac{1}{2}$ *of a 5V supply (2.5V) by 1.25V.*

Notes:

The three-dots symbol $\therefore$ means **therefore** in mathematics – you may have come across this symbol in other textbooks

When adding fractions start by looking for a common factor amongst the denominators (bottom values).

Compare the lowest common factor you can find with the result of multiplying the denominators together – then use the lower number as the ***lowest common denominator***.

When multiplying a whole number by a fraction, convert the whole number into a fraction by dividing by 1 - thus, in the example above 5 becomes $\frac{5}{1}$

Notice that a value of 1.25V is equivalent to $\frac{1}{4}$ of a 5V supply, which is not immediately clear when working with the decimal values – ***this is much more obvious when considering the values as fractions:*** $\frac{3}{4}-\frac{1}{2}=\frac{1}{4}$

Follow these steps when adding/subtracting fractions-

1. Find the lowest common denominator – *either a common factor or multiply both denominators together.*
2. Check the numerator result – *this is a very useful check: if your numerator is greater than 1 then you will need to simplify your answer later.*
3. To multiply whole numbers with fractions, convert the number to $\frac{x}{1}$ – *you can use a calculator for this to get the decimal value.*
4. Always state the answer to your question clearly – *this avoids any misunderstanding if you have made a simple error in your calculations.*

Q2: *Multiply* $\frac{1}{2}\times\frac{2}{3}$ – *what is the* **reciprocal** *of the answer?*

$$\textbf{Answer} = \frac{1}{2}\times\frac{2}{3}=\frac{2}{6}=\frac{\mathbf{1}}{\mathbf{3}}$$

$$\text{Reciprocal} = \frac{1}{\textit{Answer}}=\frac{3}{1}=\mathbf{3}$$

Answer: *The multiplication answer is* $\frac{1}{3}$. *The reciprocal of this answer is 3.*

Notes:

When multiplying fractions always multiply the numerators (top values) first, then the denominators (bottom values).

To get the final fraction, simplify your result by finding the lowest number that divides both numerator and denominator into whole numbers (this is the number 2 in the example above).

There are several ways to calculate the **reciprocal**:

1. divide 1 by your current answer – *this can be confusing when getting the reciprocal of a fraction*
2. flip the numerator and denominator – *this is the easiest method to learn*
3. find the number needed to multiply your answer with to get 1 – *this is a useful check for your answer*

Follow these steps when **multiplying** fractions-

1. Multiply the numerators and check the result – *this is a very useful check: if your numerator is greater than 1 then you will need to simplify your answer later.*
2. Multiply the denominators – *if the numerator is 1 you now have the result, but if it's greater than 1 you will need to simplify.*

3. Simplify the result – *look for a common factor between numerator and denominator, then divide both numbers by it and write the new answer (in the example above, 2 divides both 2 and 6).*
4. Check for other common factors – *if your simplified numerator is greater than 1, double check there isn't another common factor that can reduce the fraction further.*

For the **reciprocal**:

1. Get the reciprocal by flipping the numerator and denominator – *calculate the result of the division if the new denominator is greater than 1 (e.g. reciprocal of* $\frac{2}{3}$ *is* $\frac{3}{2} = 1.5$ *)*
2. Multiply your original answer by the reciprocal to check it equals 1 – *you can use a calculator for this if the numerator is greater than 1.*

Worked examples are used throughout this book to introduce the theoretical equations used in circuit theory. The notes will try to break down the steps involved in working out an answer, alongside indicating where common mistakes can be made. A common barrier to learning electronics can often be the combination of scales, symbols and equations rather than any of them as individual elements. At all stages of your learning, try to remind yourself that any difficulties you may be encountering are incremental steps forward from **what you already know** – it can often feel like electronics is a very complex discipline, but in fact there are a relatively simple set of underlying mathematical relationships involved in basic circuit theory. Thus, the important point to remember is:

If you can add, multiply and divide numbers then you can learn electronic circuit theory!

1.2 Electrical fundamentals

When learning electronics, you will come across terms like current, voltage and resistance as part of the language used to describe the behaviour of circuits. The basis of all these fundamental concepts is **electrical charge**, which defines how subatomic particles are either attracted to or repelled by one another based on their positive or negative polarity. Electrically charged particles follow simple rules:

Opposite electrical charges will **Attract** each other
Like (same) electrical charges will **Repel** each other

We are interested in **electromagnetic force**, which defines how particles actually *move* based on their charge. Although a particle can be electrically charged, it requires an electromagnetic force to move it closer to an opposite charge (or further away from a like particle). This is fundamentally important when learning about charge – *the force needed to move a charge is not the charge itself.* In a simple model of the atom, nucleons can either be a **proton** with a **positive** electrical charge, or a **neutron** that has **no charge**. An atomic nucleus is formed by a certain number of protons and neutrons that are bound to each other by the *strong nuclear force* (Figure 1.1).

The strong nuclear force is greater than the electromagnetic force at very short distances, and so it helps bind neutrons to protons to prevent the positively charged protons from *repelling* each

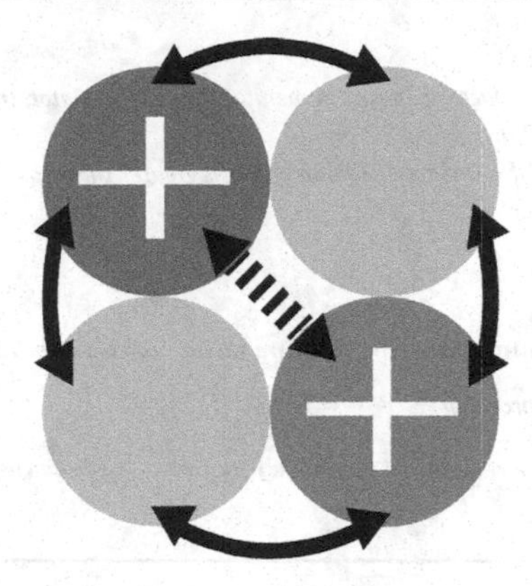

Figure 1.1 An atomic nucleus. ***This diagram shows the combination of nucleon particles within an atomic nucleus. Positively charged protons would repel each other without the presence of neutrons, which act as a spacer because they have no charge. The protons and neutrons are held together by the strong nuclear force between protons and neutrons (solid lines), which at shorter distances is greater than the electromagnetic force (dashed line) that would otherwise repel the protons from one another.***

other – this creates an atomic **nucleus**. It can help to think of neutrons as subatomic 'spacers' that hold positive protons together, without them the electromagnetic force created between these like particles would split them apart. Apart from nucleons, the other major component of atoms are **electrons**, much smaller **negatively** charged particles that orbit the nucleus within different shells (Figure 1.2).

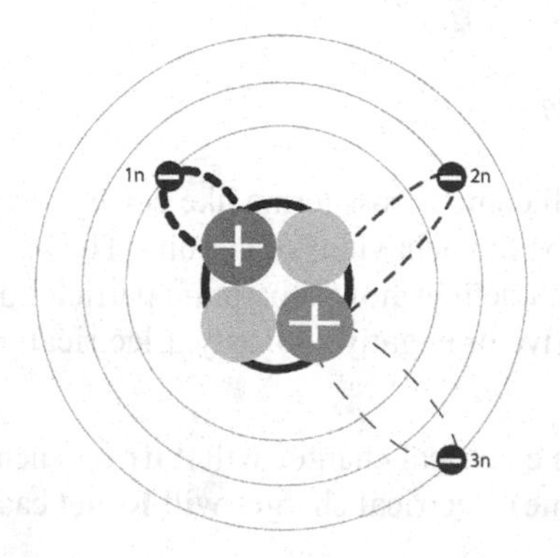

Figure 1.2 The Rutherford-Bohr atomic model. ***This diagram is an illustration of how negatively charged electrons surround the positive nucleus (of protons and neutrons) in increasing orbital paths. As the distance from the nucleus increases, the electromagnetic force (dashed lines) between protons and electrons reduces. This model is often used as it simplifies the teaching of concepts like electrical conductance and chemical bonding.***

The diagram shows each electron shell as having a principal quantum number (n), which defines the maximum number of electrons that can orbit within that specific shell using the relation $2n^2$. Although we don't need to completely derive this proof for our electronics work, it is important to note that the factor of 2 in this relationship occurs because the electrons in each shell orbit in pairs (we will return to this in chapter 2 when we look at electromagnetic induction). If the maximum number of electrons are not present within a given shell, then there are **holes** available for other electrons to move into. Electrons have more energy than holes, so when an electron fills a hole the difference in energy is released as a photon – the fundamental particle of all electromagnetic radiation (this includes light – you may already have heard about photons in this way). In an atom, the nearest shell to the nucleus has the strongest electromagnetic force between positively charged protons and negatively charged electrons. This means that the closest electron shells have the strongest attraction to protons in the nucleus. On the other hand, the **outer shell contains valence electrons** – free electrons that can be attracted to other positive charges **outside** the atom because they are not held by as strong an electromagnetic force (Figure 1.3).

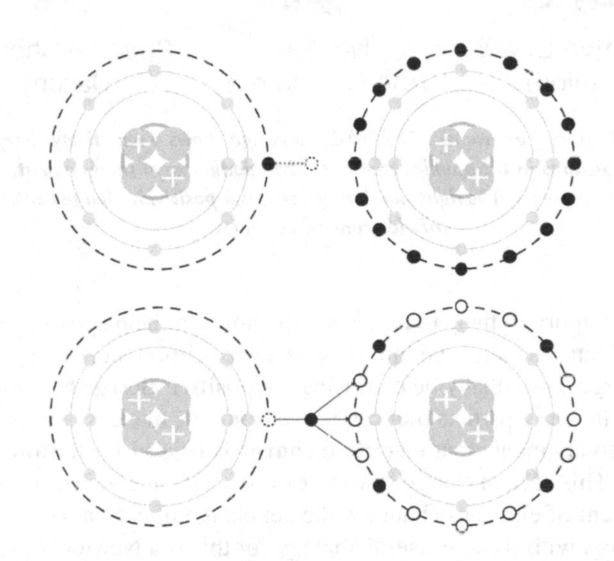

Figure 1.3 Valence electron movement. ***The diagram shows the outer valence shell of an atom, which has a weaker electromagnetic bond to the positive nucleus. In the top example, the right-hand atom has a stable outer shell, so no valence electrons can pass into its orbit as there are no holes for them to move into. In the bottom example, the less stable shell in the right-hand atom has many possible holes for a valence electron to move into.***

In Figure 1.3, valence electrons are not strongly bound to the positive nucleus of their own atom and so they can be attracted to another nucleus – but only if this nucleus has **holes** available in its own valence shell. If an atom has a full valence shell it is considered to be more stable (in chemistry they are defined as less reactive), but those with holes have the potential to accept new electrons. We define **net electrical charge** as the **difference** between the number of protons and electrons in an atom (net being the result of a sum or difference), where electrically charged atoms are known as ions:

> A **cation** is an ion with more **protons** than electrons, it has a **positive** net electrical charge
> An **anion** is an ion with more **electrons** than protons, it has a **negative** net electrical charge

The difference in net charge is measured in coulombs (C), which is a Standard Unit (SI) for the measurement of electrical charge. A coulomb is defined as a net difference of 6.24 *quintillion* electrons (6.24×10^{18}) for a negatively charged anion (or the same net difference in protons for a cation). If an electron leaves an atom, the net electrical charge for that atom becomes more positive. On the other hand, if an electron passes into the orbit of a new nucleus then the *increased number of electrons means that the net electrical charge for that atom becomes more negative* (Figure 1.4).

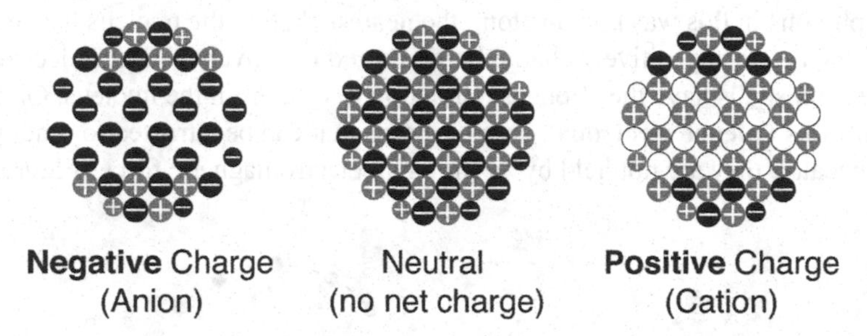

Figure 1.4 Net electrical charge examples. ***The left-hand image shows a negatively charged anion, where there are more electrons than protons in the atomic structure. The middle image shows a neutral charge, where both protons and electrons are balanced. The right-hand image shows a positively charged cation, where the absence of electrons leaves holes.***

In the diagram, the most important thing to notice are the holes in the positive cation charge: these are spaces that valence electrons can move into. This is a very important point to remember, as the concept of a positive charge can initially be confusing – a **positive charge has holes for electrons** to move into. Electrons are the only particle that can move from one nucleus to another (under normal conditions), so this effectively means that **electrical charge is passed from atom to atom by the movement of electrons**. This idea of charge flowing can be confusing at first, as we are actually talking about the movement of electrons changing the net charge of each atom they pass through – the *change* in net charge *moves* with them. A useful analogy for this is a Newton's Cradle, where the inner spheres effectively do not move, but the force of the initial collision is transferred *through* them to reach the outer sphere (Figure 1.5).

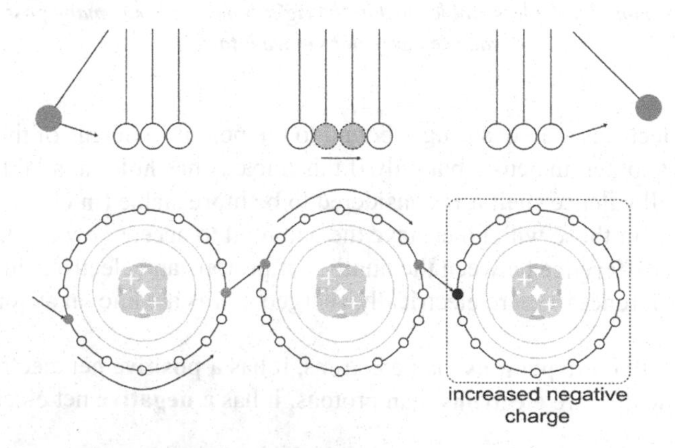

Figure 1.5 Newton's cradle analogy for electrical charge. ***In this analogy, the spheres in the cradle are similar to atoms in a material. The force of the collision between the first and second sphere is transmitted to the sphere at the other end of the cradle, but (effectively) the other spheres do not move. In a similar way, as valence electrons move from one atom to another so the net electrical charge changes for each atom. The atoms themselves do not move, but their charge changes as electrons move between them.***

Now we have a basic understanding of electrical charge and how it moves, we can define one of the fundamental units of electronic circuits:

> **Current is the flow of electrical charge over time** – *measured in amperes (A)*

One ampere of current is equivalent to one coulomb of charge flowing for one second of time. Time is the important difference between charge and current, as it is used to define the **rate** at which charge flows – 1 coulomb of charge flowing every 2 seconds would be 0.5A of current. You may already have encountered amperes (often called amps) as a common means of rating electrical and electronic equipment (Table 1.2).

Table 1.2 Listing of common current ratings. ***The table is a (broadly) indicative list of current ratings for common household and consumer equipment. We will be working with low-current circuits in this book, but many audio devices (particularly in live sound) operate at a high current rating.***

Device	Current	Notes
Light Emitting Diode	20 - 30mA	LED's can easily burn out, so checking current ratings is important. We covered measurement scales in the previous section, and the current here is measured in milliAmps (mA)- which means a factor of x10-3 must be applied to the value to scale it correctly.
Phone charger (Computer USB2)	500mA	The current rating is limited on the USB cable protocol when used with laptop or desktop computer ports- mainly so it will work with a wide range of devices such as keyboards, mice or webcams without damaging them. As we will learn, power is the product of current and voltage (), which means a dedicated mobile phone charger plugged into a wall socket will draw more power than a USB2 connection and so charge quicker.
Phone charger (Computer USB2)	900mA	
Phone charger (wall socket)	1A	
Laptop Computer	1 - 2A	Laptops are battery powered, and so they draw much smaller currents than desktops. For many applications a desktop machine is now not necessary, though many media production and rendering tasks still require more powerful hardware than found in laptops.
Desktop Computer	3A	
Microwave	4.5A	The power usage for devices like microwaves, hair dryers and kettles is very high, hence you can often see the electricity meter moving faster when they are being used. In the case of a kettle, the current is much larger that a laptop computer so it will draw much more power (but usually for a shorter period of time).
Hair Dryer	10A	
Kettle	13A	

Current ratings are very important, as supplying the wrong level of electricity to a device can easily damage or destroy it. Similarly, a high-current device will require a connection that can provide that current – thus a mobile device charger will not be able to power a kettle, but a 13A kettle plug will deliver enough current to destroy a mobile phone battery. The size of an electrical current is effectively the amount of electrons moving past a given point, and the increased energy created by more electrons moving generates heat that requires thicker cables and more insulation to protect those handling them (Figure 1.6):

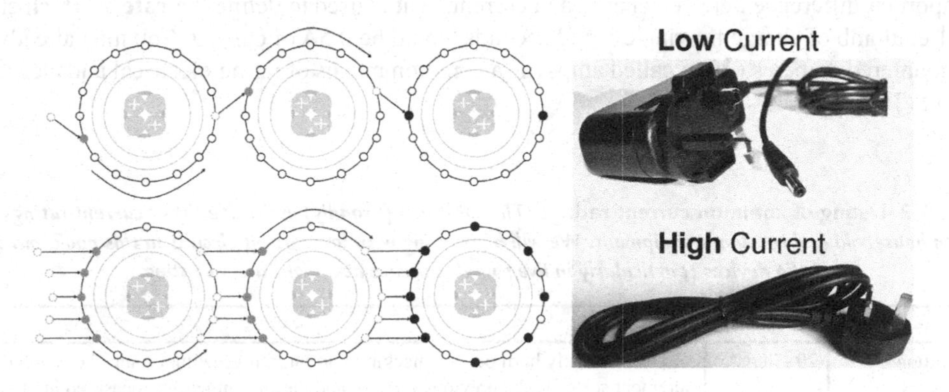

Figure 1.6 Illustration of low and high current flow. ***The upper illustration shows a low current, where a small number of electrons are displaced and thus a thinner wire can be used without overheating (a transformer is used to step down the mains supply). The lower illustration shows a high current, where the heat generated by a larger number of electrons moving requires a much thicker wire to conduct it safely.***

Conductivity defines the ability of a material to allow electrical charge to pass through it, where metals like copper have a high level of conductivity because they have less stable valence shells than other materials. This means copper wire is commonly used to conduct electricity, and thicker copper wires are able to move more electrons without overheating. We will learn more about resistivity (which is the opposite of conductivity) in the next section, but for now the next important concept to define is that of **potential difference**. You may already be familiar with the term **voltage**, but this is actually the *unit of measurement* for the potential difference between two points in an electronic circuit. We will use potential difference extensively in our audio circuit calculations, but it is often misunderstood as a concept and hence can cause confusion when learning the basics of electronic circuit theory. Potential difference is a term analogous with other areas of physics, and so we can use other forces to explain it more easily. For example, you may have learned that the potential energy of an object at a certain height is greater than that of one at a lower level – in this case the effect of gravity means that the higher object has more *potential* to move within the gravitational field by falling to the ground (Figure 1.7).

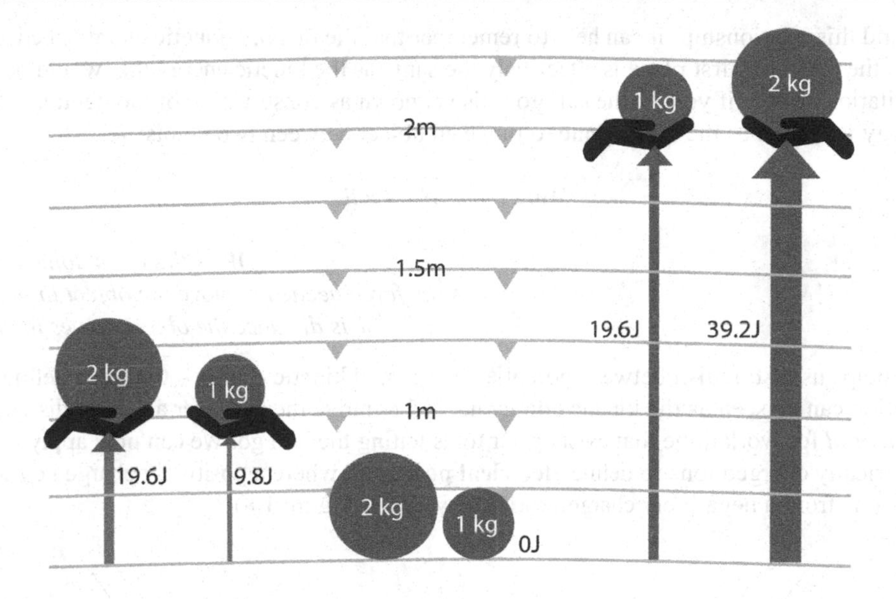

Figure 1.7 Potential energy example. ***In the diagram, the Earth's gravitational field (light grey lines) exerts a force that pulls objects towards it. The hands on the left hold two spheres of mass 1kg and 2kg at a height of 1m. The hands on the right also hold two spheres of mass 1kg and 2kg, but at a height of 2m. Notice that the spheres on the ground have no potential energy (measured in joules), and that the lower 2kg sphere has the same potential energy as the higher 1kg sphere.***

The Earth's gravity is a force that pulls objects towards it, and the further away an object is from the surface of the Earth the more gravitational energy needed to pull it back. An object at a certain height above the Earth is defined as having potential energy because there is no actual force being applied to the object yet – only the **potential** for that force to be applied if the object was free to move. Taking the diagram in Figure 1.7 as an example, if you hold a ball (sphere) in your hand you are providing the **kinetic energy** needed to cancel out the potential energy of the ball to fall to Earth. If you let go of the ball, the potential energy of the ball is now converted into kinetic energy by the force of gravity acting upon it – the ball falls to the ground. We measure potential energy in the Standard Unit (SI) of joules (J) as the product of mass, gravity and height (from the surface of the Earth) using the equation:

$$\textit{Potential energy}, P = m \times g \times h \qquad (1.2)$$

P *is the potential energy in joules (J)*
m *is mass of the object in kilograms (kg)*
g *is acceleration due to the gravity of Earth in metres per second (m/s^2) – this value is defined as a constant of 9.8 ms^2*
h *is the height of the object in metres (m)*

To understand this relationship, it can help to remember that the *original* kinetic energy needed to lift the ball into the air in the first place is effectively the same as the kinetic energy that would be created by the gravitational force if you let the ball go (this is known as conservation of momentum). This kinetic energy is known as the **work done** to move an object between two points:

$$\text{Work done, } W = f \times d \tag{1.3}$$

W is the work done in joules (J)
f is the force needed to move the object in newtons (N)
d is distance the object moves in metres (m)

Work done helps us distinguish between potential energy and kinetic energy – the force defined in the above equation can be seen as the kinetic energy needed to move the ball over a certain distance, rather than the *potential* for work done that exists prior to us letting the ball go. We can now apply the same idea to electrically charged ions to define **electrical potential**, where a positively charged **cation** will attract electrons from a negatively charged **anion** towards it (Figure 1.8).

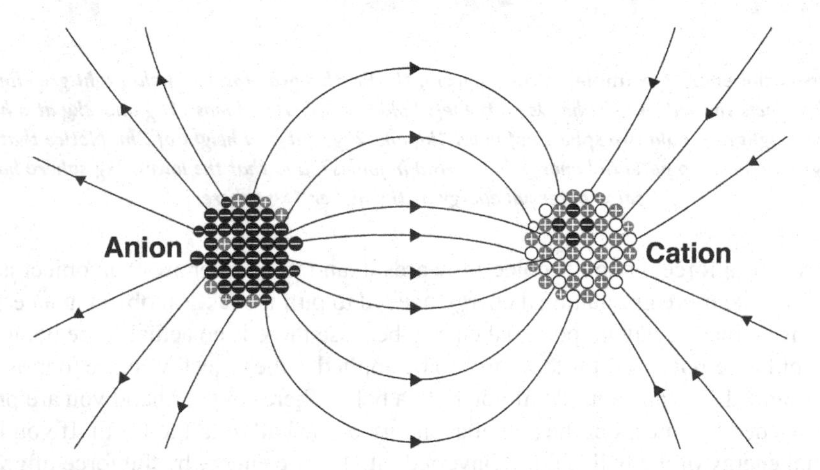

Figure 1.8 Electric potential example. ***In the diagram, the holes in the positively charged cation on the right attract free electrons from the negatively charged anion on the left. Thus, we consider the positive cations to have a higher potential to move charge – though the actual movement requires work done. The directional lines represent the electric field that exists between the two charges, which in this case is defined from negative to positive.***

In the diagram, the positively charged cation on the right has holes that can accept valence electrons from the negatively charged anion on the left. This means that the positive cation has the potential to move charge by attracting electrons towards it. The directional lines represent the resulting **electric**

field that is created by the potential difference between the two charges, showing where there is potential for the movement of electrons (to restrict the size of the diagram, many of the wider field lines have been truncated). These lines are much like those shown in Figure 1.7, which indicate the direction of force exerted on the mass of an object by the gravitational pull of the Earth. As a result, we consider the positive charge to have a higher potential than the negative, which allows us to define the **potential difference** between the two charges as the work done in moving electrons between them:

$$\textit{Potential difference}, V = W \times C \tag{1.4}$$

V is the potential difference between the two charges, in volts (V)
W is the work done in moving electrons between charges, in joules (J)
C is the number of electrons being moved, in coulombs (C)

The equation above is similar to our previous example in Figure 1.7 of potential energy – in the same way that gravity must do work to move an object that has potential energy closer to Earth, so the work done in moving a certain amount of charge is *equivalent* to the potential for that charge to move. The actual potential difference is governed both by the charge being moved (in coulombs) and the work done (in joules) to move it. This allows us to define our second fundamental concept of potential difference:

Potential difference is the difference in electric potential between two points – *measured in volts (V)*

We measure potential difference (in volts) as the work done (in joules) needed to move a charge (in coulombs) between two points – **1 volt equates to 1 joule of work to move 1 coulomb of charge**. The main problem you will encounter is that potential difference is commonly referred to as **voltage** – when actually what is meant is a potential difference *measured* in volts! To keep things simple, voltage will be used as it is the most common term you will encounter in electronics. Having said this, it can often be difficult for new learners when multiple terms are used interchangeably. We will be using voltage extensively in this book, so spend a little time reading this section again until you are comfortable with the concept of the potential to move charge.

As you now know, voltage is the **potential** for an electrical charge to move – *not* the work done by electrons when they are moving as an electrical current. The simplest practical example of potential difference is a battery, where chemical reactions between different elements are used to produce **ions** (Figure 1.9).

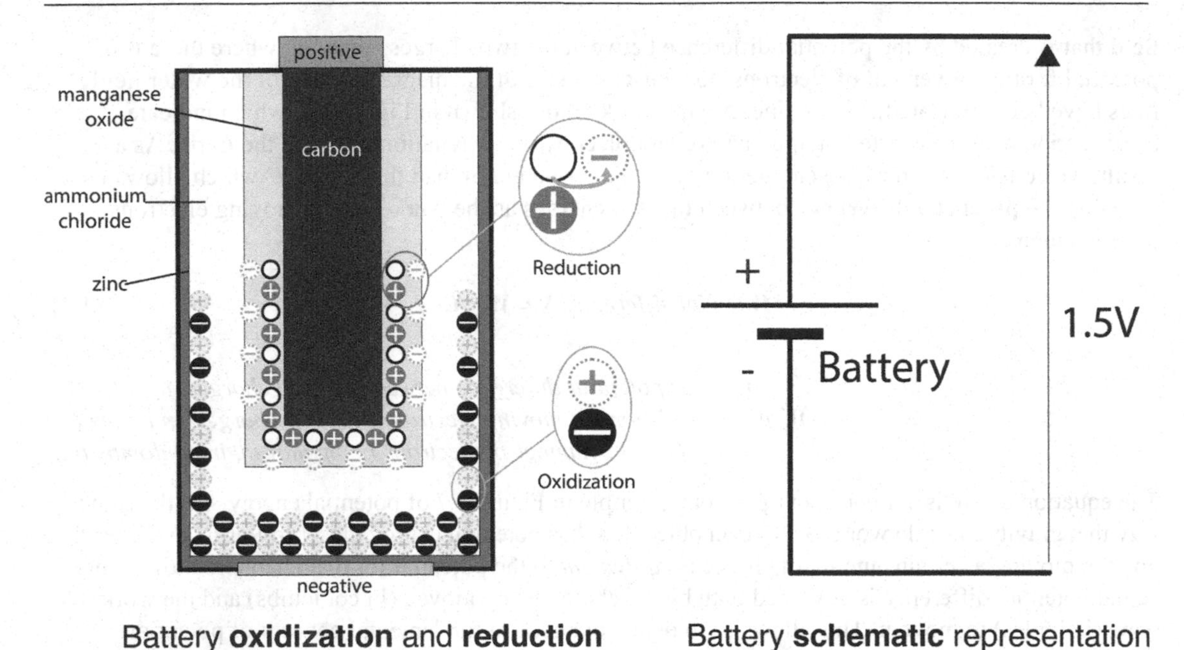

Figure 1.9 A zinc-carbon battery. ***The left-hand diagram shows how a battery uses the chemical processes of oxidization and reduction to create anions and cations (respectively). The right-hand diagram shows a schematic representation of a battery as an electronic component, where the potential difference between negative (−) and positive (+) is shown for a standard 1.5V battery.***

In the above example, zinc atoms **oxidize** (dissolve) in ammonium chloride leaving spare electrons behind, which creates an overall negative net charge of **anions** that collect at the metal base plate. At the same time, the manganese oxide and ammonium chloride **reduce** (removing electrons by using them to produce hydrogen) to leave holes and thus create an overall positive net charge of **cations** that collect around the carbon conductor (chosen because carbon does not oxidize like metals).

We now have a negative and positive terminal, and just as we saw in Figure 1.8 the imbalance between the charges at each terminal creates a potential difference between them. The right-hand diagram shows a *schematic representation* of the battery, where the arrow going from negative to positive indicates the potential difference between battery terminals. We refer to linked combinations of electronic components as circuits because they provide a *circular* path for current to flow. The schematic of the battery shown above is known as an **open circuit** – there is no path for a current to flow. If we were now to connect a single wire between the anode and cathode terminals of our battery, we would create what is known as a **short circuit** (Figure 1.10).

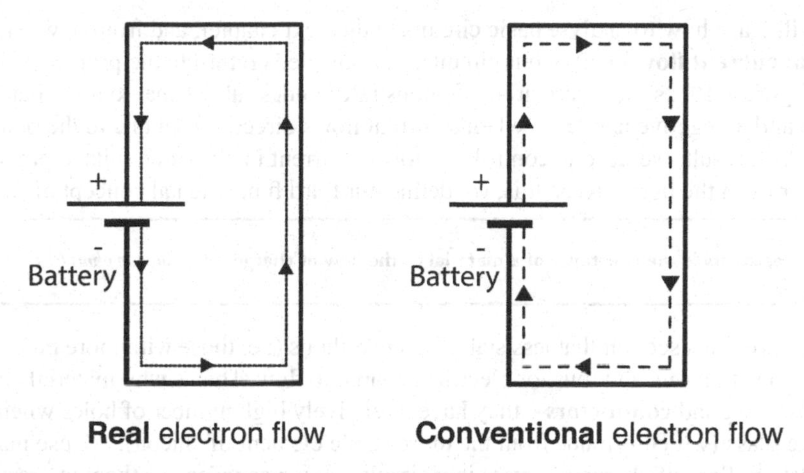

Figure 1.10 Conventional current flow in a short circuit. ***The schematic diagram shows a direct wire connection between the anode and cathode terminals of a battery, known as a short circuit. The left-hand diagram shows the actual flow of electrons from negative to positive, whilst the right-hand diagram shows the direction of conventional current flow used in circuit analysis.***

As current can now flow through the circuit from anode to cathode, the battery terminals will start to overheat due to the free movement of electrons. This can cause damage to the battery, other components or even the user of the circuit – thus we always avoid short circuits in practice! However, the biggest conceptual problem with the above diagram is the definition of **conventional current flow** on the right – how can this be conventional when electrons move in the **opposite direction** from negative to positive!

The influence of history

In **conventional current flow**, the circuit is analysed with the **current flowing from positive to negative**. This can be *very confusing* as we have just learned that the movement of electrons (which are *negative*) carries charge as a current – so why not use this direction from negative to positive?

The reason is that the concept of electricity ***predates*** *the discovery of electrons, and at that time the assumption was that current flowed from positive to negative. As with many things in life, precedence defines convention and so conventional current flow has helped to confuse students ever since!*

As we will learn in chapter 3, the total voltage and current in any electrical circuit are calculated in a single direction. In practice, the *actual* direction of the flow of charge is not important if we **consistently analyse in one direction**. You may also have noticed that Figure 1.8 refers to an electrical field that '*in this case is defined from negative to positive*' and this is again (partly) due to the convention of positive to negative – many electric field diagrams will reference the movement of a positive test charge towards a negative!

To this day, we still define circuits in terms of their positive potential difference and then analyse the flow of current going from positive to negative (or ground). You will find that every textbook (and university electronics programme) teaches in this way, so the same convention is used in this book to be consistent with the rest of the field. As our primary focus is to learn how to apply electronics theory to audio circuits, the direction of current flow is noted purely to disambiguate the concept for your own understanding. Though the flow of conventional current is effectively incorrect, in practice it makes *no difference* to the electronic calculations we will perform as long as **we always calculate in a single**

direction. We will learn how to analyse basic circuits in the next chapter, and from now on **we will use conventional current flow** in all of our circuits. For now, let's return to the problem of our battery overheating in Figure 1.10, as this short circuit demonstrates undesirable behaviour – a battery has a positive cathode and a negative anode, so when a current flows directly from one to the other it creates significant heat. As a result, we need to control the flow of current in all our circuits to prevent damage to components (or even the user). To do this, we define our third fundamental concept of resistance:

Resistivity is the resistance of a material to the flow of charge – *Measured in ohms (Ω)*

We learned in the previous section that less stable valence shells (i.e. those with more holes) make it possible for electrons to move and thus for electrical charge to flow. This is why materials like copper (and other metals) are good **conductors** – they have a relatively high number of holes when compared to **insulators** like glass (which is made from the more stable element of silicon). We use materials like copper to conduct the flow of charge (current) in a circuit, but we can also use them to create **resistance** within a circuit. To do this, we can combine some of what we learned in Figure 1.3 about current and the thickness of a wire with what we learned in Equation 1.3 about work done being related to the distance between points of electrical potential to understand how a wire wound resistor works (Figure 1.11).

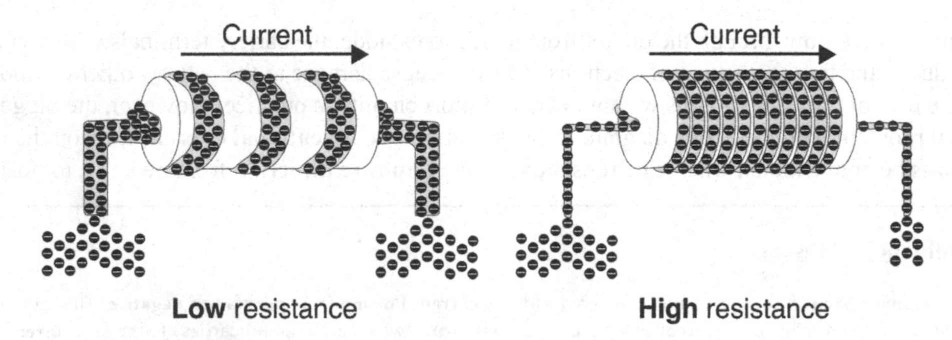

Figure 1.11 Wire wound resistor examples. ***The examples show a coil of copper wire (in grey) wound around an insulating core (in white), where the thickness of the wire and the number of turns (distance) dictate the overall resistivity of the component. In the left example, the thicker wire and lower number of turns allow more current to flow – it has low resistance. In the right example, the thinner wire and higher number of turns make it more difficult for current to flow – it has high resistance.***

In this diagram, the copper coil in the left-hand example has thicker wire – it can carry more charge. This coil also has fewer turns, and so the current has a shorter distance to travel within the resistor (recall Equation 1.3 where *Work done*, $W = f \times d$). This means it presents *less resistance* to the flow of current than the right-hand example, where a thinner wire with more turns will do more to resist the flow of charge – it will have a **higher resistance** value. There are many ways to create resistive components (e.g. carbon pile/film/printed) but wire wound resistors like those shown in the diagram above are particularly important because audio applications often use long cables for input (e.g. microphones) and/or output (e.g. loudspeaker) signals. Thus, understanding that **all conductive materials have a resistance** that varies with both length and thickness is of practical benefit when analysing the input and output requirements for amplifier circuits (we will learn more about this in chapter 2).

Among other fundamental contributions to electronics, George Ohm defined resistance as the relationship between the resistivity of the material used, the distance (length) the current must travel and the surface area (thickness) of the material involved:

$$\textit{Resistance}, \ R = \frac{\rho \times L}{A}$$

***R** is the resistance in ohms (Ω)*
ρ is the resistivity of the material used in ohm metres (Ω m)
***L** is length of the material in metres (m)*
***A** is surface area of the material in square metres (m^2)*

Resistivity varies significantly between materials, with a conductor like copper having a value of around 1.72×10^{-8} Ω m (0.0000000172), whilst an insulator like glass has a resistivity of around 1×10^{14} Ω m (100000000000000). Resistance values are defined by colour-coded bands marked on the body of the resistor as shown in Figure 1.12

Colour	1st Band	2nd Band	3rd Band	Scale	Tolerance
Black	0	0	0	1	
Brown	1	1	1	10	1%
Red	2	2	2	100	2%
Orange	3	3	3	1k	
Yellow	4	4	4	10k	
Green	5	5	5	100k	0.5%
Blue	6	6	6	1M	0.25%
Purple	7	7	7	10M	0.1%
Grey	8	8	8		0.05%
White	9	9	9		
Silver				0.01	10%
Gold				0.1	5%

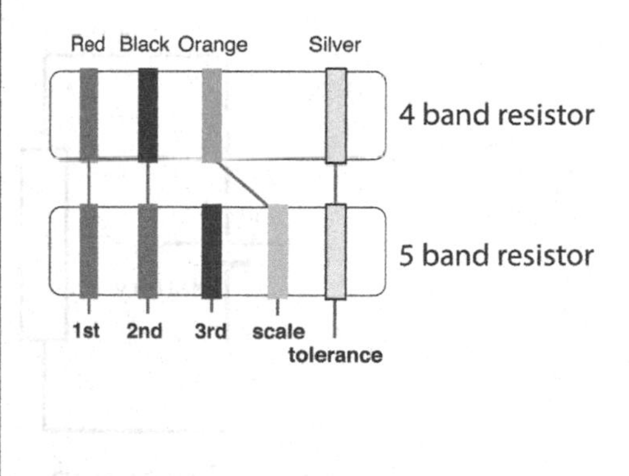

Figure 1.12 Resistance value examples. ***The colour coding of the bands on a resistor indicates both the resistance and the tolerance (accuracy) of the component involved (shown in grayscale in the right-hand images). For a 4-band resistor (top), the first two bands are the value and the third is the scaling factor. For a 5-band resistor (bottom), the first three bands are the value, whilst the fourth is the scaling factor. In both cases the last band is the tolerance value, which is often silver (10%) or gold (5%) – or omitted (20%) for less accurate resistors.***

The first example in the diagram is a 4-band resistor, where the value is defined by the first two digits (colours of red and black – giving 20) multiplied by the scaling factor (orange for 1k) to give a resistance of 20kΩ. The second example is a 5-band resistor, which uses three digits (220) multiplied by the same scaling factor (1k) to give 220kΩ. These examples are of low tolerance (silver means a value of 10%) and in practice this can often mean a cheaper component – even though it is marked with silver or gold!

Reading resistance values

One difficulty in reading resistance values is the size of the components, which can make it difficult to see the individual colours. In addition, most people have different levels of colour perception and so colours like red and brown can often be difficult to distinguish. When you move beyond breadboard circuits it is worth investing in a soldering vice that has both clamps and a magnifying glass, but a quick option when starting off is to **use the camera on your phone to magnify the component** to get a better look at the bands.

When working with lower-tolerance resistors, you can use the fact that there is a metallic silver or gold band (or none at all) to orient the component in the right direction. You will often find that the tolerance band is **sometimes slightly thicker** than the others, though this is not always true! These issues are often compounded by working long hours and/or later at night under artificial light, which can lead to mistakes that can impede your learning. It is essential to **store and label your resistors by value**, either in clearly marked tubs or even plastic bags to avoid getting confused between values when working on a circuit (you can even invest in component shelves as you progress). There is nothing more demoralizing than discovering you have used the wrong component value in a circuit (often after many hours of checking every other possible problem instead!) so try to develop good habits from the start - they will pay you back later in your studies!

In the next chapter we will learn how resistance can be used by sensors to measure changes in real-world quantities, and in chapter 3 we will learn how to calculate the resistance needed for a real circuit. For now, we can revisit our battery short circuit from Figure 1.10 to see how we can include a resistor to prevent the battery from overheating (Figure 1.13).

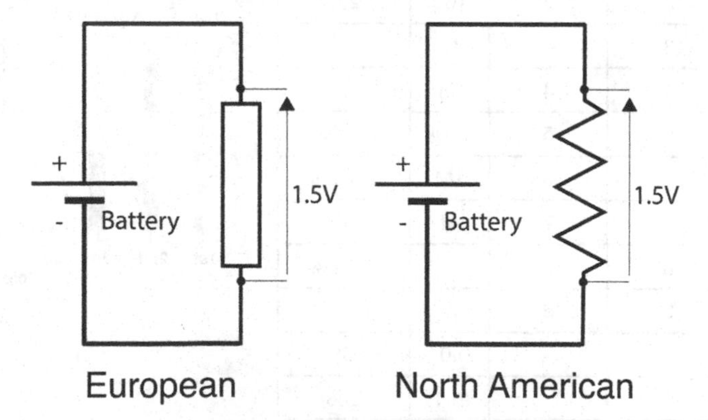

Figure 1.13 A simple battery circuit. ***The diagram shows a battery connected to a resistor, which is the simplest practical electronic circuit. The potential difference of 1.5V is now across the terminals of the resistor, which prevents the battery from overheating. The left-hand diagram shows the standard European symbol for a resistor, which will be used throughout this book. The right-hand diagram shows the equivalent North American standard symbol, which you will see in many circuit examples.***

In this circuit, the potential difference between the terminals of the battery is now 'seen' across the terminals of the resistor and so a short circuit across the battery is avoided – the potential for charge to move is now related to how much the resistor reduces the current flowing through it. It is important to remember that the resistor resists the flow of charge, and in chapter 3 we will learn how to use Ohm's Law to calculate the required resistance needed when using power either from our Arduino board or from a separate battery.

In practice, if we connected a voltmeter across the terminals of the resistor to measure the actual voltage the connections between the resistor and battery will also present a (small) resistance to the flow of current (remember that **all wires have resistance**) and thus we will see a slightly lower voltage reading across the resistor than across the battery itself. This is particularly relevant in audio circuits, where we will often work with signal cables that are several metres long – even longer in live sound systems. These cables will have an increased resistance (and also a capacitance) due to their length, and in practice it is important that we bear this in mind when designing our audio systems to work with them. For now, we have a circuit where the flow of current is controlled by the resistor and we can use this as the basis of our first practical electronic circuit to light up a light-emitting diode (LED) (Figure 1.14).

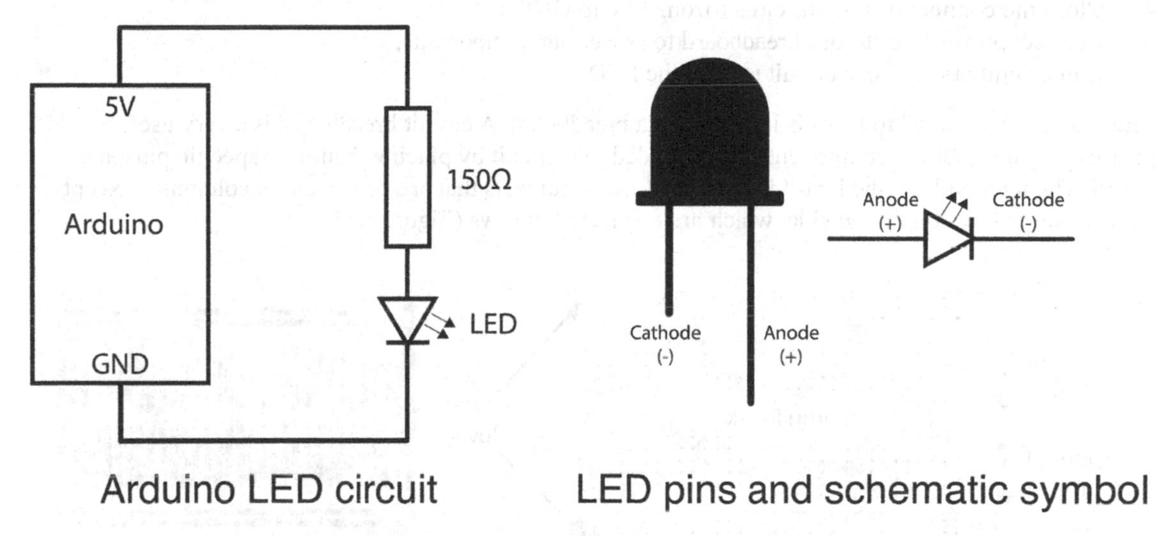

Figure 1.14 A simple light-emitting diode (LED) circuit. ***The left-hand schematic diagram uses the 5V and ground (GND) pins of an Arduino to replace the battery power supply in our previous example. A 150Ω resistor is connected in a series circuit with an LED light to prevent the current (in blue) from burning out the LED component. The right-hand diagram shows the positive anode (long leg) and negative cathode connections of an LED, alongside its schematic symbol.***

In this circuit, the 150Ω resistor is used to limit the current flowing into the LED to prevent it from burning out. Without the resistor controlling the amount of current, the Arduino would supply too much current for the LED and it would burn out (creating a short circuit). Notice that LEDs have a specific polarity, where the **longer** leg is the **positive** terminal of the component. You may also notice that the words anode and cathode seem similar to anion and cation – this is another example of the confusion that can be caused by conventional current flow! We will learn more about how this circuit works in chapter 3, but for now the tutorial below will use it to introduce the Tinkercad application that we will use for some of our early circuits.

1.3 Tutorial – introduction to Tinkercad

Tutorials are used throughout this book to learn how to **apply** the concepts we have learned. All tutorials follow a series of steps that should allow you to successfully complete the task involved. It is **vital** that you **do not skip** any steps, as the tutorials in later chapters are designed to guide you through some of the common problems and mistakes that can occur when learning about audio electronic circuits. In this tutorial, we will do the following:

1. Set up a Tinkercad account and create a new Arduino circuit for simulation.
2. Learn how to read a breadboard layout to plan our circuit before building it.
3. Add components to our breadboard circuit by rotating and placing them correctly.
4. Close the connections in our circuit from +5V to GND.
5. Connect power lines to our breadboard to power our components.
6. Run a simulation of our circuit to light the LED.

The first thing we need to learn is how to read a breadboard. A circuit breadboard is a very useful prototyping tool, where components can be added to a circuit by placing them into specific pins in the board. The underside of the board is a series of metal contacts that are connected in columns – except for the power lines on the outside, which are connected in rows (Figure 1.15).

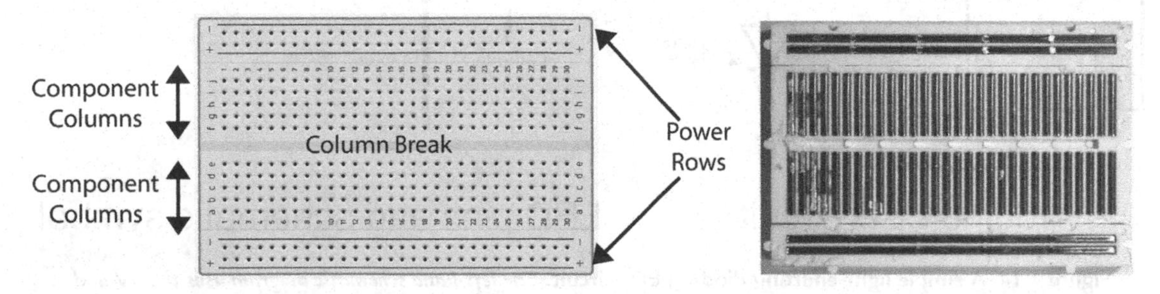

Figure 1.15 Breadboard layout. ***The left-hand diagram shows how power rows (marked +/−) are connected horizontally across the two top (and bottom) rows of the breadboard to provide access to positive and negative (or ground) terminals at multiple points on the board. Component columns are connected vertically (right-hand image) and numbered in sequence, with each pin in a column having a different letter reference. The middle of the board breaks each component column, creating two separate work areas.***

It can initially be confusing to work with a breadboard, as the power lines create a visual cue that suggests all other pins are also connected on rows. It can help to think of a breadboard like a crossword puzzle, where **power** (is) **across** and **signals** (are) **down**. The breadboard is also split along the middle of each column **(a–e, f–j)** to create **two separate columns** on the board – in later projects we will see how this allows us to add a lot more components to a single board. The best way to learn breadboard layouts is by practising with them, but there are some initial guidelines you can follow. Whenever you are designing a breadboard circuit layout, the best approach is to start with the first positive component connection in the circuit and place it somewhere near a power row (leaving one pin free to connect to the power rail itself). Then, add each subsequent component by placing its positive terminal on the same column as the negative terminal of the previous component – they are now linked because every pin in each column is already connected (Figure 1.16).

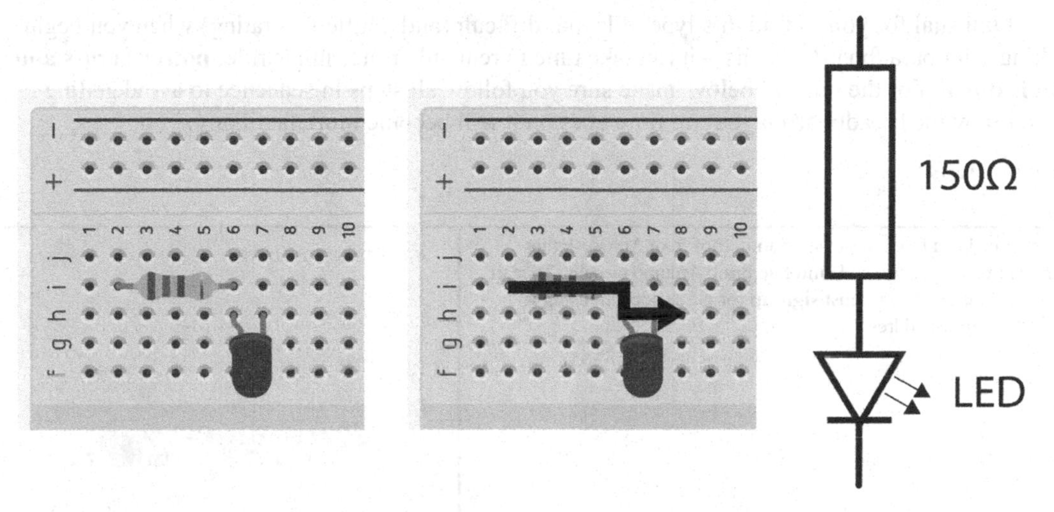

Figure 1.16 Breadboard component layout. ***The left-hand image shows the component layout, the middle image shows the current path and the right-hand image shows the schematic. The two components from the circuit in Figure 1.14 (a resistor and an LED) are connected to each other because both of their pins are on the same column (column 6) of the board – we will use this layout in our circuit below.***

Notice the resistor in this figure is connected on pins **[i2–i6]**, which then allows a connection from the power line above to be added to pin **j2** – because the resistor is also connected to column 2 it will now receive power. The final step is to add a ground connection to column 7, which will link the negative terminal (short leg) of the LED to ground (Figure 1.17).

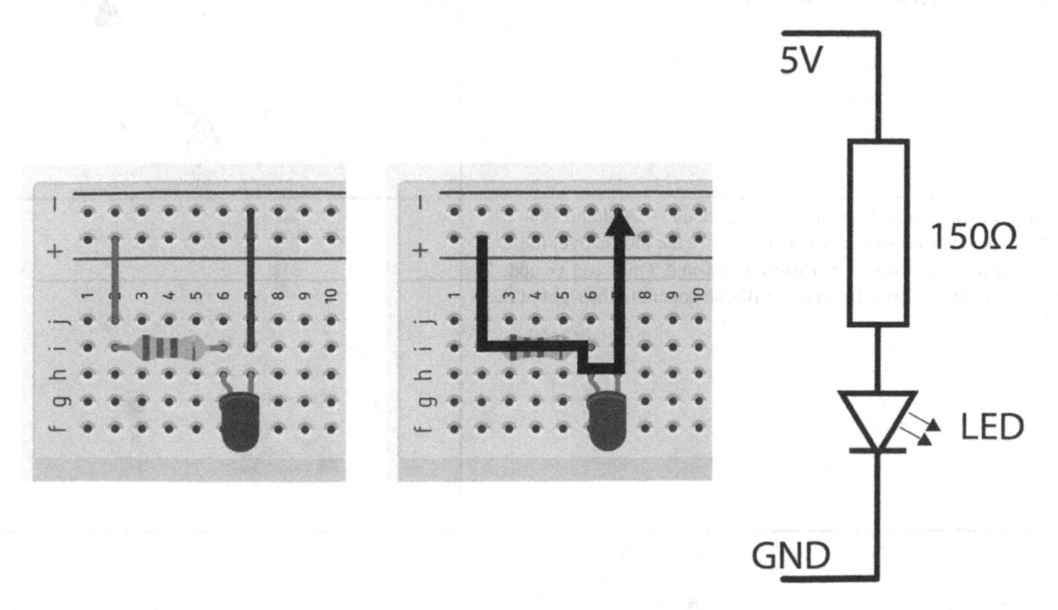

Figure 1.17 Adding power and ground connections. ***The left-hand image shows the component layout, the middle image shows the current path and the right-hand image shows the schematic. The power row is connected to all pins in column 2, which includes the resistor. The ground row is connected to all pins in column 7, which includes the negative terminal of the LED.***

It is not unusual for you to find this type of layout difficult (and a little frustrating) when you begin working with breadboard circuits – it can take time to remember the simple rule: **power across and signals down**. For the tutorial below, make sure you follow all steps in sequence to avoid getting mixed up by the breadboard layout; as time goes on it will become more familiar to you.

Tutorial steps

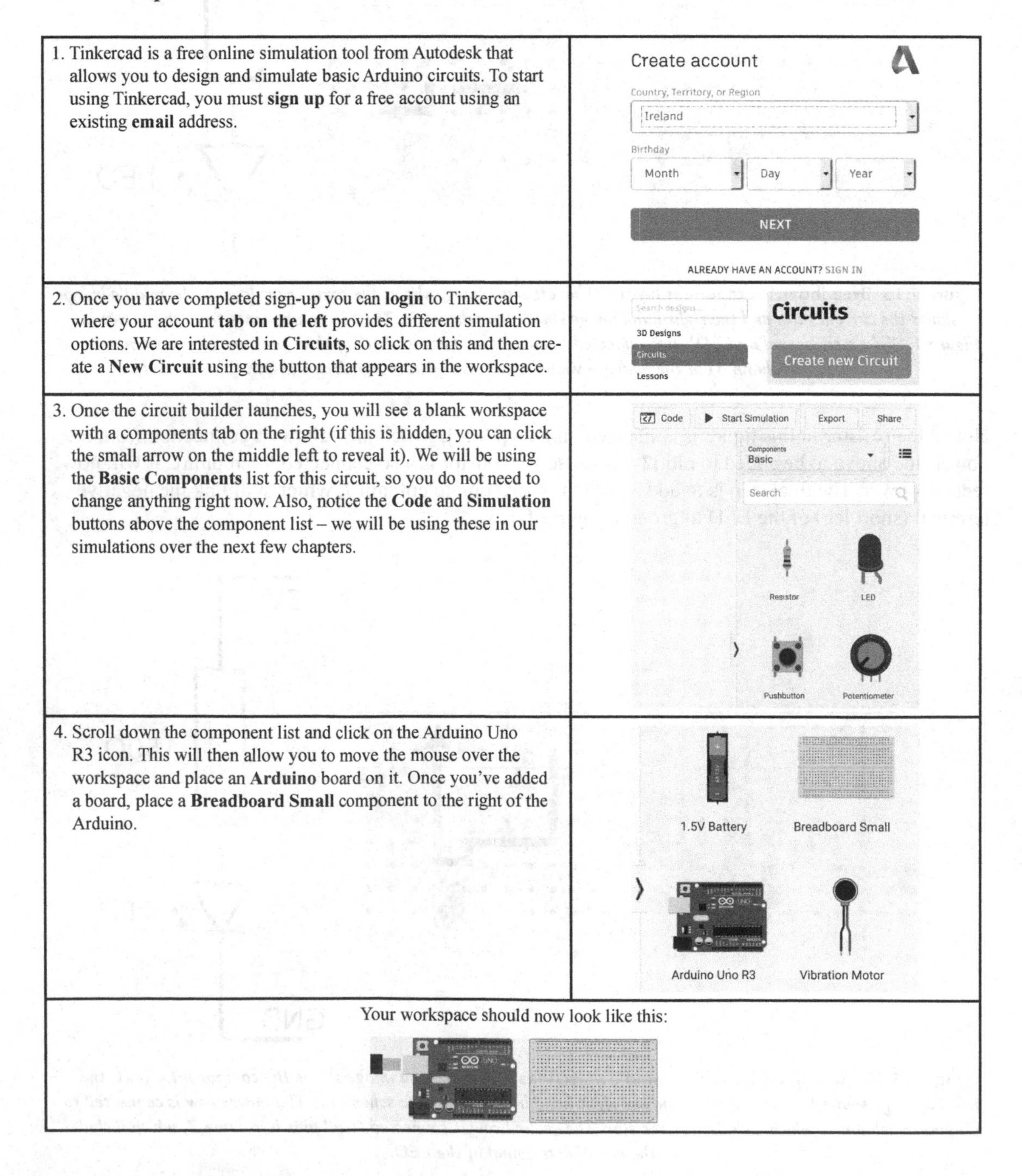

Step	
1. Tinkercad is a free online simulation tool from Autodesk that allows you to design and simulate basic Arduino circuits. To start using Tinkercad, you must **sign up** for a free account using an existing **email** address.	
2. Once you have completed sign-up you can **login** to Tinkercad, where your account **tab on the left** provides different simulation options. We are interested in **Circuits**, so click on this and then create a **New Circuit** using the button that appears in the workspace.	
3. Once the circuit builder launches, you will see a blank workspace with a components tab on the right (if this is hidden, you can click the small arrow on the middle left to reveal it). We will be using the **Basic Components** list for this circuit, so you do not need to change anything right now. Also, notice the **Code** and **Simulation** buttons above the component list – we will be using these in our simulations over the next few chapters.	
4. Scroll down the component list and click on the Arduino Uno R3 icon. This will then allow you to move the mouse over the workspace and place an **Arduino** board on it. Once you've added a board, place a **Breadboard Small** component to the right of the Arduino.	
Your workspace should now look like this:	

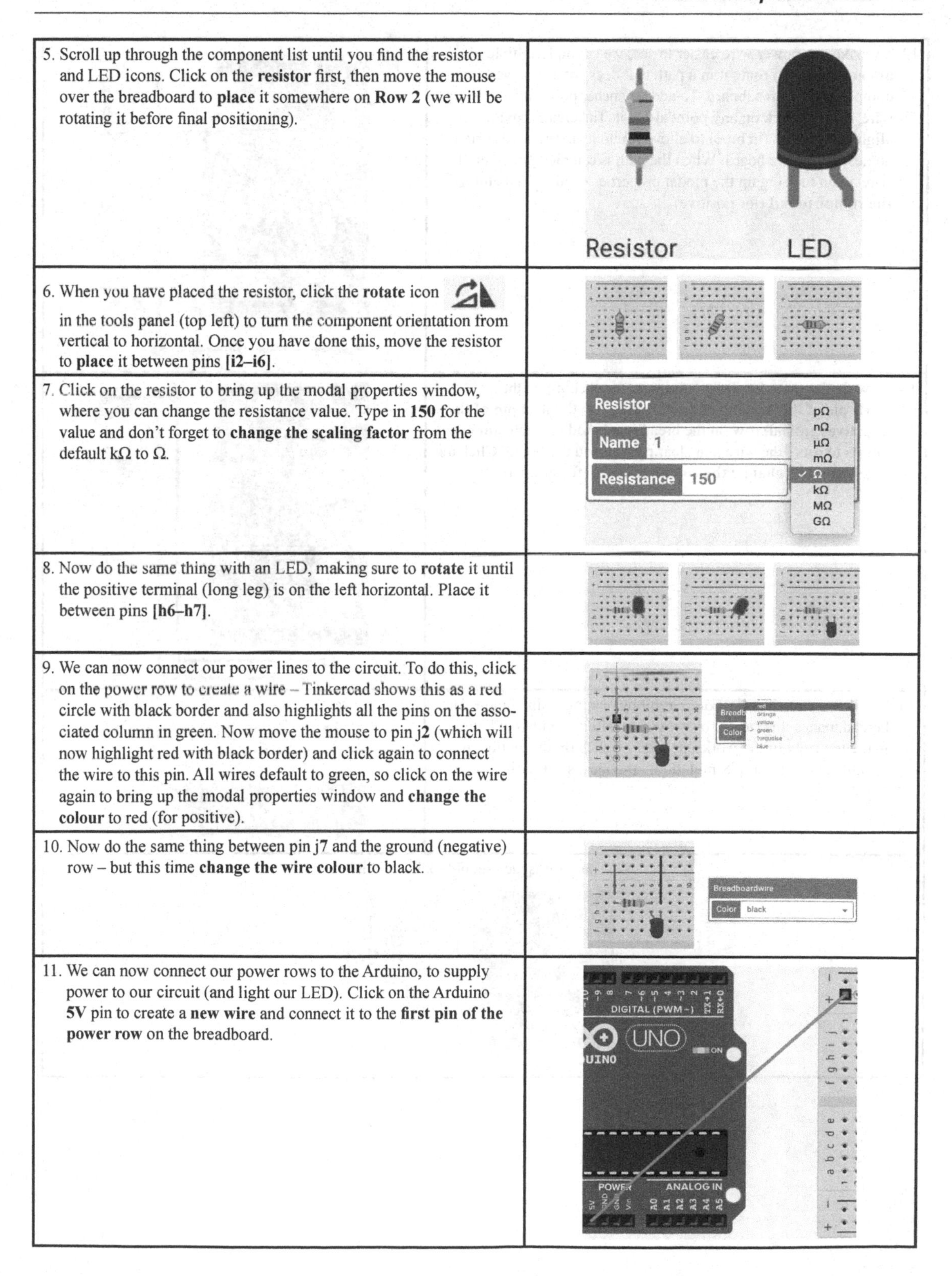

5. Scroll up through the component list until you find the resistor and LED icons. Click on the **resistor** first, then move the mouse over the breadboard to **place** it somewhere on **Row 2** (we will be rotating it before final positioning).	
6. When you have placed the resistor, click the **rotate** icon in the tools panel (top left) to turn the component orientation from vertical to horizontal. Once you have done this, move the resistor to **place** it between pins **[i2–i6]**.	
7. Click on the resistor to bring up the modal properties window, where you can change the resistance value. Type in **150** for the value and don't forget to **change the scaling factor** from the default kΩ to Ω.	
8. Now do the same thing with an LED, making sure to **rotate** it until the positive terminal (long leg) is on the left horizontal. Place it between pins **[h6–h7]**.	
9. We can now connect our power lines to the circuit. To do this, click on the power row to create a wire – Tinkercad shows this as a red circle with black border and also highlights all the pins on the associated column in green. Now move the mouse to pin j2 (which will now highlight red with black border) and click again to connect the wire to this pin. All wires default to green, so click on the wire again to bring up the modal properties window and **change the colour** to red (for positive).	
10. Now do the same thing between pin j7 and the ground (negative) row – but this time **change the wire colour** to black.	
11. We can now connect our power rows to the Arduino, to supply power to our circuit (and light our LED). Click on the Arduino **5V** pin to create a **new wire** and connect it to the **first pin of the power row** on the breadboard.	

12. To make our power wire easier to see, we can add multiple **anchor points** to route it in a path that does not cross over our components or breadboard. To add an anchor point to an existing wire, **double click** on any point along it. Tinkercad provides alignment guides (in blue) to allow you to route the wire at right angles around the board. When the path is complete, click on the wire again to bring up the modal properties window and **change the colour to red** (for positive).

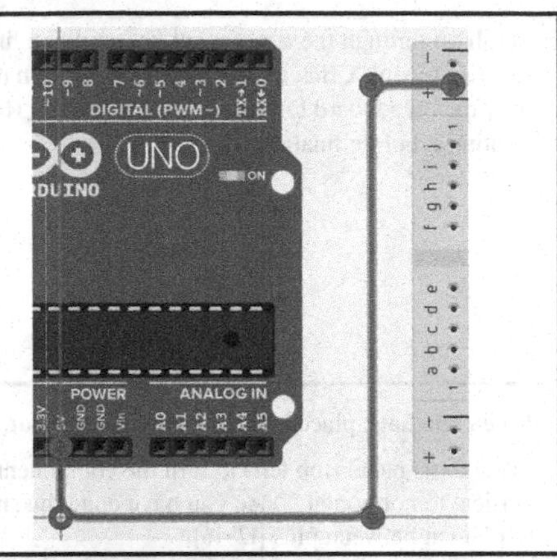

13. Now do the same for the ground wire, by clicking on the **GND** pin of the Arduino and connecting it to the **first pin of the negative** (ground) row on the breadboard. Add multiple **anchor points** to route the wire in a clear path around the board. Click the wire again and **change the colour to black** (for ground).

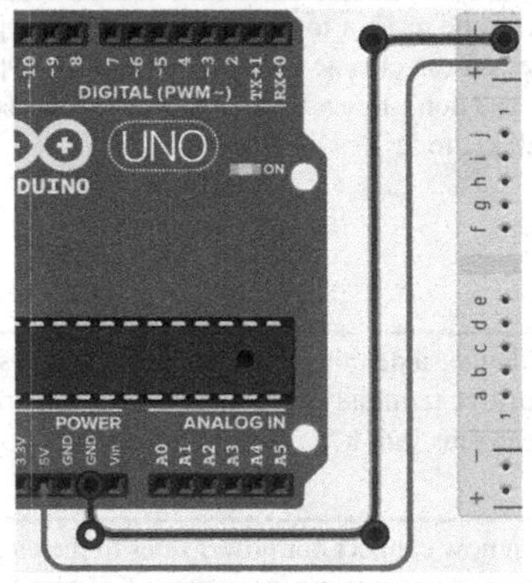

14. We are now ready to test our circuit by running a simulation. Tinkercad makes this simple by providing a Run Simulation button in the top right of the workspace. If you click on this button, it should change to Stop Simulation – this means it is now running.

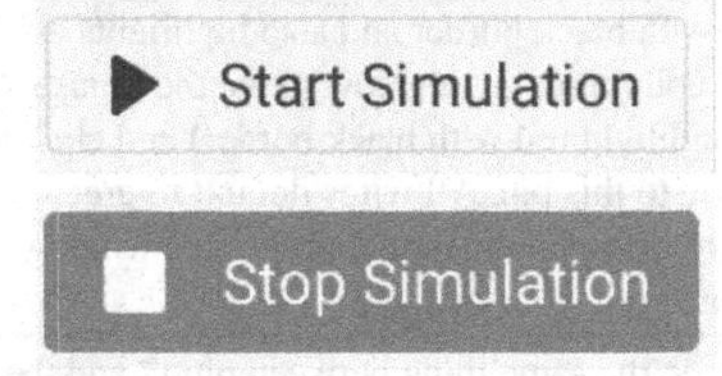

Your workspace should look like this:

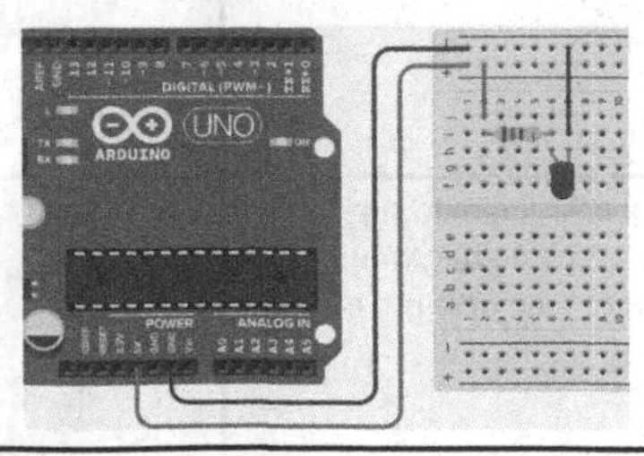

If you have followed all steps in the tutorial and have a lit LED in your simulation window, then **well done** – you have completed your first electronic circuit! We will use Tinkercad for other circuits in later chapters (particularly those that involve Arduino code), so it is important to learn how to use it as quickly as possible to avoid the tool becoming a stumbling block in your audio electronics learning.

At this point, we have now covered all three fundamental elements needed to begin learning about electronic circuits:

Current is the flow of electrical charge over time – *measured in amperes (A)*

Potential difference is the difference in electric potential between two points – *measured in volts (V)*

Resistivity is the resistance of a material to the flow of charge - *measured in ohms (Ω)*

We will be using current, voltage and resistance throughout this book so it can be useful to refer back to this section if you find some later concepts difficult to grasp. At the fundamental level, everything in electronics is about the movement of electrical charge – we measure either this level of movement or the potential for it to move (most of our circuits will use voltage changes rather than current). Resistance is used to control the movement of electrical charge, and this is how all circuits operate – even more advanced components like capacitors and semiconductors are all resistive in some way, so these three elements will form the core of all your learning in audio electronics. The example project below will take you through the steps needed to build the tutorial circuit – one of the aims of this book is to have at least one practical project at the end of each chapter that covers what you have learned. In later projects we will focus more on audio circuits, but the example project aims to use the minimum amount of components to get you up and running with a practical circuit straight away.

1.4 Example project – getting started: an Arduino-powered LED light

The example project will use an Arduino with a 150Ω resistor, LED and breadboard to build the circuit we have just simulated in Tinkercad – which again highlights its usefulness as a prototyping tool for Arduino. To build this project, you will need the following components (Figure 1.17):

1. An Arduino Uno board (or equivalent).
2. A breadboard.
3. A 150Ω resistor.
4. An LED (ideally red, but any colour will do for this project).
5. 4 wires (ideally 2 red and 2 black, to make it easier to follow the circuit path).

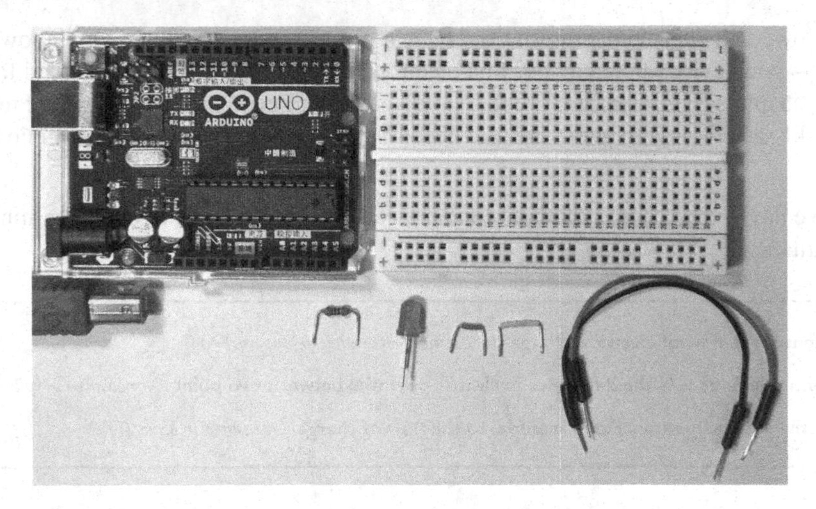

Figure 1.18 Project components. ***The image shows the Arduino, resistor, LED, connector wires and connector cables that will be used to create the project circuit on the breadboard.***

We will build our first circuit powered by the Arduino based on the schematic in Figure 1.19 (used in the previous tutorial):

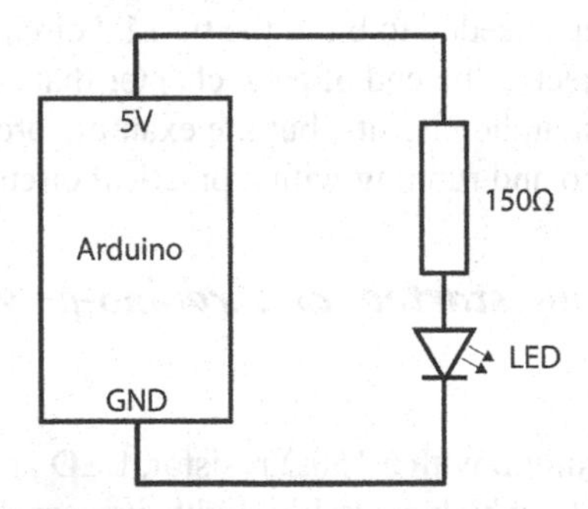

Figure 1.19 Project schematic. ***The image shows the schematic introduced in Figure 1.14 that was used to build the Tinkercad tutorial circuit in the previous section. We will now build the same circuit on breadboard.***

In each project in the book, every build stage begins by showing the schematic for that stage and then its breadboard layout. In the opening chapters, we will also include the current flow for this layout to help you follow the breadboard layout process, and also as a check to ensure you have set everything out correctly (Figure 1.20).

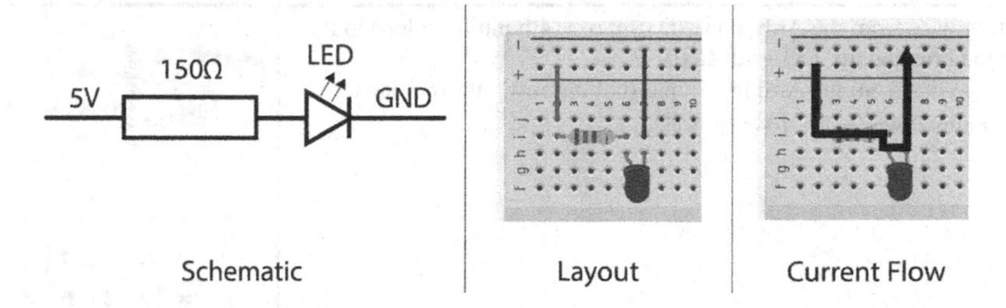

Figure 1.20 Translating a schematic to breadboard layout. ***The first image shows part of the schematic from Figure 1.19, redrawn with inline components to help you follow the circuit path. The second image shows the breadboard layout for the components involved, and the third image shows the current flow through these components.***

Now we have a plan for our component layout, we can begin to build the circuit on breadboard.

Project steps

Start by placing the components (resistor and LED) on the breadboard. 1. Add the 150Ω resistor between pins **[i2–i6]** 2. Add the LED between pins **[h6–h7] – long leg (cathode) on pin h6**	
1. Add a connecting wire (or cable if not available) from pin **j7** to the ground rail on the breadboard. **[j7–GND]** 2. Add a connecting wire from pin j2 to the positive power rail. **[j2–V+]**	

1. Add a connecting wire from the Arduino GND (top row, 4th pin from left) to the ground rail on the breadboard. **[ADgnd–GND]**
2. Add a connecting wire from the Arduino 5V pin (bottom row, 5th from left) to the power rail on the breadboard. **[ADv+–V+]**

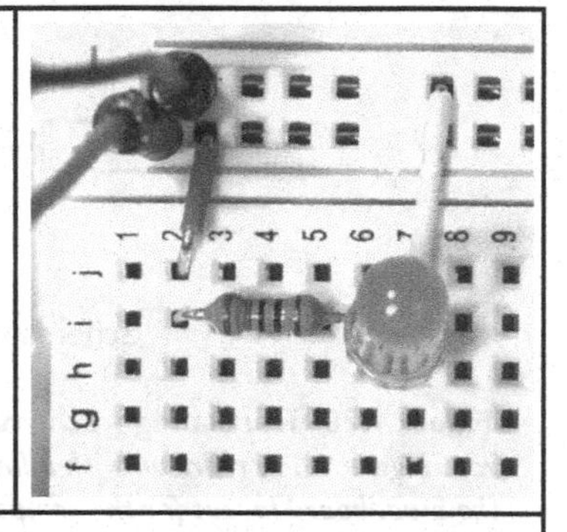

If you have connected everything correctly, the LED should light up (see next page):

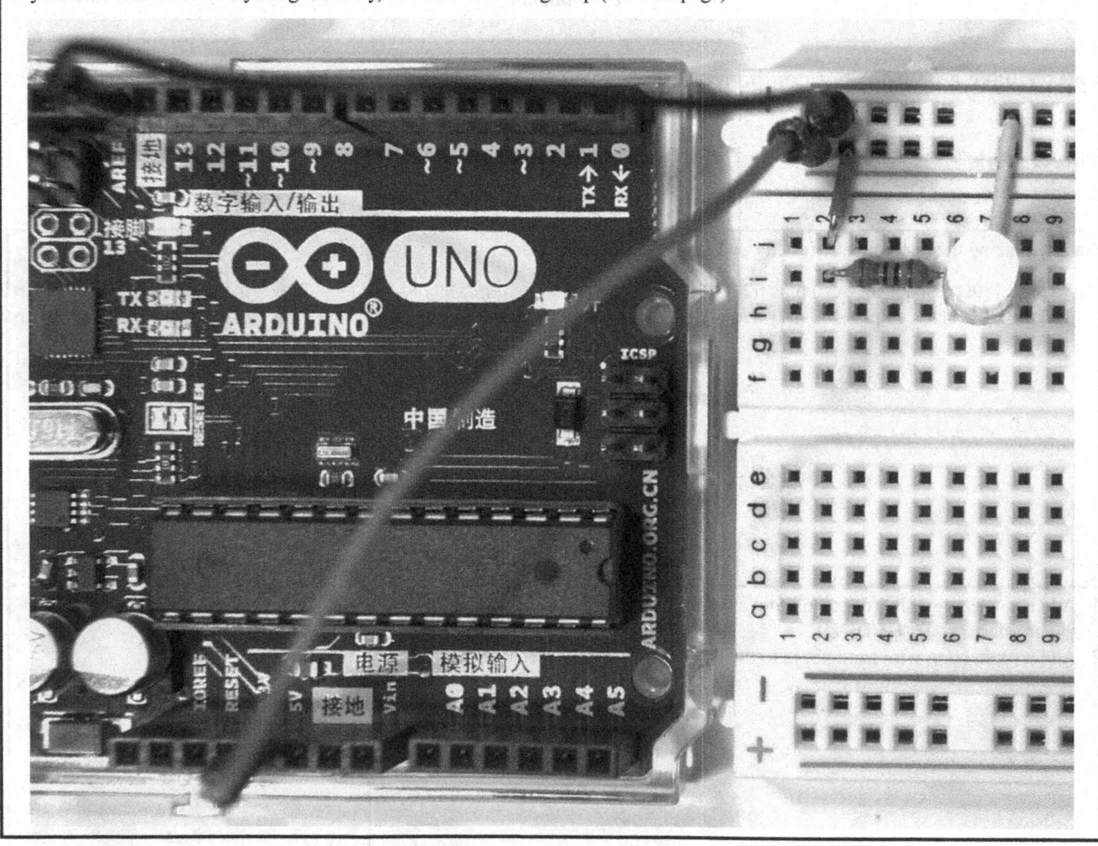

1.5 Conclusions

In this chapter, we have begun learning some of the fundamental methods and concepts of electronics. We found that scales, symbols and equations are effectively the language of electronics – all languages take time to learn. We also covered three fundamental concepts that we will use throughout this book: current, voltage and resistance. Voltage can be a difficult concept to understand, and we will return to it again in chapter 3 to reinforce what we have learned. The tutorial introduced the Autodesk Tinkercad online application for testing Arduino code and circuit designs – we will use this application again in the following chapters. The final project built a prototype of the tutorial circuit, a first practical circuit that lights an LED.

In the next chapter, we will learn the basic elements of an electronic system that combines inputs from sensors with outputs to actuators. This will help us understand the requirements of our audio circuits and what components we need to design them. When we learn how to program the Arduino controller in chapter 4 we will use our knowledge of systems to consider how it can be used to control the processes carried out by our circuits, so it is important to begin with a conceptual understanding of how this is implemented.

1.6 Self-study questions

In each chapter, a short set of questions are provided that aim to enhance your learning of the topic. Such questions can often be ignored because they are either considered by the learner to be too easy (or too difficult), or sometimes the learner is keen to progress onto more interesting topics and doesn't want to be slowed down by what they believe to be yet more examples of the same things they have just learned. One of the reasons for self-study questions is to develop your **practice** of electronic theory, to make sure you build on a set of solid examples that will help you to **reduce the mistakes** you make when learning. As mentioned at the beginning of the chapter, learning can be significantly impeded by an increasing belief that '*this is too difficult*' based on errors or mistakes in your work.

> *If you think you didn't understand the chapter, use the questions to focus on what you still need to learn.*
> *If you think you did understand the chapter – prove it by getting 100% in the questions!*

Scales

The important point with scales is not to skip the steps – even in simple addition this can lead to errors.

1. Write out all the quantities
2. Convert all units to include scaling factors
3. Count the number of digits from the decimal point
4. State the answer to 2 decimal places

Q1: *What is the total current I_{tot}, as a result of adding 100mA and 3mA?*

Q2: *What is the total voltage V_{out} as a result of adding 10µV and 7mV?*

Equations

For adding/subtracting fractions:

1. Find the lowest common denominator
2. Check the numerator result - *if your numerator is greater than 1 then you will need to simplify*
3. Always state the answer to your question clearly

For multiplying fractions:

1. Multiply the numerators and check the result
2. Multiply the denominators
3. Simplify the result – *look for a common factor between numerator and denominator*
4. Check for other common factors – *double-check if your simplified numerator is greater than 1*
5. Always state the answer to your question clearly

For the reciprocal:

1. Flip the numerator and denominator
2. Multiply your original answer by the reciprocal to check it equals 1

***Q1**: Add* $\frac{2}{3}+\frac{1}{4}$*: what is this fraction (as a voltage) of an Arduino 5V power supply?*

***Q2**: What is the reciprocal of* $\frac{1}{3}$ *and what is the reciprocal of* $\frac{3}{4}$ *?*

Electrical charge

As a quick memory test, see if you can answer the following as an explanation in your own words (either to yourself or someone else) without looking back through the chapter.

***Q1**: What is current?*

***Q2**: What is potential difference?*

***Q3**: What is resistivity?*

All answers are provided for each chapter at the end of the book.

Chapter 1 summary

The following images may help you to remember some of the key points from this chapter:

Current is the flow of electrical charge over time – *measured in amperes (A)*

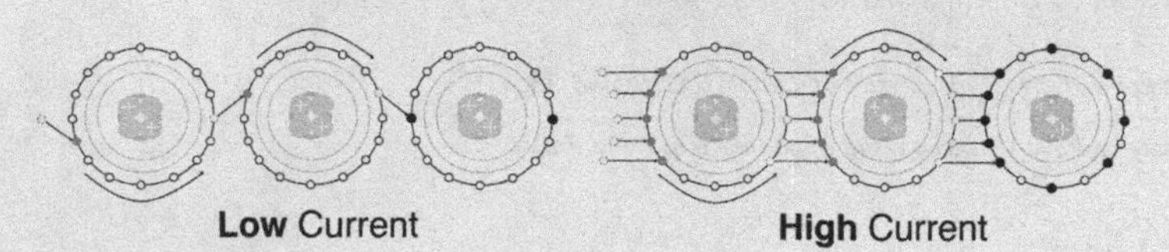

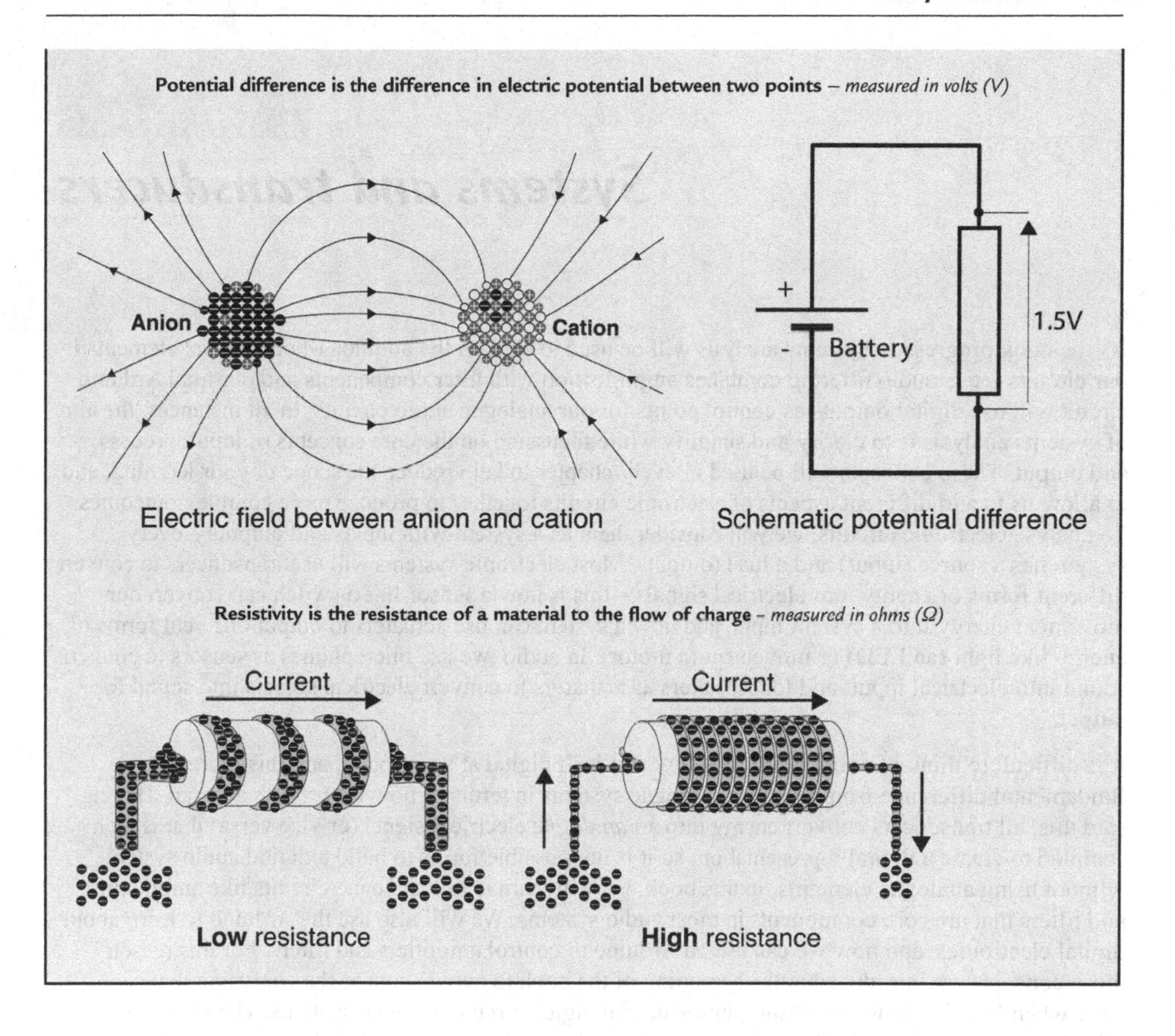
Potential difference is the difference in electric potential between two points – measured in volts (V)
Anion
Cation
+
-
Battery
1.5V
Electric field between anion and cation
Schematic potential difference
Resistivity is the resistance of a material to the flow of charge – measured in ohms (Ω)
Current
Current
Low resistance
High resistance

CHAPTER 2

Systems and transducers

As the book progresses, systems analysis will be used to explain the addition of conceptual elements in our circuits – e.g. audio filtering combines amplification with filter components and our final Arduino circuit will use digital outputs as control points for our analogue audio circuits. In all instances, the aim of systems analysis is to clarify and simplify while focussing on the core concepts of input, process and output. These concepts will be used in every chapter to help reduce the scope of your learning, and to allow us to add different aspects of electronic circuits together to produce more complex outcomes. To analyse electronic circuits, we will consider them as a system with inputs and outputs – every system has a source (input) and a load (output). Most electronic systems will use transducers to convert different forms of energy into electrical signals – this is how a sensor like a switch can convert our movement energy into a system input, and how a system can use actuators to output different forms of energy like light (an LED) or movement (a motor). In audio, we use microphones as sensors to convert sound into electrical input, and loudspeakers as actuators to convert electrical signals into sound for output.

It is difficult to think of a modern technology that isn't digital at some point, and this represents a fundamental difference from previous analogue systems in terms of how we *process* a signal. Having said this, all transducers convert energy into an *analogue* electrical signal (or vice versa) that is then sampled to create a digital representation, so it is not possible for us to build a digital audio system without using analogue elements. In this book, we will learn about analogue circuits like amplifiers and filters that are core components in most audio systems. We will also use the Arduino to learn about digital electronics, and how we can use an Arduino to control amplifiers and filters. For this reason, these concepts are introduced at the beginning of the book to avoid some of the confusion that can arise when learning how to combine analogue and digital circuits in audio systems. The tutorial in this chapter covers some basic component preparation, to allow you to place them on a breadboard more easily. Our project extends the simple LED circuit from chapter 1 to include a switch sensor that controls the flow of current into the LED – this is a basic example of an open-loop system. Although systems thinking may appear fairly straightforward when applied to the basic examples used in this chapter, it is a powerful analysis technique that can significantly reduce the complexity involved when designing electronics circuits. As mentioned in the previous chapter, most equations in electronics are fairly simple to work with but when you combine them with scales, symbols and now sensor and actuator requirements the sum of these elements can quickly become complex. It is recommended that you devote some time to familiarizing yourself with all aspects of the material covered in this chapter; these ideas will be used throughout the book to explain more advanced concepts.

What you will learn

How to understand electronic systems in terms of input/process/output
How digital systems like the Arduino work
How analogue signals are converted to digital data
How MIDI can control audio devices
How transducers convert sensor inputs to electrical signals
How transducers convert electrical signals to actuator outputs
How common types of microphones and loudspeakers work
How to prepare components for breadboard use
How to use build an open-loop system to control an LED actuator output

2.1 Electronic systems and transducers

The Oxford English dictionary[1] defines the word system as '*A set of things working together as parts of a mechanism or an interconnecting network*'. Systems are used throughout this book to analyse and explain each project, and also the associated Arduino code (see chapters 4 and 5) that is needed to control them. In engineering, a distinction is made between the *systematic* (elements that make up the whole) and *systemic* (properties that may relate to more than one part of the system), and this helps us to combine both of these to consider a system as '*any closed volume for which all the inputs and outputs are known*'.[2] This leads to a an *open-loop* system (Figure 2.1).

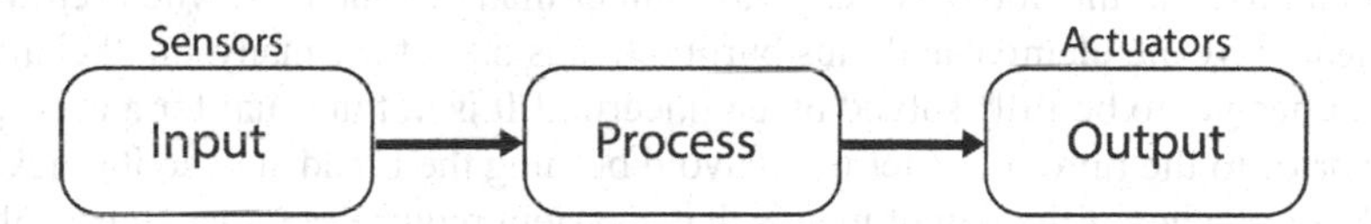

Figure 2.1 Conceptual diagram of an open-loop system. ***In the above diagram, the system processes inputs from sensors to create outputs to actuators. In this way, input to the system controls its output.***

Any system can be modelled in terms of three core elements of input, process and output – **a system processes inputs to create outputs**. The inputs and outputs to our system are provided by **transducers** – something that **converts** one physical quantity to another. You will have encountered many different types of transducers already: a weighing scales converts physical mass to a numeric value, a digital thermometer converts temperature to a digital value, a lightbulb converts electricity to light and a motor converts electricity to mechanical force. We will be using electrical transducers in all of our circuits, where a physical quantity like sound is converted either to or from electrical energy. The important distinction to make is that **input** transducers are called **sensors**, whilst **output** transducers are called **actuators** – we will learn more about them in the following sections. The other part of an open-loop system is the central **process** that uses sensor input to control actuator output.

A simple open-loop system for a toaster can help us better understand the relationship between input, process and output (Figure 2.2).

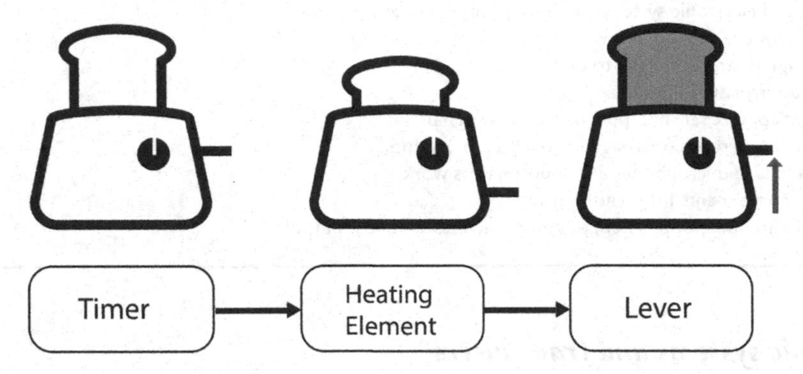

Figure 2.2 Open-loop system example. ***A toaster is an example of a simple open-loop system, where the timer (input) determines how long the heating element (process) will be active before the lever (output) lifts the toast out of the heating element.***

The above diagram is known as an *open-loop system* because the output of the system has no effect (control) on the input, while **inputs** to the system **control** the output. This is an important point, as the logical flow of a system is always from input to output and the need to change inputs because of outputs is often a source of confusion in system design (particularly when a user is responsible for this change). If the bread used in the above system was thinner than the one used when setting the timer, it could be overheated by the element and thus burnt – this is one of the many small challenges of modern living that has yet to be fully solved by engineering! It is not unusual for a user to manually engage the lever prior to the timer completing to avoid burning the bread, and so it quickly becomes clear that lack of knowledge of the output means that a system requires another source of control (other than the user) to avoid producing undesirable outputs. To relate this problem to our own domain, consider a simple example of an open-loop audio system (Figure 2.3).

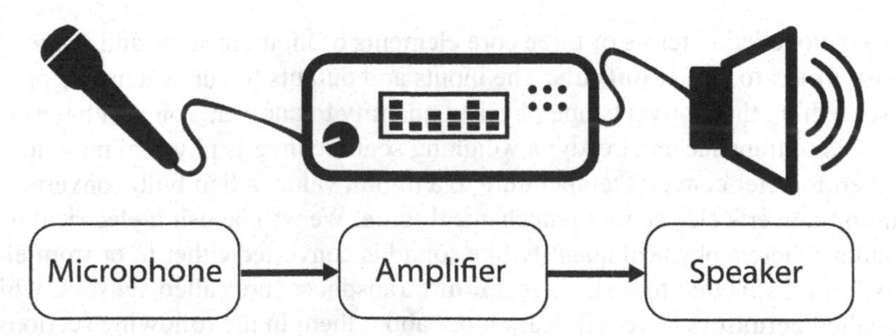

Figure 2.3 Open-loop audio system example. ***The example shows a microphone that creates an electrical signal as the input to an amplifier, which then processes this electrical signal to increase it to drive a loudspeaker that creates sound as an output.***

In this system, input from a microphone (sensor) is amplified to drive a loudspeaker (actuator) and thus create sound. The problem here is one of control – how do you prevent a microphone signal to the amplifier from distorting the output to the loudspeaker (which could blow the speaker cone, or even damage the amplifier)? In the above open-loop system, the only way of keeping loudspeaker output within safe limits (other than controlling the amplifier settings in advance) is to rely on the input from the microphone being within a range that will not distort when amplified. This type of distortion is not an unusual occurrence, indeed most professional vocalists learn to control their distance from the microphone when performing live to avoid distortion – the singer Sinead O'Connor often holds the microphone at the side of her head to reduce the volume, and other famous singers like Patti Labelle are known for letting the microphone fall to their side when hitting certain notes (this is also part of the performance, to emphasize the volume that they can achieve!). Even in the studio, vocal legends like Aretha Franklin have released recordings where their voice has distorted the microphone at certain points because of the unusually large volume range they are capable of producing – not due to a lack of competence by the engineer or the artist, but rather an example of an audio system that does not have sufficient control over its output to respond to the input conditions involved.

Sound engineering

A live sound engineer will aim to set up all audio levels during a soundcheck (if one is possible!) and will then have to monitor them throughout the performance. Part of their job is to try and avoid feedback being introduced due to changes in the initial setup, which can include:

- *A microphone moving during the performance* – either intentionally (such as a piano vocal mic) or by accident (someone trips over a cable under low light)
- *One of the artists changing their own equipment setup* – this includes the well-known problem of a lead guitar player turning up their amplifier during a gig to ensure they can be heard
- *A singer begins to struggle with their vocal delivery and asks for more monitor level* – the combination of heat, fatigue and noise exposure (a stage can get very loud!) makes it more difficult for a singer to hear themselves as the gig progresses
- *The room acoustics have changed* – a soundcheck is usually carried out in a largely empty venue, but when a crowd are present during the performance they will not only significantly change the absorption coefficient of the room but will also increase the room temperature (which changes the speed of sound)

The practical application of audio systems for recording or performance will not be covered in any depth in this book, but it is crucial to gain a thorough understanding of the requirements for audio systems – and how rapidly these can change. Practical experience is the best way to develop this, either by going to gigs, playing in a band, asking to assist a live engineer or volunteering in a recording studio (this will also help build your CV). Whether you are a musician or not, there is no better way to expand your knowledge of using audio systems than by practice – you will need this knowledge in the future to design and build effective audio electronics circuits.

Systems can create significant problems when they are not controlled – a distorted vocal track can completely ruin a recording. This highlights why we usually need to design a *closed-loop system* – a system where the output **controls** the input in some way. Using a simple non-audio system again to demonstrate the concept, we can look at home heating as an example (Figure 2.4).

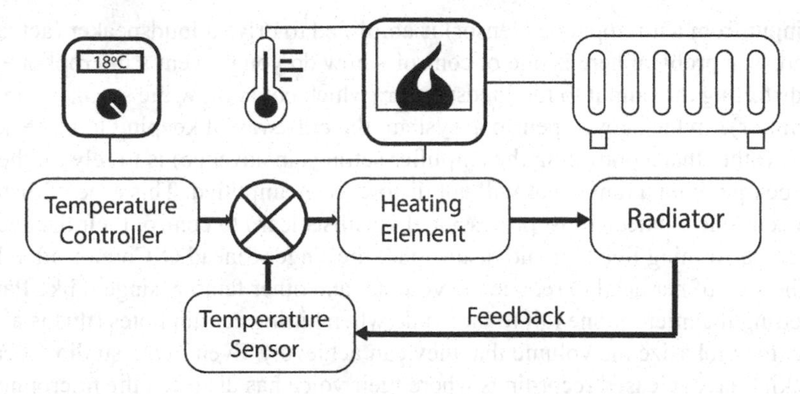

Figure 2.4 Closed-loop system example. ***In this example, a temperature controller (input) sets the required temperature that the heating element (process) needs to generate to feed heat to the radiator (output). A temperature sensor acts as a control input to provide feedback to the heating element on the current temperature, preventing it from overheating the room.***

The closed-loop system in this diagram allows the output (radiator) to control the input (temperature level) by including a temperature sensor – a type of transducer that converts heat to electricity. Note the circle with a cross that links the two inputs – this summing junction represents a **decision** point in the system, where the circle represents some evaluation of the inputs relative to each other. In our example, the decision point takes the temperature value specified by the temperature controller and *compares* this to the temperature of the room (as reported by the sensor) and *evaluates* the result. The input from the sensor is known as **feedback**, which can be defined as '... *control ... by reinserting into the system the results of its performance*'.[3] Feedback is a crucial part of electronics systems and also computer programming – we need to build systems that can be controlled to perform as required. These decision points are a fundamental element of all systems, and this is partly why microcontrollers like the **Arduino** have become so popular – they **control** systems. We will look at decisions in more detail in chapter 5 when we learn how to program selection instructions on the Arduino – they are one of the core elements of computer programming.

We could consider the open-loop audio example from Figure 2.3 as being closed loop if we include the user in the system; in this case the LED volume indicator on the amplifier would act as a form of feedback to tell the user to turn down the gain control. Nevertheless, we want to try and keep our systems simple to begin with, so the heating example above allows us to quickly grasp the idea of feedback in a closed-loop system without performing extensive user modelling (which is outside the scope of this book). Having said this, we can use our previous open-loop audio system example to demonstrate the concept of **system boundaries** (Figure 2.5).

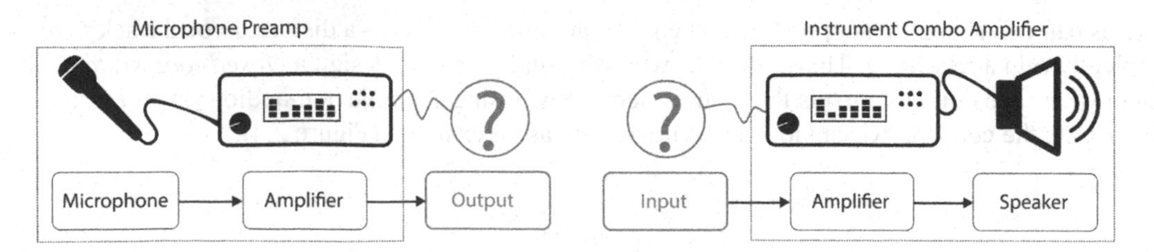

Figure 2.5 System boundary examples. ***In the above diagram, the first system on the left represents a microphone with integrated preamplifier, whilst the second system on the right is an instrument combo amplifier (so called because it combines an amplifier and loudspeaker in a single unit).***

In this diagram, the point at which we encapsulate the *closed volume* of the system can completely change the final system we obtain. In a microphone with preamplifier (such as a USB or Bluetooth microphone) our system diagram might not necessarily drive a loudspeaker, but could instead be connected to an analogue to digital (A/D) converter that creates a digital representation of the sound detected at the microphone input (we will look at this in more detail in the next section). In a guitar (or keyboard) 'combo' amplifier the *specific* input is not known, but in chapter 7 on amplifiers we will learn about different inputs like mic and line level that audio electronic equipment is designed to work with. This means that our second system example in the diagram above must be able to accommodate a specific **type** of input signal (you may have seen a mic/line switch on certain musical equipment), even though the *actual* input is not known during the design process. This is also true for the first system example – you may have come across loudspeaker impedance ratings like 4Ω and 8Ω (specified in ohms), where loudspeakers and amplifiers should have the same rating to get the best power transfer from the system (and in some cases avoid damaging the equipment).

As we learned in chapter 1, electrical **resistance** is defined in **ohms** and in the next chapter we will learn more about how it changes the flow of electrical current in a circuit. Loudspeakers are often defined by their impedances, which is a more complex way of measuring resistance (we will learn about impedance for AC circuits in chapter 6). For now, the word resistance will be used to avoid confusion when learning about electronic systems. The resistance of a component defines how much it **stops** electrical **current** from flowing, and so it is important that our system plans for the resistances of all **possible** inputs and outputs involved. To do this, we consider inputs as sources and outputs as loads on the system (Figure 2.6).

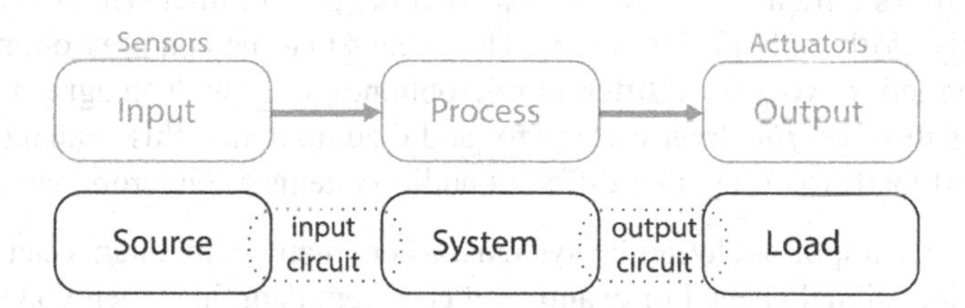

Figure 2.6 Representing a system as source and load. ***In electronic circuits, every component has a resistance, and so the system should model the external resistances it will encounter as either a source (input) or load (output). This helps to plan how a system should respond when connected to other elements – it must accept a source and drive a load.***

The idea that **a system must accept a source and drive a load** is crucial to electronic circuit design – you always have to plan for the type of input and output your system will be connecting to. Knowing the sources and loads in larger systems is critical to their effective use – if you've ever plugged a guitar into a microphone input you will have experienced unexpected distortion because the signal level of the guitar is much higher than that of a microphone (the source is different from the one the system was designed for). In audio, we will often connect several components together to create a larger overall system – even a small recording studio may contain separate preamps, compressors, monitors (loudspeakers) and an HDD recording system. We can think of all the components as **subsystems** of the larger overall system, where each subsystem must have a source and load that are compatible with the others it is connected to. To understand how components like those in our example studio are interconnected, we can break the overall recording studio system down into a set of subsystems by **partitioning** (Figure 2.7).

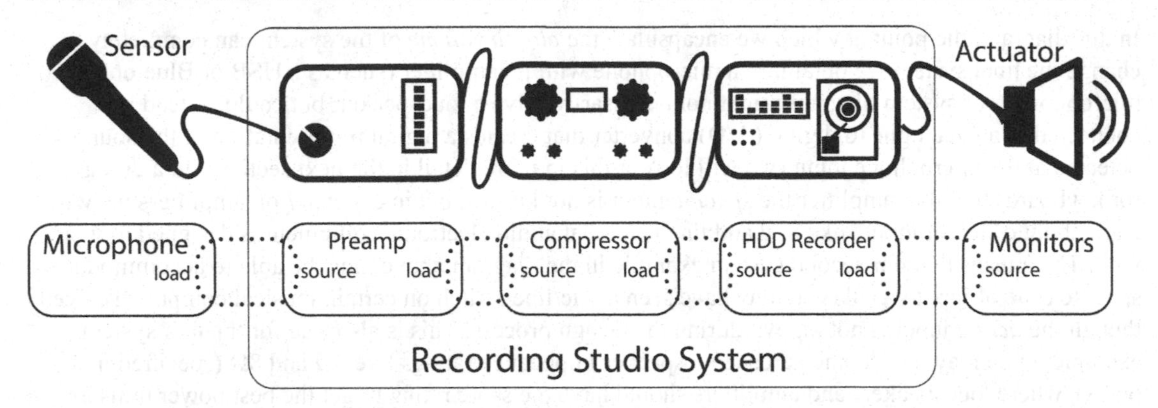

Figure 2.7 Partitioning an example recording studio system. ***In the diagram, different parts of a simple recording studio setup are shown as subsystems within the signal path from microphone to monitors. The microphone and loudspeakers are not considered part of this system, as in practice these elements can vary.***

Partitioning can help to reduce the complexity of a system, as it allows you to focus on each subsystem individually without having to include all the details of the rest of the overall system. In each subsystem, the source and load are the elements that define how that subsystem must interface correctly with the other parts of the system that they are connected to. In our example system above, we may not know the impedance of the microphone (or loudspeakers) involved when designing the compressor or hard disk (HDD) recorder – the only time this is important is when we are designing the preamplifier subsystem. As long as we know the load that the preamplifier will create, we can design a compressor to correctly work with this resistance. This means that the compressor and HDD recorder do not need to be designed to work with different microphones, only with an agreed load at input. You may have come across the term *line level voltage* for audio equipment – this is an agreed standard for connecting sources and loads to ensure that different audio systems are interoperable with one another.

Partitioning is particularly important for audio systems, where many components can be connected together in a much longer signal chain. For example, it can seem daunting when you first encounter a mixing desk with 48 channels, but once you realize that each one of these input channels is a duplicate example of the same subsystem it becomes much less difficult to learn – you really only have to learn how to operate one channel! We will not be building complex audio systems in this book, but as the chapters progress we will learn how to combine processes like amplification and filtering to create more useful circuits. All the project circuits we will build in this book will be based around open-loop analogue systems with digital control, focussing on the concepts of input/process/output and how each of these can be modelled as resistive sources and loads. We will begin each of these projects with a system diagram to help focus our analysis of the accompanying circuits that implement them, so spend a little time re-reading this section to ensure you are comfortable with the concepts we have introduced.

2.2 Digital systems and Arduino control

Digital systems arguably represent the greatest advance in modern electronics, where their adoption is now virtually universal at some stage within all systems. We will be using the Arduino as a digital controller throughout this book, so it is important to know the basics of how a digital system works – particularly when learning to program the Arduino in later chapters. As we learned in chapter 1, the flow of electrical charge is the fundamental element of all electronic circuits – we use current (amps) and voltage (volts) to measure our electronic signals. In any analogue circuit, the changes in the flow of charge create a continuous-time (CT) signal, which **continuously varies its output over time** without stopping (Figure 2.8).

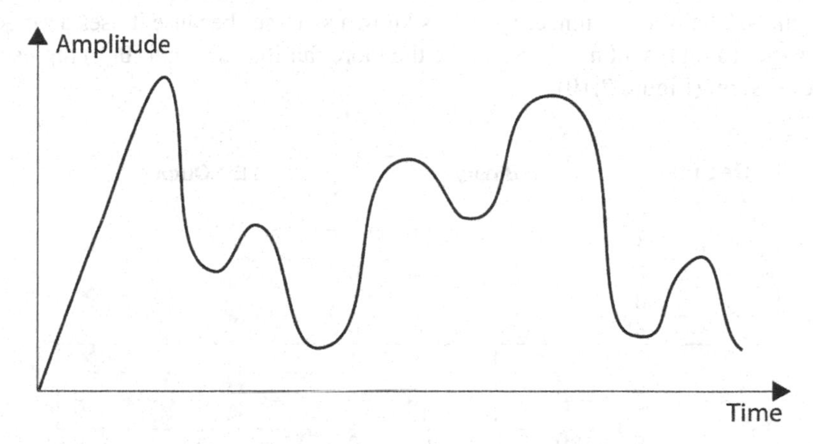

Figure 2.8 An analogue signal. ***The diagram shows an example of an analogue signal. The output amplitude level (y-axis) of the signal continuously changes over time (x-axis). This is a continuous-time (CT) signal, as it can conceivably be at any output level at any time.***

Systems that work with CT signals take a continuous input to produce a continuous output. Examples include many guitar/keyboard amplifiers and public address (PA) systems, where a CT signal obtained from an input transducer is processed (e.g. filtering, amplification) for output in a similarly continuous manner. A CT signal is often called **an analogue signal**, and we will return to the example above in the next section when we learn about analogue to digital (A/D) conversion. In contrast, a discrete-time signal is a signal that changes between **specific output levels at specific time intervals** – this is often called a **digital signal** (Figure 2.9).

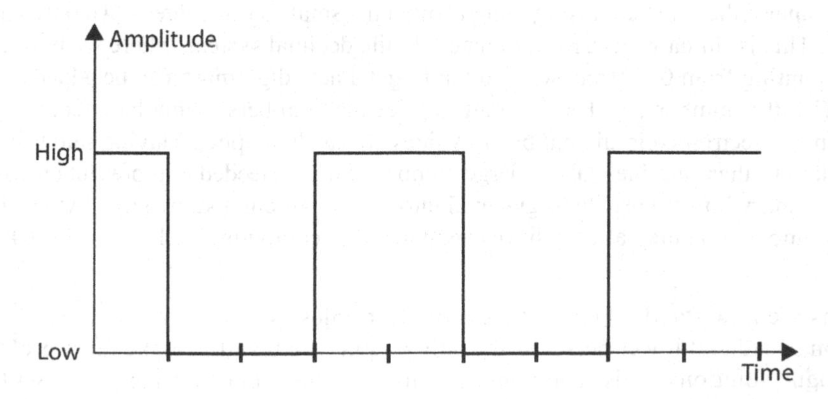

Figure 2.9 A digital signal. ***The diagram shows an example of a digital signal. The output amplitude levels (y-axis) are defined as either low or high, and the signal is measured (sampled) at specific time intervals (tick marks on x-axis). This is a discrete-time (DT) signal, as it can only be a specific output level at certain defined times.***

Digital signals have specific output levels measured at defined times, which allows changes in electrical charge to be quantified into discrete information that can then be sent through a circuit. One of the many advantages of using digital signals is that because there are only two output levels (low-high) in the circuit, it is easy to map these output values to binary numbers (0–1) – the fundamental unit of digital signals. The binary number system is known as Base_2 because it uses a range of only two possible values (0–1) to represent numbers, unlike the more familiar decimal (0–9) representation of values in a Base_9 system (Figure 2.10).

Decimal		Binary				LED Output			
	0				0				
	1				1				
	2			1	0				
	3			1	1				
	4		1	0	0				
	5		1	0	1				
	6		1	1	0				
	7		1	1	1				
	8	1	0	0	0				
	9	1	0	0	1				
1	0	1	0	1	0				

Figure 2.10 Binary data representation. ***The binary number system can use two possible values (0 and 1) to represent quantities, whilst the decimal number system uses a larger range of values (0–9). Notice that in both number systems a new digit must be added (highlighted in grey) every time we reach the end of its numeric range. A visualization of binary values as LED outputs is also shown to demonstrate how digital systems can easily use on/off to represent binary numbers.***

In the diagram in Figure 2.10, a mapping of binary values to LED outputs has been included to show how the binary values of **0** and **1** can be **mapped** to **on** and **off** using electronic components – this is the key to understanding digital systems and why binary is so essential to using them. When you first encounter binary, the most confusing thing is often the small set of values – we only have 0 and 1 to work with. This is (in part) because we relate it to the decimal system, where we have become familiar with counting from 0 to 9 and so we often forget a new digit must also be added at the end of this range too (i.e. the number 10). The 0–9 range of decimal numbers cannot be directly represented as on/off within an electronic circuit, but binary values are easily mapped. Having said this, with only 0 and 1 to work with there are inevitably a large number of digits needed to represent any useful data. For this reason, binary bits are usually segmented into a group of eight known as a byte (this factor of 8 scales up to numbers you may already have encountered in computing such as 16, 32, 64, 128, 256, 512, 1024).

Digital systems are now standard in most areas of electronics, where factors such as communication speed and lower noise (as the only outputs are 0 and 1) make them preferable to previous analogue solutions. This is an important distinction for our learning, as we will be using the Arduino (which is digital) to control analogue audio circuits like filters and amplifiers. Some of the reasons for using analogue audio signals will be discussed in the next section, but for now we can look at the basic digital system used in microcontrollers like the Arduino – a system known as the Harvard architecture (Figure 2.11).

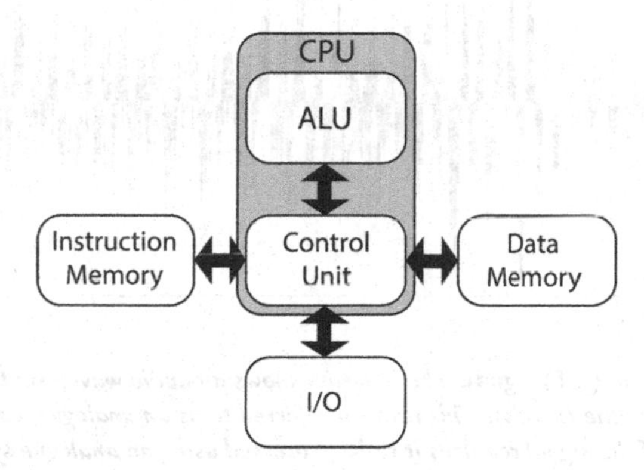

Figure 2.11 Harvard computer architecture. ***The Harvard architecture combines an arithmetic logic unit (ALU) and a control unit within a central processing unit (CPU). Data in the system is stored in specific memory, whilst the instructions for the control unit have their own separate memory storage. Input and output connectivity (I/O) allows sensors and actuators to interact with the system.***

A microcontroller is effectively a small computer: the arithmetic logic unit (ALU) can perform calculations on binary data that is stored in memory, whilst the CPU executes instructions that determine how **inputs** and **outputs** to the system will be **processed** – much like the open-loop system in Figure 2.1. Also note that the Harvard architecture distinguishes between CPU instructions and system data by placing them in separate memory, but in practice most modern microcontrollers like the Arduino mix both within what is known as a *modified* Harvard architecture. The important point for our understanding is that the **CPU** executes instructions to **process inputs and outputs** to the system – when we learn about Arduino control in later chapters we will be using the C language to program these instructions.

In section 2.5, we will see how a switch is a sensor that is either on or off – we can use the Arduino to map these values to a binary 1 or 0 as a control input to the system. Things become more complex when the sensor input is not simply an on/off switch, as we have to convert the continuous signal from the sensor (e.g. a potentiometer) to a representative set of data points. To do this, we must **sample** the signal at regular intervals and then store these values as binary data.

2.3 Analogue to digital conversion – sampling

Analogue and digital signals were introduced at the beginning of the previous section. We can now formally define an analogue signal as a continuous-time (CT) signal, which, as the name suggests, varies its output over time without stopping (Figure 2.12).

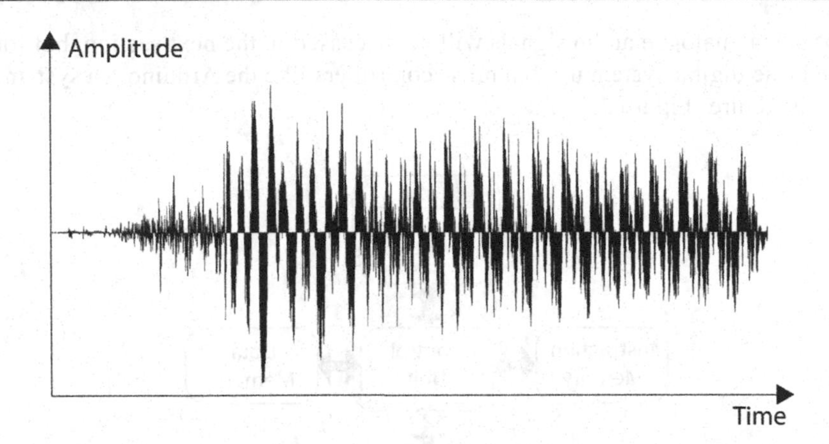

Figure 2.12 A continuous-time (CT) signal. ***The diagram shows an audio waveform that continuously changes its output amplitude (y-axis) over time (x-axis). This is often referred to as an analogue waveform, as the time-varying nature of the signal requires it to be processed using an analogue system.***

Systems that work with CT signals take a **continuous** input to produce a continuous output. Examples include many guitar/keyboard amplifiers and public address (PA) systems, where a CT signal obtained from an input transducer is processed (e.g. filtering, amplification) for output in a similarly continuous manner. These are often called analogue signals as they are processed by circuits with components like capacitors that vary their response over time (and hence frequency) – we will learn more about capacitors later in this chapter. In contrast, a discrete-time signal is a **representative sample** of a continuous-time signal, where the output is measured at specific time intervals (Figure 2.13).

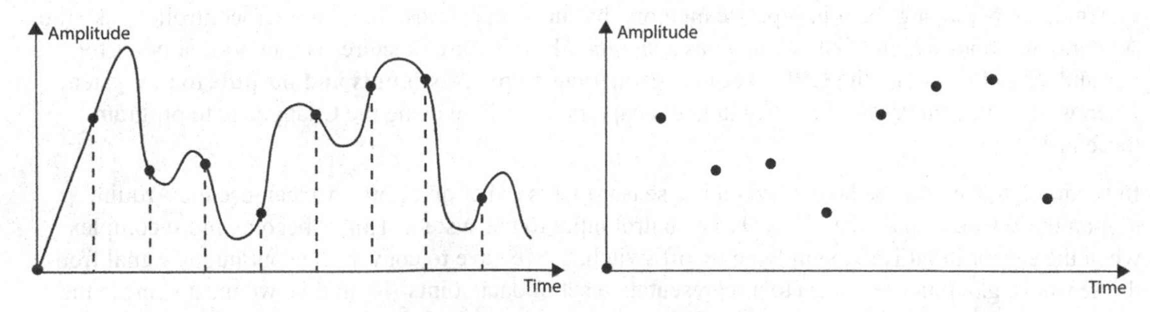

Figure 2.13 A discrete-time (DT) signal. ***The diagram shows how a continuous-time signal can be sampled at discrete time intervals, with the measurement data then being used to create a representation of that signal. The left-hand image shows the CT signal being sampled, whilst the right-hand image shows the actual data points obtained by sampling.***

A DT signal is a set of data points that have been **sampled** at regular intervals. These data points can then be used to **describe** a CT signal – the original signal is no longer present, but if you have enough data to adequately describe it then you can recreate it when required. This type of signal is more commonly known as a **digital signal**, because the sampled data points are stored and/or transmitted as **binary digits**. The concept of data sampling is core to all modern electronics and computing, and in audio electronics the storage, transmission and rendering of audio sample data is now the de facto standard. Whilst we will not be working directly with digital audio in this book (we will use the Arduino to digitally control analogue audio circuits), there are some fundamental aspects of audio sampling that are useful to know. The first factor to be aware of in sampling is the rate at which you sample – how **often** you measure the signal (Figure 2.14).

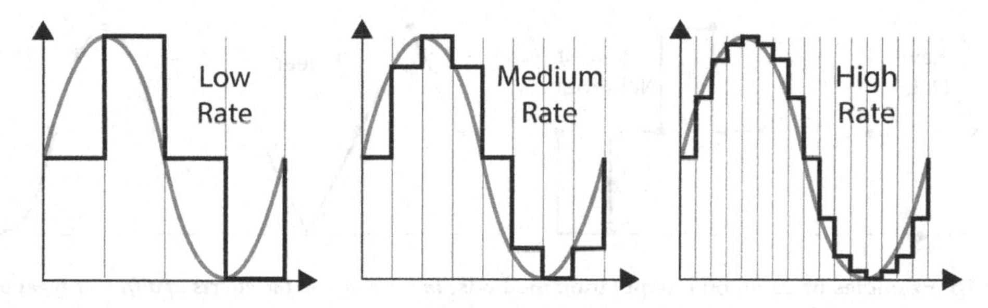

Figure 2.14 Varying sampling rate. ***The diagram shows DT representations of a waveform at three different (conceptual) sample rates, with low, medium and high sample rates shown from left to right. As the sample rate increases, so the distance between the data points reduces.***

In the diagram in Figure 2.14, the **higher** the **sampling rate** the **smaller** is the **error** in the DT representation of the signal. This is particularly important for higher frequencies, where Nyquist's theorem states that the number of data points required must be a **minimum** of **double** the **frequency** of the signal being sampled. The proof of this is not particularly complex, but also requires some discussion of Fourier analysis which is outwith the scope of this book. We will return to the topic of audio frequencies when we learn about AC signals, but for now you should note that the second factor of importance in sampling is the **range** of values available to describe the output signal being measured – this is known as the bit depth (Figure 2.15).

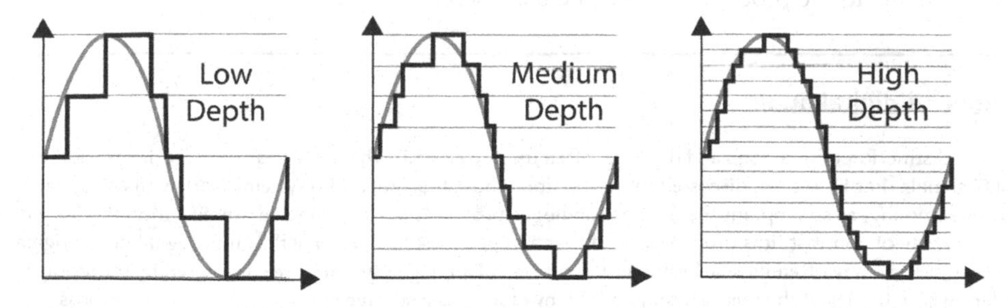

Figure 2.15 Varying bit depth. ***The diagram shows DT representations of a waveform at three different bit depths (ranges of measurement), with low, medium and high bit depth shown from left to right. A higher bit depth means there are more available values in the range used to measure the signal output, so the distance between the values of the data points is reduced.***

In this diagram, as with sample rate, the **higher** the **bit depth** the **smaller** is the **error** in the DT representation of the signal. Bit depth describes the number of binary bits available to measure the data points with – we saw in Figure 2.10 that the more bits you have the greater the range of data values you can describe. As with sampling rates, the proof of this is not particularly complex but requires a discussion of binary arithmetic that is again outwith the scope of this book. We will briefly look at binary again in the following section on MIDI, but for now we can focus on the third factor of importance in sampling – how we can **interpolate** between the data points to fill in the missing information (Figure 2.16).

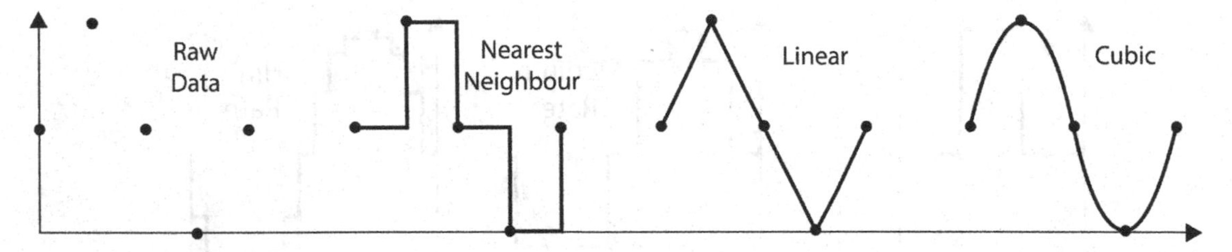

Figure 2.16 Examples of common interpolation methods. ***In the diagram the effects of different types of data interpolation are shown. The nearest-neighbour method is arguably the simplest to implement, whilst linear and cubic interpolation methods require extra mathematical processing.***

Once again, as with sample rate and bit depth the topic of data interpolation is also outwith the scope of this book! Having said this, we can note its impact on the resulting shape of the reconstructed signal, which varies significantly based on the method used. At first glance cubic methods appear to be the most useful as the cubic curve effectively describes a sine wave (which is fundamental to all audio signals) – but what if this was not the shape of the original CT signal being sampled? In practice, every form of interpolation leads to a possible error as the information from the original signal is no longer available – interpolation is a **best guess** for the missing data. Although nearest-neighbour methods are often the least accurate, they are relatively simple to implement electronically and do not require additional mathematical processing. Nearest neighbour was used in the previous diagrams to visually illustrate the variance of sample rate and bit depth on discrete signals, though in practice there are many other elements to the process that must be considered.

Analogue and digital audio

In audio, the distinction of analogue and digital is often used to describe the somewhat broader delineation of CT and DT signals that have been discussed in this section. Though digital audio systems work with DT signals obtained by sampling, at some point they still use analogue processes like filtering and amplification that require CT signals. By way of contrast, it is often proposed that analogue systems preserve the audio signal in its original form and whilst they may often have a far higher resolution, all analogue systems are still limited by the storage and rendering capabilities of the medium they use. Many older audio storage methods such as vinyl records and magnetic tape are used in analogue systems (and thus perceived as CT signals) but at some level they are arguably still discrete representations of a signal – just at a much **higher resolution** than digital sampling can currently achieve. For example, a vinyl record stores audio information in a physical groove that has an amplitude resolution related to the size of the needle used to etch it – this would equate to a very high digital bit depth. In addition, the physical movement of this needle can conceivably achieve a very high time resolution, but the width of the groove prevents lower frequencies from being accurately represented – you may encounter RIAA compensation for low frequencies if you have a record player. Whilst vinyl thus has a very high data resolution, it can also introduce significant errors (as noise) on playback due to factors such as motor stability, arm movement and the presence of dust (there are many DSP plugins that will add these 'authentic' sounds to your digital signals!).

Similarly, when early digital formats like compact discs (CDs) were first commercially introduced the obvious benefits included ease of storage and stability of rendering, but the implicit assumption of higher audio quality was never as straightforward to argue (much of the improvement was actually lower noise). Whilst a 44.1kHz sampling rate may be higher than the minimum Nyquist value needed for audible frequencies, this specific sampling frequency was not used for reasons of audio quality – it was to allow the audio signal to be embedded in a National Television System Committee (NTSC) signal for broadcast in North America. Similarly, low amplitude resolution (bit depth) due to the limitations of 16-bit address busses (which have now become outdated in most computing systems) means that broadcast wave files (BWF) often still contain a form of broadband digital noise called dithering, which is added to raise the overall volume level of the signal to a region that can be more accurately described by a 16-bit amplitude range.

Although digital audio is not directly covered in this book, it is important to highlight this distinction between analogue and digital to better clarify that analogue audio circuits are still essential to all audio systems, but digital audio has made massive advances in both storage and rendering quality. The debate on audio quality will continue regardless, but as electronics engineers our goal is to design, build and test the best audio systems possible – rather than making decisions based on what are in many ways commercial distinctions. A high fidelity (HiFi) vinyl system can produce an amazing sound, but so can a high data rate digital streaming system – neither has an exclusive claim to audio quality.

Our discussion of CT and DT signals may initially appear somewhat superficial, as we have only distinguished between them in broad terms whilst introducing some of the main factors relating to DT signals. The primary reason for covering CT and DT signals, other than as they relate to our understanding of systems, is to make an initial definition of these signal types prior to working with analogue audio components and digital Arduino control. We will be building electronic circuits that process CT signals, but will also use the Arduino to control certain parameters of these circuits. Most modern audio systems employ analogue-digital (A/D) conversion to sample and store a digital representation of the input audio as a DT signal, and whilst it is technically possible to use an Arduino for this purpose its slow processor speed and low on-board memory preclude any useful demonstration of these concepts.

Thus, digital audio circuits are not covered in this book, but we will be using the **Arduino for digital control** of our audio systems. This is an important distinction to make when defining systems, as the control of a system can in some ways be more complex than the system itself. As an example, consider how the closed-loop heating system of Figure 2.4 is implemented with modern temperature controllers that learn heating patterns over days and weeks to improve the living environment and reduce energy costs – the algorithms involved incorporate machine learning (ML), artificial intelligence (AI) and cloud computing for data analysis and storage that are much more complex than a simple thermostat reading. In audio, this distinction is arguably best demonstrated by giving a brief overview of MIDI control systems.

2.4 MIDI control systems

Musical Instrument Digital Interface (MIDI) is widely used in music production and represents a truly impressive example of digital system control. MIDI combines a simple hardware interface with a digital communications protocol for musical information that has been a standard in audio production for nearly 40 years since the original MIDI 1.0 specification was written in 1982. One of the reasons for its longevity is the simplicity with which this protocol can be adapted to other interfaces, and it is now common to find USB, Ethernet and Bluetooth MIDI interfaces that all use the same MIDI protocol for audio control. The original MIDI design incorporated a 5-pin DIN

(Deutsches Institut für Normung) connector that was used to establish communication between devices – a connection that is also used an **opto-isolator** to convert MIDI signals to light to avoid noise problems (Figure 2.17).

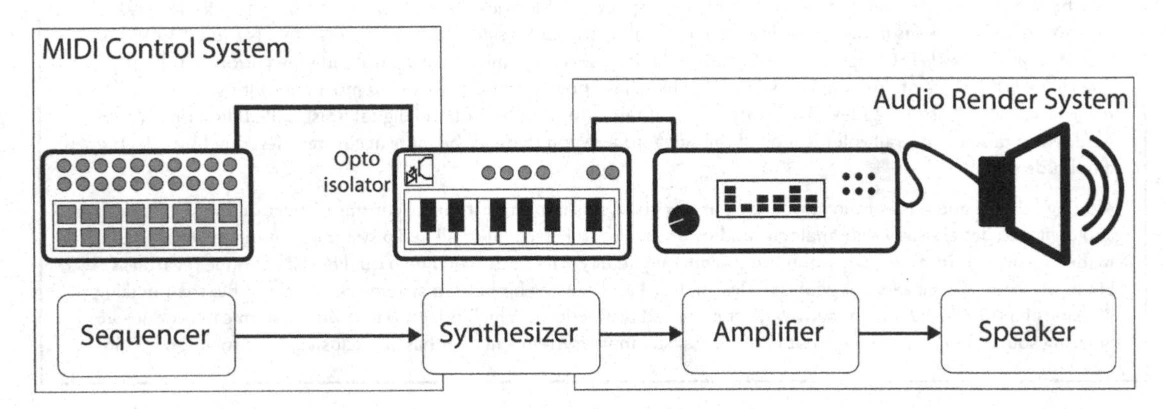

Figure 2.17 A Basic MIDI System. ***The MIDI control system on the left consists of a sequencer connected to an audio synthesizer using a MIDI cable (which passes through an opto-isolator circuit). The audio render system on the right defines the analogue signal connections between the synthesizer, amplifier and loudspeaker that generate sound output.***

In the diagram in Figure 2.17, a MIDI sequencer is used to control an audio synthesizer – so called because it synthesizes simple audio signals (like sine waves) to create more complex musical timbres. The sequencer sends MIDI messages (in binary format) that tell the synthesizer what notes to generate (and when), but it does not generate any audio signals – the synthesizer produces these separately as part of the audio render system. This is an important point: the MIDI control system tells the synthesizer what to do, but it **does not create any sound** – it simply sends MIDI information as a digital signal in binary form. Each MIDI message consists of a status byte (8 bits) that defines the type of message (e.g. Note On, Note Off), the channel it will apply to (1–16), accompanied by one or more data bytes that provide specific information related to this status (e.g. Note Number, Note Velocity). All bytes are then combined into a single binary message for digital transmission using a MIDI interface (Figure 2.18).

Status				Channel				Data 1								Data 2							
Note On (9)				1 (binary 0)				Middle C (60)								Velocity (100)							
1	0	0	1	0	0	0	0	0	0	1	1	1	1	0	0	0	1	1	0	0	1	0	0

Figure 2.18 MIDI message example. ***The diagram shows the composition of a MIDI Note On message, with its binary equivalent listed underneath. The first byte defines Status, which combines the message type (always starting with a binary 1) with the MIDI channel that the message applies to. The second and third bytes provide Data (which always starts with a binary 0).***

Notice that the channel number in the message is 1, but its binary equivalent is 0 – this is because computers always count from zero, even though Western mathematical systems usually begin from

1 when counting in decimal! This shift in counting can often confuse new programmers, and will be discussed again when learning how to program the Arduino in later chapters. For now, the definition of a specific MIDI channel for each message allows up to 16 receiving devices to be connected in a single MIDI system, which significantly increases its capacity for audio device control (Figure 2.19).

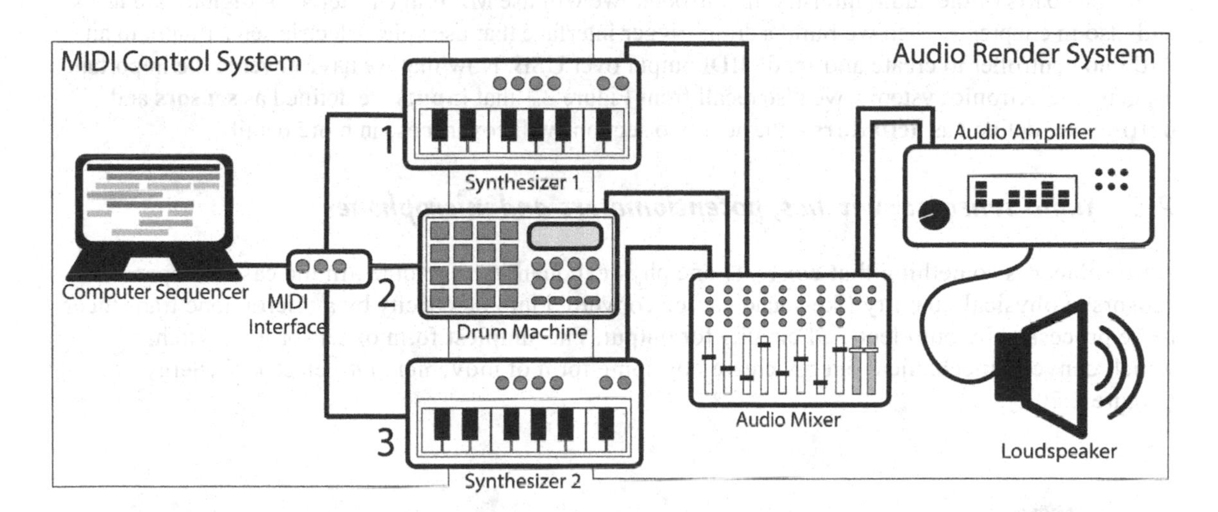

Figure 2.19 MIDI channel control. ***In the diagram, a computer running a sequencer application is connected to a MIDI interface using a USB cable. This interface is then linked to two synthesizers and a drum machine using MIDI DIN connectors – this is the MIDI control system. The three devices are then connected to an audio mixer for output – this is the audio render system.***

In this diagram, the MIDI interface allows multiple audio devices to be connected together in a single MIDI system. If each audio device is now set up to respond to a different MIDI channel (listed as 1,2,3 in the diagram) then the sequencer can send specific control messages to any/all three as needed. This is the real power of MIDI: a simple digital system with multiple possible hardware interfaces (USB and DIN in the diagram) can be set up to control multiple audio devices at the same time. It is fair to say that many styles of music that stem from the 1980s were directly shaped by the possibilities and capabilities of MIDI control – styles as diverse as Techno, House, Contemporary R&B and Hip Hop all have significant audio synthesis elements that could only be rendered as part of a predefined sequence.

Although MIDI use is still widespread, it inevitably has some limitations (partly due to the evolution of some other technologies it now operates with). For example, in the MIDI 1.0 system described in Figure 2.19, there is no feedback loop from the drum machine or synthesizers to the sequencer – the sequencer merely outputs MIDI messages that may or may not be received correctly. This means that a device may respond incorrectly (or not at all) to the message being sent, and if the device is not set up for the correct MIDI channel it may respond to messages intended for another device. Although this does not seem to be a major issue in abstract, you can mimic this behaviour in any software sequencer by assigning a grand-piano sound to a drum machine plugin to get a sense of why many synthesizers were often abruptly disconnected when things went wrong in the MIDI system – another reason the DIN plug was a good choice! Further, any MIDI 1.0 device can respond to incoming MIDI messages, but this does not mean that they are capable of doing so. A bass synthesizer could mistakenly be assigned a sweeping arpeggio part that spans multiple octaves, well outwith the range the instrument was designed for.

To address this, the MIDI Manufacturers Association (MMA) is currently introducing MIDI 2.0 – a major revision of the MIDI protocol. Although at time of writing the protocol is still in prototyping, the main advances of device profiling and bi-directional communication (in the form of property exchange) have already been announced. If adopted, MIDI 2.0 could provide answers to many of the problems of modern audio device control that have led to other options (e.g. OSC, HID) being used in various parts of the audio industry. In this book, we will use MIDI in chapter 4 for digital output, and also in chapter 5 when we build a drum trigger interface that uses piezoelectric sensor input to an Arduino controller to create and send MIDI output over USB. Now that we have covered the important aspects of electronic systems, we also recall from Figure 2.1 that **inputs** are defined as **sensors** and **outputs** are defined as **actuators** – the next two sections will cover these in more detail.

2.5 Audio sensors: switches, potentiometers and microphones

A transducer is something that **converts** one physical quantity to another. In the case of input sensors, a physical quantity like sound can be converted into electricity by a microphone transducer to be processed by our electrical circuits for output. The simplest form of sensor is a switch, which converts mechanical energy created by some form of movement into electrical energy (Figure 2.20).

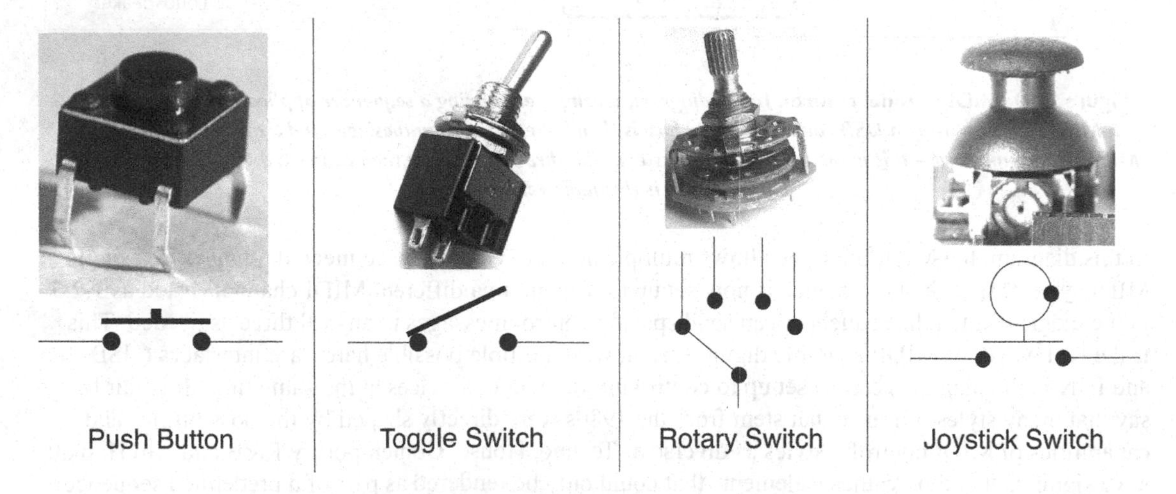

Figure 2.20 Common switch examples. ***The diagram shows some of the most common types of switch, and the circuit symbols used to represent them. Push and toggle switches are based around a single contact that completes a circuit, whilst rotary and joystick switches are more advanced combinations of contacts that generate multiple possible outputs.***

There are many types of switch, but they all effectively change an open circuit into a closed circuit by closing some form of electrical contact (we learned about open circuits in the previous chapter). The push-button switch is often referred to as a *momentary action* switch, because electrical contact is only made during the moment that mechanical force (e.g. finger pressure) is applied. This type of switch is commonly used in computer keyboards, and at a basic level the same trigger is used in synthesizer keyboards and other MIDI controllers (though they also incorporate more advanced velocity and pressure sensors made from piezoelectric materials). Toggle switches, as the name implies, toggle between open and closed states – such as the on/off switch on an amplifier. Rotary and joystick switches effectively combine more than one switch within a single component, allowing either multiple outputs (rotary) or multiple switches (joystick) to be engaged with a single movement. We can also describe a switch by the number of poles (circuits) and throws (output paths) it has (Figure 2.21).

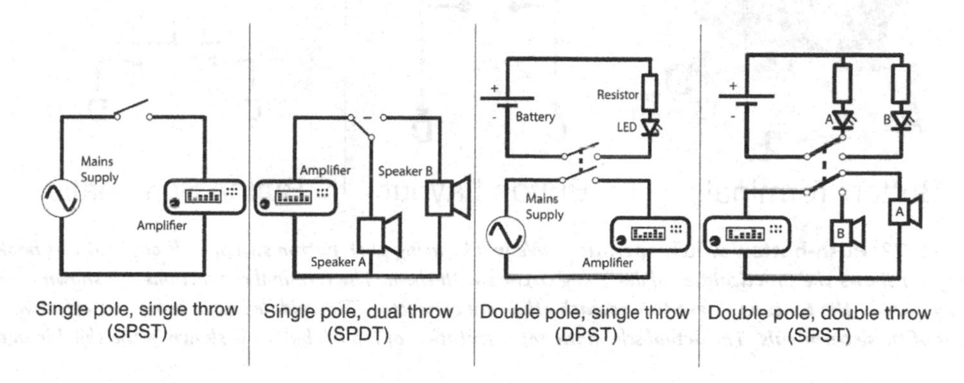

Figure 2.21 Common switch configurations. ***A switch can be defined by the number of poles (circuits) and throws (paths) it has. Each pole is a separate circuit that can be closed by the switch, whilst each throw is a possible output path that the switch can connect to. The example audio circuits shown are described in more detail below.***

The example circuits in Figure 2.21 show how we can use these different switch configurations in our audio circuits:

- **SPST switch:** 1 circuit, 1 output path – *an amplifier on/off power switch (off is no connection, which is not a path!)*
- **SPDT switch:** 1 circuit, 2 output paths – *an A/B loudspeaker switch (each circuit connects to a different speaker)*
- **DPST switch:** 2 circuits, 1 output path (each) – *an amplifier on/off power switch that also lights an LED (1 circuit for amplifier power, 1 circuit for LED power). Both circuits are on/off (off is no connection, so not a path)*
- **DPDT switch:** 2 circuits, 2 output paths (each) – *if the A/B loudspeaker switcher above also had LED indicators for each speaker, we could use a DPDT to light them (in practice, we should avoid mixing audio and power signals)*

As we learned with scales, symbols and equations in chapter 1, electronics has its own language that takes time to learn. Switch configurations can be confusing to begin with as there are four different acronyms to remember – unfortunately we will encounter many more acronyms during this book! One way to simplify your learning is to think of **single throw** (ST) switches as being on/off – you only have **one output path** for each circuit involved. Double pole (DP) switches effectively do two things at once, and it can help to think of them being used in a closed-loop system where LED indicators give user feedback (though DP switches are also used in many other applications).

Push button switches will be used as sensors in many of our Arduino projects, so it is important to spend a little time learning how to work with them in practice. The first thing to note is the terminal **layout** of a push-button switch and how this translates into an actual circuit schematic (Figure 2.22).

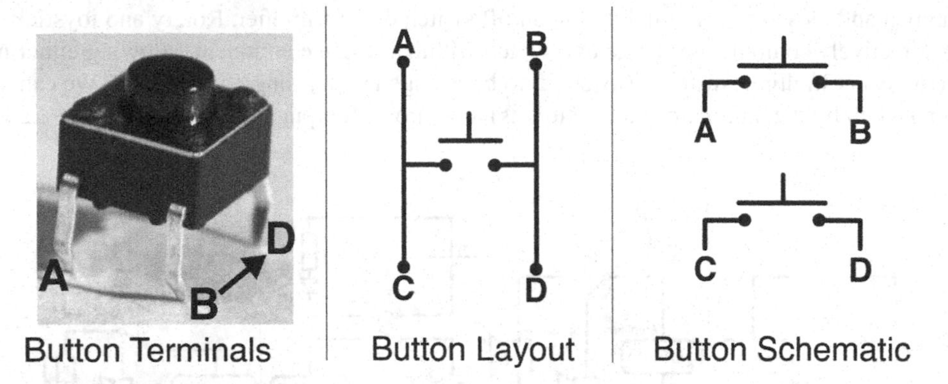

Figure 2.22 Push-button switch operation. ***We will be using push-button switches throughout this book, so this diagram shows the practicalities of designing circuits with them. The terminal connections are shown in the left-hand image, where the terminals bent toward each other are connected. The middle image shows the effective circuit layout of these terminals. The actual schematic representation of a push button is shown in the third image.***

A simple push-button switch has four contacts which are already connected to each other in pairs (A-C, B-D). As we can see in the button layout in Figure 2.22, by pressing the button we effectively connect all four terminals to each other. Although this makes the switch simple to work with (and cheap to manufacture), we are **effectively only able to use two of the four terminals** in our circuits. We know that the pins which are **physically bent towards each other are already connected** together (see left-hand image) so we need to orient our switches accordingly. In our project circuits, we will be using one half of each switch to connect two terminals together (either A-B or C-D). The specific half depends on the rotation of the components on the breadboard, where some components are set out 'bottom up' and others 'top down' to maximize the board space available.

In the previous chapter, we learned about using **resistivity** when we added a resistor to our LED circuit to prevent the current from burning out the component. In simple terms, a switch is effectively an on/off resistor – the circuit is switched from open (infinite resistance) to closed (no resistance). Switches are very useful (particularly for digital electronics), but what if we wanted to measure a quantity in more detail than simply on or off? The answer is to use a component where the quantity we want to measure **changes** its resistivity (Figure 2.23).

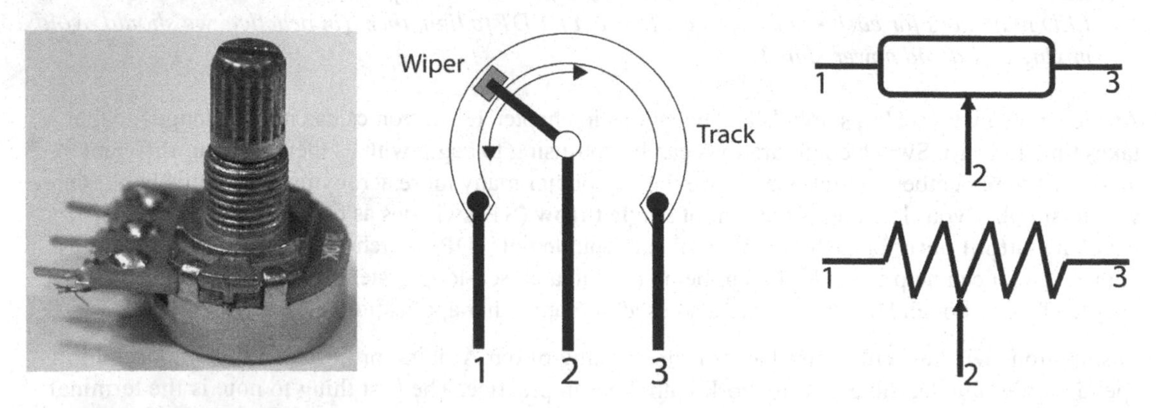

Figure 2.23 The variable resistor. ***The left-hand image shows an example of a variable resistor (or potentiometer), whilst the middle image shows the internal workings of the component. The right-hand images show the standard European (top) and North American (bottom) circuit symbols, where the second terminal is the variable resistance output.***

Variable resistors (also known as **potentiometers** or pots) are sensors that convert mechanical energy (**movement**) into changes in **resistance**. The basic construction of a variable resistor is a length of copper track with some form of wiper or other contact connected to it. As the position of the **contact** is varied (by moving it) so the length of wire between input and output terminals changes. As we saw in Figure 1.10 in chapter 1, changing the **length** of a wire will increase its **resistivity**, so by moving the contact we can increase or decrease the resistance that the component presents to the flow of current in the circuit. Variable resistors are used extensively in audio for all manner of volume, equalization, panning and effects controls – they are the most common type of sensor you will encounter. Whilst switches are used for power circuits and digital input, audio signal requirements mean that a variable resistor volume control will often be the preferred means of engaging an amplification circuit to prevent pops and clicks from damaging loudspeakers and other equipment in the system. We will use variable resistors in some of the audio circuits in this book, mostly as part of a voltage divider (see chapter 3) that controls the size of the signal being sent to output.

We will not be building high-precision circuits in this book, but nevertheless it is important to understand the capabilities and *limitations* of sensors so we can use them correctly. For example, a cheaper potentiometer may not be accurate enough for use in an audio filter circuit (and may also introduce significant noise) – knowing this can help you design more effective circuits. All sensors will measure values over a given **range**, and their **resolution** is the smallest measured difference between each value (Figure 2.24).

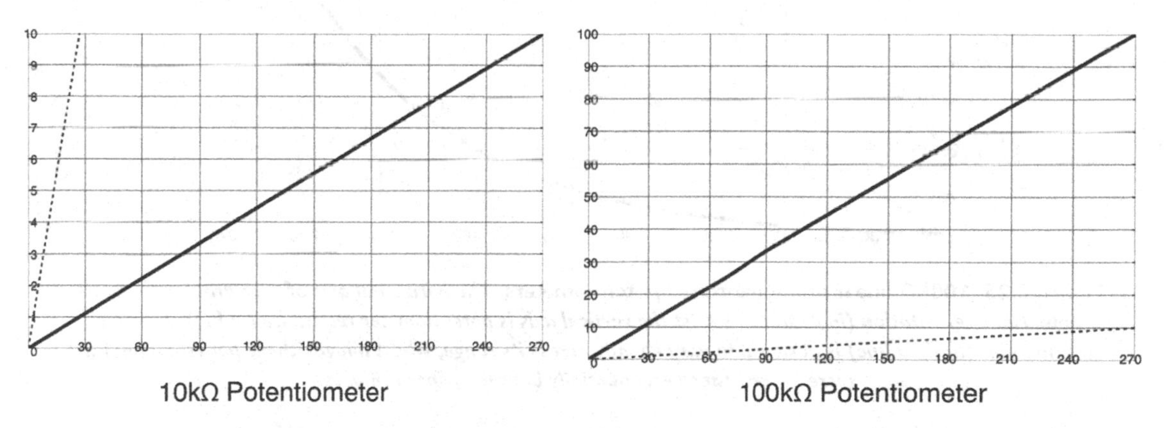

Figure 2.24 Sensor range and resolution examples. ***The horizontal axis of each graph represents potentiometer rotation (in degrees), whilst the vertical axis is potentiometer resistance (in kΩ). The sensor on the left is a potentiometer with a smaller range (10kΩ) but a higher resolution (1kΩ increments). The sensor on the right measures a much larger range of values (100kΩ) but at a lower resolution (10kΩ increments). In both graphs, the dashed line represents the other potentiometer range, to give a sense of proportion between the two sensors.***

The potentiometer measurements in the graphs in Figure 2.24 show how both range and resolution can vary between sensors. It is common to work with values like 10kΩ and 100kΩ for potentiometers in audio circuits, and so it is important to remember that a component with a larger range may not necessarily have the same resolution as one that works with a smaller range of values. In the case of the potentiometers above, we have labelled measurements at various rotation angles from 0 (wiper fully left) to 270 (wiper fully right). In practice, we may find that this potentiometer is not completely **linear** across its range of values – a small movement at one or other end of the wiper may create a greater change in resistance than the same movement in the middle. This is particularly true of cheaper potentiometers, where poorly fabricated wiper contacts can not only create non-linearity but also introduce noise, which is a type of **error**. All sensors are prone to errors at some level, which leads to the definition of **accuracy** as the maximum error that may occur for that sensor – the more accurate the sensor, the less likely it is to give a false reading. In audio, it is also important to note the *taper* of a potentiometer, which is usually either linear or logarithmic (often known as an audio taper). Even though a linear taper pot is *designed* to increment resistance in line with movement of the wiper, in practice it may not be completely **linear** at all points across its range – the taper does not guarantee quality! You may already have encountered logarithmic (or audio) taper pots, which are used in many audio circuits specifically because they *do not* produce linear values (Figure 2.25).

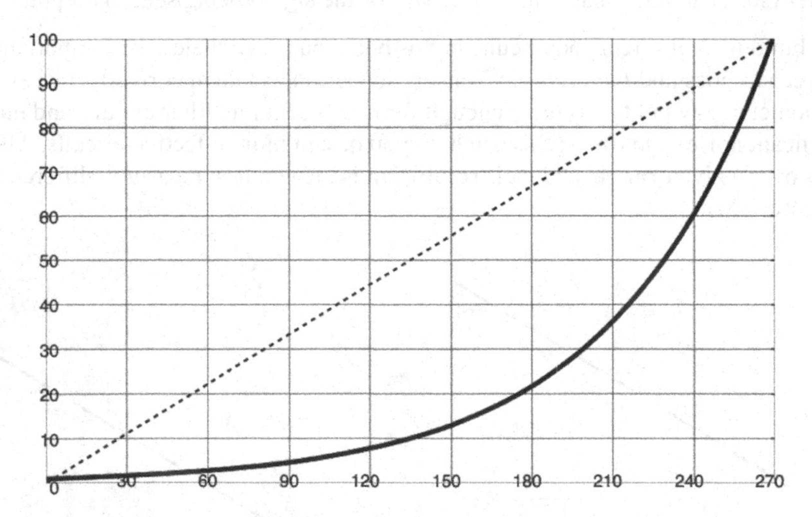

Figure 2.25 100kΩ linear and logarithmic potentiometers. ***The horizontal axis of each graph represents potentiometer rotation (in degrees), whilst the vertical axis is potentiometer resistance (in kΩ). A linear potentiometer (dashed line) is designed to vary equally across its range, whilst a logarithmic pot (solid line) will increase resistance exponentially between wiper contacts.***

A logarithmic pot has a resistance curve that starts and ends slowly, with most change occurring in the middle of the range. The human ear can detect a huge range of intensities and frequencies, and often a linear scale cannot represent these ranges effectively – we will look at equal loudness curves later in chapter 8. Although a thorough discussion of psychoacoustics is outside the scope of this book, you will often encounter logarithmic pots being used in audio for volume and tone controls due to their non-linear behaviour that better aligns with human hearing.

Another form of sensor you may have encountered is the **piezoelectric** sensor – they are commonly used in microphones, acoustic instrument pickups and even as accelerometers in mobile phones. These types of sensor exploit the *piezoelectric effect* that occurs in certain materials like crystals and ceramics, where an increase in mechanical stress (pressure) changes the net electrical charge of that material (Figure 2.26).

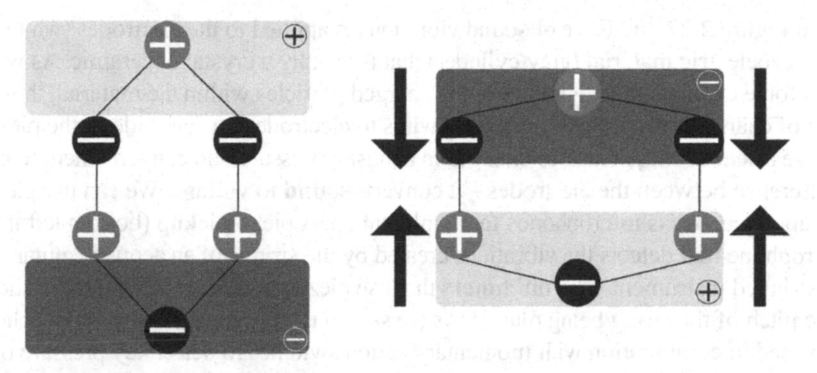

Figure 2.26 The piezoelectric effect. ***In the above diagram, a quartz crystal (silicon dioxide) that combines positive silicon and negative oxygen atoms is sandwiched between two copper plates. The left-hand example shows a single positive and negative atom at each plate, whilst the right-hand example shows how mechanical stress pushes the positive silicon and negative oxygen atoms closer together. This alters the net charge at each plate and hence the overall potential difference between them.***

This example shows the effect of piezoelectric compression on a quartz crystal, where pressure applied to the copper plates changes the alignment of the silicon dioxide – moving its silicon and oxygen atoms closer together. This changes the net electrical charge (overall sum or difference of all charges) at each plate – the top plate becomes overall negative because of the addition of two negative charges, whilst the bottom plate becomes overall positive because of the addition of two positive charges. We learned in chapter 1 that voltage (potential difference) is the potential for a charge to move between two points. Thus, the outcome of applying mechanical stress to quartz is to change its potential difference – a change we can use as part of our electronic circuits. You should also note that the polarity of each plate changes when it is compressed, and this variation creates what is known as an alternating current (AC). We will look at AC circuits in more detail from chapter 6 onwards, but for now we will focus on how the piezoelectric effect can be used to construct a microphone (Figure 2.27).

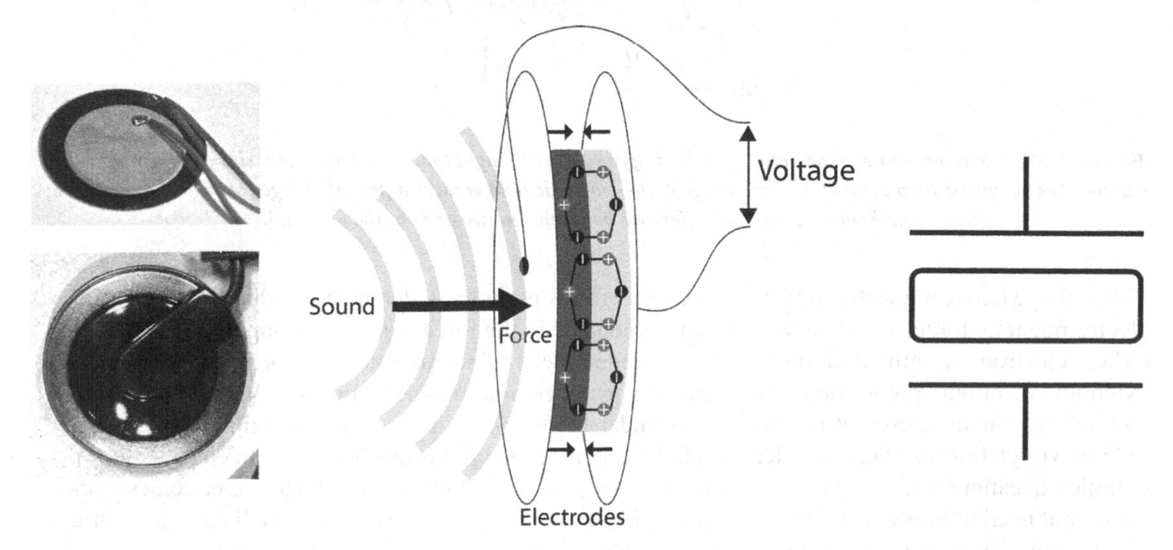

Figure 2.27 The piezoelectric microphone. ***The left-hand images show examples of piezoelectric sensors (bottom left is the piezo used in the chapter 5 project), whilst the middle diagram shows how a piezoelectric sensor converts sound to voltage by reacting to the acoustic pressure of the sound waves on the material. The right-hand image shows the standard circuit symbol for a piezoelectric transducer.***

In the diagram in Figure 2.27, the force of sound vibrations is applied to the electrodes (white discs) on either side of a piezoelectric material (grey cylinder) that is usually a crystal or ceramic. As we saw in Figure 2.26, this force changes the alignment of the charged particles within the material, thus changing the overall level of charge within it. By connecting wires to electrodes on each side of the piezoelectric material we create a **sensor** that measures changes in acoustic pressure and converts them to changes in the potential difference between the electrodes – it converts **sound** to **voltage**. We can use piezoelectric sensors (often called piezos) as microphones for amplification. A piezo **pickup** (bottom left in the diagram) is a type of microphone that detects the vibrations created by the strings of an acoustic guitar – you may also have seen stringed instrument 'clip-on' **tuners** that use piezos to detect and analyse sound vibrations to determine the pitch of the string being played. As we saw in the previous section on switches, piezo sensors are also used in combination with momentary action switches to detect key pressure on electric pianos and other synthesizers. Another common use of piezo sensors is for **drum triggers**, where the vibrations created by a stick or beater hitting the drum head are converted into MIDI signals that control the triggering of other audio samples (we will learn how to build a drum trigger project in chapter 5).

Although piezoelectric sensors are widely used in audio, electromagnetic **induction** is still the most common way of converting air pressure waves (sound) to and from electricity. Induction exploits the relationship between a magnetic field and an electric field, where changes to the magnetic field *around* a conductor will also change the charge flowing *through* it (Figure 2.28).

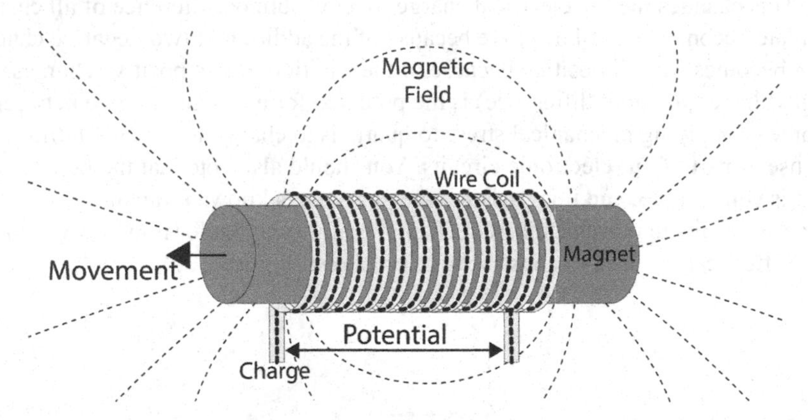

Figure 2.28 Electromagnetic induction. ***In the diagram, as the magnet moves through the wire coil (conductor) it changes the magnetic field around it. The change in the magnetic field creates a flow of charge through the conductor, and hence a potential difference between the two ends of the wire coil.***

This linkage between electric and magnetic fields forms part of a wider understanding of the electromagnetic force, which as we learned in chapter 1 is the force that attracts opposite charges (where electrons are attracted to holes in the valence shells of positive atoms). The problem with extending electromagnetic force to include the phenomenon of magnetism is that although the effect is well known and measured, it is still not fully understood! To avoid confusion, we will simply exploit the known relationship between electrical fields and magnetic fields rather than try to consider the complex questions (and incomplete answers) that follow them. For our introductory electronics work, we do not need to know why magnets attract and repel one another or why they are linked to electric fields – we only need to know how to use these phenomena as part of our audio systems.

Understanding electromagnetism

Whilst a full explanation of electromagnetism is beyond the scope of this book, the lack of any explanation is always problematic so some brief notes are provided here to avoid creating a significant gap in your understanding. At the beginning of chapter 1 we learned that atomic particles are moved by the **electromagnetic** force, which will either attract or repel charges from one another based on their respective net charges. We also recall from that chapter that electrons are paired in an orbital shell based on the Pauli Exclusion Principle (the relation $2n^2$), where each electron in a pair has a different direction of axial **spin** (either up or down). This difference in spin directions effectively allows both of these negative charges to exist in the same shell orbit without repelling one other – the **overall direction of electromagnetic force has changed because of the difference in axial spin.**

Magnetic materials are usually conductors that have unpaired electrons, and these electrons can potentially have either direction of axial spin – this is known as paramagnetism. In some magnetic materials (e.g. iron, nickel, cobalt) the extra electrons also align their spins in parallel to one other (known as ferromagnetism) to create a much greater magnetic field. It is believed that this alignment of axial spins may be part of how magnetism works – the extra electrons are spinning in a single direction in the first material, which may create a force of attraction to those spinning in the other direction on the second material – even though they both have a negative charge! We will see a related effect when we encounter **capacitors** later in this section – if you've ever rubbed a balloon on a material like wool to generate static to stick it to a wall then you have created an electrostatic field, which in many ways is the same electron-driven force as magnetism. Whilst the full answer to this phenomenon is still a matter of considerable debate, for our understanding we note the common factor of free electrons as being fundamental to both electrical charge and magnetism.

For a more detailed explanation of electromagnetism, see '*The Feynman Lectures on Physics, Vol. II: Mainly Electromagnetism and Matter*', Basic Books, 2001. ISBN: 978-0465024940.

Electromagnetic induction is used widely throughout electronics, from proximity sensors and microphones to actuators like motors and loudspeakers (as we will see in the next section). In audio, electromagnetic **induction** is used by the **dynamic microphone**, where changes in air pressure move a thin diaphragm that is connected to a copper coil wound around a magnet (Figure 2.29).

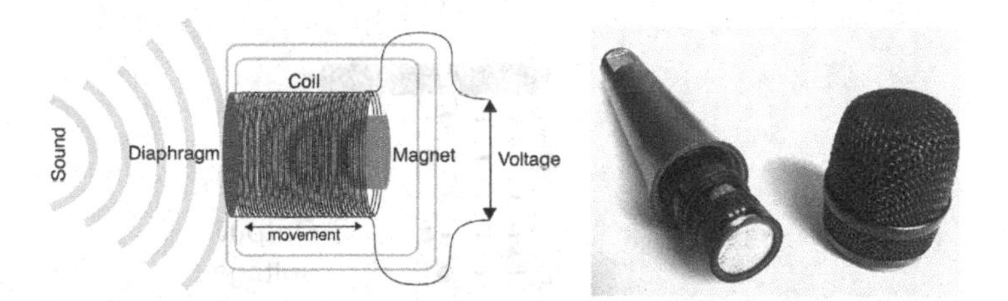

Figure 2.29 The dynamic microphone. ***The left-hand diagram shows how a microphone diaphragm is used to detect changes in air pressure caused by sound. As the diaphragm moves the conductor around the magnet, the voltage across the conductor changes due to electromagnetic induction. The right-hand image shows an example of a dynamic microphone capsule (with protective grille removed).***

In a dynamic microphone the copper coil acts as a conductor, surrounding *but not touching* the magnet that now creates a magnetic field around this coil. As the diaphragm moves with changes in air pressure (due to sound), the copper coil is moved across the magnet – thus **changing the magnetic field** around the conductor. The principle of electromagnetic induction means that this change in the magnetic field surrounding the coil will also create a **change in the voltage** across its terminals, and in this way a dynamic microphone converts sound into electrical energy. Dynamic microphones are often used in live music systems, partly due to their simplicity and also their low construction cost.

Another type of microphone, the condenser, works by converting sound pressure into changes in **capacitance**. Capacitors (formerly called condensers) exploit the occurrence of an **electrostatic field** between static charges, which is used to store electrical charge (Figure 2.30).

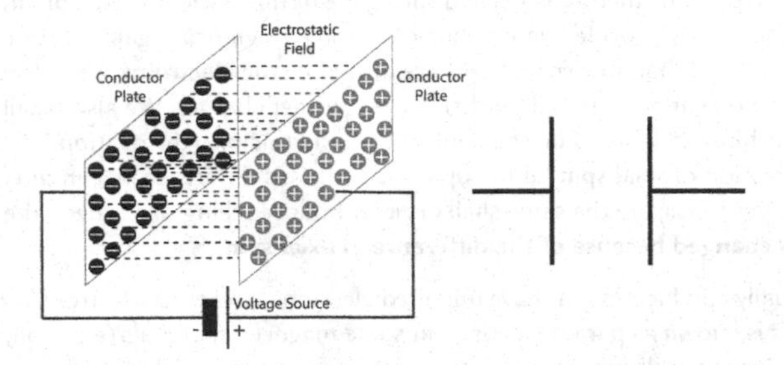

Figure 2.30 Capacitor example. ***In the left-hand diagram, a voltage source is used to supply electrical charge to the conductive plates (one positive, one negative). The build-up of charge between the plates is stored electrostatically in the air (or other insulator) between them. The right-hand image shows the standard schematic symbol for a capacitor.***

An electrostatic field occurs when **stationary charges** are brought into close proximity with one another – no electromagnetic force is present to move the charges, but an electrostatic field is created because they are **physically** close. This is how static electricity works, where the negative charge on the surface of one material (e.g. shoes can be charged by walking on thick carpet) can transfer to a conductor (e.g. a metal door handle) due to their relative physical proximity. In a capacitor, the air between the conductive plates is capable of storing electrostatic charge. The **capacitance** of this component is the amount of charge it can store, and this varies (partly) based on the distance between the plates. Capacitors are very important electronic components, particularly in audio circuits where their varying frequency response (a capacitor takes **time** to charge) makes them very useful for filtering. We will be looking at capacitors in more detail later in this book when we learn about AC circuits, but for now we can see how one of the capacitor plates could be connected to a diaphragm that is moved by air pressure (Figure 2.31).

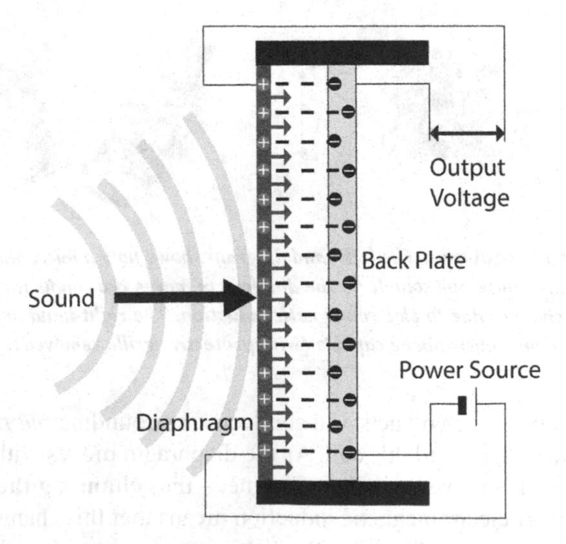

Figure 2.31 A condenser microphone. ***In the diagram, one of the microphone capacitor plates is a flexible diaphragm that can be moved by air pressure waves due to sound. As the diaphragm moves closer to the other plate, the capacitance changes and thus so does the output voltage from the microphone. The power source (either battery or dedicated +48V signal) is needed to charge up the capacitor.***

In the diagram above the movement of the diaphragm changes the distance between the conductive plates, which in turn changes the capacitance between them. This change in capacitance changes the output voltage across the plates of the microphone, and so in this way a condenser microphone converts sound into electrical energy. Notice the requirement for a power source to provide the charge for the capacitor plates, which is why condenser microphones will either have a dedicated battery or require a power signal (known as +48V phantom power) to be provided from a mixing desk or other audio interface. Condenser microphones are often found in higher-end audio recording applications, where the sensitivity of the diaphragm is typically much higher than in a dynamic microphone because it is not connected to a copper coil (which increases the weight that the diaphragm must displace). This higher sensitivity was also one of the reasons that condensers were often considered to be less robust than dynamic microphones, though in recent years new construction methods have led to them being used in a much more versatile range of recording applications.

Microphone types

There are many different makes and models of microphone, all designed with different applications in mind. Dynamic microphones, being the most common, can often be conflated with low quality but as noted above their robust design and relatively simple construction make them very popular for live music performance. You may also have come across another form of dynamic microphone called the ribbon microphone, where the diaphragm itself is a very thin strip of conductive material (usually aluminium) that moves within the magnet. Because this strip (ribbon) of conductive material is much lighter than a coil and diaphragm apparatus, ribbon microphones are known for their accuracy – the downside being their fragility when compared to a moving coil and diaphragm. As technology improves, it is becoming more common to find ribbon microphones using materials that are robust enough for live use.

For more information on microphones, see '*Sound and Recording*', 6th edition, by Rumsey and McCormack, Focal Press, 2009. ISBN: 0240521633.

Microphone technologies are rapidly evolving, partly due to the demands of new mobile, wearable and now hearable systems that often have primary applications in other areas (e.g. health, fitness, augmented reality). The change in domain (from dedicated audio production) can sometimes be interpreted as an impediment to better quality, as commercial factors often significantly impact on the performance of audio systems (e.g. installing audio speakers behind the screen on flat panel televisions to reduce the form factor). The positive side of this evolution is that new domains often provide much of the investment needed to innovate in new ways – note that microphone technology began with the earliest telephony systems, not for music production. It is important for us to consider both innovation and application in our electronics learning, to avoid discounting new ideas from other domains that may add value to our systems. As voice interfaces become more prevalent, so microphone technology will advance – this will also benefit audio production as a result. Now we have looked at the basic operation of some audio sensors, we will turn our attention to the outputs of audio systems – actuators.

2.6 Audio actuators: LEDs and loudspeakers

All systems have outputs to actuators, where this actuator converts electrical energy into another physical quantity. We are primarily interested in actuators that create sound, but it is also useful to consider those that convert electrical energy into light – specifically the **light-emitting diode** (LED). We used an LED in the previous chapter to build a simple circuit where the output is light, and audio systems will often use LEDs to communicate information such as power indicators, input level meters and even frequency analysis (a graphic EQ meter was considered very impressive in 1980s consumer hi-fi equipment!). LEDs are very versatile, and even if a circuit does not need a visual indicator they can still be very useful for prototyping and testing purposes.

We recall from chapter 1 that electrons carry charge, and in order for electrons to move there must be holes in the valence shells of other atoms for them to move into. We also learned that some materials (e.g. copper) are conductors that allow electrons to move easily, whilst others (such as pure silicon, i.e. glass) are insulators with stable valence shells that prevent the flow of electrical charge. There are also materials that can be chemically altered to behave as **both** conductor and insulator – these are known as **semiconductors**. The materials used in semiconductors are created by chemically **doping** the element **silicon** (Si) to change its atomic structure – this is part of the reason why silicon is such a crucial element in electronics. The simplest form of semiconductor is a **diode**, which combines an **n-type** (negative) material that contains a surplus of **electrons** with a **p-type** (positive) material that contains extra **holes** for those electrons to move into. In this way, a diode acts like a **one-way valve** for current, which is very useful for electronics (particularly for digital operations). We recall from chapter 1 that when an electron moves into a hole in a valence shell a **photon** is emitted – a photon being the fundamental particle of electromagnetic radiation, including light (Figure 2.32).

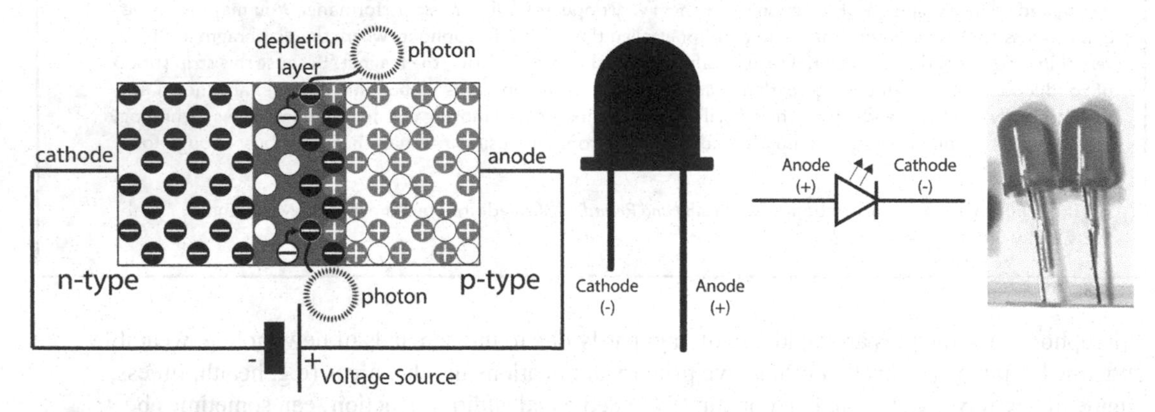

Figure 2.32 LED operation. ***The diagram shows how electrons from the power supply cathode (negative) cross the diode depletion layer (in grey) to fill holes in the side that's nearest the p-type region. At the same time, these moving electrons leave behind holes in the side of the depletion layer nearest the n-type region. Every time a hole is filled by an electron, the change in energy is emitted as a photon of light. The schematic symbol and pin connectors for an LED are also shown for reference.***

A diode exploits the movement of electrons around the junction between n-type and p-type materials known as the depletion layer. We remember from chapter 1 that when an electron fills a hole the change in energy is emitted as a photon – these photons create the light in an LED through a process called electroluminescence. The colour of an LED is determined by the *band gap* of the diode, which broadly speaking is the specific difference between the energy levels of electrons and holes. We will look at diodes in more detail in chapter 6, alongside other semiconductors like transistors and operational amplifiers. We will use diodes extensively in our circuits, both as an indicator of certain information in our system and also as a quick prototyping tool for testing. For now, let's move on to consider the most important actuator in audio electronics – the **loudspeaker**.

As we learned in the previous section, electromagnetic **induction** is a change in the magnetic field *around* a conductor that creates a corresponding change in the voltage *across* the conductor. For dynamic microphones, we use the movement of a thin diaphragm due to changes in air pressure (sound) to move a coil across a magnet, which creates a change in voltage across that coil. In the case of loudspeakers, we simply reverse this process by applying a voltage to the coil which is again wrapped around a magnet (Figure 2.33).

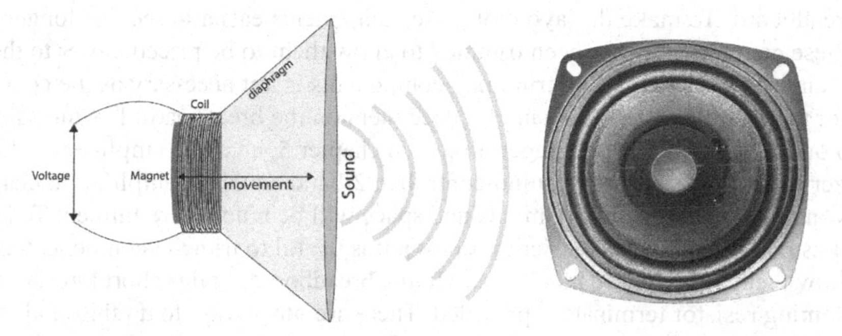

Figure 2.33 The dynamic loudspeaker. ***The diagram shows how a wire coil (known as a voice coil) is wound around a magnet. The coil is connected to the loudspeaker diaphragm (often referred to as the speaker cone), which is usually a paper or plastic composite with a large surface area designed to move air. When a voltage is present at either end of the coil, it will move relative to the magnet (which is fixed) and thus move the attached diaphragm.***

Dynamic loudspeakers move a diaphragm (known as a speaker cone) to create changes in air pressure, which creates sound. This is the most common form of electroacoustic transducer, where electrical energy is converted into sound by the movement of a conductor (voice coil) around a magnet. Audio quality loudspeakers (usually known as drivers) focus on precision of construction and refinement of the materials used to achieve high-performance results. It is not uncommon to find multiple drivers being used in a commercial loudspeaker enclosure, where the audio signal is first passed through a crossover network to split it into different frequency ranges. This allows each driver used to be constructed specifically for certain frequencies, most commonly for bass (low), middle and treble (high) sounds. We will learn more about different frequency ranges, and how to design circuits to process them, when we cover audio filters later in this book.

It is worth noting that whilst dynamic loudspeakers are overwhelmingly the most common form of audio actuator, there are many other types of loudspeaker. As many of these designs are either experimental (i.e. not generally commercially available) and/or of lower audio quality, they have not been included in this book because we are primarily interested in understanding the role of actuators in audio systems, rather than studying any specific transducer in great detail. Having said this, piezoelectric transducers (where changes in voltage create changes in the pressure in the piezoelectric material) are now used in some laptops and mobile devices because of their very small surface area (they do not require a magnet or a cone, which both take up space). Electrostatic speakers (which exploit the variance of an electrostatic field to move a flat diaphragm) also have great potential due to their highly linear frequency response in certain ranges (there are none of the resonances created by the cone or enclosures used in dynamic loudspeakers) but in most cases they do not deliver bass frequencies well. It is possible that the loudspeakers of the future will not be solely dynamic designs, and arguably both piezoelectric and electrostatic transducers have some commercial potential in this regard.

2.7 Tutorial – working with components

This book provides an introduction to audio electronics and Arduino control, where prototype circuits built at the end of each chapter aim to summarize what has been discussed in a practical context. To do this, all project circuits will combine components such as resistors, LEDs and capacitors on a single small breadboard. To make the layout of these components easier to see, the longer connecting terminals for these components have been trimmed to allow them to be placed closer to the board. For the project circuits in chapters 1–4, trimming components is not necessary as the connections involved are not too complex – you can simply place them on the breadboard. Having said this, as we progress to building a MIDI drum trigger project in chapter 5, an audio amplifier in chapter 7, a 2-band equalizer in chapter 8 and an Arduino-controlled 2-band equalizer amplifier in chapter 9 the component layouts will become more complex and space will be much more limited. To keep your circuits as neat as possible (and thus easier to follow), it is useful to trim the component-connecting terminals to allow them to be placed much closer to the breadboard. In this short tutorial, a simple method for trimming resistor terminals is provided. There are other ways to do this, and so you should consider how to use the tools you have available to achieve consistently trimmed terminals for use in your projects.

Tutorial steps

Every project circuit begins with a summary of the components required – in the previous chapter you will already have seen a resistor with trimmed terminals. LEDs and capacitors can also be trimmed in a similar way, but in all cases the aim is **not to remove all of the terminal wire** – only to trim it.	
The first step in trimming a resistor is to ensure that the length of each trimmed terminal is equal. In this case, a pair of needle-nosed pliers is used to provide a suitable length for the terminal, where the excess wire is bent at 90° around the pliers ready for clipping. If you do not have these, any flat surface of 1 cm or less can be used – e.g. an old cutlery knife handle or even a house key will provide a suitable surface to bend the terminal wires.	
Once the wires are bent to 90° at a distance of around 1 cm from each side of the resistor, the excess wire can now be clipped at the bend points so that both terminals are now the same (shorter) length. This can be done with any pair of pliers or snips that are designed for cutting (scissors are not recommended as they are not designed to cut metal and can slip).	Clip

Once trimmed, the shorter resistor terminals can now be bent downwards at an angle of 90° to each side of the resistor – if you have tweezers then they can be placed at the side of the resistor to provide a pivot. The component is now ready to be placed much closer to the surface of the breadboard, making it easier to see and to work with in more complex projects.	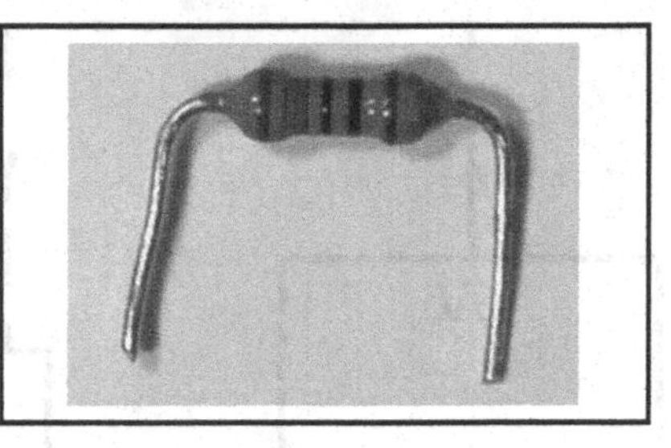

2.8 Example project – sensor control of LED output

In our example projects from now on, we will begin with the system diagram for the project involved, followed by the schematic representation of the circuit used to build it. We will then work through each stage of the schematic to discuss how best to lay out our components on breadboard to create the actual circuit that can either be simulated in Tinkercad or built using an actual Arduino and some components. In all instances, the system diagram is crucial to both analysing and building effective circuits, as it focusses on the core elements of input, process and output that we will return to throughout this book. To learn about working with sensors (specifically switches), we will now build two circuits based on the open-loop systems in Figure 2.34.

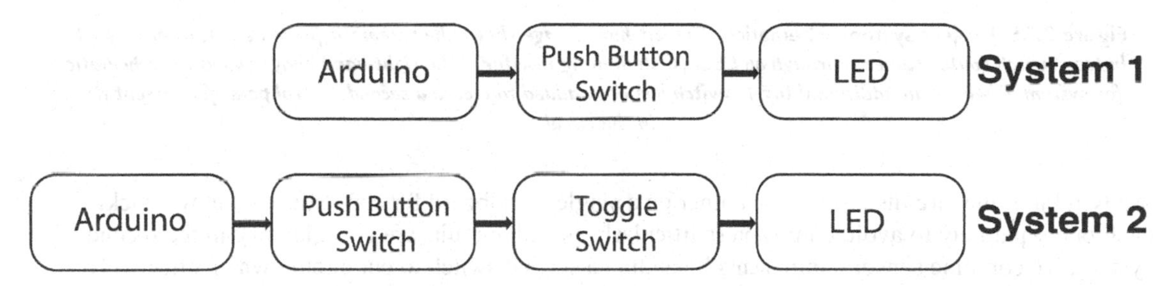

Figure 2.34 Systems used in the example project. ***The left-hand image shows an open-loop system that uses a switch to route current from the Arduino through an LED. The right-hand image routes current in the same way, but combines two switches to provide separate control points for current flow within the system.***

These systems build on the example circuit from chapter 1, where we used the Arduino to provide power to an LED. This was effectively an output to an actuator, so we will now look at adding sensor input to create the simplest level of open-loop system control. As with the project in chapter 1, neither of these systems is particularly interesting from an audio perspective, but we are using these early projects to consolidate our learning and apply the systems thinking we have covered in this chapter to further our understanding of how to break a circuit down. We will quickly be building much more complex projects that combine analogue circuits with digital control (such as the MIDI drum trigger project in chapter 5), all based on the same systems thinking introduced in this chapter. For this project, we will build each of these systems in turn using the following two schematics in Figure 2.35.

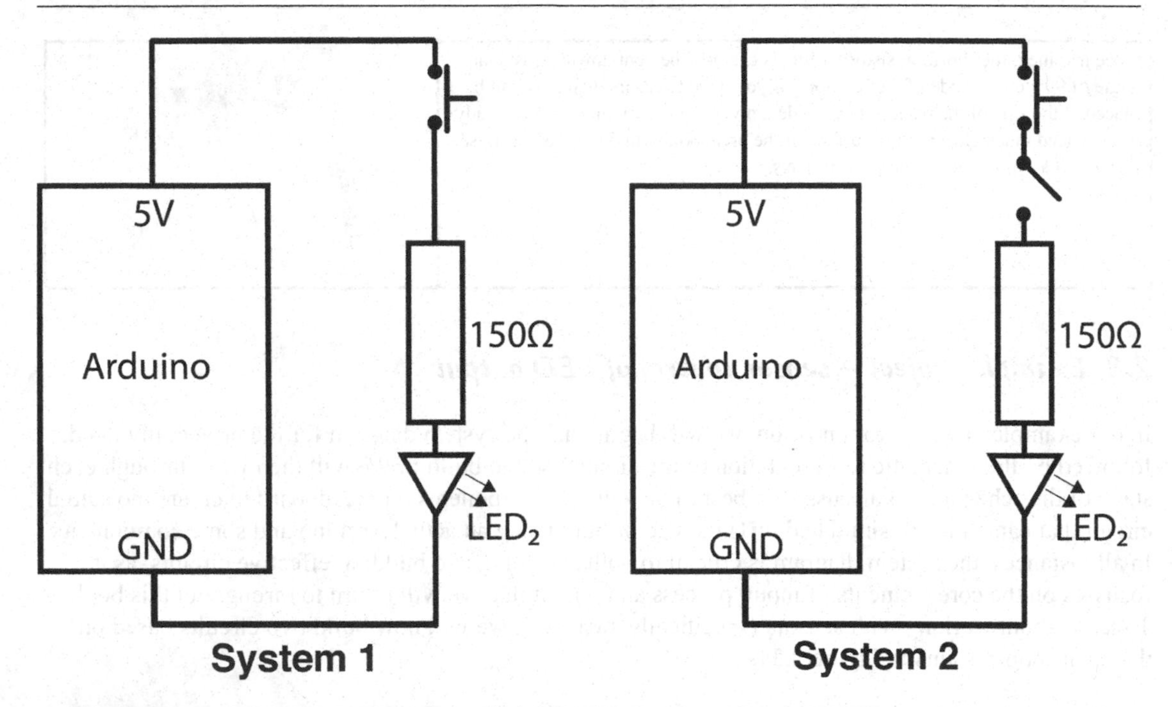

Figure 2.35 Project system schematics. ***The left-hand image shows the schematic for system 1, where a push-button switch routes current through an LED (with limiting resistor). The right-hand image shows the schematic for system 2, where an additional toggle switch has been added to create a second control point for current flow in the circuit.***

We will build the circuits next to each other on a single breadboard for convenience, but will tackle each one separately to avoid confusion (particularly as both circuits are very similar). In the second system, we combine sensor components by adding a second switch to our circuit, which effectively provides a second control point for current flow. This is a very typical configuration, where a power switch is used to provide overall current flow to the system, but then further switches are added to route current in specific ways. We will learn more about routing current in the next chapter when we look at series and parallel circuits, but for now we will look at the specific design requirements involved when working with push-button switches.

Circuit design

In this project the first thing to consider is the layout of the switch, which we previously discussed earlier in this chapter. The next thing to address is the layout of the resistor and LED, which must be connected in series with the output button terminal (either B or D). In Tinkercad, LEDs are oriented with the anode on the left so they must be rotated 180° in order to be linked correctly in series with a resistor. In addition, the size of the component means it needs to be placed on a higher row in

Tinkercad to allow other connections (in this case GND) to be added underneath. We can set out the components so that each input terminal is connected on the same column as the output terminal of the previous component – in the case of a resistor input and output are interchangeable. The system 1 schematic has also been redrawn to rearrange the three components in a straight line, to help us focus on the layout based on a series current flow (Figure 2.36).

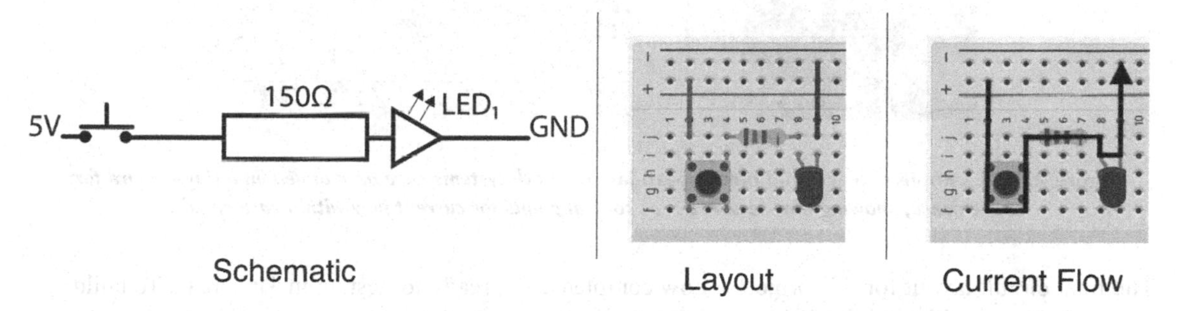

Figure 2.36 System 1 component layout. ***The schematic on the left is used to plan where the components will be situated on a breadboard. The power lines have been included to show why the switch and LED are situated on [row i] to allow 5V and GND connections to be added.***

Notice in the schematic in Figure 2.36 that we have marked the 5V and GND connections at each end of our components. In the breadboard layout, we have added wires to connect these components to the power rails – the Arduino will eventually supply +5V and GND to these rails. The circuit for system 2 replicates the same structure as our previous schematic, but with an additional switch being added in series to provide an extra control point for current flow (Figure 2.37).

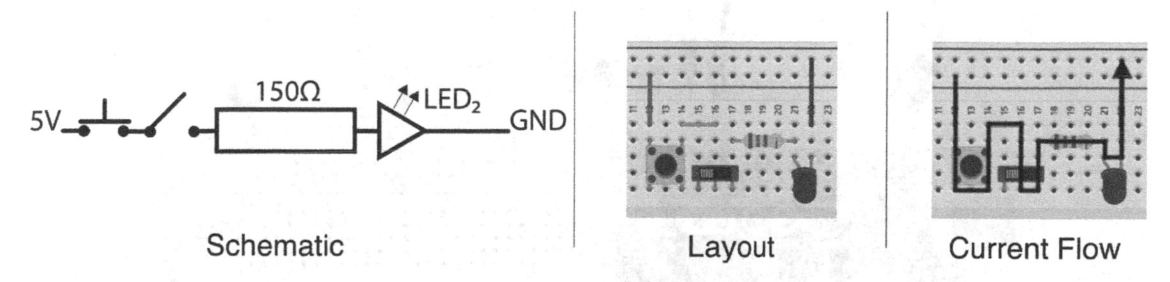

Figure 2.37 System 2 component layout. ***The schematic on the left is used to plan where the components will be situated on a breadboard. The addition of the second toggle switch (a sliding switch in Tinkercad) can be replaced with a second push-button switch on an actual breadboard if necessary.***

Now we can add power connections from the breadboard to the Arduino 5V and GND pins (Figure 2.38).

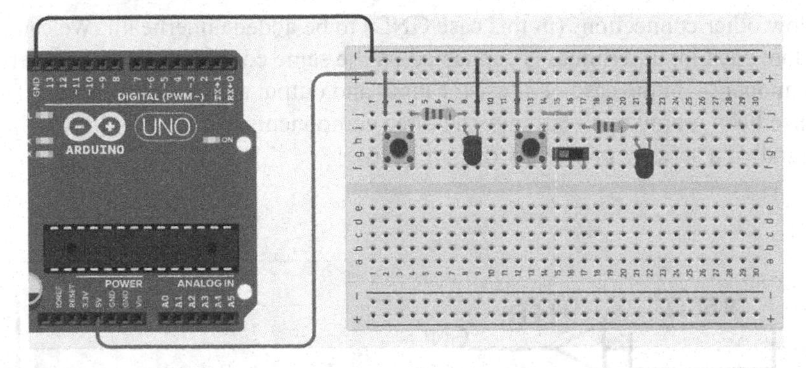

Figure 2.38 Example project final breadboard layout. ***Both systems have been added on a single board for convenience, showing how sensors act as control points for current flow within each circuit.***

The final circuit layout for this project is now completed and ready for testing in Tinkercad. To build this project, you will need the following components (Figure 2.39).

1. An Arduino Uno board (or equivalent).
2. A breadboard.
3. Three × push-button switches (you can use a toggle switch in system 2 if you have one).
4. Two × 150Ω resistors.
5. Two × LEDs (ideally red, but any colour will do).
6. Six wires (ideally three red and three black, to make it easier to follow the circuit path).

Figure 2.39 Project components. ***The image shows the switches, resistors, LEDs and wires used in building both systems in this example project. The toggle switch in system 2 has been replaced with a second push-button switch to reduce build costs.***

To build the project on a breadboard, we will now work through the specific pin playouts for each system in the steps below.

Project steps

1. Add a push-button switch so that the common terminals (A-C, B-D) are connected between pins **[g2–i2]** and **[g4–i4]** – common switch terminals are bent towards each other 2. Add a 150Ω resistor between pins **[j4–j8]** 3. Add the first LED between pins **[i8–i9]** – long leg (cathode) on pin i8	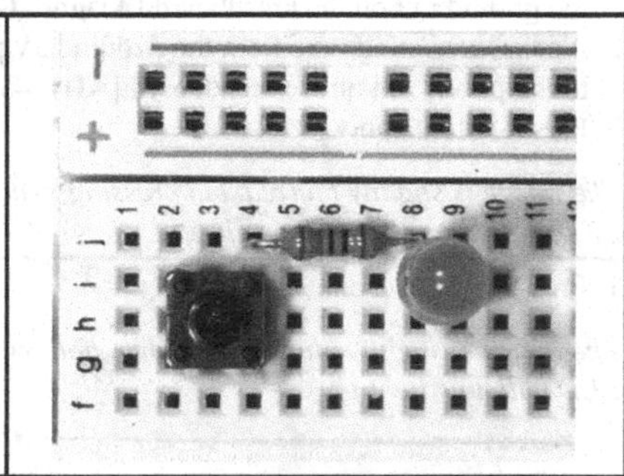
1. Add a connecting wire from pin j2 to the power rail on the breadboard **[j2–V+]** 2. Add a connecting wire from pin j7 to the ground rail on the breadboard **[j7–GND]**	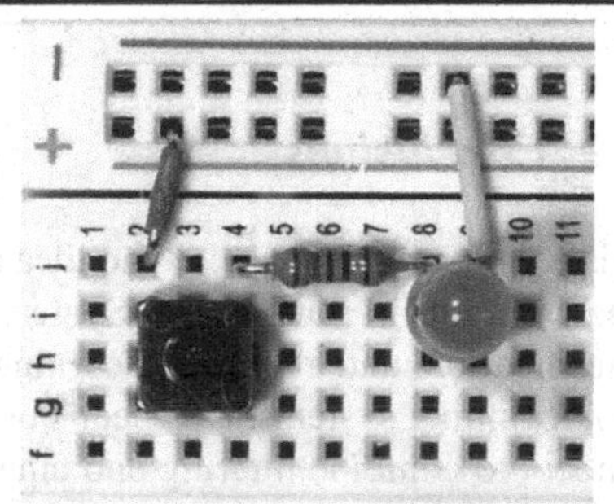
1. Add a connecting wire from the Arduino GND (top row, 4th pin from left) to the ground rail on the breadboard **[ADgnd–GND]** 2. Add a connecting wire from the Arduino 5V pin (bottom row, 5th from left) to the power rail on the breadboard. **[ADv+–V+]** *If you have connected everything correctly, the LED should light up*	
Project steps – system 2	
1. Working from the previous circuit for system 1, add a second push-button switch with the common terminals (A-C, B-D) connected between pins **[g12–i12]** and **[g14–i14]** 2. Add a third push-button switch with the common terminals connected between pins **[g17–i17]** and **[g19–i19]** 3. Add a connecting wire in series between the push-button switches on pins **[j14–j17]** 4. Add a 150Ω resistor between pins **[j19–j23]** 5. Add the second LED between pins **[i23–i24]** – long leg (cathode) on pin i8	
1. Add a connecting wire from pin j12 to the power rail on the breadboard **[j12–V+]** 2. Add a connecting wire from pin j24 to the ground rail on the breadboard **[j24–GND]**	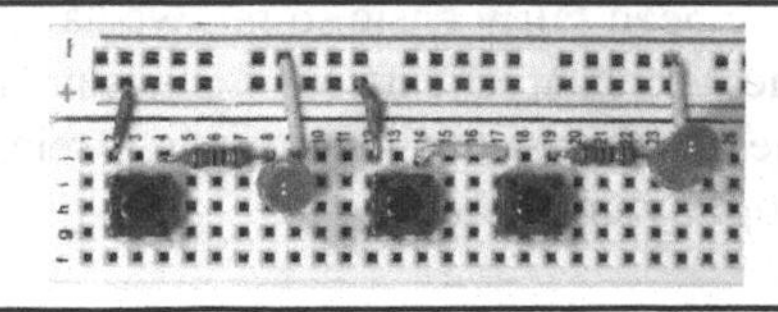

<table>
<tr><td>1. Add a connecting wire from the Arduino GND (top row, 4th pin from left) to the ground rail on the breadboard [ADgnd–GND]
2. Add a connecting wire from the Arduino 5V pin (bottom row, 5th from left) to the power rail on the breadboard [ADv+–V+]
3. Press switch 2 only

The image is slightly blurred, but pressing switch 2 will not light the second LED – the circuit is not completed</td><td>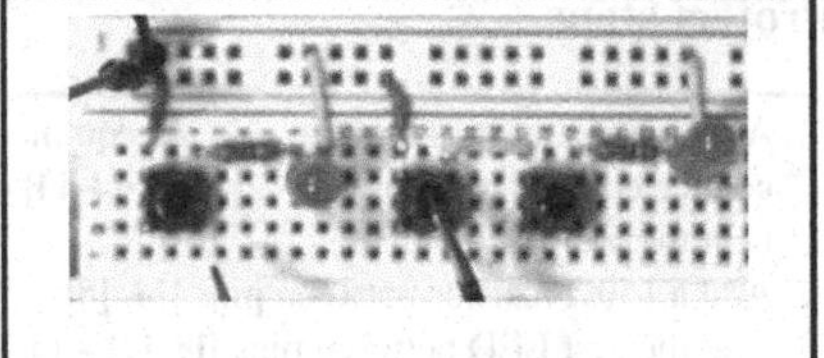</td></tr>
<tr><td>1. Press switches 2 and 3 together

The image is slightly blurred, but when both switches are pressed the second LED will now light up</td><td>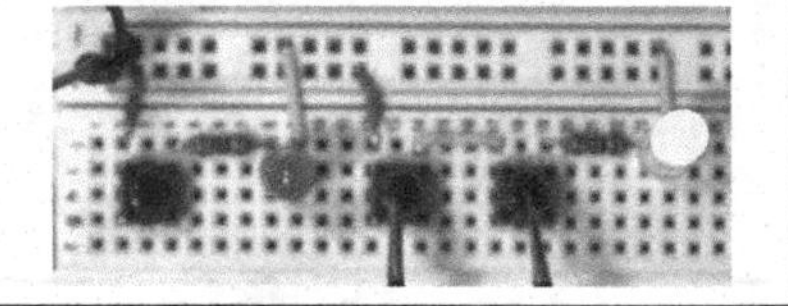</td></tr>
</table>

2.9 Conclusions

This chapter has introduced the idea of an electronic system that processes input from sensors to create outputs to transducers. System analysis is important when considering source and load issues in amplification, and also when learning how to control a circuit by programming the Arduino. We looked at partitioning systems as a means of breaking down complexity, and in later chapters we will use this concept to combine filtering and amplification stages in our circuits. In addition, MIDI systems have been a mainstay of audio electronics for over 30 years, and some of the basic elements of the MIDI protocol were introduced in this chapter as preparation for our chapters on Arduino programming that will use it more extensively.

We looked at sensors like switches and potentiometers – we will be using them with the Arduino throughout this book to control our audio circuits. We also looked at other sensors that can be used as audio transducers like piezos, condensers and dynamic microphones and noted that the dynamic loudspeaker effectively operates using the same electromagnetic principles. Analogue and digital signals were also briefly introduced to discuss the conceptual differences between continuous time (CT) and discrete time (DT) representations. Although we will not be covering digital audio in this book, it is nevertheless important to understand the fundamental distinction between signal types to better inform our future learning with the Arduino. We will be using the Arduino as a microcontroller for our audio circuits, but you will encounter many examples that include audio – the Arduino is not powerful enough for us to use it in this way.

In the next chapter, we will begin to learn how to analyse direct current (DC) circuits. You have already been working with DC circuits in the example projects in chapters 1 and 2, and we will return to the circuit from the project in chapter 1 to show how a figure of 150Ω was arrived at for the resistor that limits the current flowing through the LED. We will learn about Ohm's Law and Kirchoff's Laws for voltage and current – the fundamental analysis tools used throughout electronics. Although our projects have been fairly basic so far, we have covered a lot of conceptual ground in a short space of time. The next chapter carries on in a similar way, so please take some time to work through the self-study questions provided below to help reinforce your learning – systems analysis is a powerful technique if properly understood.

2.10 Self-study questions

The following questions aim to help you remember some of the core concepts in this chapter. The questions are not particularly difficult, but focus on stimulating your memory of what you have read, so try to avoid flicking back through the pages to find the answer – research on studying and the testing effect shows that any attempt at recalling an answer helps to reinforce the question in your mind (even if wrong, once you see the correct answer you are then more likely to retain it).

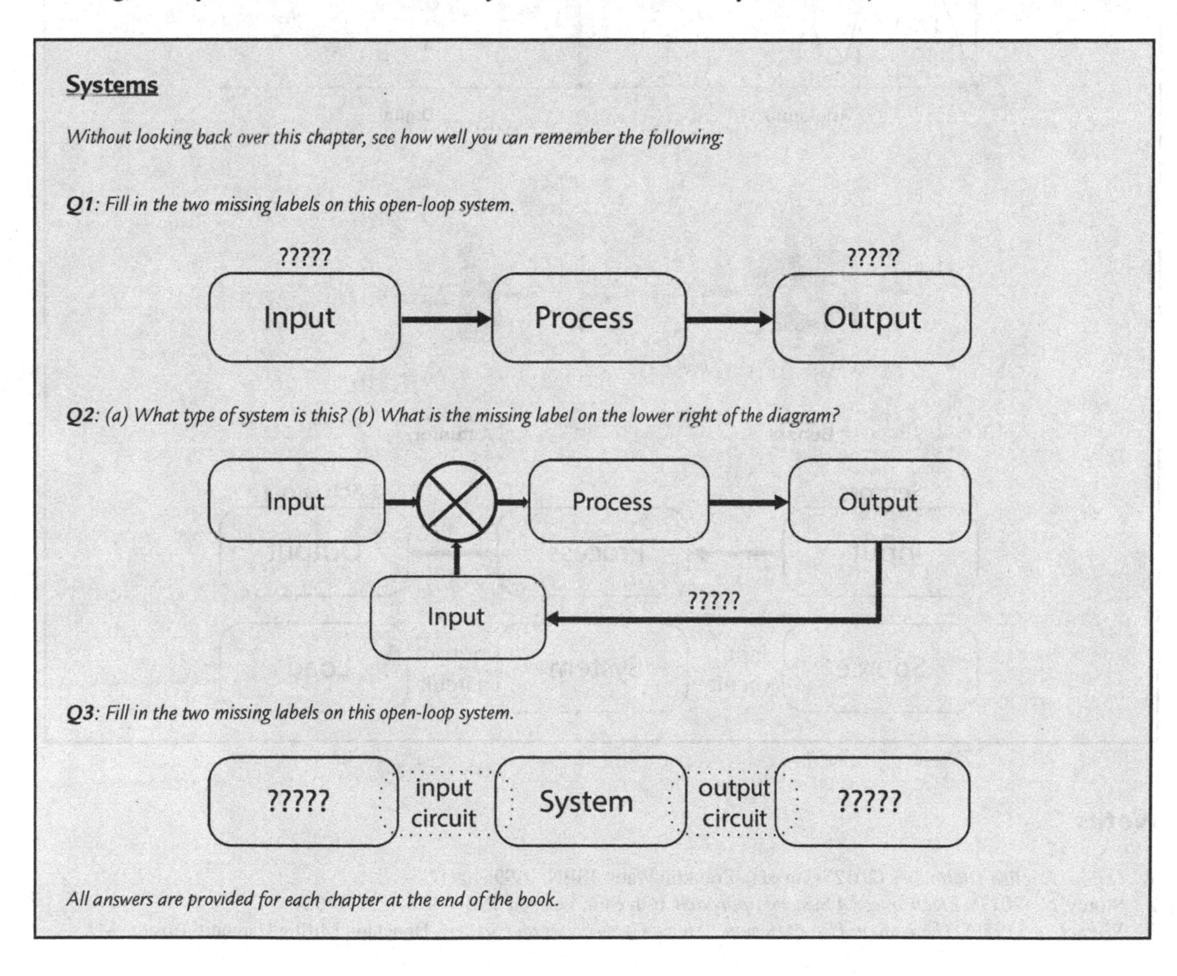

Chapter 2 summary

The following images may help you remember some of the key points in this chapter:

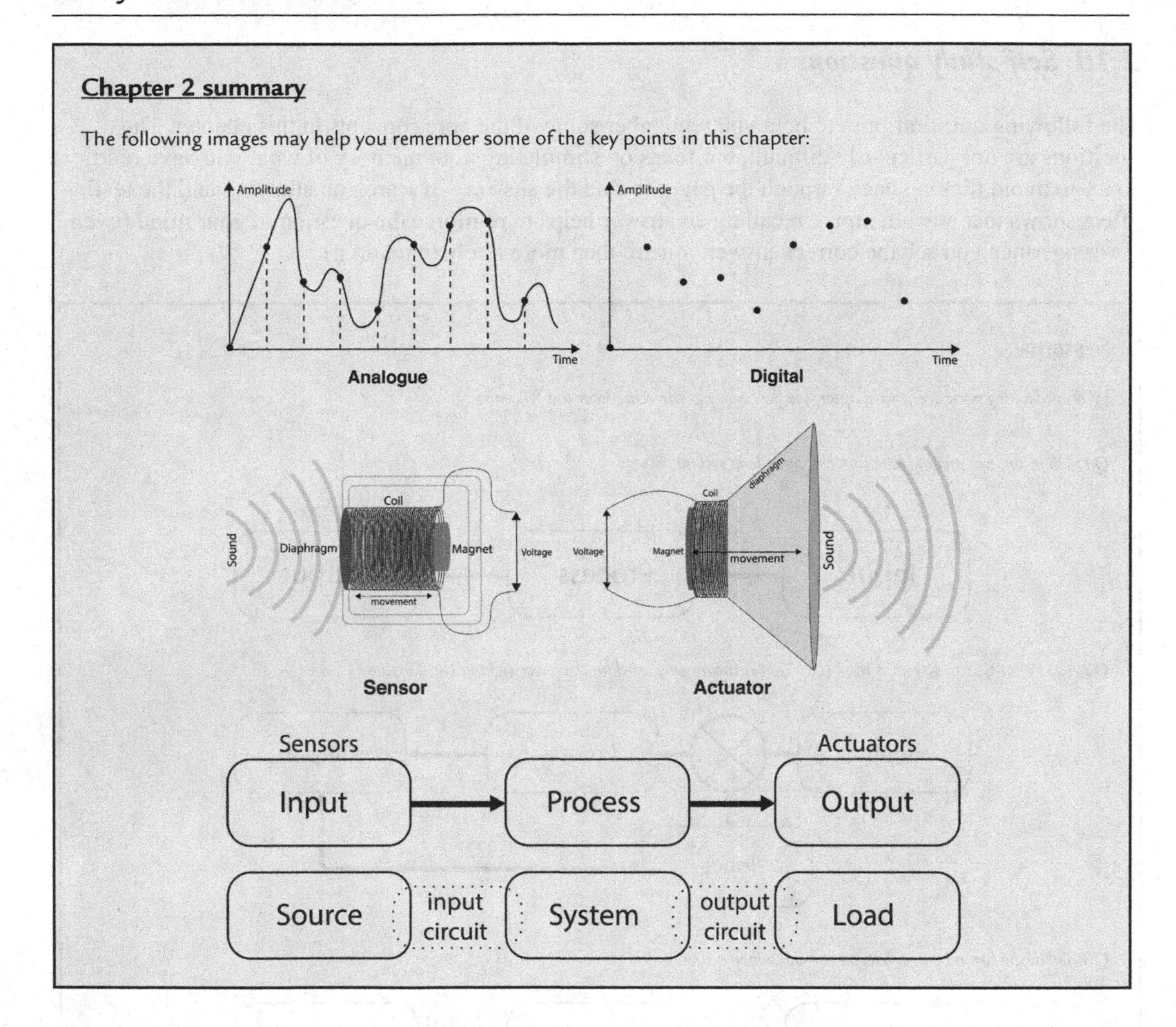

Notes

1 *Oxford English Dictionary* (2012) (7th edn). Franklin Watts. ISBN: 0199640947.
2 Storey, N. (2017). *Electronics: A Systems Approach* (6th edn). Pearson. ISBN: 978-1292114064.
3 Wiener, N. (1950). *The Human Use of Human Beings: Cybernetics and Society*. Houghton Mifflin Harcourt, Boston, MA.

CHAPTER 3

DC circuits

In this chapter, we will learn how to analyse direct current (DC) circuits. The basic principle behind DC circuits is that they are time-invariant – they do not change. This means we can specify voltage and current values and assume that these will remain the same as long as the power supply to the circuit is connected, which differs from the alternating current (AC) circuits we will encounter later in the book. This chapter represents our first real introduction to electronic circuit analysis, where we will learn the basic equations and rules of series and parallel circuits. You may find the maths in the chapter examples takes a little time to work through (particularly for parallel circuits) but remember that the equations themselves are quite straightforward – it's the scales and symbols that take some time to get used to. We will begin by learning about Ohm's Law, a fundamental equation that defines the relationship between voltage, current and resistance in a circuit. Ohm's Law was briefly introduced in chapter 1 as an example of a relatively straightforward equation used in electronics, and now we will learn how to work with it in practical terms. We will also encounter two other fundamental aspects of DC circuit analysis – Kirchoff's Voltage Law (KVL) and Kirchoff's Current Law (KCL). These two laws are used in all electronic circuits, so this chapter will spend a little time deriving them and explaining their use. We will use KVL to analyse series circuits (where all components are connected in a single loop) and KCL to analyse parallel circuits (where current can travel down multiple branches within the circuit) demonstrating the following rules:

series circuits – **voltage varies** and current is constant
parallel circuits – **current varies** and voltage is constant

A short tutorial is provided showing how Ohm's Law and KVL were used to derive the resistor value for our LED circuit in chapter 1 – explaining why the value of 150Ω was used in that circuit. The final projects compare series and parallel LED circuits to show the practical differences between them. Unfortunately they are not particularly exciting (nor audio-related) projects, but they will give you more practice with Tinkercad, circuit design and breadboard layout to get familiar with the practicalities of building real circuits.

What you will learn

Ohm's Law to relate voltage, current and resistance
Kirchoff's Voltage Law (KVL) for series circuits
Kirchoff's Current Law (KCL) for parallel circuits
How to use voltage dividers to create output voltage ratios
For **series** circuits, why **voltage varies** and current is constant
For **parallel** circuits, why **current varies** and voltage is constant
How to apply Ohm's Law with KVL and KCL to analyse circuit examples
How to design and build series and parallel circuits on breadboard

3.1 Ohm's Law and direct current

Chapter 1 introduced the three fundamental elements needed to begin learning about electronic circuits:

> **Current is the flow of electrical charge over time** – *measured in amperes (A)*
> **Potential difference is the difference in electric potential between two points** – *measured in volts (V)*
> **Resistivity is the resistance of a material to the flow of charge** – *measured in ohms (Ω)*

We also used the Arduino in our first project to power a simple circuit with a resistor and an LED as output, intending to look at this circuit in more detail in this chapter. As a first step, we can reduce this circuit to its minimum functional level by removing the LED output, so we can learn how to use circuit analysis techniques to determine the value of the resistor in the circuit (Figure 3.1).

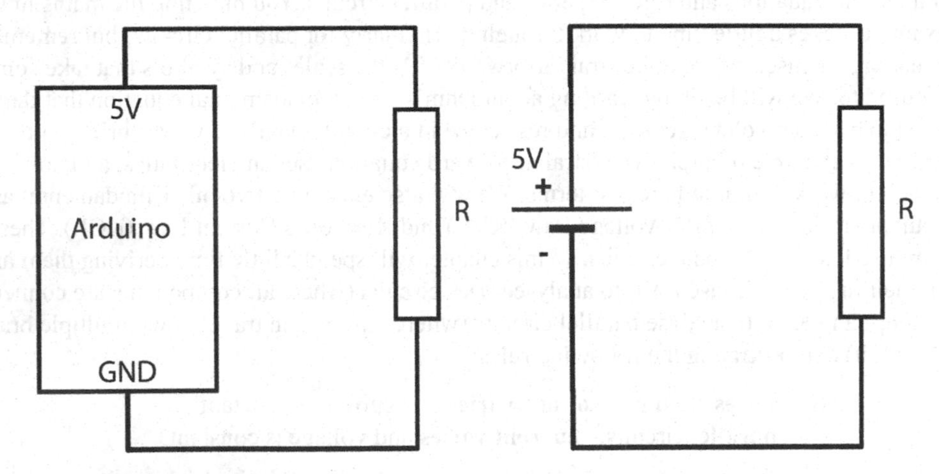

Figure 3.1 A simple Arduino circuit. ***The left-hand diagram shows an Arduino providing 5V to power to a single resistor. The right-hand diagram shows the schematic representation of the Arduino as a standard 5V DC voltage source. This is the simplest circuit we can design, though in practice it does not provide any inputs or outputs and so is not a practical system.***

In the example circuit in Figure 3.1, a single resistor is connected to the Arduino 5V output to create a circuit path to ground (GND). The right-hand schematic shows the same circuit with the standard symbol for a DC voltage source, which includes the polarity (+/−) of the circuit for conventional current flow (see next section). The following examples will use a DC voltage source unless we are building a project circuit, in which case the left-hand representation will show the Arduino as a power supply. Although this circuit does not do anything useful, it allows us to consider a practical constraint of Arduino output when connected over USB. As we saw in chapter 1 (Table 1.2), the USB protocol defines a maximum current of 500mA for connected devices so our Arduino (which will draw power from the computer that programs it) will only be able to supply a current up to this value. If the Arduino can output a maximum current of 500mA through its 5V pin, the resistor in our circuit should limit the current to this value (though in practice the computer will limit the supply regardless). To determine a suitable value for this resistor, we use an equation introduced in chapter 1 – **Ohm's Law**:

$$\mathbf{\textit{Voltage}}, V = IR \tag{3.1}$$

V is the potential difference in volts (V)
***I** is the current in amperes (A)*
***R** is the resistance in ohms (Ω)*

George Ohm developed the equation above in 1827 to show that the potential difference (V) between two points on a wire is equal to the magnitude of the current (I) flowing through it multiplied by the resistance (R) of the wire itself (Figure 3.2).

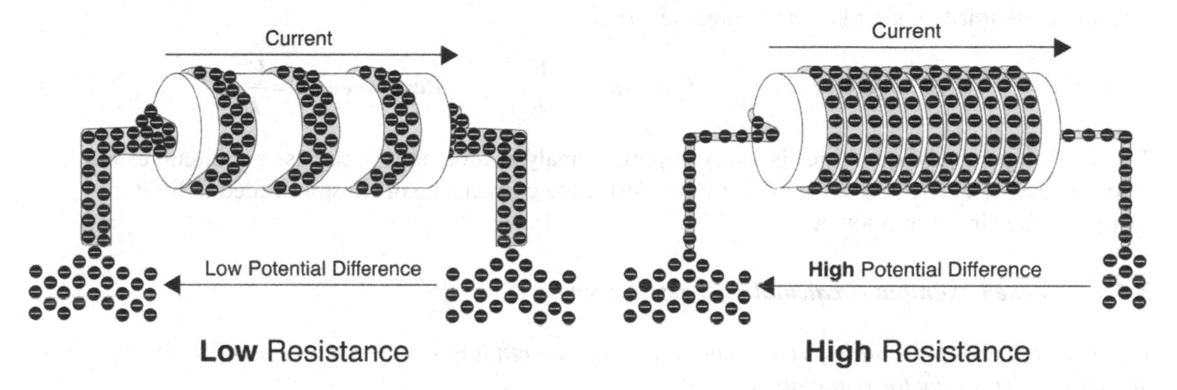

Figure 3.2 Resistance and Ohm's Law (taken from chapter 1, Figure 1.11). ***In the left-hand example, the lower resistance will not significantly impede current flow, so the difference in charge (potential difference) between the ends of the wire will be lower. In the right-hand example, the higher resistance prevents charge from flowing easily, creating a greater difference between the number of free electrons at each end of the wire.***

In Figure 3.2, because the wire resists the flow of current, more charge builds up at one end of the wire than the other. As a result, the imbalance in electrical charge between the two ends of the wire creates a **potential difference.** For a lower resistance value, the imbalance is not as significant. as more current can flow through the resistor and so both points have similar levels of charge. For a higher resistance value, the reduction in current flow means more charge is now concentrated at one end of the resistor as it cannot flow easily to the other end. This increased resistance creates an increased imbalance in charge, which creates a higher potential difference between the two ends of the wire as a result. We know that the **current does not change** because the number of electrons stays the same, the only difference is **where** the electrons are located. Thus, we can see that increasing R increases V, so we can derive $\mathbf{\textit{Voltage}}, V = IR$.

Scales, symbols and equations

In chapter 1, we learned that fundamental electronics equations are not mathematically complex, but the scales and quantities involved can take time to learn. Ohm's Law defines the simple mathematical relationship between the three fundamental elements of voltage (V), current (I) and resistance (R), but the symbols involved can initially appear a little confusing. The letter (I) is used for current partly because it was originally discovered by André-Marie Ampére, who as a French speaker named the quantity Intensité du Courant (Current Intensity). In addition, electrical charge is measured in coulombs (C), which would also create a confounding between terms! As you progress through the book you will find that **scales and symbols often take more time to learn** than equations, so terms will be explained whenever they are first introduced.

Ohm's Law is a simple equation to work with and forms the basis of all electrical circuit analysis. We can apply the equation in different ways, where we can use **any two of the quantities** involved to determine the third by simply rearranging the terms:

$$\textit{Voltage}, V = IR \qquad \textit{Current}, I = \frac{V}{R} \qquad \textit{Resistance}, R = \frac{V}{I} \tag{3.2}$$

This rearrangement of quantities is a very important analysis tool, and we can use our circuit example from Figure 3.1 to show how Ohm's Law can derive the correct size of resistor needed to limit the current in the circuit to 500mA.

3.1.1 Worked example – calculating a resistor value

This example uses the simple circuit from Figure 3.1, though it has no practical use other than to introduce Ohm's Law for calculation.

Q1: *A single resistor is connected in a circuit powered by a 5V voltage source (from an Arduino). As the Arduino has a maximum output current of 500mA, a resistance value is needed to limit the output current to this level – what would be a suitable value for this resistor?*

$$V = 5V, \qquad I = 500mA$$

$$\textit{Resistance}, R = \frac{V}{I} = \frac{5}{0.5} = 10\Omega$$

Answer: A suitable resistor value for the circuit would be **10Ω**.

Notes:

The only issue in this example is the scaling factor of milliamps for current, which requires 500mA to be scaled to 500×10^{-3} = **0.5A**.

The example above is simple enough, though of no actual practical use – the Arduino is limited by the USB protocol of the computer supplying it, so it cannot exceed 500mA even if the resistance in the circuit was lowered to 5Ω (giving a theoretical current of 1A). It is also important to note that **all Arduino output pins are limited to 40mA** and driving them beyond this current will damage the Arduino (the +5V power pin is not directly connected through the microcontroller so it is not limited in this way). The purpose of this circuit is to show how straightforward Ohm's Law is to use, providing you know two of the three quantities and **pay attention to any scaling factors** involved. As you progress through the book, we will use Ohm's Law extensively in our circuit analysis so it will help to spend some time looking at equation (3.2) to become more comfortable with rearranging the terms involved. Now we have learned how to apply Ohm's Law, we can move on to another set of fundamental laws that define the behaviour of electronic circuits – Kirchoff's Circuit Laws.

3.2 Kirchoff's Voltage Law: series circuits

Chapter 1 discussed analysing circuits using **conventional current flow**, where it is (incorrectly) assumed that charge flows from the positive to negative terminal in the circuit. We learned that this is a result of historical precedent, where electrical charge was originally believed to be positive long before the negatively charged electron was discovered. Aside from the reason for the difference in current direction, note that in electronic circuit analysis the direction of the current does not actually make a difference – as long as we calculate in a **single** direction (Figure 3.3).

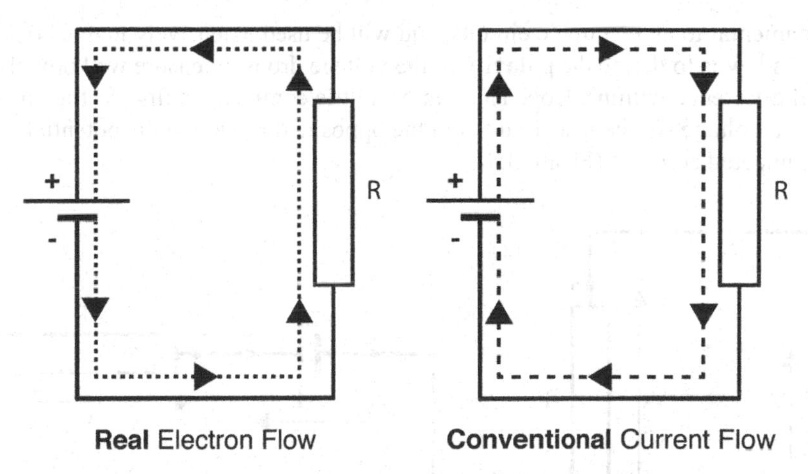

Figure 3.3 Real and conventional current flow in a simple Arduino circuit. ***The schematic shows an Arduino providing current to a single resistor. The left-hand diagram shows the actual flow of electrons from negative (GND) to positive, whilst the right-hand diagram shows the direction of conventional current flow used in circuit analysis.***

Conventional current flow

The idea of maintaining a single direction for analysis can seem either obvious or arbitrary (or both!), but in many ways it is not unlike adding (2+3) or (3+2) – regardless of the direction in which we add the quantities, the result is still 5. In practice, the flow of electrons may be in the opposite direction to our analysis but the answer to our equations will remain the same if we are consistent in the direction used.

This seems to be quite straightforward: as we have noted the reason for this historical precedent so we can then move on to its application. The problem occurs when trying to reinforce our conceptual understanding of voltage, current and resistance whilst also learning to use Ohm's Law and Kirchoff's Laws – it is then we may discover that thinking in reverse can be more difficult than it first appears! The solution is to **proceed slowly to reiterate the concepts** learned in chapter 1, while learning to use the new analysis methods introduced in this chapter. If you remember to do this, the conceptual gaps in your understanding will not become wider as we progress.

We measure current as the flow of charge over time in a **single direction** – this is known as direct current (DC). In a DC circuit, the current and potential difference at any place in the circuit are constant – they **do not change** over time. In chapter 2, we learned about continuous time (CT) signals (also known as analogue) that carry input from sensors like microphones and also drive output to loudspeakers – CT signals **vary over time**. In chapter 8, we will learn about alternating current (AC) as a means of building circuits for CT audio signals, but for now we will focus on DC circuits to help us understand and apply basic circuit analysis techniques.

In 1847, Gustav Kirchoff defined two fundamental laws that govern the behaviour of all electronic circuits. We will look at each of these in turn as they apply to **series** and then parallel circuits, beginning with Kirchoff's **Voltage** Law (KVL):

> **Kirchoff's Voltage Law**
>
> *The sum of all voltages around any closed loop in a circuit is equal to zero*

This law is fundamental to all electronic circuits, and will be used extensively in this book. The key to understanding this law is to define the **polarity** of the voltage drops, to ensure we know the **direction** of each potential difference within a loop. This can be a little confusing at first, as the potential difference across a voltage source in a circuit is in the opposite direction to the potential difference of the resistance connected across it (Figure 3.4).

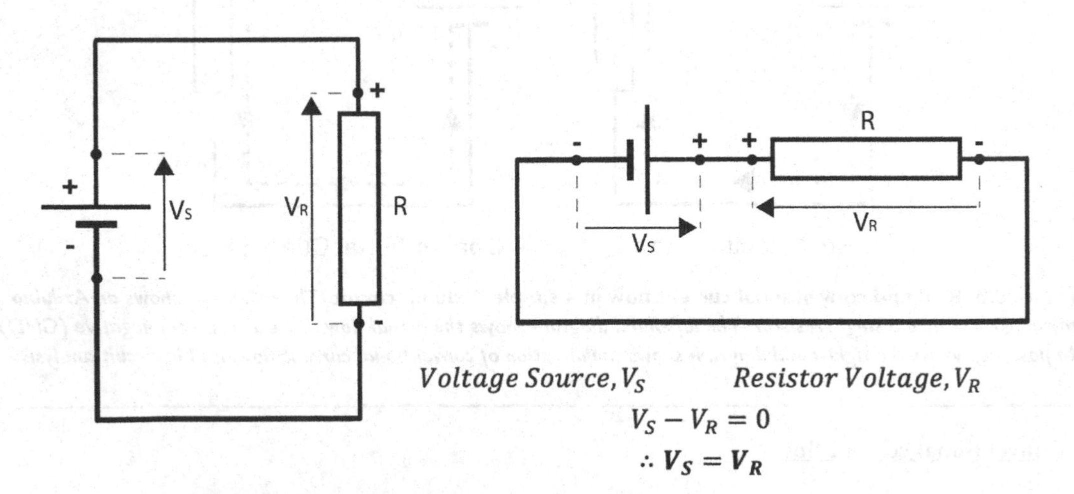

Figure 3.4 Voltage polarity in a series circuit. ***The diagram on the left shows the direction of the potential difference (low to high) across a voltage source and the subsequent direction of the voltage drop through a series resistor connected in this circuit. The diagram on the right shows a rearranged version of the same circuit, to highlight the opposite direction of the potentials involved.***

This diagram shows a series circuit: **a single loop in which all components are connected together in series**. We have already used Ohm's Law in the previous section to determine the value of a resistor in a series circuit but as noted in the previous sidebar, at this point you are synthesizing several concepts at the same time: conventional current flow, voltage (potential difference) polarity and Kirchoff's Voltage Law (KVL). For this reason, it is useful to step through each of these elements in turn to be clear about how they relate to one another – you may understand them all individually, but combining them can lead to comprehension problems when you are beginning circuit analysis:

1. *Conventional current flow* – we know the direction does not impact on circuit analysis but it does require us to reverse our thinking in terms of the flow of charge. Although we know electrons flow from negative to positive, we must now think of **current as flowing from positive to negative**.
2. *Voltage polarity* – We define the direction of potential difference as being from low to high, to show the difference in polarity between the voltage source and resistor (second image in Figure 3.4). Consider that the arrowhead indicates the direction of movement of charge, where a **resistor opposes the flow of charge** from the voltage source. This can be confusing when you try to combine your thinking with conventional current flow to imagine how electrons would move, so take some time to consider this point as being both conceptual (polarity) and practical (conventional current).
3. *Kirchoff's Voltage Law* – as mentioned above, the key to understanding KVL is to see voltage polarity as being the direction of potential for charge to flow. If you are comfortable with step 2 then realize that the potential difference of the resistor is being subtracted from that of the voltage source – they will **cancel each other out** to give a **total of zero** volts.

From KVL, we can now see that because all the voltage drops in a loop must equal 0, the voltage source (V_S) is effectively being opposed by the resistor voltage (V_R). This means that $V_S - V_R = 0$ and by rearranging the equation $\boldsymbol{V_S = V_R}$. To think of this another way, the voltage source provides the potential difference needed to move charge around a loop and other circuit components (i.e. the resistor) reduce it – they cancel each other out.

Current in a series circuit

We have learned that different voltage polarities determine how components resist the flow of charge from a source by cancelling each other out. In addition, if a voltage source is the only component in a loop providing free electrons then the **current in that loop will not change** – it will remain constant at all points. This can initially be a little confusing, as we have learned that current (amps) is the rate of electrical charge (coulombs) per second. If this is the case, then it can seem logical that the electrons 'slow down' after entering a resistor and so the current would thus vary on either side of the resistor.

In fact, **electrons do not change speed** – we are simply counting their number at a particular point in the loop for a specific length of time. If **fewer electrons can flow out of a resistor, then fewer electrons will flow into a resistor** – the resistor has reduced the number of electrons moving throughout the circuit. Thus, the current in any series loop is **limited by the total resistance** within it – all electrons will move the same overall distance, regardless of whether they are currently inside (or outside) any particular resistor:

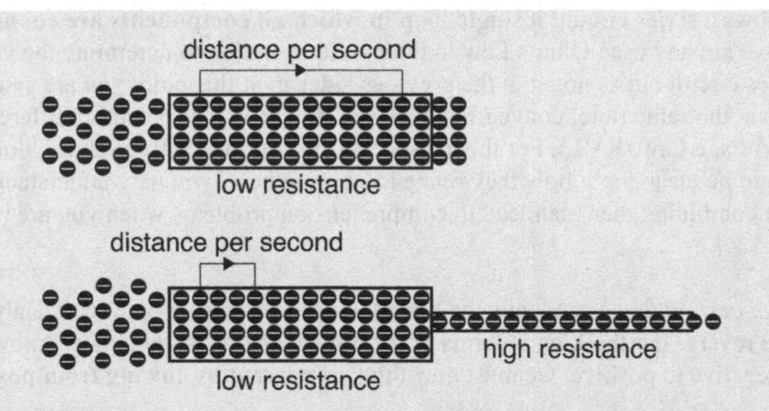

The image above shows how combining a low and high resistance effectively reduces the overall current flowing in a circuit, where current is measured as the amount of charge (in coulombs) flowing past a given point in one second. As there is less area in the higher resistor for electrons to move, the number of electrons that can pass any given point will be reduced. Thus, in the second example the distance an electron can travel within the lower resistance is also now reduced – there is nowhere else for it to go. A simple analogy would be to think of electron flow through multiple resistors like a traffic jam, where cars on a multiple-lane road move less overall distance because the nearest exit ramp only allows single-lane traffic to leave.

As free electrons can only enter through the voltage source, in **series circuits voltage varies and current is constant.**

Now we know that all voltage drops in a loop will cancel each other out to zero, and also that the current in that loop will be constant. By combining these two elements with Ohm's Law, we can develop a simple equation for determining the total resistance in a series circuit (Figure 3.5).

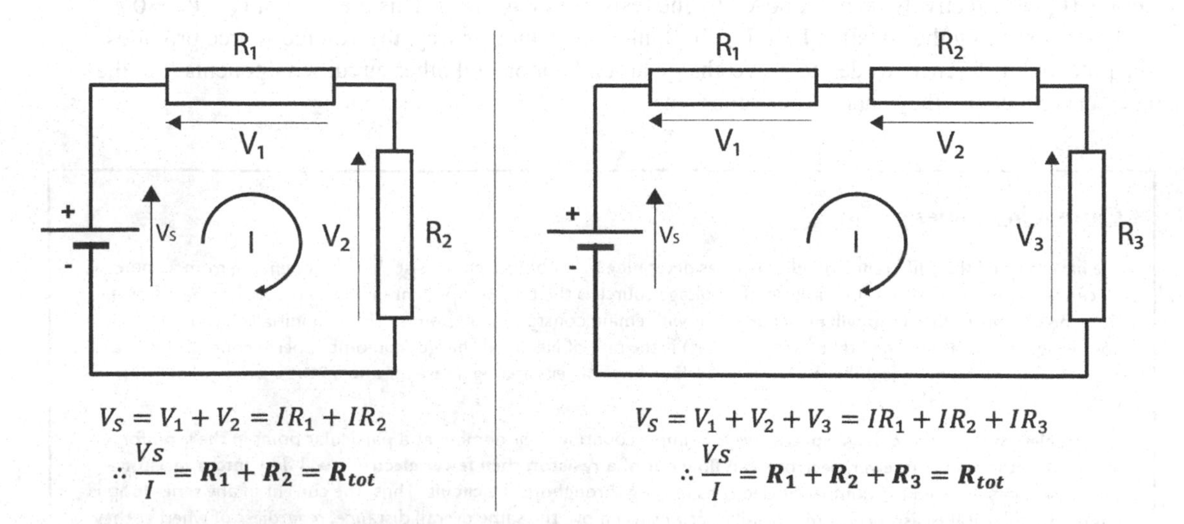

Figure 3.5 Calculating total resistance in a series circuit. ***In the diagrams, examples of series circuits combining two and three resistors are shown. We know that the supply voltage (VS) is equivalent to all component voltage drops, and that current in the loop is constant. By rearranging the terms, we can derive the total resistance (Rtot) for each example.***

In Figure 3.5, KVL is combined with Ohm's Law to determine the total resistance for each series circuit. Taking the left-hand example of a two-resistor series circuit, we can now state the following:

1. That the supply voltage (V_S) has the opposite polarity to the voltage drops across the resistors ($V_S = V_1 + V_2$).
2. That current (I) is constant throughout the circuit, so ($V_S = IR_1 + IR_2$).
3. Ohm's Law can be used to get the total resistance ($R_{tot} = \frac{Vs}{I}$).

This allows us to derive the general equation for **resistance in a series circuit:**

$$
\begin{aligned}
V_S = V_1 + V_2 &= IR_1 + IR_2 \\
\frac{Vs}{I} &= R_1 + R_2 \\
R_{tot} &= \frac{Vs}{I} \\
\therefore R_{tot} &= R_1 + R_2 \\
\textbf{\textit{Total Series Resistance}}, R_{tot} &= R_1 + R_2 + \ldots R_n
\end{aligned}
\tag{3.3}
$$

R_{tot} is the total resistance in ohms (Ω)

n is the number of resistors in series

The derivation of this equation has been provided to make sure that the concepts of voltage polarity and constant current do not confuse your understanding of series resistance in a circuit. In practice, the equation is a simple addition – **for any number of resistors in series, the total resistance is their sum**. This makes series circuit analysis quite straightforward, as the following examples will show:

3.2.1 Worked examples – calculating series resistance

Taking the circuit examples from Figure 3.5 to replace the resistors with actual values.

Q2: *A 5V voltage source is connected in series with two resistors* $R_1 = 10\Omega$ *and* $R_2 = 20\Omega$ *:*

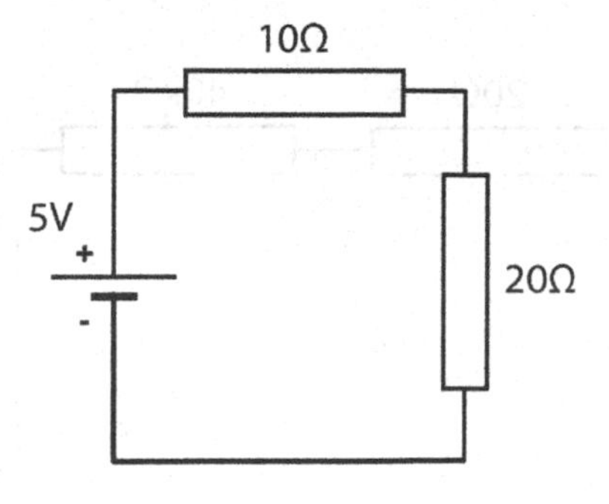

(a) *What is the total resistance of the circuit?*

$$V = 5V, \quad R_1 = 10\Omega, \quad R_2 = 20\Omega$$

$$Total\ Series\ Resistance,\ R_{tot} = R_1 + R_2 = 10 + 20 = \mathbf{30\Omega}$$

Answer: The total resistance in the circuit is **30Ω**.

(b) *What is the total current flowing through the circuit?*

$$V = 5V, \quad R = 30\Omega$$

$$\text{Current, } I = \frac{V}{R} = \frac{5}{30} = \mathbf{0.17A}$$

Answer: The total current in the circuit is **0.17A**.

Notes:

The mathematical equations used in this example are straightforward, and we have not used any quantities that require scaling factors to be applied. The only note is that $\frac{5}{30} = 0.166666666666666\dot{6}A$ is rounded to 2 decimal places – the dot above the last 6 means recurring in mathematics, because there is no **exact** decimal value for $\frac{5}{30}$.

Q3: *A 5V voltage source is connected in series with three resistors* $R_1 = 20\Omega$, $R_2 = 40k\Omega$ *and* $R_3 = 1M\Omega$*:*

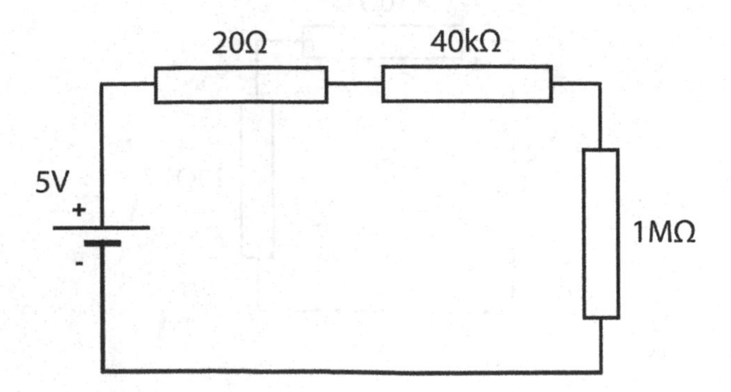

(a) *What is the total resistance of the circuit?*

$$V = 5V, \quad R_1 = 20\Omega, \quad R_2 = 40k\Omega, \quad R_3 = 1M\Omega$$

$$Total\ Series\ Resistance,\ R_{tot} = R_1 + R_2 + R_3 = 20 + 40{,}000 + 1{,}000{,}000 = \mathbf{1{,}040{,}020\Omega}$$

Answer: The total resistance in the circuit is **1.04MΩ**.

(b) *What is the total current flowing through the circuit?*

$$V = 5V, \quad R = 1.04M\Omega$$

$$\text{Current},\ I = \frac{V}{R} = \frac{5}{1{,}040{,}000} = 0.0000048 = \mathbf{4.8\mu A}$$

Answer: The total current in the circuit is $4.8\mu A$.

Notes:

The inclusion of a proportionally large resistor (1MΩ) makes the other resistances negligible within a series circuit – although we have rounded to 2 decimal places, 1.04MΩ is effectively 1MΩ for practical purposes.

Because the overall resistance is now much higher than the first example, the current flowing in the circuit has significantly decreased. Whenever we see a number of zeros in a result, we should always be looking to find the nearest scaling factor that will represent the quantity for ease of writing. To scale a result, follow these steps:

1. Count the zeros and determine which factor of 3 is nearest (i.e. 3/6/9/12) to the answer digits (first non-zero)
2. If the number of zeros is not a factor of 3, either move the answer digits up (add a zero) or down (add a decimal point) – in our example, the answer digits are 48 but the scaling factor is 5 (zeros) so we move the answer down by one decimal point to 4.8 to get a 10^{-6} result
3. Write the answer in the correctly scaled units – 0.0000048 becomes 4.8μ
4. To check your answer, multiply it by the divisor (in this case total resistance) to ensure it gives you the same numerator (i.e. voltage) you started with

By combining KVL with Ohm's Law we have now developed a practical means of analysing a series circuit, allowing us to restate all the things we now know at this point.

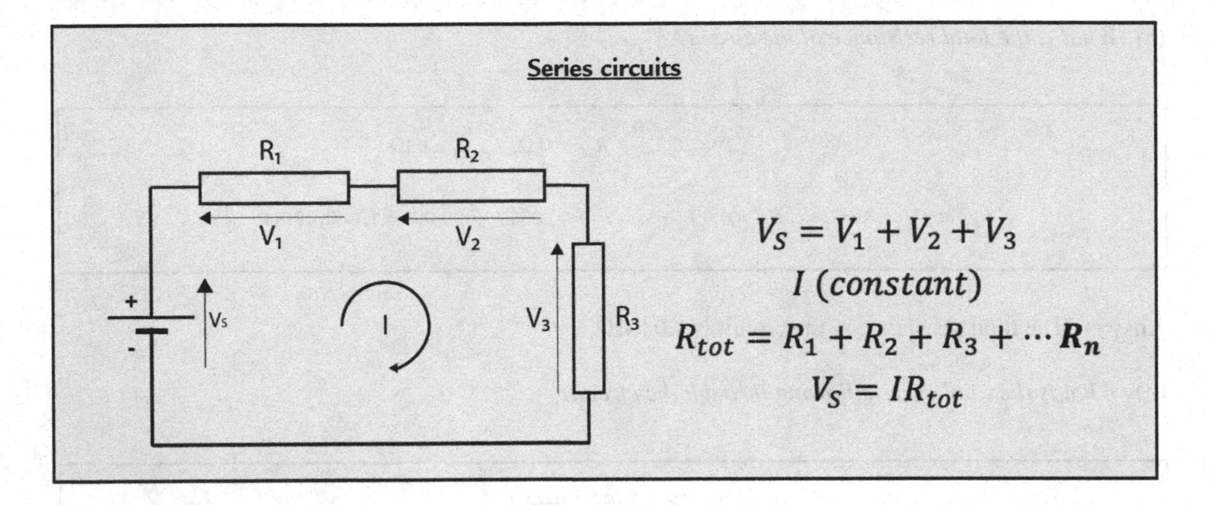

Let's now use these equations to calculate the specific voltage drops across resistors in a circuit – a very common circuit analysis task.

3.2.2 Worked example – calculating series resistor voltages

In this example, similar resistance values are used to avoid scaling factors in the calculations of results.

Q4: *A 5V voltage source is connected in series with three resistors* $R_1 = 10\Omega$, $R_2 = 20\Omega$ *and* $R_3 = 30\Omega$ *:*

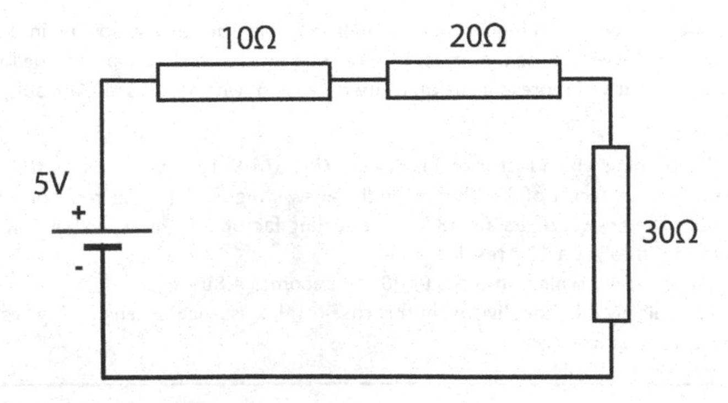

(a) *What is the total resistance of the circuit?*

$$V = 5V, \quad R_1 = 10\Omega, \quad R_2 = 20\Omega, \quad R_3 = 30\Omega$$

$$\textit{Total Series Resistance}, \ R_{tot} = R_1 + R_2 = 10 + 20 + 30 = \mathbf{60\Omega}$$

Answer: The total resistance in the circuit is **60Ω**.

(b) *What is the total current flowing through the circuit?*

$$V = 5V, \qquad R = 60\Omega$$

$$\text{Current, } I = \frac{V}{R} = \frac{5}{60} = \mathbf{0.083333333333333 = 83.33mA}$$

Answer: The total current in the circuit is **83.33mA**. (*the choice of mA is explained in the notes*).

(c) *What are the voltage drops across the three resistors in the circuit?*

$$I = 83.33A, \qquad R_1 = 10\Omega, \qquad R_2 = 20\Omega, \qquad R_3 = 30\Omega$$

$$V_{R1} = IR_1 = 0.08333 \times 10 = 0.8333 = \mathbf{0.83V}$$

$$V_{R2} = IR_2 = 0.08333 \times 20 = 1.6666 = \mathbf{1.67V}$$

$$V_{R3} = IR_3 = 0.08333 \times 30 = 2.4999 = \mathbf{2.5V}$$

Answer: The voltage drops across the three resistors are $V_{R1} = \mathbf{0.83}V, V_{R2} = \mathbf{1.67V}, V_{R3} = \mathbf{2.5V}$.

Notes:

In example 1.1.2, $\frac{5}{30} = 0.166666666666666\dot{6}A$ was rounded to 2 decimal places giving a truncation of the final value. In this case, rounding $\frac{5}{60} = 0.08333333333333\dot{3}A$ to 2 decimal places would give a loop current of **0.08A**.

The problem is that this introduces an **inaccuracy** when we calculate the voltages across the loop, because we are now scaling the value up by factors of 10, 20 and 30. If we had simply rounded the current value to 0.08A, we would arrive at voltage drops of $0.8 + 1.6 + 2.4 = \mathbf{4.8V}$, which is less than the source voltage of 5V!

There is no single solution here, as rounding a fraction to state an answer will always truncate the actual value. The previous example noted that when encountering zeros in a result a scaling factor should be considered, but with only a single zero in the value this may not initially spring to mind when working through the calculations.

In actual fact, the current is stated as 83.33mA to retain some of the resolution in the answer by using a scaling factor of (10^{-3}) instead, which highlights another reason why scaling factors are useful in electronics – we are dealing with ratios that often have **no exact decimal equivalent**. Although there is no standard rule, a good sanity check in the example above would be to add up the loop voltages and notice the error of 0.2V relative to the source of 5V – this would suggest that the truncated current value may not be accurate enough for the calculation.

The impact of truncating values is shown in the example above, but it is also important to highlight the ratio of the resistances in the circuit. Resistor values in simple proportions were deliberately chosen to show how the voltage drops scaled accordingly – the voltage drop across the 30Ω resistor is three times that of the 10Ω, but also the same as the combined voltage drop across the other two resistances (10+20 = 30Ω, 0.83+1.67 = 2.5V). This use of resistor ratios is core to circuit analysis and is best demonstrated by a very common electronics concept – the voltage divider.

3.3 Voltage dividers

Voltage dividers are a fundamental circuit building block, and we will encounter them in different forms at various points throughout this book. A voltage divider uses the ratio between two series resistors to divide a source voltage (Figure 3.6).

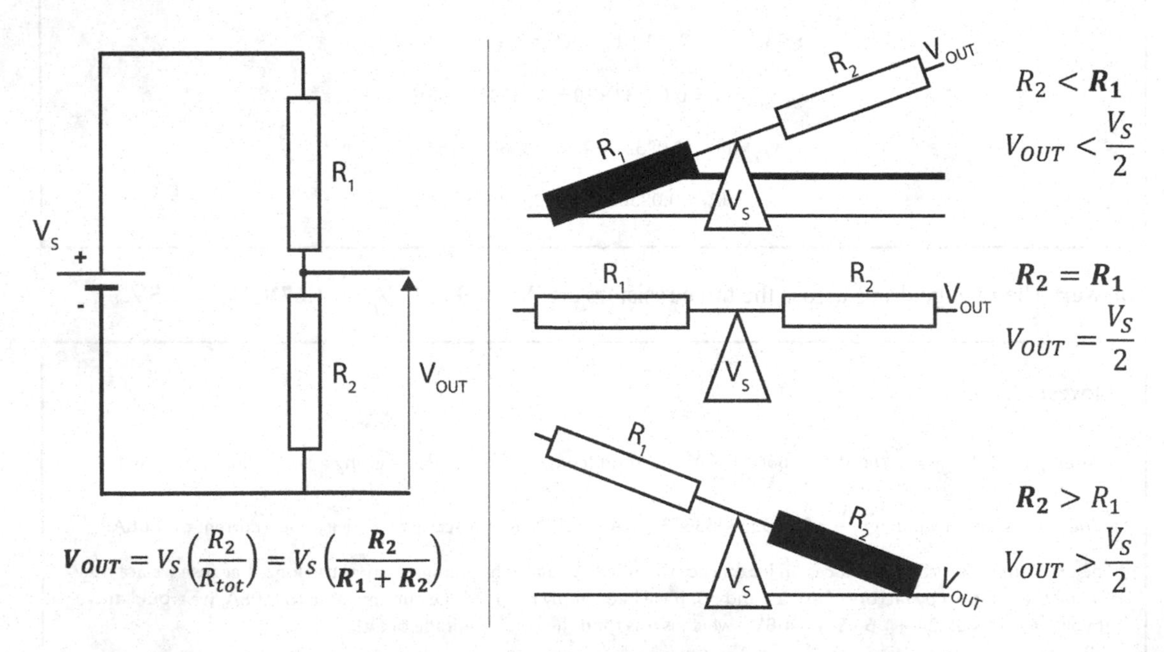

Figure 3.6 A voltage divider. ***The diagram above shows one of the most common building blocks in electronic circuits. Connecting the two resistors in series allows the source voltage (V_S) to be divided in proportion to the ratio between them. The second diagram shows a simple analogy with a seesaw, where the ratio of the resistances is represented by weights balanced on a pivot.***

In Figure 3.6, the standard equation for resistor ratios is shown underneath a voltage divider circuit (you may see the divider drawn in different configurations, but it will always be a voltage source with two series resistors where the output is measured across one of them). The right-hand diagrams show a simple analogy with a seesaw, where the resistances are equivalent to weights on either side of a pivot. This analogy is used to illustrate how a voltage divider is based on the principle of the ratio between two quantities – the fact that we use them in electronics is simply their application. To illustrate this, consider the three seesaw examples:

1. R_2 is less than R_1 ($R_2 < R_1$) – the heavier weight is on the left of the pivot (R_1) so the output proportion is less than half of the total ($V_{OUT} < \frac{V_S}{2}$).
2. R_2 is equal to R_1 ($R_2 = R_1$) – the seesaw is balanced so the output proportion is half of the total ($V_{OUT} = \frac{V_S}{2}$).
3. R_2 is greater than R_1 ($R_2 > R_1$) – the heavier weight is on the right of the pivot (R_2) so the output proportion is greater than half of the total ($V_{OUT} > \frac{V_S}{2}$).

The mathematical explanation of the voltage divider equation is that the divider exploits the ratio between two quantities, where we are interested in the proportion of R_2 relative to the total resistance ($R_1 + R_2$) – hence we measure $\frac{R_2}{R_1 + R_2}$. By multiplying this proportion by the source voltage (V_S) we get the proportion of the voltage (V_{OUT}) dropped across the output R_2. We can also derive the proof of this equation using a combination of Ohm's Law and KVL as follows:

$$
\begin{aligned}
\textit{Total Series Resistance}, R_{tot} &= R_1 + R_2 \\
\textit{Source Voltage}, V_S &= IR_{tot} = I(R_1 + R_2) \\
\therefore I &= \frac{V_S}{R_{tot}} = \frac{V_S}{R_1 + R_2} = V_S\left(\frac{1}{R_1 + R_2}\right) \\
\textbf{\textit{Output Voltage}}, V_{OUT} &= IR_2 = \left(V_S\left(\frac{1}{R_1 + R_2}\right)\right)R_2 = V_S\left(\frac{R_2}{R_1 + R_2}\right)
\end{aligned}
\tag{3.4}
$$

Note: rearranging the loop current to $V_S\left(\frac{1}{R_1 + R_2}\right)$ *allows the final equation to be stated in terms of the proportion* $\left(\frac{R_2}{R_1 + R_2}\right)$

If you find the use of ratio (one quantity relative to another) and proportion (one quantity relative to the total of all quantities involved) confusing it may be because these are mathematical relationships that were often taught using fractions, unlike the decimal system where percentages are much more common. Although fractions are no longer as widely used as they were when these equations were first derived, the rounding error created when moving to decimal in the previous example highlights that fractions, ratios and proportions are exact mathematical relationships and so are useful to know. Having said this, the aim of this book is to give you a practical understanding of audio electronics, so if you are feeling a little less confident after working through the proof in equation 3.4 then you can simply remember the following:

Voltage dividers

$$V_{OUT} = V_S\left(\frac{R_2}{R_1+R_2}\right)$$

as R_2 gets bigger – so does V_{OUT}

The following examples aim to demonstrate that in practice voltage dividers are relatively easy to work with:

3.3.1 Worked examples – voltage dividers

Simple ratios will be used to apply the proof of the voltage divider equation.

Q5: *A 5V voltage source is connected to a voltage divider with resistors* $R_1 = 10\Omega$, $R_2 = 20\Omega$:

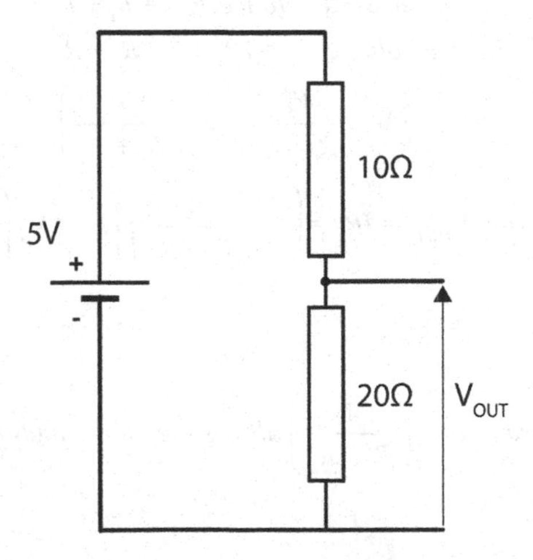

(a) *What is the output voltage (V_{OUT}) across R_2?*

$$V_S = 5V, \quad R_1 = 10\Omega, \quad R_2 = 20\Omega$$

$$V_{OUT} = V_S\left(\frac{R_2}{R_1+R_2}\right) = 5\left(\frac{20}{10+20}\right) = 5\times\frac{2}{3} = \frac{10}{3}$$

$$V_{OUT} = \mathbf{3.3\dot{3}V}$$

Answer: The voltage divider output voltage is **3.33V**.

Q6: *A 5V voltage source is connected to a voltage divider with resistors* $R_1 = 50\Omega,\ R_2 = 50\Omega$:

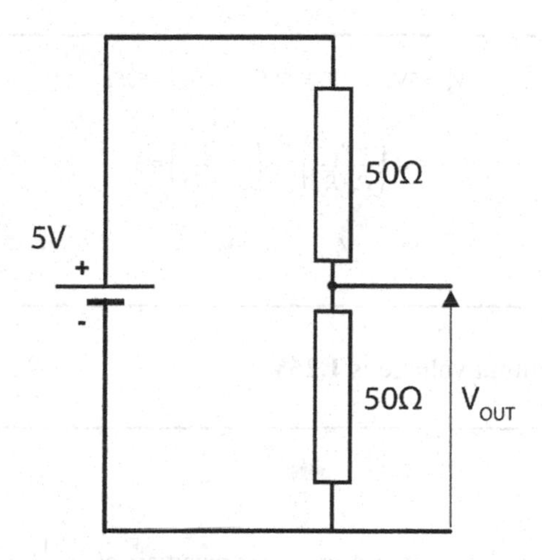

(a) *What is the output voltage* (V_{OUT}) *across* R_2?

$$V_S = 5V, \quad R_1 = 50\Omega, \quad R_2 = 50\Omega$$

$$V_{OUT} = V_S\left(\frac{R_2}{R_1 + R_2}\right) = 5\left(\frac{50}{50+50}\right) = 5 \times \frac{1}{2}$$

$$\mathbf{V_{OUT} = 2.5V}$$

Answer: The voltage divider output voltage is **2.5V**.

Q7: *A 5V voltage source is connected to a voltage divider with resistors* $R_1 = 20\Omega,\ R_2 = 60\Omega$:

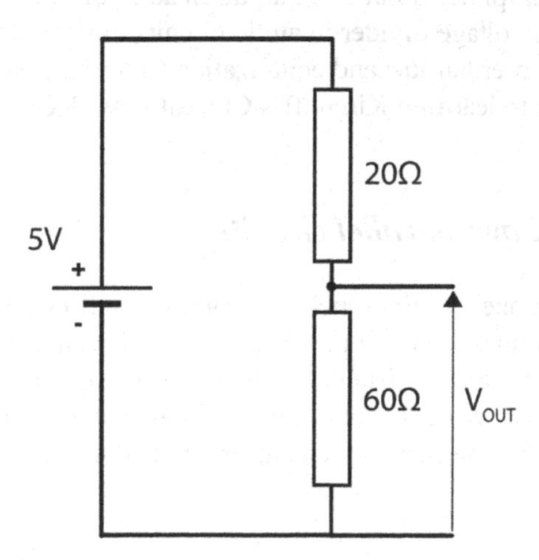

(a) *What is the output voltage (V_{OUT}) across R_2?*

$$V_S = 5V, \quad R_1 = 20\Omega, \quad R_2 = 60\Omega$$

$$V_{OUT} = V_S\left(\frac{R_2}{R_1+R_2}\right) = 5\left(\frac{20}{20+60}\right) = 5\times\frac{1}{4}$$

$$\mathbf{V_{OUT} = 1.25V}$$

Answer: The voltage divider output voltage is **1.25V**.

Notes:

In these examples, resistor values were chosen to create simple proportions of $\frac{2}{3}, \frac{1}{2}$ and $\frac{1}{4}$. The reason for doing this is to highlight how a voltage divider uses the ratio of the two series resistors to output a proportion of the source voltage (V_S).

The important thing to remember when calculating the **ratio** between R_1 and R_2 is that it is **not the proportion** between R_2 and R_{tot}, so a voltage divider with resistors of 20Ω and 60Ω will have:

a **ratio** of $\frac{20}{60} = \frac{2}{3}$

a **proportion** of $\frac{20}{20+60} = \frac{1}{4}$

Voltage dividers are used widely throughout electronics in both analogue and digital systems. We will learn how to use voltage dividers in DC circuits in this chapter and later in AC circuits as filters (where we replace a resistor with a frequency-dependent capacitor). We will also learn how to use a voltage divider as feedback to control operational amplifier gain, and to help explain the practicalities of source and load in terms of amplifier input and output. In addition, a potentiometer (variable resistor) is commonly used as a voltage divider in audio circuits to control parameters like gain (amplifier input), volume (amplifier output) and equalization (filter balance). Now we have covered series circuits, we can move on to learning Kirchoff's Current Law (KCL) and how this applies to parallel circuit analysis.

3.4 Kirchoff's Current Law: parallel circuits

A parallel circuit has more than one circuit branch that current can flow through. When learning parallel circuits, one of the first questions you may ask is why this is necessary – particularly when working with more complex reciprocal resistance calculations as we will see below. The simplest answer is that providing multiple circuit branches give a circuit some level of redundancy, so if one branch becomes open circuit for some reason then current will continue to flow down the others.

This makes most sense when you think of household appliances – if your toaster and television were connected in series then damaging the toaster would also impact on your media consumption! To begin learning about parallel circuits, Kirchoff's second law relates to the conservation of charge in an electrical circuit:

Kirchoff's Current Law

The charge entering a junction is equal to the charge leaving that junction

Kirchoff's Current Law (KCL) states that the total current (I) entering any junction (or node) in a circuit will be the same as the current leaving that junction – charge is conserved. The first step towards understanding KCL is to remember that there will always be the samc level of charge present in a DC circuit – free electrons cannot enter or leave the circuit other than through a voltage source (Figure 3.7).

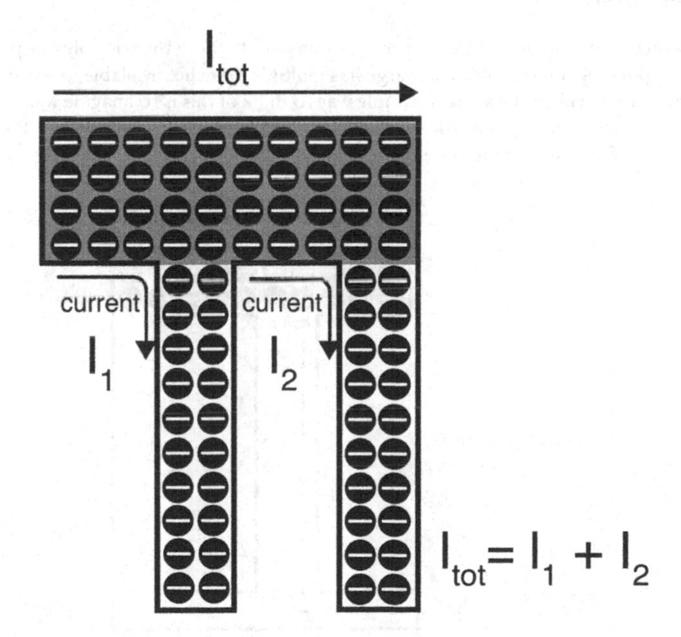

Figure 3.7 Conservation of charge at circuit junctions. ***In the diagram, a total charge of 40 electrons (in grey) can flow down one of two branches in the circuit, and thus 40 electrons (in total) must leave through both of these junctions. No other sources of charge are present, so the total current must be the sum of the two currents flowing through the branches.***

When we looked at KVL for series circuits, it was noted that the current flowing in a single loop does not change – a resistor will reduce the flow of electrons throughout the entire loop, not just through it. This idea makes sense in a series circuit, but it becomes more complex when we add other **parallel branches** to our circuit (Figure 3.8).

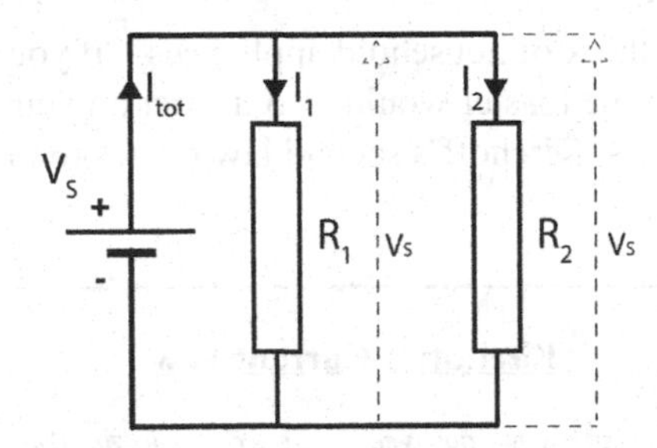

Figure 3.8 A simple parallel circuit. ***In the schematic diagram, current (I_{tot}) can flow through two possible branches (either I_1 or I_2) into either of two resistors (to avoid any branch being a short circuit). The amount of charge flowing through each branch is governed by the resistor (either R_1 or R_2) – thus the current in each branch is different. The potential for charge to flow through each branch is equivalent – the voltage in the entire circuit remains constant.***

Voltage in a parallel circuit

In series circuits, we learned that voltage varies and current is constant because there is only one path for it to take within the circuit. In **parallel** circuits, because charge has multiple branches available, the **voltage (potential difference) across each branch will be the same**. A simple way to think of this is to imagine what the potential difference would be across each branch just as the voltage source is connected (i.e. when time= 0 seconds) - **before** any current has been able to flow through the circuit:

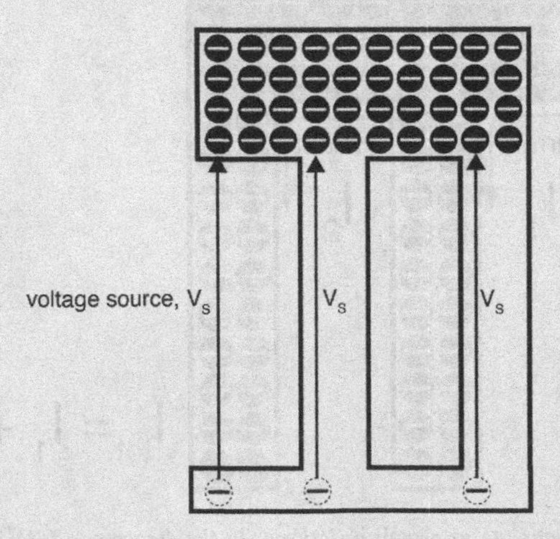

The image above shows how the charge from the voltage source has equal **potential** to move down either branch – it can go in more than one direction. Although the current in the branches may not be the same, the potential for current to flow between the top and bottom of each branch is **equivalent to the potential created by the voltage source.**

This means that parallel circuits have the opposite relationship between voltage and current. As free electrons only enter through the voltage source, in **parallel circuits current varies and voltage is constant.**

In Figure 3.8, charge from the voltage source (V_S) provides the total current (I_{tot}) flowing through the circuit:

1. Using Ohm's Law with the total circuit resistance (R_{tot}) we can define $V_S = I_{tot} R_{tot}$.
2. As current can go down multiple branches, we define two currents (I_1 and I_2) – one for each branch.
3. Using KCL, we know that the total current in the circuit must be equal to the sum of the two branch currents, so $I_{tot} = I_1 + I_2$.
4. We also know that voltage is constant throughout the circuit: this allows us to define each of the branch currents using Ohm's Law ($I_1 = \frac{V_S}{R_1}, I_2 = \frac{V_S}{R_2}$).

From here, we can derive the equation for total resistance in a parallel circuit using a combination of Ohm's Law, KVL and KCL as follows:

$$\mathrm{I_{tot}} = \mathrm{I_1} + \mathrm{I_2}$$

$$I_1 = \frac{V_S}{R_1} \text{ and } I_2 = \frac{V_S}{R_2}$$

$$\text{substitute these terms } I_{tot} = I_1 + I_2 = \frac{V_S}{R_1} + \frac{V_S}{R_2}$$

$$V_S = I_{tot} R_{tot}$$

$$\text{rearrange these terms } I_{tot} = \frac{V_S}{R_{tot}}$$

$$\therefore I_{tot} = \frac{V_S}{R_{tot}} = \frac{V_S}{R_1} + \frac{V_S}{R_2}$$

$$\text{divide by } V_S \text{ throughout } \frac{1}{\boldsymbol{R_{tot}}} = \frac{1}{\boldsymbol{R_1}} + \frac{1}{\boldsymbol{R_2}}$$

Note: for each additional branch resistor we simply add another term, so:

$$\textbf{\textit{Total Parallel Resistance,}} \frac{1}{\boldsymbol{R_{tot}}} = \frac{1}{\boldsymbol{R_1}} + \frac{1}{\boldsymbol{R_2}} \cdots + \frac{1}{\boldsymbol{R_n}} \tag{3.5}$$

$\boldsymbol{R_{tot}}$ *is the total resistance in ohms (Ω)*
n *is the number of resistors in series*

There is an aspect of this equation for parallel resistance that consistently causes problems for new electronics students. The equation actually defines the **reciprocal** of the total parallel resistance, where **reciprocal means the result of dividing 1 by the value** involved (in this case, $\frac{1}{R_{tot}}$). We must remember that ($\frac{1}{R_{tot}}$) is not the actual resistance value – the mathematical statement for this would be

$$R_{tot} = \frac{1}{\frac{1}{R_{tot}}} = \frac{1}{\frac{1}{R_1} + \frac{1}{R_2} \cdots + \frac{1}{R_n}}$$ (this is not an easy equation to decipher!).

In practice, finding the reciprocal of a fraction is quite straightforward, where you have to invert the terms of the fraction to divide the denominator by the numerator:

fraction	$\frac{1}{R_{tot}} = \frac{2}{3}$	$\frac{1}{R_{tot}} = \frac{3}{4}$	$\frac{1}{R_{tot}} = \frac{1}{2}$
reciprocal	$R_{tot} = \frac{3}{2}$	$R_{tot} = \frac{4}{3}$	$R_{tot} = \frac{2}{1}$

Thus, in parallel resistance circuits we must always remember this final step:

> When calculating parallel resistance always find the **reciprocal of the result** - you must **invert the terms.**

By combining KCL with Ohm's Law we have now developed a practical means of analysing a parallel circuit, allowing us to restate all the things we now know at this point.

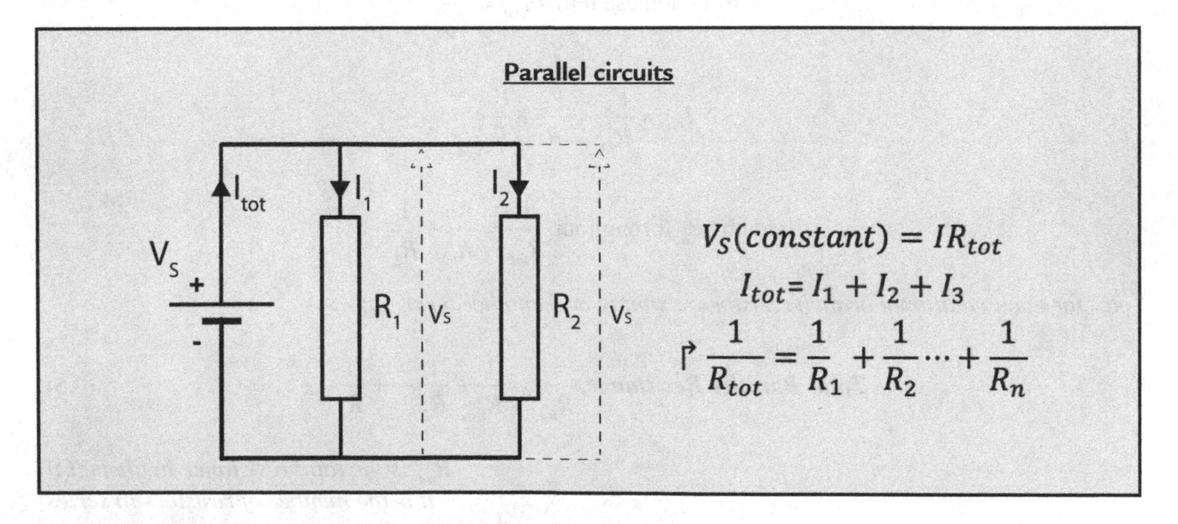

Let's now use these equations to calculate the total resistance in a parallel circuit, to highlight the importance of the reciprocal in determining the final result:

3.4.1 Worked examples – calculating parallel resistance

We will calculate the total parallel resistance in each circuit, to show how the reciprocal must be found as the last step.

Q8: *A voltage source is connected in parallel with two resistors* $R_1 = 10\Omega$ *and* $R_2 = 20\Omega$:

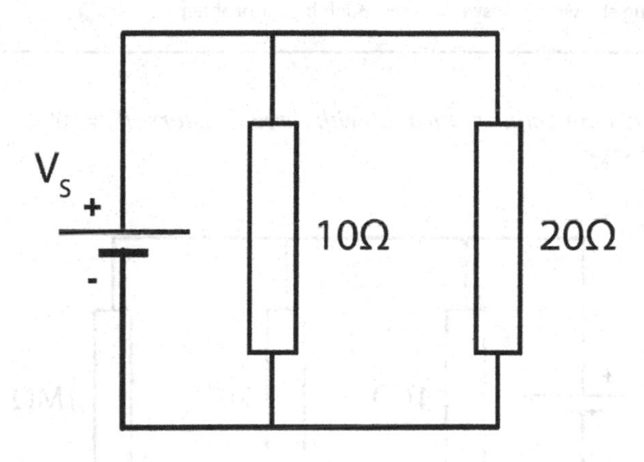

(a) *What is the total resistance of the circuit?*

$$R_1 = 10\Omega, \quad R_2 = 20\Omega$$

$$\textit{Total Parallel Resistance}, \frac{1}{R_{tot}} = \frac{1}{R_1} + \frac{1}{R_2}$$

$$= \frac{1}{10} + \frac{1}{20} = \frac{2}{20} + \frac{1}{20}$$

$$\frac{1}{R_{tot}} = \frac{3}{20}$$

$$\mathbf{R_{tot}} = \frac{20}{3} = \mathbf{6.67\Omega}$$

Answer: The total resistance in the circuit is **6.67Ω**.

Notes:

Simple resistor values were chosen to show how the fractions must be added. The first thing to do is state the values R_1 and R_2, and find a common denominator (in this case 20). Then, both terms must be scaled to this same denominator to allow them to be added. The final step is to find the reciprocal of this addition, which involves inverting the terms of the fraction (numerator and denominator) and then calculating the final result. The key thing in parallel resistance calculations is to carry out each step to avoid making simple arithmetic mistakes - it may seem like this takes longer, but getting the wrong answer causes much bigger problems!

Q9: *A voltage source is connected in parallel with three resistors* $R_1 = 10\Omega$, $R_2 = 20\Omega$ *and* $R_3 = 1M\Omega$:

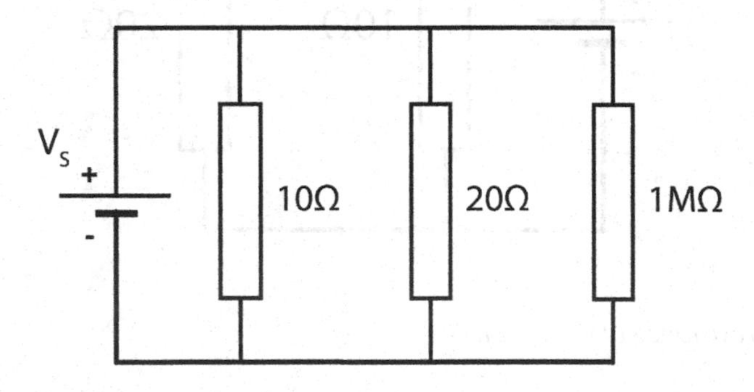

(a) *What is the total resistance of the circuit?*

$$R_1 = 10\Omega, \quad R_2 = 20\Omega, \quad R_3 = 1M\Omega$$

$$\text{Total Parallel Resistance}, \frac{1}{R_{tot}} = \frac{1}{R_1} + \frac{1}{R_2} + \frac{1}{R_3}$$

$$= \frac{1}{10} + \frac{1}{20} + \frac{1}{1{,}000{,}000} = \frac{100{,}000}{1{,}000{,}000} + \frac{50{,}000}{1{,}000{,}000} + \frac{1}{1{,}000{,}000}$$

$$\frac{1}{R_{tot}} = \frac{150{,}001}{1{,}000{,}000}$$

$$R_{tot} = \frac{1{,}000{,}000}{150{,}001} = \mathbf{6.67\Omega}$$

Answer: The total resistance in the circuit is **6.67Ω**.

Notes:

In this example, a third resistor $R_3 = 1\text{M}\Omega$ is added as another parallel branch in the circuit. A much larger value of $1\text{M}\Omega = 1{,}000{,}000\Omega$ (compared to $R_1 = 10\Omega$ and $R_2 = 20\Omega$) was chosen - the reason for doing this is to show how a large resistor has a negligible impact on a parallel circuit. The final result is effectively the same, and we would have to go to 5 decimal places to find the difference between this answer and the result for example 1. The resistor calculations also require some focus, as the common factor of 1,000,000 makes the arithmetic more complex. In all instances, taking a little more time to check a parallel calculation is recommended to avoid errors.

As the above examples show, calculating total parallel resistance takes a little more time because of the use of fractions. In addition, you may forget to calculate the reciprocal as the final step when you begin working with DC circuits – the only solution to this is practice. Knowing the total parallel resistance for a circuit can prove very useful when working with multistage circuits – we looked at the idea of source and load resistance in chapter 2, and this allows us to partition a system into discrete elements (e.g. filters, amplification). In these cases, knowing the overall resistance of each part of the circuit allows it to be replaced with an equivalent resistor, which is known as Thevenin and Norton equivalence (calculating equivalent circuits is outwith the scope of this book). For now, the following examples calculate branch currents for a parallel circuit, using the total resistance as a check for the results. This is a useful approach in electronics, where you can quickly verify a set of values by using Ohm's Law.

3.4.2 *Worked examples – calculating parallel current*

In these examples, KCL is used to calculate all the currents flowing through the circuit – the value for total resistance can then be used to check the results.

Q10: *A 5V voltage source is connected in parallel with three resistors* $R_1 = 10\Omega$, $R_2 = 20\Omega$ *and* $R_3 = 40\Omega$:

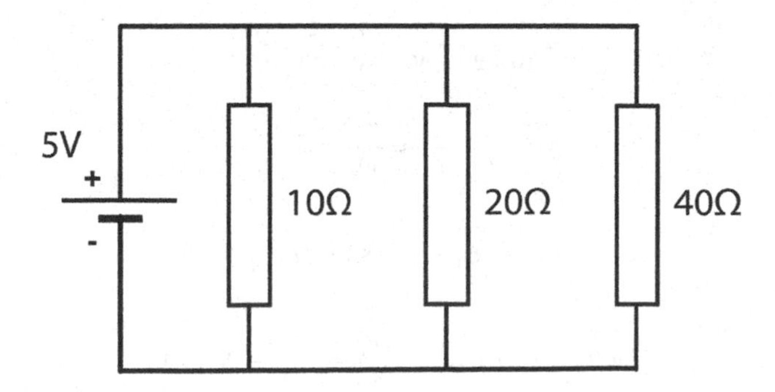

(a) *What are the branch currents (I_1, I_2, I_3) and total current (I_{tot}) of the circuit?*

$$R_1 = 10\Omega, \quad R_2 = 20\Omega, \quad R_3 = 40\Omega$$

$$I_1 = \frac{V_S}{R_1} = \frac{5}{10} = \mathbf{0.5A} \quad I_2 = \frac{V_S}{R_2} = \frac{5}{20} = \mathbf{0.25A} \quad I_3 = \frac{V_S}{R_3} = \frac{5}{40} = \mathbf{0.125A}$$

$$I_{tot} = 0.5 + 0.25 + 0.125 = \mathbf{0.875A}$$

Answer: The branch currents are $I_1 = \mathbf{0.5A}$, $I_2 = \mathbf{0.25A}$, $I_3 = \mathbf{0.125A}$. The total circuit current is $I_{tot} = \mathbf{0.875A}$.

Notes:

In this example, resistor $R_3 = 40\Omega$ in the third branch of the circuit is changed to make the fraction calculations easier for total resistance. For each branch in the circuit, we know that the voltage (V_S) will be constant at 5V so knowing the resistance in each branch means Ohm's Law can be used to find the associated current (I_1, I_2, I_3). We can then simply add these three values to get the total current (I_{tot}) in the circuit.

(b) *What is the total resistance (R_{tot}) of the circuit? Use Ohm's Law to check against the previous current (I_{tot}) value.*

$$R_1 = 10\Omega, \quad R_2 = 20\Omega, \quad R_3 = 40\Omega$$

$$\textit{Total Parallel Resistance}, \frac{1}{R_{tot}} = \frac{1}{R_1} + \frac{1}{R_2} + \frac{1}{R_3}$$

$$= \frac{1}{10} + \frac{1}{20} + \frac{1}{40} = \frac{4}{40} + \frac{2}{40} + \frac{1}{40}$$

$$\frac{1}{R_{tot}} = \frac{7}{40}$$

$$\therefore \mathbf{R_{tot}} = \frac{40}{7} = \mathbf{5.714\Omega}$$

$$\textit{Total Circuit Current}, \mathbf{I_{tot}} = \frac{V_S}{R_{tot}} = \frac{5}{5.714} = \mathbf{0.875A}$$

Answer: The total resistance in the circuit is $R_{tot} = \mathbf{5.714\Omega}$. Checking using Ohm's Law, the total current $I_{tot} = \mathbf{0.875A}$.

Notes:

For this question, we know the three branch resistors and also the parallel equation for total resistance. The first step is finding the common factor between the resistors (40) and then scale the terms to this common denominator ($\frac{1}{10} = \frac{4}{40}$, $\frac{1}{20} = \frac{2}{40}$). Once the terms are added, the reciprocal of the answer can be found by flipping ($\frac{7}{40}$) to get the final result ($\frac{40}{7} = 5.714\Omega$). This value can be used for total circuit resistance with Ohm's Law to check the current total from the previous question – if the branch currents are added correctly then we should get the same answer (0.875A). This is quite a common way to check a circuit, and the more you practise with Ohm's Law the easier it will become.

This chapter has now covered a few worked examples, and you may be feeling a little fatigued (particularly if it's been a while since you last calculated fractions!). To consolidate your knowledge of DC circuits, some practical examples of circuit analysis can now be considered – starting with the example circuit from chapter 1 where a resistor was connected in series with an LED to limit the current.

3.5 Tutorial: limiting current to protect components

In the chapter 1 project, we created a simple circuit that used power from the Arduino to light an LED (we learned in chapter 2 that this LED is an output **actuator**). We learned in chapter 1 that we needed a 150Ω resistor to prevent the flow of current from the Arduino from damaging the LED – the resistor resists the flow of charge to reduce the current level in the circuit to a safe level for semiconductor operation. Now we have built and tested this simple circuit, we can analyse how this resistance value of 150Ω was determined (Figure 3.9).

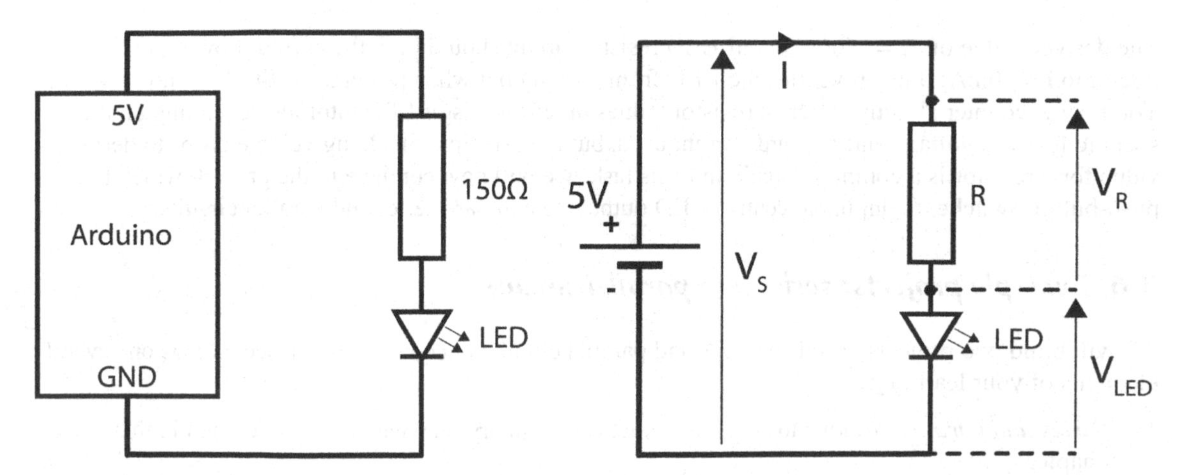

Figure 3.9 Simple LED circuit from chapter 1. ***The left-hand schematic diagram uses the 5V and ground (GND) pins of an Arduino to provide a voltage source for the circuit. A resistor is connected in series with an LED light to prevent the current from burning out the semiconductor component. The right-hand diagram shows a standard schematic representation, including both the circuit current (which is constant) and the voltage drops in the circuit with an unknown value for resistance (R).***

In this circuit, the resistor is used to limit the current that can flow through the LED to prevent it from overheating. This is a very common use of resistors, and we can now learn how to use Ohm's Law and KVL to calculate the correct resistance value for our circuit. To do this, the first thing we need to consider is the maximum **forward current** of the LED – the current it can safely handle. As we will see in chapter 7, we call this *forward* current because a diode is designed to only allow current to flow in one direction. All LEDs have a maximum forward current, and you can find the value for the one you intend to use by looking at its data sheet (which is usually available online). For our purposes, we will use a general value of $\mathbf{I = 20mA}$ which sits broadly within the range of forward current values for most LEDs.

The next thing we need to take into account is the maximum **forward voltage** of the LED, which is based on the colour of the LED being used. Once again, the data sheet for the LED you have purchased will provide specific values, but we will work from a general value of $\mathbf{V_{LED} = 2V}$ for a red LED. In practice, actual red LED values vary from around 1.7 to 2.5V, whilst blue and white LEDs (which must generate a lower wavelength) can reach a forward voltage of 4V. Like forward current, the forward voltage drop created across a diode is related to the depletion layer that was discussed briefly in chapter 2. Now we have values for LED current and voltage, we can use Ohm's Law and KVL to find the value of V_R (and hence R) in the circuit above:

$$V_S = 5V, \qquad V_{LED} = 2V, \qquad I = 20mA$$

$$from\,KVL: V_S = V_R + V_{LED}$$

$$5 = V_R + 2$$

$$\therefore V_R = 5 - 2 = 3V$$

$$R = \frac{V_R}{I} = \frac{3}{0.02} = 150\Omega$$

The derived value of $R = 150\Omega$ for our LED resistor circuit should limit the current flowing in the circuit to $I = 20mA$, thus preventing the LED from burning out when powered by the Arduino +5V pin. You may encounter slightly different resistor values in other resistor/LED tutorials depending on the specific forward voltage and forward current used, but the principle of adding voltage drops to derive a value for a resistor is a common circuit analysis task. We will now continue to the project, which uses push-button switches as inputs to control LED outputs to compare series and parallel circuits.

3.6 Example projects: series and parallel circuits

We will build two projects based on series and parallel circuit concepts. These projects focus on several elements of your learning:

1. *Series and parallel circuits*: to provide a practical summary on what you have learned in this chapter.
2. *Translating schematics to breadboard*: to work through the process of building circuits.
3. *Expanding breadboard connections*: to consider ways to extend our use of a single breadboard.
4. *Working with Tinkercad*: this was introduced in chapter 1, but more practice will accelerate your learning.

At this point, it is assumed that you have completed the projects for chapters 1 and 2. Whilst you may not have made anything particularly relevant to audio yet, these projects are designed to act as building blocks for later tasks so please try to devote some time to fully completing them. In each project, the schematic and the breadboard connections are provided for you to follow. It is recommended to begin by prototyping the circuit in Tinkercad, as building actual circuits can be tricky when you are learning (the actual components are small and must be correctly placed or the circuit will not work). You can complete these projects using only Tinkercad for simulation, but it is also assumed that you have (or will have) access to some basic electronic components and an Arduino board to begin building the projects in this book.

3.6.1 Series circuit project

For this project, we will build a circuit powered by the Arduino that combines the three schematics in Figure 3.10.

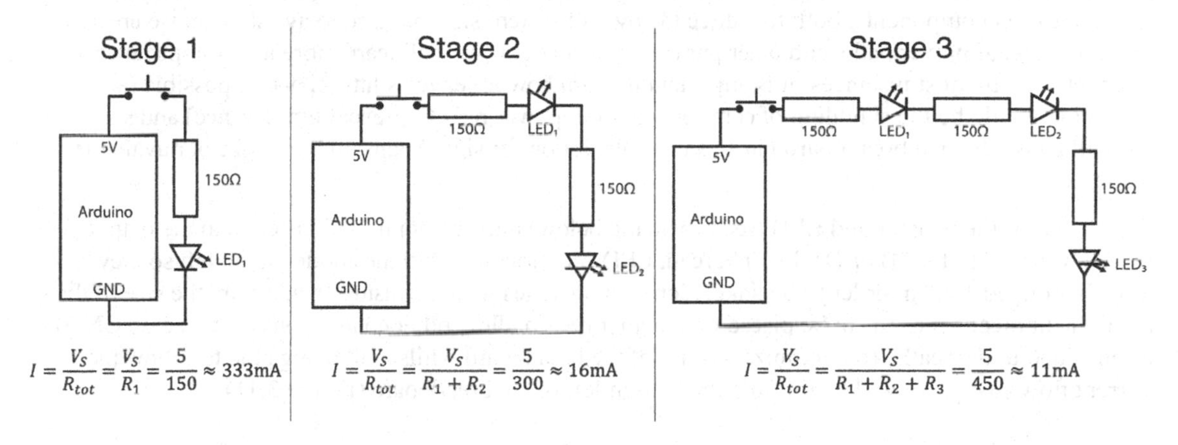

Figure 3.10 Series circuit schematics. ***The diagram shows three schematics, one for each stage of the project. In each stage, a push-button switch is used to control the flow of current through the loop. A resistor and LED are added to each stage to increase the number of components introducing voltage drops around the loop.***

This circuit has three stages, where each stage is a separate series loop connected to a common voltage source (Arduino) that can be switched on/off by a push-button switch. Therefore, the overall system has three sensor inputs (push-buttons) and multiple LED outputs (1, 2 and 3 respectively). The purpose of the push-buttons is to give you a visual comparison of the effect of adding multiple resistors in series, where each loop will be dimmer than the previous one due to the additional resistors reducing the overall current (from $\approx 333\text{mA}$ down to $\approx 11\text{mA}$). We know that resistances in a series circuit add up, and by using Ohm's Law in the equations above we can show that more resistors in series will reduce the overall current flowing. If you don't have six red LEDs then you can use other LED colours if you have them, but blue and white LEDs will be much dimmer than red, orange or green because of their higher forward voltage (around 4V). You can also leave out the last stage of this project if you do not have enough LEDs and/or resistors – or you can simply build stages 1 and 2 then break them down to build stage 3 separately afterwards. If you do so, you will still see a reduction in LED brightness in the second-stage circuit, but it will not be as marked as the difference between stages 1 and 3.

Circuit design

In this project, we are effectively replicating the same component linkages in three stages, so there are some practical steps we can take when designing our circuit layout. Consider the component structure for schematic 1 – notice how the switch/resistor/LED combination is common to all three diagrams. If we can determine how to set these components out on breadboard then we can simply extend it for diagrams 2 and 3. The first thing to note is the layout of the switch, which we previously discussed in chapter 2. A push-button switch has four contacts which are already connected in pairs (A-C, B-D) and by pressing the button we connect all four terminals to each other. In our circuit, we will be using one half of each switch to connect two terminals together (either A-B or C-D).

In this project, we will be designing the circuit from **both** top down and bottom up to help expand our understanding of the practicalities of breadboard circuits. As we progress to more complex filter and amplification circuits in later chapters, we will be working within the constraints of the space available on the breadboard. It may seem simpler to just use a bigger breadboard (or add a second one) but in practice a significant amount of effort in circuit design (particularly in audio) goes into the reduction of space between components, both to reduce the overall system size and also to avoid the noise created by longer signal/power paths and other parasitic impedances (we will learn more about impedance in chapter 6). In most instances, it is important to learn how to leave as little space as possible (an exception would be the shielding of components from power rails to prevent interference) and so we will consider each breadboard pin as a possible option for signal paths to maximize our available resources.

The layout of the resistor and LED (see schematic below) must be connected in series to the output button terminal (either B or D). In Tinkercad, LEDs are oriented with the anode on the left so they must be rotated 180° in order to be linked correctly in series with a resistor. In addition, the size of the component means it needs to be placed on a higher row to allow other connections (in this case GND) to be added underneath. To maximize our use of the breadboard whilst still being able to follow the current flow easily, we will begin from the bottom left of the breadboard (Figure 3.11).

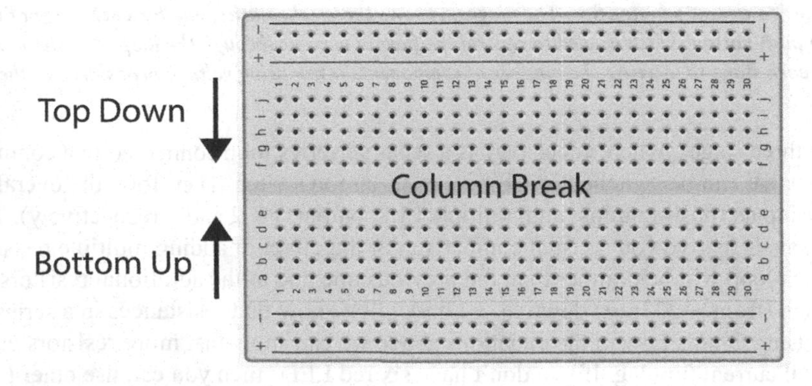

Figure 3.11 Breadboard layout. ***In many of our projects from now on, we will maximize the breadboard space available by working from both bottom up and top down to create multiple circuit branches using the column break to separate them.***

We can now set out the components so that each input terminal is connected on the same row as the output terminal of the previous component – in the case of a resistor input and output are interchangeable. The stage 1 schematic is also redrawn to rearrange the three components in a straight line, to help us focus on the layout based on a series current flow (Figure 3.12).

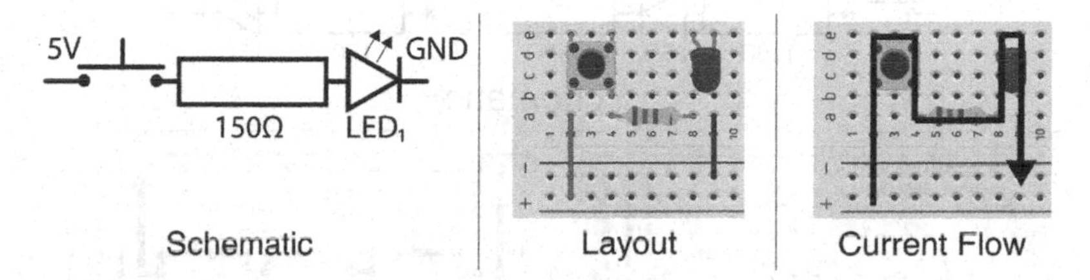

Figure 3.12 Series circuit project stage 1. ***In the above diagram, the original stage 1 schematic has been redrawn horizontally in the left-hand image to act as a reference for the layout in the middle image. The current flow through this component layout is also shown in the right-hand image.***

In the layout, connections have been added to the power rails on the breadboard to complete the circuit loop – the Arduino will eventually supply +5V and GND to these rails when we connect it later in the build. Stage 2 of the circuit replicates this structure, but with an additional resistor/LED pair being added in series (Figure 3.13).

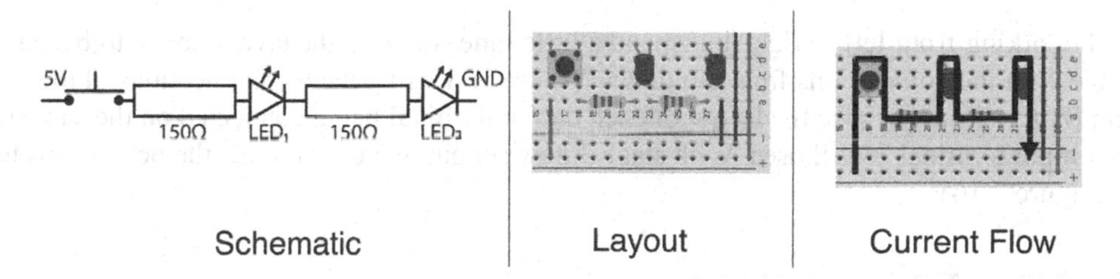

Figure 3.13 Series circuit project stage 2. ***In the diagram, the stage 2 schematic/layout/current flow structure is shown. The additional wires on columns 29 and 30 are for power bridging, which will be discussed below.***

Now we are ready to add stage 3, which requires us to use the top half of the breadboard. To do this, the first step is to bridge the power rails on the board (you will already see these connections in the layout diagram above), to allow a single input connection from the Arduino to provide +5V and GND to both power rails (Figure 3.14).

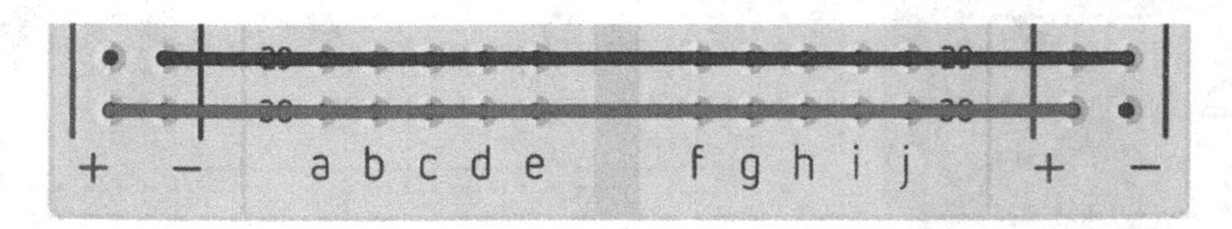

Figure 3.14 Bridging breadboard power rows. ***The image above shows how the two sets of power rows on a breadboard can be bridged by connecting a wire between them. This allows power connections to the Arduino to provide 5V and GND rails to the entire board. Note that the board has been rotated to show you where to connect your wires.***

The bridging connections mean that both +/− rails are now connected, which allows us to use the top half of the breadboard for stage 3. The first thing that may confuse you is the reverse layout of the components as we are now working **top down** rather than bottom up as in the previous stages (Figure 3.15).

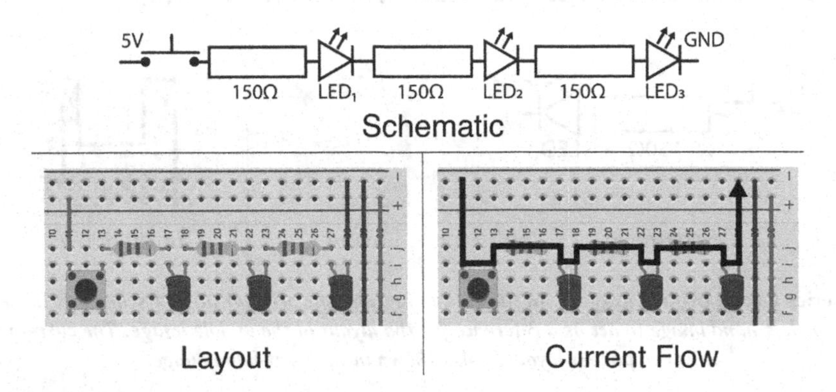

Figure 3.15 Series circuit project stage 3. *In the diagram, the stage 3 schematic shows where an additional resistor and LED have been added. Below this, the layout and current flow structure are shown, where we are now working top down from row 11 onwards to maximize the breadboard space available. The additional wires on columns 29 and 30 bridge power to both rails on the breadboard, allowing a single Arduino connection for 5V and GND to be used.*

We are still working from left to right, but spend a little time studying the layout above to become comfortable with the connections from top down. We are now using the top connections of the switch to connect [A-C], but otherwise the layout of each output terminal being connected on the same row as the next input terminal is still used. With stage 3 now set out, we can connect the power rails to the Arduino (Figure 3.16).

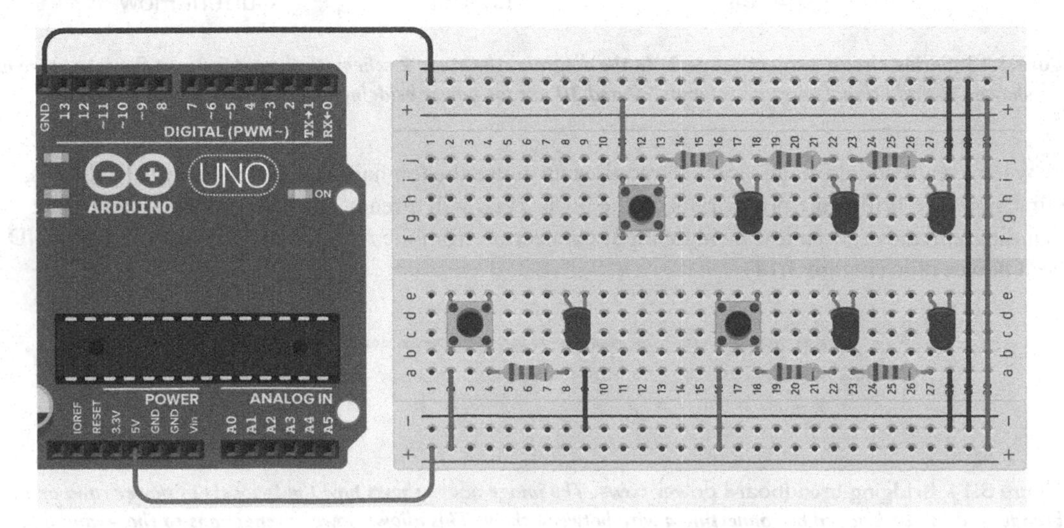

Figure 3.16 Series circuit project final breadboard layout. *The image shows all three stages of the project laid out on a single breadboard. This allows all stages to be tested simultaneously in Tinkercad.*

The final circuit layout for this project is now completed and ready for testing in Tinkercad – we will now work through the specific pins for each stage in the steps below. If you do not have a lot of components, you can build each stage of the project in turn with a minimum of three resistors and LEDs. If this is the case, you can use the Tinkercad simulation to see the effect of adding voltage drops to a series circuit – the LEDs become dimmer as you add a greater total resistive load to the circuit and hence reduce the overall current flowing. For this project you will need:

1. Arduino Uno board or other 5V power supply.
2. One small breadboard (30 rows).
3. Six × red LEDs (minimum three).
4. Six × 150Ω resistors (minimum three).
5. Six × push-button switches (minimum two).
6. Ten wires (ideally five red for +5V, five black for GND).

Project steps

For this project, you will need: 1. Arduino Uno board 2. One small breadboard (30 rows) 3. 6 × red LEDs (minimum 3) 4. 6 × 150Ω resistors (minimum 3) 5. 6 × push-button switches (minimum 2) 6. 4 × connector cables and 6 × connector wires	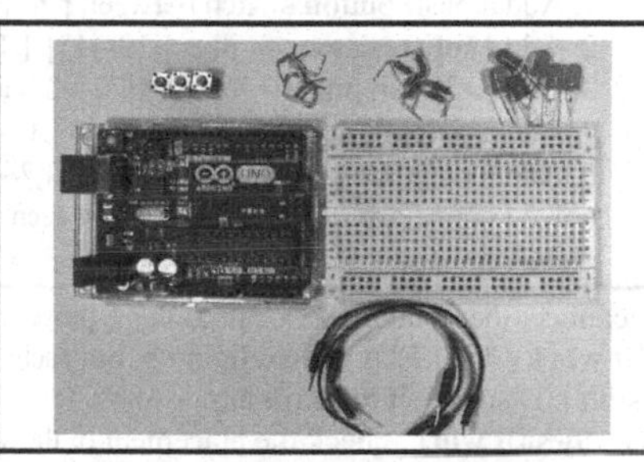
Stage 1: 1. Add a push-button switch between pins [d2–b4] (top left/bottom right) 2. Add a 150Ω resistor between pins [a4–a8] 3. Add an LED – anode pin (e8) to cathode (e9) 4. Add a positive connector wire from [a2–V+] 5. Add a negative connector wire from [a9–GND]	
Add a connector cable between the GND pin and the – power rail. Add a connector cable between the Arduino +5V pin and the + power rail. Test the circuit by pushing the button. **If it works:** the LED should light up. Disconnect the Arduino power connector cable and continue to the next build stage. **If it doesn't work**: check the placement of the resistor, check the push-button switch orientation and check the direction of the LED (long-leg cathode is positive).	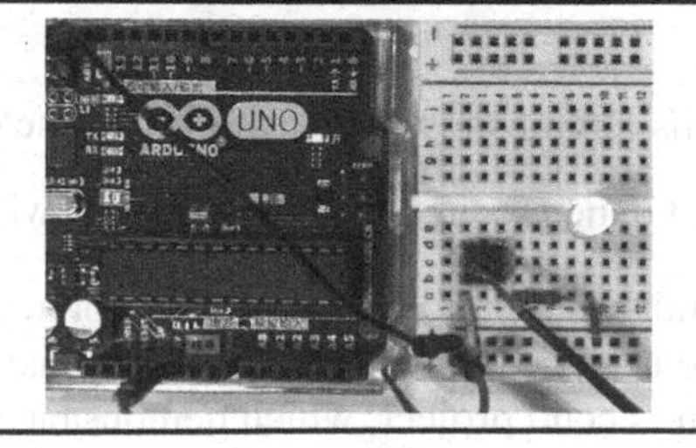

Stage 2: 1. Add a push-button switch between pins [d16–b18] 2. Add 150Ω resistors between pins [a18–a22] and [a23–a27] 3. Add LEDs between pins [e22–e23] and [e27–e28] 4. Add a positive connector cable from [a16–V+] 5. Add a negative connector cable from (a28–GND]	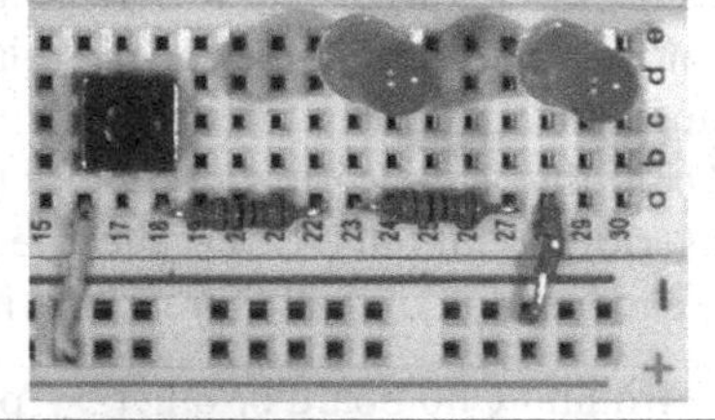
Reconnect the Arduino +5V pin to the + power rail and test the circuit by pushing each button in turn. **If it works:** the LEDs should light up, but the LEDs in the stage 2 circuit should be dimmer. Disconnect the Arduino power connector cable and continue to the next build stage. **If it doesn't work**: check the placement of the stage 2 resistors, check the push-button switch orientation and check each LED's direction.	
Stage 3: 1. Add a push-button switch between pins [i11–g13] 2. Add 150Ω resistors on pins [j13–j17], [j18–j22] and [j23–j27] 3. Add LEDs on pins [i17–i18], [i22–e23] and [i27–i28] 4. Add a positive connector wire from [j11–V+] 5. Add a negative connector wire from [j28–GND] 6. Add connector cables to bridge between both +/− power rails (bottom rails not shown)	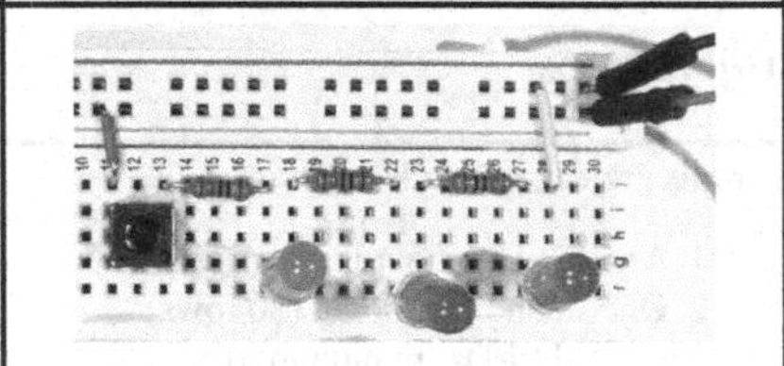
Reconnect the Arduino +5V pin to the + power rail and test the circuit by pushing each button in turn. **If it works**: the LEDs should light up, but each stage of the circuit should be dimmer (in the image below, stage 3 is barely visible). Disconnect the Arduino. **If it doesn't work**: check the placement of the stage 3 resistors, push-button switch orientation and check each LED's direction. 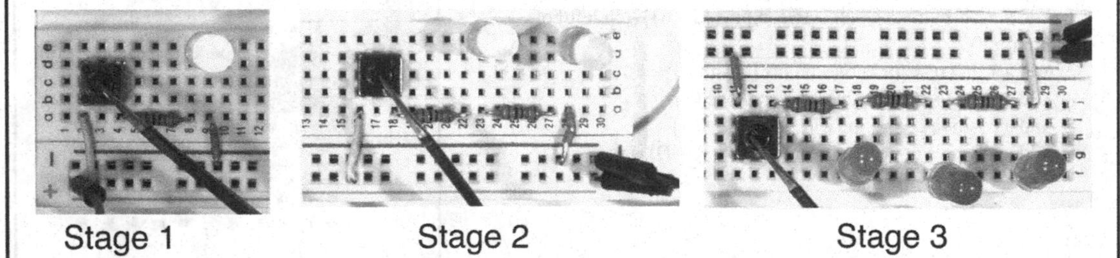	

This circuit shows how LEDs become dimmer because the overall resistance in the circuit increases and hence reduces the total current, where Ohm's Law shows that $I = \frac{V_S}{R_{tot}}$. Although the project gives a fairly simple illustration of the principles of a series circuit, it is important to note how many steps are involved in translating the schematics into an actual breadboard circuit. We will now move on to our second project, which demonstrates how parallel circuits can exploit constant voltage to create redundancy.

3.6.2 Parallel circuit project

For the second project, we will build a parallel circuit powered by the Arduino based on the schematic in Figure 3.17.

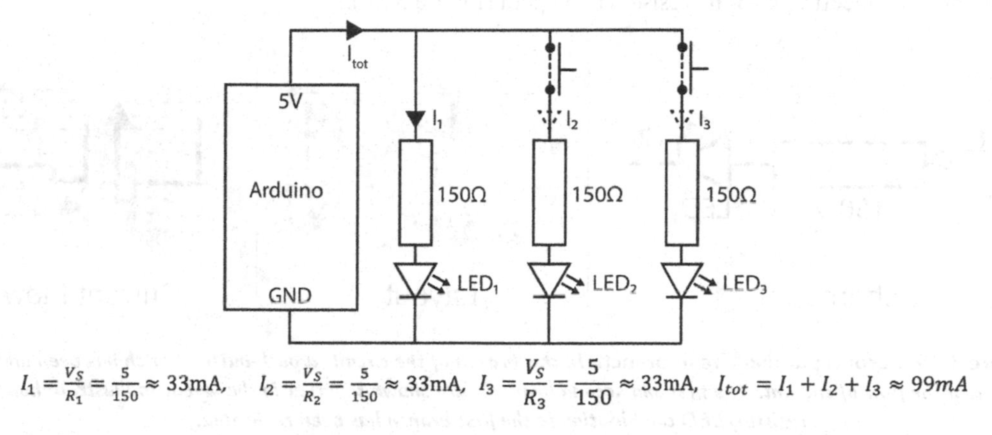

Figure 3.17 Parallel circuit schematic. ***The schematic shows three branches connected in parallel with an Arduino power source, where the second and third branches are controlled by push-button switches. In each branch, a resistor controls the current flowing through an LED. The branch currents I_1 and I_3 are indicated with dashed lines for closed switch positions.***

We recall that voltage is constant in a parallel circuit, so if the resistances are all equal to 150Ω then each branch current will also be equal (approximately 33mA). If the currents flowing through each branch are equivalent, then all three LEDs should light up with the same brightness. The previous project covered a lot of the practicalities of switch/resistor/LED component combinations, so we will move onto circuit design to see how we can use the breadboard to lay out our components in the most effective way possible.

Circuit design

In some respects, the circuit involved in this project will be a lot simpler to design and construct than the previous example. Other than the omission of the push-button switch from the first branch, each parallel branch of the circuit is the same so we can focus on designing a single section and then replicating it (Figure 3.18).

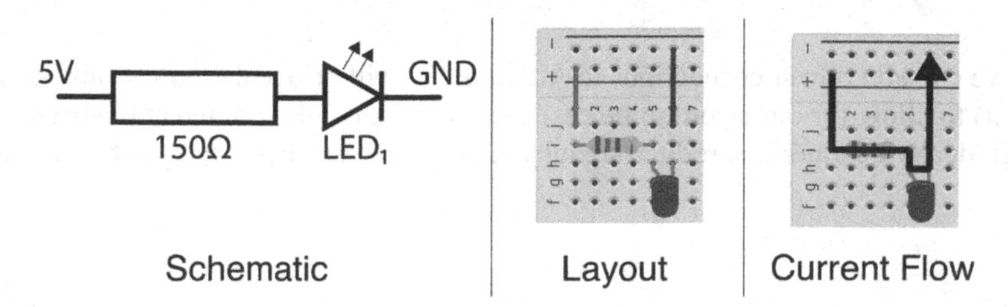

Figure 3.18 First parallel circuit branch. ***This branch of the circuit is replicated throughout with the addition of push-button switches to create the full parallel circuit. The resistor LED combination should be familiar to you from the previous series circuit project design.***

In the layout in Figure 3.18, the connections to the power rails are included to show another practicality of circuit design – you must always lay out each section of a circuit in relation to the signal or power pins that it will connect to. This becomes more apparent when we add one of the branches that also contains a push-button switch – the power rail must now be connected to the input of that switch before connecting it to the resistor/LED path (Figure 3.19).

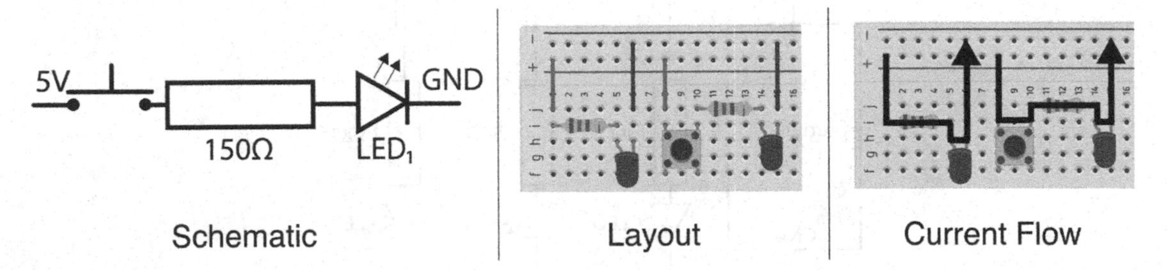

Figure 3.19 Second parallel circuit branch. ***In this branch of the circuit, a push-button switch has been added to control the flow of current. The first and second branches are shown together in the layout to illustrate how the resistor/LED combination in the first branch has been replicated.***

Now we can simply replicate this branch to create the third branch of our circuit (Figure 3.20).

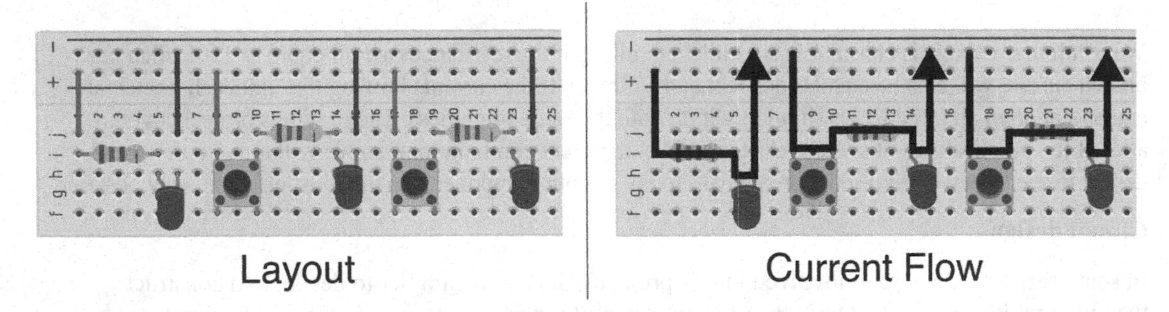

Figure 3.20 Combined parallel circuit branches. ***This diagram shows the combined layout for all three parallel branches in the project circuit. The right-hand image shows the current flow for the entire circuit, where each parallel branch is connected to the common 5V and GND rails powered by the Arduino.***

At this point, we can add power connections to Arduino. We will bridge the power rails (as in the previous circuit) to allow for our layout to be easily followed, though as noted at the beginning of this project we will ideally be aiming to reduce signal and power paths whenever possible (Figure 3.21).

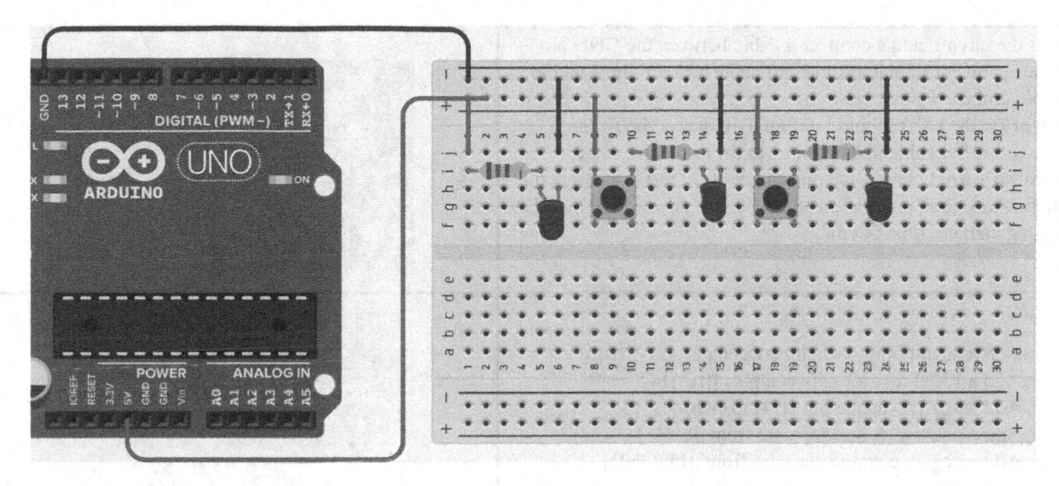

Figure 3.21 Final parallel circuit layout. ***All three branches of the circuit are shown connected to an Arduino for simulation in Tinkercad. This layout combines bottom-up and top-down components to maximize the space available on a single breadboard.***

The final circuit layout for this project is now completed and ready for simulation testing in Tinkercad. Once this testing is complete, we can work through the specific pins needed to build the actual circuit in the steps below.

Project steps

For this project, you will need: 1. Arduino Uno board 2. One small breadboard (30 rows) 3. 3 × red LEDs 4. 3 × 150Ω resistors 5. 2 × push-button switches 6. 4 × connector cables and 6 × connector wires	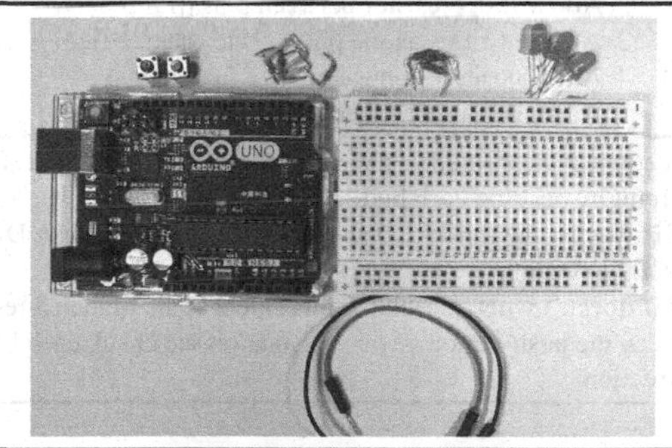
Branch 1: 1. Add a 150Ω resistor between pins [i2–i5] 2. Add an LED – anode pin (h5) to cathode (h6) 3. Add a positive connector wire from [j1–V+] 4. Add a negative connector wire from [j6–GND]	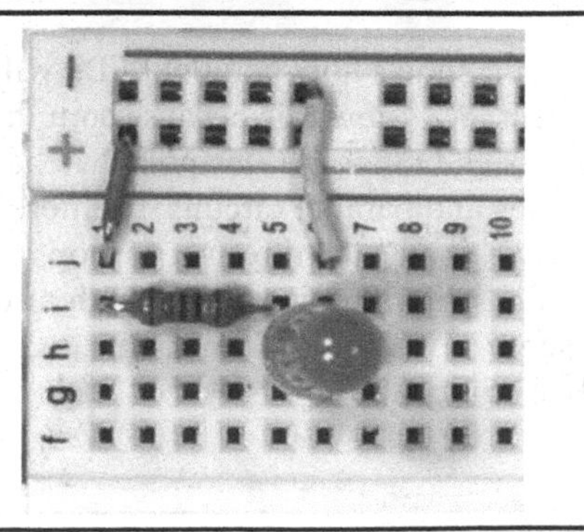

To test the circuit, add a connector cable between the GND pin and the – power rail and a connector cable between the Arduino +5V pin and the + power rail. **If it works:** the LED should light up. Disconnect the Arduino power connector cable and continue to the next build stage. **If it doesn't work**: check the placement of the resistor, check the push-button switch orientation and check the direction of the LED (long-leg cathode is positive).	
Branch 2: 1. Add a push-button switch between pins [i8–g10] 2. Add a 150Ω resistor between pins [j10–j14] 3. Add an LED – anode pin (i14) to cathode (i15) 4. Add a positive connector wire from [j8–V+] 5. Add a negative connector wire from [j15–GND]	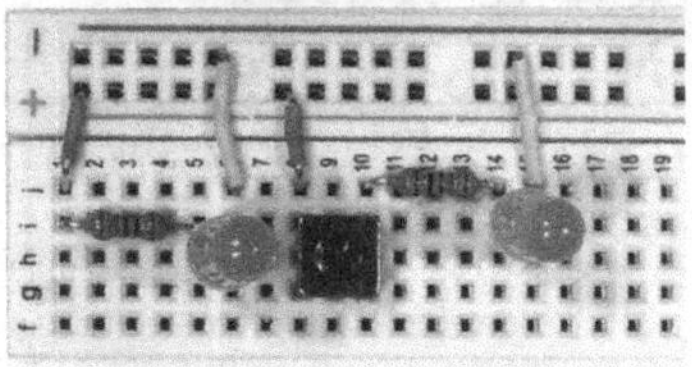
Reconnect the Arduino +5V pin to the + power rail and test the circuit by pushing the button. **If it works:** the second LED should light up with the first. Disconnect the Arduino power connector cable and continue to the next build stage. **If it doesn't work**: check the placement of the branch 2 resistor, check the push-button switch orientation and check each LED's direction.	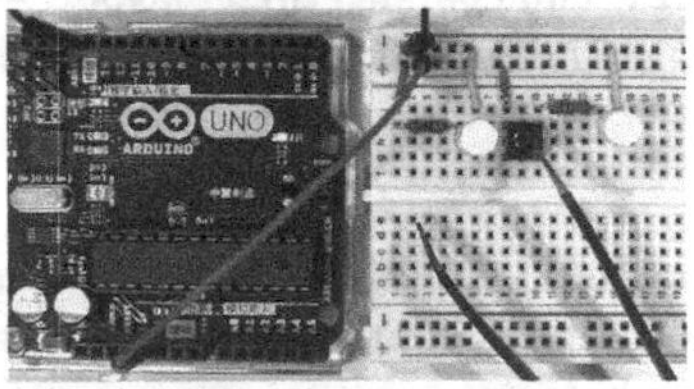
Branch 3: 1. Add a push-button switch between pins [i17–g19] 2. Add a 150Ω resistor between pins [j19–j23] 3. Add an LED – anode pin (i23) to cathode (i24) 4. Add a positive connector wire from [j17–V+] 5. Add a negative connector wire from [j24–GND]	
Reconnect the Arduino +5V pin to the + power rail and test the circuit by pushing each button in turn. **If it works:** the third LED should light up with the first. Disconnect the Arduino. **If it doesn't work**: check the placement of the branch 3 resistor, check the push-button switch orientation and check each LED's direction.	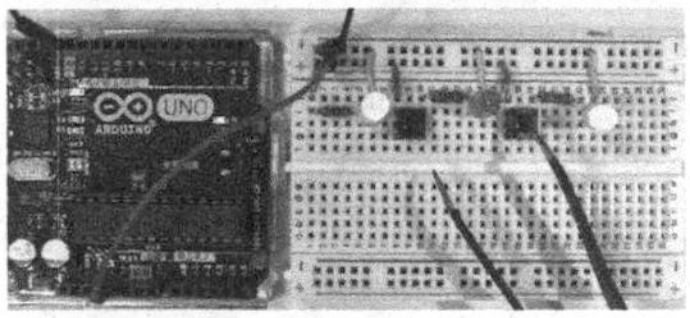

This circuit shows how parallel LEDs will have the same brightness because each branch current is equivalent. We know this because both the voltage (constant in a parallel circuit) and resistance are equal in each branch. At this point, we have now worked through the practical steps needed to build both series and parallel circuits, but there is now one last point to note about our circuits. In the first project, we connected three series loops to the same supply rails – this is also what we have done in the parallel circuit project. Thus, both circuits are **effectively connected in parallel**, which makes more sense if we now redraw the entire circuit from project 1 including the common Arduino power supply (Figure 3.22),

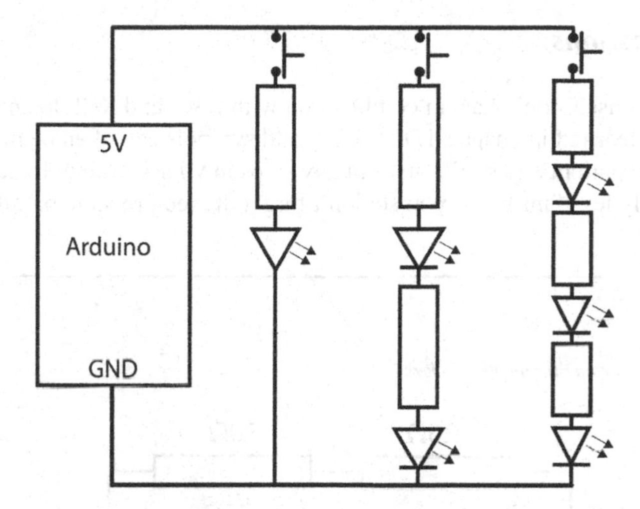

Figure 3.22 Full project 1 schematic. ***This diagram shows how the three series stages of the first project are effectively parallel branches within the overall Arduino circuit. The resistor and LED labels are omitted for clarity.***

This may initially seem slightly confusing, but the practical point is that we can design different circuit stages and power them from a common supply (the Arduino). For your own learning, the important point is not whether you see the first project circuit as series or parallel but rather that you understand how and why the circuit was divided into different stages. This is a very common circuit design task and you will see it again in later circuits in this book.

3.7 Conclusions

This chapter has covered a lot of ground, introducing Ohm's Law, Kirchoff's Laws (KVL, KCL) and also working through more involved project circuits. It is recommended that you **spend some time going back through this chapter** to make sure you are comfortable with the calculations involved – as noted in chapter 1, the mathematics may be fairly straightforward but working with symbols and scales can introduce a lot of pitfalls when you are beginning to learn electronics. If you find yourself struggling with some of the material in the next few chapters, it is highly likely that the root of the problem lies somewhere within the concepts and calculations introduced in this chapter – they may appear fairly simple in isolation, but we have worked through most of the core concepts of fundamental DC circuit analysis in a very short time and it is to be expected that you will take a little longer to synthesize everything you have just learned.

The next chapter will look at configuring and programming the Arduino for input processing and output control. This represents a **significant shift in focus** from DC circuit analysis, so be prepared to find the combination of new concepts more difficult to work through than any of the individual parts in isolation. As noted in chapter 1:

If you can add, multiply and divide numbers then you can learn electronic circuit theory!

Applying these basic mathematical concepts requires practice – similarly, learning introductory code is not complex, but can be very confusing!

3.8 Self-study questions

The following questions use Ohm's Law in combination with KVL and KCL to analyse series and parallel circuits. As we learned in chapter 1, the scales and symbols can often be much more difficult than the equations so be patient and don't miss out any steps in your working. In addition, working with fractions is possibly less familiar to you so don't forget the reciprocal in parallel circuits!

Series circuits

Use the following schematic to answer the questions below:

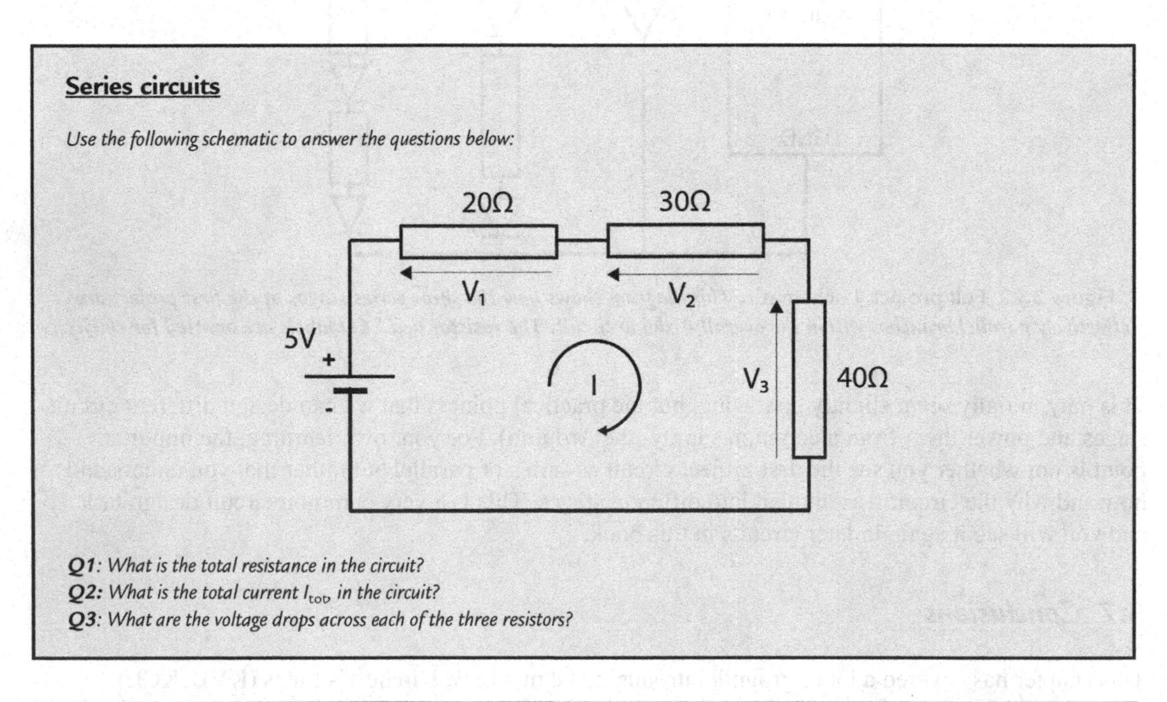

***Q1**: What is the total resistance in the circuit?*
***Q2**: What is the total current I_{tot}, in the circuit?*
***Q3**: What are the voltage drops across each of the three resistors?*

Parallel circuits

Use the following schematic to answer the questions below:

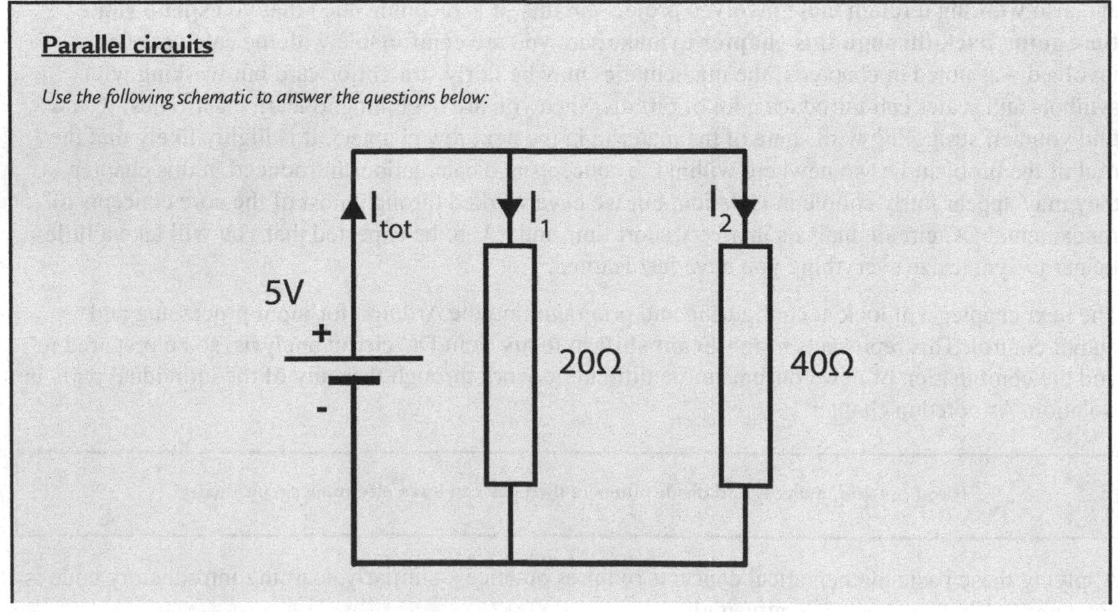

Q1: What is the total resistance in the circuit?
Q2: What is the current I_{tot}, in each branch of the circuit?

Voltage dividers

Use the following schematic to answer the questions below:

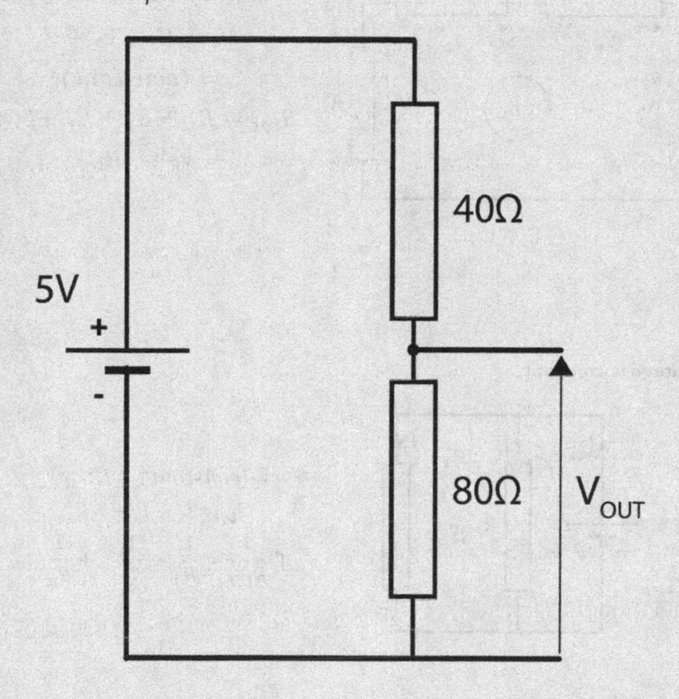

Q1: What is the voltage V_{out}?
Q2: What is the voltage across the first resistor, V_{R1}?

All answers are provided for each chapter at the end of the book.

Chapter 3 summary

Ohm's Law

Voltage, $V = IR$

Kirchoff's Voltage Law

The sum of all voltages around any closed loop in a circuit is equal to zero

Kirchoff's Current Law

The charge entering a junction is equal to the charge leaving that junction

Series circuits

Voltage varies and current is constant

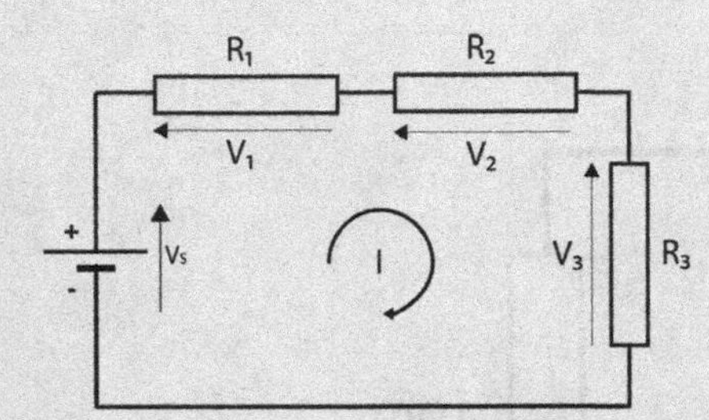

$$V_S = V_1 + V_2 + V_3$$
$$I\ (constant)$$
$$R_{tot} = R_1 + R_2 + R_3 + \cdots \boldsymbol{R_n}$$
$$V_S = IR_{tot}$$

Parallel circuits

Current varies and voltage is constant

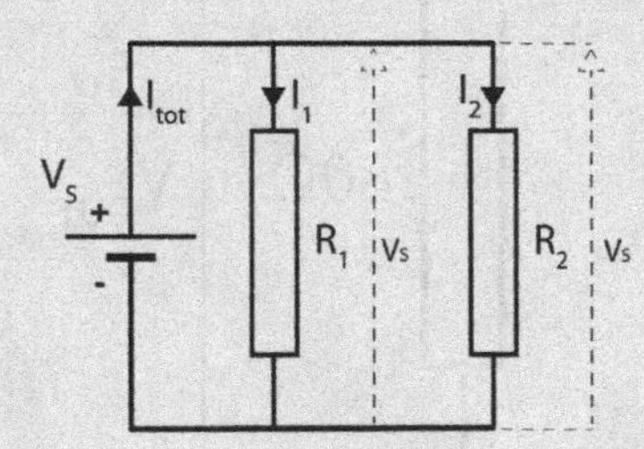

$$V_S(constant) = IR_{tot}$$
$$I_{tot} = I_1 + I_2 + I_3$$
$$\frac{1}{R_{tot}} = \frac{1}{R_1} + \frac{1}{R_2} \cdots + \frac{1}{R_n}$$

Voltage dividers

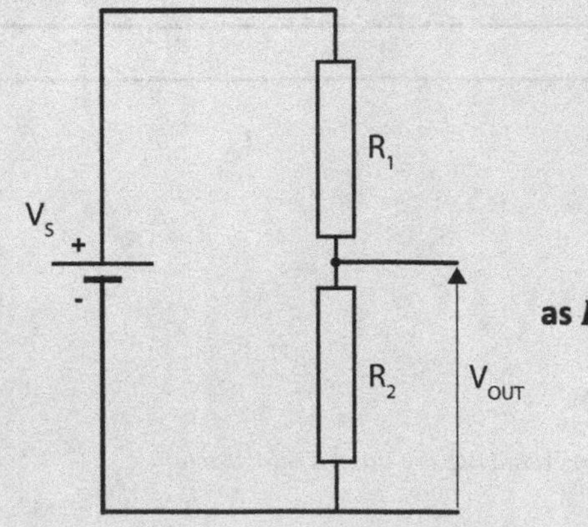

$$V_{OUT} = V_S\left(\frac{R_2}{R_1 + R_2}\right)$$

as R_2 gets bigger- so does V_{OUT}

CHAPTER 4

Digital systems 1 – Arduino output

In this chapter, we will learn about the Arduino microcontroller and how to program it. Until now, we have simply been using the Arduino as a 5V power source for our DC circuits, so the next two chapters (Digital Systems 1 and 2) will focus on how to use the Arduino as a microcontroller. We begin by returning to the system concepts introduced in chapter 2, as a means of separating analogue inputs and outputs from the digital processing and control performed by the Arduino. Over the next two chapters, we will learn how to use the Arduino to create actuator outputs, and how input from sensors can be used by the Arduino to execute processes.

To do this, we must learn the basics of computer programming, as the Arduino executes code compiled in the C/C++ languages. This book aims to provide an introduction to computer programming, covering core concepts and principles to help you get started with digital control, Having said this, these two chapters are **not a full computer programming course**. In this chapter, we will use code to configure Arduino outputs to drive actuators in the form of LEDs and a piezo loudspeaker. In so doing, we will learn about the three instructions types (sequence, selection and iteration) as the verbs of our code, and also how a basic Arduino sketch is configured. We will then learn about variables (named locations that store our data) and constants as the nouns of our code – you cannot have a verb without a noun! We then progress to declaring and using functions as a means of encapsulating our code instructions. Functions are the key to reducing and reusing our code, so in the tutorial some pointers on coding technique are provided, as computer programming requires both diligence and accuracy (particularly in a strictly typed language like C).

This chapter uses several example projects to apply the programming concepts involved. The examples aim to show you how code can be used to tell the Arduino to control our electronic circuits – this is where the Arduino is most powerful. The final project will use the Arduino to light a sequence of LEDs in tandem with emitting a series of musical pitches through a piezo loudspeaker. Although this is by no means a full musical project, the next chapter will build on these concepts to construct a MIDI drum trigger.

What you will learn

How the Arduino system is configured and how we will use it
The three basic programming instruction types (there is no fourth instruction!)
How variables (and constants) are used to give named access to data in memory locations
How functions encapsulate code instructions for reuse
How we can use Arduino PWM output to generate audio signals that drive a loudspeaker
How we can approach problems by debugging our code

4.1 Microprocessor control systems

Chapter 2 introduced the idea of a basic digital system as used in microcontrollers like the Arduino – a system known as the Harvard architecture (Figure 4.1).

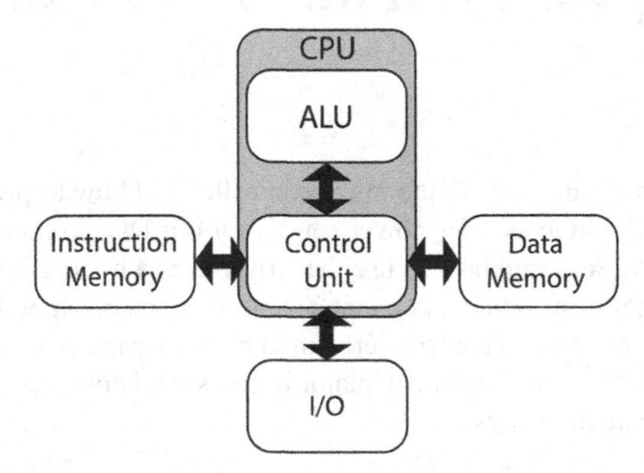

Figure 4.1 Harvard computer architecture. ***The Harvard architecture combines an arithmetic logic unit (ALU) and a control unit within a central processing unit (CPU). Data in the system is stored in specific memory, whilst the instructions for the control unit have their own separate memory storage. Input and output connectivity (I/O) allows sensors and actuators to interact with the system.***

A microcontroller like the Arduino is effectively a small computer, where the central processing unit (CPU) executes instructions that process sensor inputs to create actuator outputs. In this way, the Harvard architecture functions like an open-loop system (Figure 4.2).

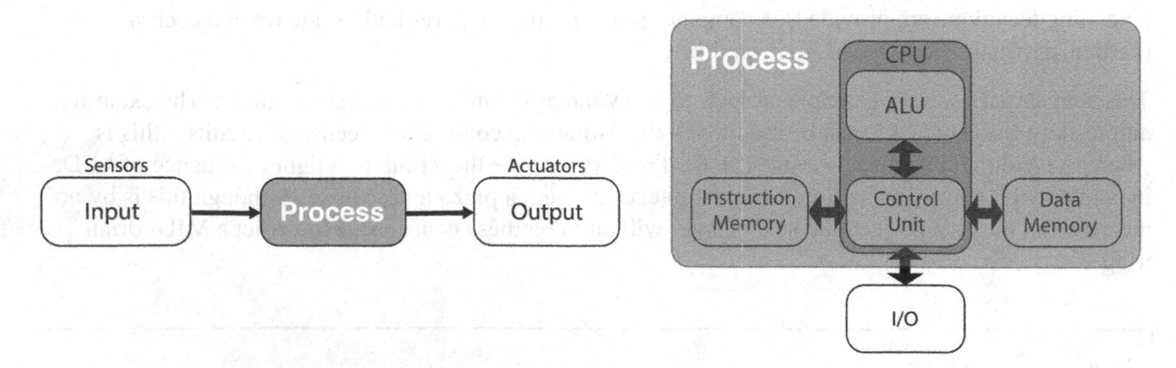

Figure 4.2 Basic Harvard architecture system diagram. ***The left-hand image defines the core elements of an open-loop system as being input/process/output. The right-hand image shows how the Harvard architecture (as discussed in Figure 4.1) can be viewed in the same way for coding purposes.***

This concept of processing inputs to create outputs is core to our understanding of systems, and in this chapter we will learn how to use the **C programming language** to program the Arduino with instructions that carry out these processes. The Arduino also incorporates some elements of C++, which in some ways is a more advanced version of C. In the first three chapters, we effectively used the Arduino as a power supply for the circuits in our projects. Now we will learn how to compile instructions written in C and upload them to an Arduino, but before doing this we need to spend a little time familiarizing ourselves with the basic layout and configuration of a common Arduino board – the Uno.

At time of writing, the Arduino Uno is one of the cheapest fully featured boards available, which makes it a good choice for those learning electronics. In addition, there are many Uno hobby kits available for relatively low cost that provide a range of input and output components alongside breadboard and wires – everything you need to build basic Arduino projects. The Arduino Uno is based around an ATMega328P microcontroller and is considered to be the reference model for future boards (hence Uno, which is Italian for 'one'). The layout of the input and output pins in the Uno is shown in Figure 4.3.

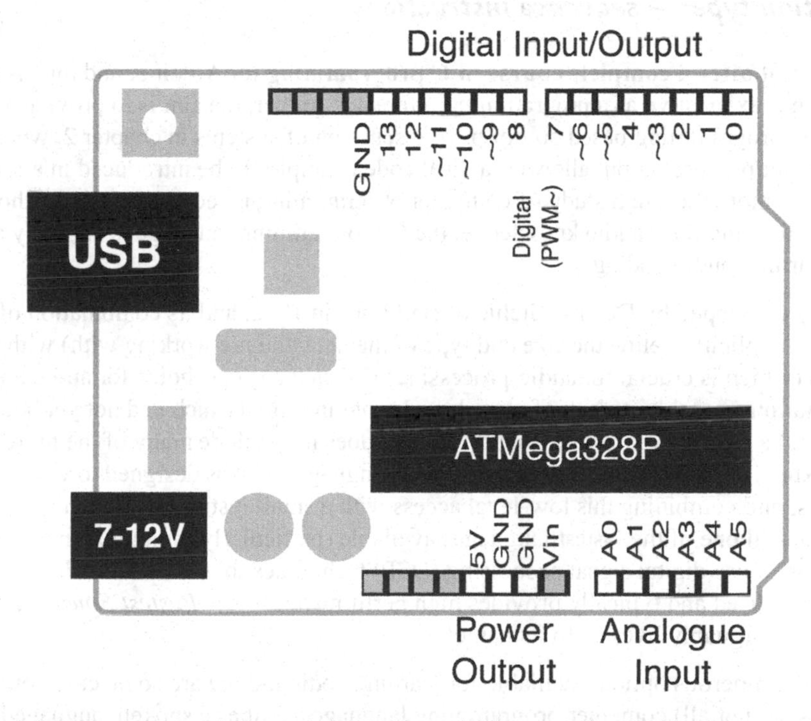

Figure 4.3 Arduino Uno layout. ***The image indicates the main elements of an Arduino Uno, starting with power supply options (either USB or 7–12V DC) on the left. The ATMega328P microcontroller is connected to 6 analogue inputs and 13 digital input/outputs (all pins marked white). Power output is provided via the 5V and GND pins at the bottom (also 1 GND pin at the top). Note the use of a tilde (~) to indicate which digital pins can produce pulse width modulation (PWM) output.***

The Arduino can be powered by USB (5V) or from a 7–12V DC power supply. The board can use either of these supplies to power its microcontroller (the ATMega328P) and also to provide a power supply output of 5V and GND (pins marked white in Figure 4.3), which were used in the previous three chapters to power our project circuits. This makes the Arduino a very adaptable board, as it can be powered using the same USB bus that is used to upload C programming instructions to control the ATMega328P. For more permanent circuit installations, the Arduino can also be powered by a dedicated 7–12V DC supply or even by a battery connected to the Vin pin next to the power output section (this is also where other Arduino shields can supply power to the Uno board).

The Arduino Uno is a very versatile board that has 14 pins (0–13) available for either digital input and/or output. By allowing all digital pins to be configurable for either input or output, the Arduino can be adapted to a wide range of digital system designs. In addition, digital pins 3,5,6,9,10,11 can be configured for pulse width modulation (PWM) output, which will be covered in more detail later in this chapter (the Arduino can also receive sensor input from up to six analogue inputs). Now we have a broad sense of the layout and configuration of the Arduino Uno, we can begin to learn how to use C programming instructions to create digital actuator outputs.

4.2 Instruction types – sequence instructions

This **book does not offer a complete course in C programming** for Arduino, and thus is neither comprehensive nor exhaustive as a programming reference. Rather, the aim is to provide an initial introduction to C programming based on our prior discussion of systems in chapter 2, where the delineation of input/process/output allows practical code examples to be introduced in a focussed manner. Though a more thorough study of computer programming is recommended for those who are interested in progressing their audio knowledge, the C programming language is arguably a very good place to start learning audio coding.

C was originally developed by Dennis Ritchie of Bell Labs in 1972, and its combination of strict typing (where you must explicitly define the size and type of the data you are working with) with low-level memory access (which is crucial for audio processing) makes it a good choice for audio applications. C also has limitations, notably its lack of class-based programming (which had not yet been invented in the 1970s!) and a relatively small instruction set that does not include many of the more modern features (e.g. extended data structures) found in other languages. C was designed to work closely with microprocessors, and combining this low-level access with a small instruction set (thus a small compile time) means it is still one of the fastest languages available (particularly for audio rendering). If you plan to learn how to use digital signal processing (DSP) techniques to work with audio data, the C language is widely used and typically provides high performance – the '*Fastest Fourier Transform in the West (FFTW)*'[1] arguably lives up to its name!

Though there are numerous options available for learning coding, there are some core concepts that are common to most (if not all) computer programming languages. Just as a spoken language has **nouns and verbs**, so computer programming languages have **data and functions** that broadly mimic these concepts. More fundamentally, computer programming instructions can be defined as one of three basic types.

Programming instruction types

Sequence instructions – do something **once**

Iteration instructions – **repeat** something more than once

Selection instructions – make a choice based on a **condition**

It is important to remember is that **there is no fourth instruction type**! When learning to code, it can be difficult to map the instruction to the process it will carry out and this can lead to the perception of 'other' instructions that may be needed. At its core, programming is very much an exercise of diligence and accuracy and so it is important to frame our learning within the context of simple systems that involve clear unambiguous processes. Rather than puzzling over complex lines of code that can appear very confusing (at least to begin with), we will be using a systems analysis approach to always frame our code in terms of input/process/output.

The first instruction type is the **sequence instruction** – do something once. Extending our previous analogy of nouns and verbs, we can think of a sequence *instruction* as defining the *verb* of a process. The example below introduces specific instructions to begin the process of reading code, where the first thing to note is the use of comments (**denoted by two forward slashes** //). Comments are ignored by the C compiler (a compiler builds code for upload to the Arduino) so they are a very useful way of marking our code for reference and ease of understanding:

```
// a configuration instruction that initializes pin 13 for output
pinMode(13, OUTPUT);
// an output instruction that sends a 5V signal (HIGH) to pin 13
digitalWrite(13, HIGH);
// an assignment instruction that assigns the value 8 to the integer
//variable x - notice that each comment line needs forward slashes
int x = 8;
// an arithmetic instruction that adds 5 and 4 and assigns the result
//to the integer variable y
int y = 5 + 4;
```

In the above examples, each sequence instruction is **terminated by a semicolon** (;). You must terminate every instruction in C, to prevent the compiler from continuing to read the next line as a continuation of the previous instruction. We will work through the lines of code introduced above as we progress our learning, but for now the other major point is to note the use of the **keyword** int which defines the data **type** of the **variable** x. **Keywords are reserved** as part of the C/C++ languages – you must use them, and cannot change them.

C programming does not recognize carriage returns as part of the language, so the following examples show what does and does not work when terminating an instruction:

```
// a comment on an instruction - the compiler will ignore this
pinMode(13, OUTPUT);
// the previous instruction was terminated, so this is a new instruction
pinMode(12, OUTPUT)
// the previous instruction was not terminated, so the compiler now reads the
// next line as part of the same instruction
int x = 8;
// Thus, the compiler reads pinMode(12, OUTPUT)int x = 8;
// Needless to say, this instruction does not make sense to the compiler!
```

These examples are only intended to show you what sequence instructions look like as C code, as the only elements explained thus far are comments (//) and terminating instructions (;). To avoid simply introducing elements of the language in abstract (i.e. why should variable x= 8) this chapter will **focus on practical examples** – the best way to learn to write is to read, so the best way to learn how to write code for the Arduino is to read examples of existing programs. To do this, we will adapt the example circuit first introduced in chapter 1 to light an LED – but this time using an Arduino digital output pin to control the flow of current through the LED.

4.3 Example project 1 – Arduino digital output

The first coding example in this chapter will implement the simplest possible system – one that uses processes to create digital actuator outputs (Figure 4.4).

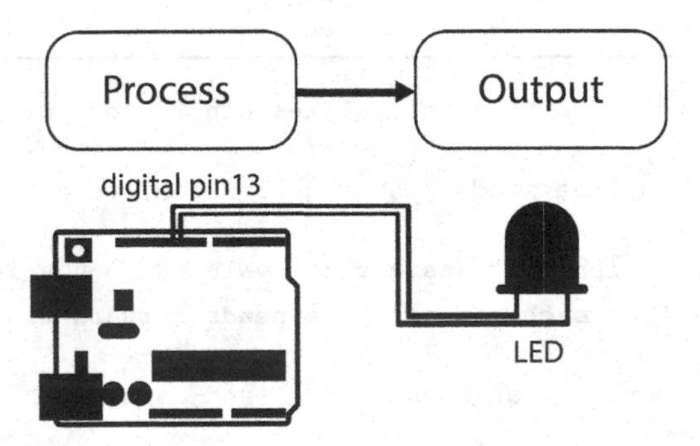

Figure 4.4 Simple digital output system. ***In the diagram, the Arduino executes processes based on coding instructions that will create an actuator output by lighting an LED. Note that the diagram does not include all pin connections, as this system is used to highlight the logical connections from Arduino process to LED output.***

The electronic component circuit in this system can be simulated using the schematic in Figure 4.5.

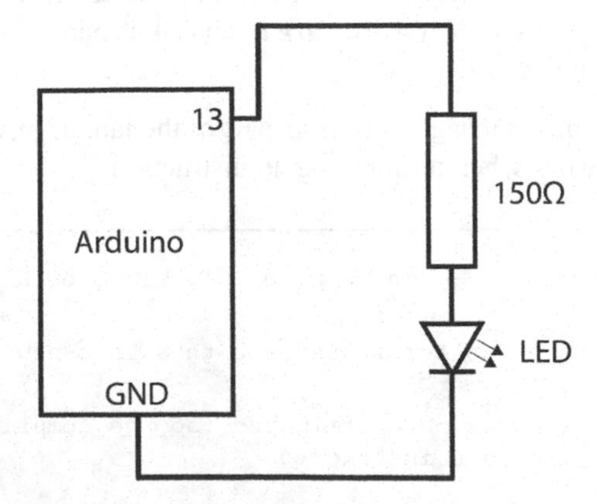

Figure 4.5 Arduino digital output schematic. ***The schematic in the diagram is very similar to the example circuit first introduced in chapter 1. The fundamental change is in the way power is supplied to the LED – this will now be provided by digital pin 13, which will be controlled by our C code instructions.***

In this schematic, the only change in the circuit is the pin providing power to the circuit. In our previous circuits, we have been using the 5V power output pin to provide power, but now we have connected the resistor/LED combination to digital pin 13 instead. This exploits the core principle of digital signals – they are **either off or on** (0 or 1). Thus, if a digital pin on the Arduino is set up for output then it can provide 5V to our circuit. In so doing, the digital pin is effectively acting as a switch to turn our LED on – the only difference is that we can now control the flow of current on each digital pin using C code. In previous chapters, step by step examples on circuit analysis and design were provided to help you learn the fundamentals of breadboard layouts. Now we can move a little faster by using what you have learned to create the final layout required in Tinkercad (Figure 4.6).

Figure 4.6 Arduino digital output breadboard layout. ***The layout shown is similar to that employed in chapters 1–3, where a resistor/LED combination is used. The power for this series loop will now be provided by digital pin 13, which replaces the 5V connection in previous examples. Tinkercad will be used to run code simulations for programming examples.***

If you have followed the examples in chapters 1–3 then the layout for this digital output example should be familiar to you. The process for building this circuit will not be covered in detail, but rather a list of connection instructions are provided for you to refer to when building the circuit for yourself (when referring to **A**rduino **D**igital pins the prefix [**AD**XX] is used):

1. Add a 150Ω resistor between pins [i1–i5].
2. Add an LED – anode pin (h5) to cathode (h6).
3. Add a connector to the negative (ground) rail (i6-GND].
4. Add a connector between the negative (ground) rail and the Arduino top left GND pin.
5. Add a connector between Arduino digital pin 13 and breadboard pin j1 [AD13–j1].

Now the circuit is connected, we can use another very powerful Tinkercad feature – coding simulation. Tinkercad allows you to write code that is then executed by a simulated Arduino Uno board, with digital outputs being assigned a voltage level of either 0V (off) or 5V (on). The coding simulator can be accessed from the menu bar on the right of the Tinkercad window (Figure 4.7).

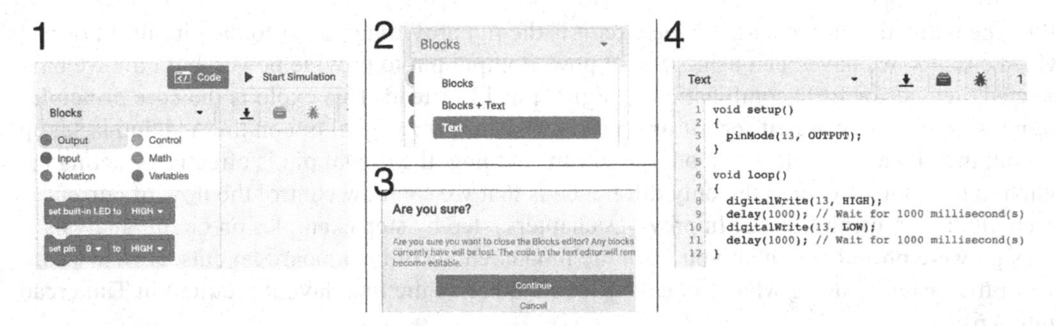

Figure 4.7 Tinkercad coding simulation window. ***The diagram shows the stages required to setup the coding editor in Tinkercad. Clicking on the Code button (1) opens the blocks window, which will not be used. The next steps involve selecting text (2) and then confirming the selection (3) in order to launch the C code editor (4).***

Although Tinkercad provides a logical blocks editor as the default option, a systems approach to learning C programming will be taken in this book so the text editor should be set as the default. It can initially seem tempting to use the blocks editor to set up code (and then convert it to text), but please **do not to use the blocks editor** as it will interfere with the method used to teach you how to write code in this book. Instead, the system diagram from Figure 4.4 can be extended to include the processes that the Arduino will execute to use a digital pin to light the LED (Figure 4.8).

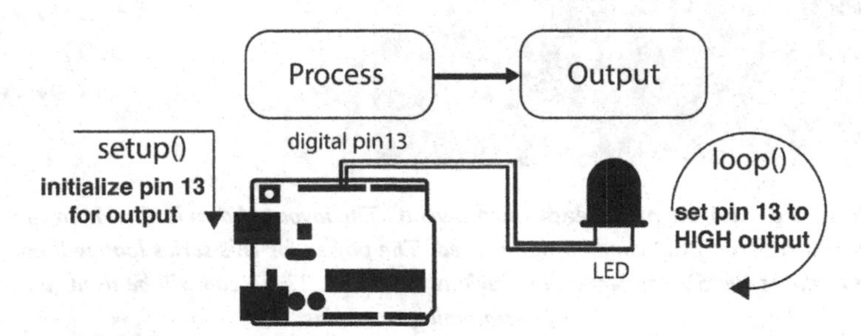

Figure 4.8 System code processes diagram. ***This diagram includes the setup() and loop() functions that are provided by the Arduino. The setup() function will be used to configure digital pin 13 for output, whilst loop() will be used to light the LED.***

With the text editor now open, you will find that Tinkercad has already provided you with some initial code to run on the Arduino. This is very useful in many respects, but this code will be edited to reduce the instructions and add some extra commenting, to take a more structured approach that more closely mimics the Arduino Integrated Development Environment (IDE) application introduced later in the chapter. In this example, replacing the default instructions provided by Tinkercad allows the inclusion of extra comments to explain what is happening in each section:

```
void setup() {
  // put your setup code here, to run once:
  //  initialize digital pin 13 for output
  pinMode(13, OUTPUT);
```

```
}//end of setup()

void loop() {
  // put your main code here, to run repeatedly:
  //set digital pin 13 to output HIGH (5V) - rather than LOW (0V)
  digitalWrite(13, HIGH);

}//end of loop()
```

The code example above contains several important structural elements that you should aim to become familiar with:

1. The **keyword** *void* is reserved by the C language to define a *function return type* – **functions are verbs**: we will look at them in section 1.6.
2. The *brackets* () hold *function input* parameters – there must **always be a pair** of opening and closing brackets.
3. The *braces* {} delineate a coding *structure* – **braces define structure** in C, and again they come in pairs!
4. The *double forward slash* // indicates a *comment* – comments are **not instructions** so they are **not executed** by the Arduino.
5. The use of *block capitals* usually indicates a **predefined** value – **constants are nouns**, data that can't change.
6. The *semicolon* defines the end of a **sequence instruction** – you must **terminate every instruction** in C.

There are two **sequence instructions** in the example above: pinMode(13, OUTPUT) and digitalWrite(13, HIGH). These instructions are actually function calls – they **execute other code** contained inside predefined functions called pinMode() and digitalWrite(). Though functions are covered in more detail in section 4.6, it is difficult to introduce sequence instructions in any practical way without using function calls to demonstrate them. More abstract mathematics calculation examples could be used, but these would not directly progress your electronics systems learning. Instead, the general forms of each of these functions can be stated to introduce their use (Figure 4.9)

Function	Description
pinMode(pinNumber, mode)	an initialization function that defines a specific digital pin (pinNumber) as being either INPUT or OUTPUT (mode)
digitalWrite(pinNumber, value)	executes a process that assigns an output voltage level to a specific digital pin (pinNumber). You can set the pin (value) to HIGH at 5V or LOW at 0V

Figure 4.9 Digital output function definitions. ***The figure shows two functions: the pinMode() function initializes a digital pin for either input or output, whilst digitalWrite() sets an output pin to 5V(HIGH) or 0V(LOW). The digitalWrite() function cannot operate on a pin that has not been initialized.***

Arduino programming requires you to setup the elements you intend to work with in advance – so you must **configure** a digital pin **before** you **use** it. These functions effectively act in tandem, where we initialize a pin and then use it in a process to create output to actuators (like our LED). This example will be used again in section 4.6 when learning about functions, but for now note that the functions setup() and loop() are predefined by the Arduino development environment to perform certain types of tasks at specific times during the lifecycle of the code (Figure 4.10).

Function	Description
setup()	runs once when the ATMega328P boots up to initialize the Arduino. This is where we set up our input/output pins and also initialize any data we may need (we introduce variables in the next section)
loop()	runs continuously while the Arduino is powered up as the main entry point for our code. This is where we execute coding processes that change input and/or outputs. The function loop() runs at a portion of the Arduino clock speed of 16MHz. In theory, this means 16,000,000 operations per second but in practice the amount of instructions in the loop() increases the time needed for each execution.

Figure 4.10 Core Arduino functions. ***These functions are used in all Arduino programs to initialize and execute processes. The setup() function runs once when the board is powered up, whilst the loop() function executes continuously while the Arduino has power.***

The setup() function, as the name implies, contains any initialization instructions needed to configure the Arduino for operation. This usually involves setting digital pins for input and/or output as required, so that they can then be used by functions like digitalWrite(). The loop() function is effectively the main entry point for Arduino code, and in some ways it can help to think of this function as the heartbeat of the microcontroller – it is continuously looping and executing code while power is available to the Arduino. Setup() and loop() will be used in all Arduino projects to configure inputs/outputs and execute our processes, and though they have not been covered in detail at this point it is important to know their overall role within Arduino development. As a first Arduino code simulation, we can run the previous example code to configure digital pin 13 for output and then write a 5V (HIGH) to this pin. The resistor/LED components are connected to pin 13, so it should now light up once the Arduino code runs. To enter the code into the Tinkercad editor, you can simply copy the text from the example above and paste it into the Tinkercad editor to run a simulation (Figure 4.11).

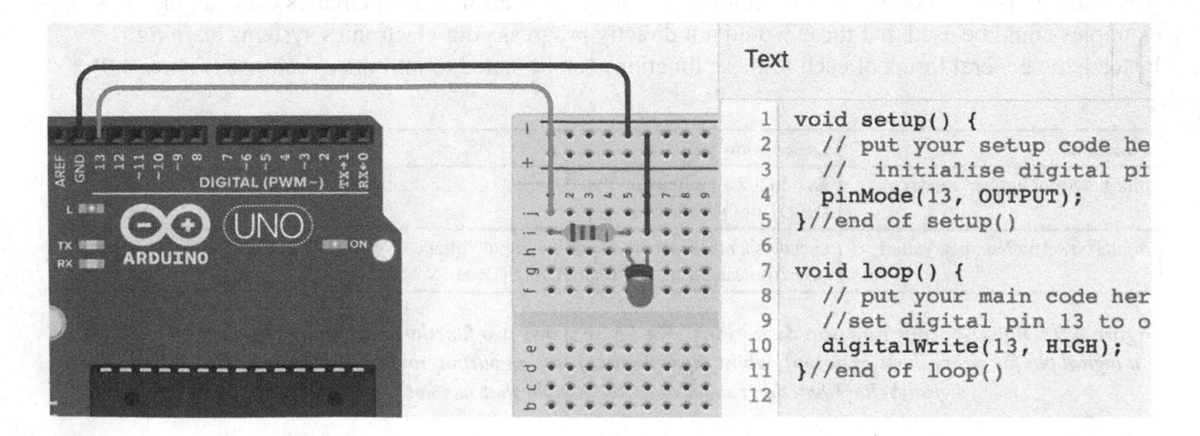

Figure 4.11 Running a code simulation in Tinkercad. ***Once we have entered our code (in this case by copying from the text provided) we can press the Start Simulation button that we also used for our previous Tinkercad circuits. If the circuit and code are correct, the LED should light up.***

This example uses a simple circuit that we have already encountered in previous chapters, to show how the Arduino can be programmed to execute processes that control that circuit. Whilst lighting an LED may not be new, we have now created a digital control point for our system that can be extended to create much more powerful circuits. Now we have covered the first (and simplest) instruction type, we can progress to learning about data storage using variables.

4.4 Data types – variables

The previous example used two sequence instructions to configure and execute an output process, and in both cases those instructions required data (e.g. pin number, output, signal level) to define them. Section 4.2 introduced the analogy of programming instructions being verbs **and data being nouns**. As in natural language, you cannot have a verb without a noun (running only makes sense if *I am* running, *you are* running, etc.) and so we will now spend some time learning about the types of data than can be used when programming the Arduino to better understand how to execute the instructions that use them. The ATMega328P microcontroller in the Arduino stores data in binary format, which is very difficult for us to read (Figure 4.12).

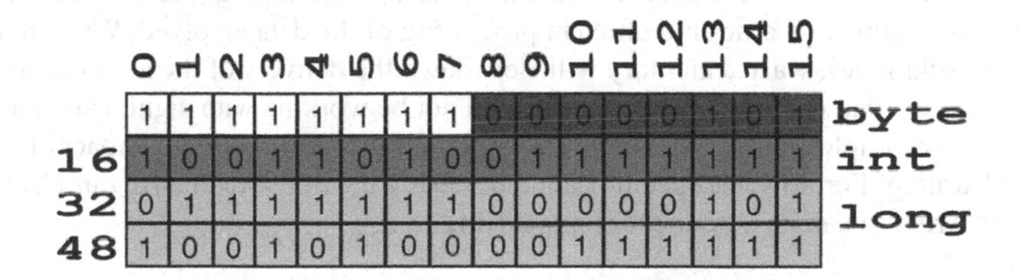

Figure 4.12 Microcontroller data memory. ***In the diagram, each memory location on the ATMega328P is either binary 0 or 1. The memory locations are grouped in multiples of an 8-bit byte (dark grey) – a 16-bit integer (grey) and a 32-bit long integer (light grey) are shown to illustrate this. Each group of memory has an address (the first bit) and can be updated over time, hence the data they contain can vary.***

In this diagram, a simplified version of microcontroller memory shows how each memory location can be either binary 0 or 1. Data values are then assigned to groups of these locations based on multiples of the 8-bit byte, where the address of the group is the location of its first data bit. The ATMega328P is an 8-bit processor, so it can only work with 8 binary bits at any one time. This limitation means that the Arduino must link larger data types like a 16-bit integer (int) and a 32-bit long integer (long) for processing in a sequence, with the final result requiring extra processing due to the need to move data around in 8-bit chunks. In addition, the Arduino has no native processing capability for floating-point numbers (numbers that contain a decimal point). This means that when working with the Arduino, data stored in bytes will be the easiest (and quickest) to process, whilst larger data types like long integers will take much more processing time. Chapter 2 noted that the Arduino is not well suited to audio data processing, which typically uses floats and doubles (higher-precision floats) to represent audio data. Thus, the 8-bit processing limitation of the ATMega328P is a significant part of the reason for using our Arduino as a microcontroller (as it was intended) rather than as an audio data processor.

Returning to our data memory diagram, we can see an 8-bit byte starting at memory address 8, with a 16-bit integer (int) starting at address 16 and a 32-bit long integer (long) starting at memory address 32. The size of the memory allocated dictates how large a data value can be stored there, and this is a crucial point for programming with microcontrollers – you must always **allocate sufficient space for your data**. If you do not allocate enough space for it, the data will overflow the storage location and cause significant problems in your code – a simple example of this would be the odometer in a car, which will eventually overflow once it runs out of storage locations (Figure 4.13).

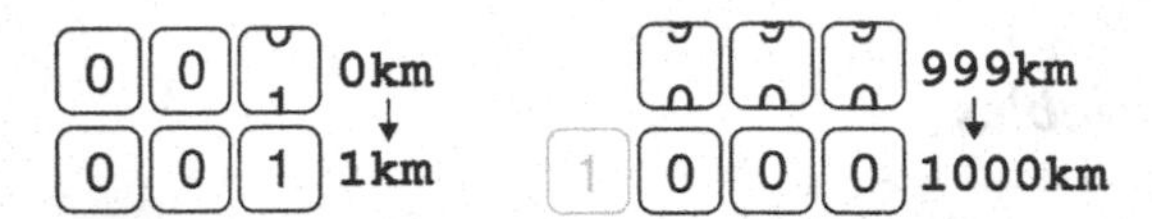

Figure 4.13 Odometer overflow example. ***In the diagram, the transition from 0 to 1km distance causes no problems for the odometer, as it is within the data range. The transition from 999 to 1000km is not possible, however, as the data range of 3 decimal digits cannot express the larger quantity.***

The odometer example illustrates the problem with data overflow in computing – if you don't have enough space for the data then you cannot store it. It would initially appear that the easiest solution to this problem is to allocate the biggest possible storage location for all data values, but in so doing significant amounts of memory are wasted. In more advanced programming (such as audio DSP), proper memory management is crucial to efficient processing of the data involved. When rendering audio using DSP techniques, wasted memory will slow down the delivery of the audio data (and potentially create leaks that can crash the code). We will not be working with significant amounts of data in our Arduino examples, but it is nevertheless important to know about the impact of data storage for your future learning. For now, the specific data types that we will use to control our electronic systems with the Arduino can be introduced (Figure 4.14).

Data type	Size	Explanation
Byte	8 bits	A whole number between 0 and 255
Int	16 bits	A signed whole number between −32768 and 32767
Unsigned int	16 bits	An unsigned whole number between 0 and 65535
Word	16 bits	The same as unsigned int – easier to read and write
Long	32 bits	A signed whole number between −2,147,483,648 and 2,147,483,647
Unsigned long	32 bits	An unsigned whole number between 0 and 4,294,967,295
Boolean	8 bits	A Boolean represents a condition – either true or false

Figure 4.14 Core Arduino data types. ***The C language defines many different fundamental data types, based on both their role and also the potential storage requirements involved. Note that the data types word and unsigned int are effectively the same for our purposes. The types listed are those that will be used when programming the Arduino. Other types (e.g. float, char, double) are more commonly used with more powerful processors, and so are omitted in this introductory text.***

In this table, the listing covers the most common data types used in this book. Other than numbers, Booleans are included (which equate to the logical values of true or false) as they are a core part of selection instructions. Selection instructions will be discussed in the next chapter when learning how to read sensor input to a system, where Boolean logic is the key to building flexible code that can respond to user input. Although the C language has a much wider set of types that include floating-point numbers (numbers with a decimal point) alongside characters and strings (numerical representations of alphabetical symbols), the Arduino is not particularly well suited to working with them. For this reason, data is focussed around the 8-bit byte, which is a whole number of sufficient size to define things like input/output pin numbers and other simple quantities. When extended data processing is needed (e.g. when reading sensor input in the next chapter) the integer (int) type will be used – a 16-bit whole number that provides more capacity to perform calculations with larger values.

Now we know more about data types, we need to learn how to use them in our code. Figure 4.12 gave a simplified representation of microcontroller memory, as in practice the allocation of memory storage is a complex and dynamic task. Although you will not need to know much detail about things like stack and heap memory for the coding work in this book, the Harvard Architecture diagram in Figure 4.1 makes a distinction between the memory needed for data and that needed to store the actual computing instructions that will tell the ATMega328P what processes to execute. This is a useful way of explaining the need for dynamic memory management – the data in memory may change over time and hence need to be moved to optimize storage.

In addition, when the output pin (13) was declared in the previous example we simply typed this as a value. This is fine for output on a single pin, but later examples will use more complex input/output configurations and it will become more difficult to keep track of pin allocations as a result. Examples will often be working with a quantity that may change over time, such as an analogue output signal that changes the pitch of a piezo speaker (which is the final project in this chapter) – how can we keep track of where this data is stored in memory, when the microcontroller may have to move it to optimize storage? The solution is to use a **variable** – a **memory location referenced by a unique name** (Figure 4.15):

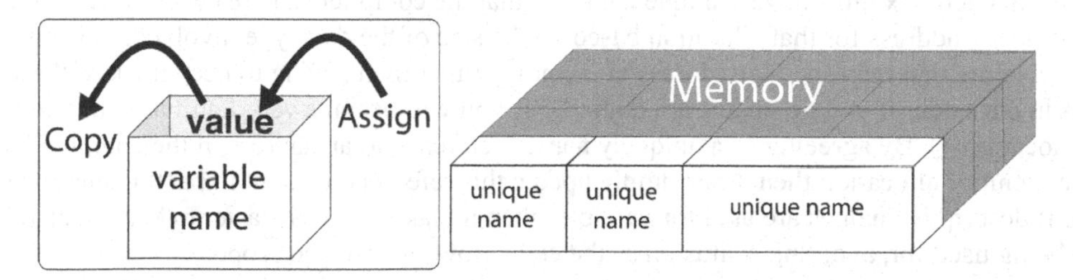

Figure 4.15 Variable storage example. ***The diagram shows a conceptual example of a variable, where its unique name references a particular storage location in memory. The analogy of a labelled box helps to show that a value can be assigned to a variable, then retrieved by copying it at a later time. Notice that the boxes on the right are of different sizes: this is no different than organizing memory storage locations.***

In this diagram, a common analogy is made between memory allocation and storing something in a labelled box. Data must be stored in boxes otherwise the processor cannot find it, thus we must first have a box ready to hold our data (this is known as **declaring a variable**). Each **box must have a unique name**, to avoid confusion between different boxes and their contents. Also, the name of the **box must not contain any spaces**, as the compiler would read each word as separate instructions. Once a box is prepared, we can store a data value to it (this is known as **assigning a value** to a variable) and because **C does not allow us to use an empty box** (mainly to keep memory use optimized) both operations must be performed at the same time. Variables are declared outside of functions to give them global scope – so they can be accessed from anywhere in the code:

```
// declare a variable to hold our output pin number 13
//a byte is more than enough storage for output pin values of 0-13
byte outputPin = 13;

void setup() {
  // put your setup code here, to run once:
    //now we can access the outputPin variable to initialize the pin
pinMode(outputPin, OUTPUT);
}//end of setup()
```

In the listing above, the **keyword byte** is used to declare a new **variable** with the **unique name** outputPin. The **value** 13 is then **assigned** to this variable using a **single equals sign**. Once a variable name is declared (remember there cannot be spaces in the name) and a value assigned to it, a copy of that value can then be retrieved for later use (this is known as accessing a variable). A new value can also be reassigned to a variable at a later point in the code, hence the name **variable** (where the data can **vary**). A crucial point to note is that the assignment operator is a single equals sign **(=), which is not the same as equivalent to**. In selection statements in the next chapter, the equivalence of data or expressions will often be tested by using two equals signs (==). It can be very confusing for new programmers to learn that a single equals sign means assignment (you are putting a value into a storage location), but you must be aware of the difference to understand what is happening whenever a variable is declared.

We know that each box must have a unique name, so that the compiler can create a reference to a specific memory address for that allocation based on the size of the data type involved. This allows us to declare a name that references an address in memory, rather than having to keep track of the specific locations in our code (if you go on to learn audio DSP, you will become very familiar with pointers to memory locations!). By agreeing on a uniquely named reference to an address, if the compiler has to move the memory allocation then it can simply update this reference without impacting our code. In addition, if descriptive names are used for variables then it makes it much easier to keep track of what they are being used for, avoiding confusion as the code grows in size and scope:

```
// you will often encounter tutorials or sample code like the following:
int myInt1 = 200;
int myInt2 = 100;

//now assign the result of adding these integers to a new integer named result
int result = myInt1 + myInt2;
//now assign the result of subtracting to a new integer named result2
int result2 = myInt2 - myInt1;
```

In the listing above, the result of performing simple arithmetic operations on two integers (declared using the **keyword int**) is assigned to two new variables (both called a variation of the word result!).

These are simple arithmetic examples, and there is nothing wrong with the code shown. Having said this, it becomes apparent that result2 tells us nothing about the contents of that variable, any more than myInt1 tells us anything descriptive about the data it holds. When you are learning to code (and even when you have a lot of coding experience) hours can be spent staring at instructions trying to figure out if something has been wrongly typed (or followed the wrong process). If the variable names from the listing above were replaced with more descriptive ones, the processes involved can be followed much more easily:

```
// declare two arbitrary x coordinates (relative to origin 0);
int beginXLocation = 200;
int endXLocation = 100;

//the result of adding these integers gives the X distance from the origin
int distanceFromOrigin = myInt1 + myInt2;

//the result of subtracting beginXLocation from endXLocation is the X distance
int distanceXTravelled = myInt1  - myInt2;
```

In this listing, the variable names show that we are working with two points in 2D space relative to an origin (assumed to be x = 0). Although the names do not enforce the data (i.e. simply calling a variable myIntegerX does not make it an integer, only the type declaration counts) they do help to give descriptive meaning to operations – even simple mathematical calculations like the addition and subtraction examples in the listing above. You may also notice that the first letter of each word in these variable names is capitalized – this is called Camel Case. This makes descriptive variable names easier to read, alongside adding comments that relate to what the instruction is doing with the data. Another aspect of variables is the use of **constants** for **values that do not change**. C allows us to use the **const** keyword to declare **read-only** variables – once assigned they **cannot be reassigned**, only copied:

```
// declare a constant byte to hold output pin number 13
const byte outputPin = 13;

// use the constant to initialize the output pin
pinMode(outputPin, OUTPUT);

//trying to reassign another value to outputPin will cause an error
outputPin = 12;
```

Constants can be used for many things in Arduino code, particularly in input/output configuration where digital pins will be connected to specific electronic components – the components will not change, so the **pin configurations shouldn't change** either. In the first code listing it was noted that block capitals usually indicate predefined constants (e.g. HIGH/LOW), and we can also define other constants for ease of reuse – though block capitals are avoided to ensure they don't either clash with an existing Arduino definition or suggest that they are part of existing Arduino libraries. To illustrate this, the second example extends the first LED light example by adding other LEDs on several output pins and flashing them on and off in sequence. Constants are used to define the output pins in advance (to match the electronics circuit), and also the **delay()** function to switch the output to the LEDs on and off over time.

4.5 Example 2 – multiple digital outputs

The second coding example extends our previous system to add multiple actuator outputs (Figure 4.16).

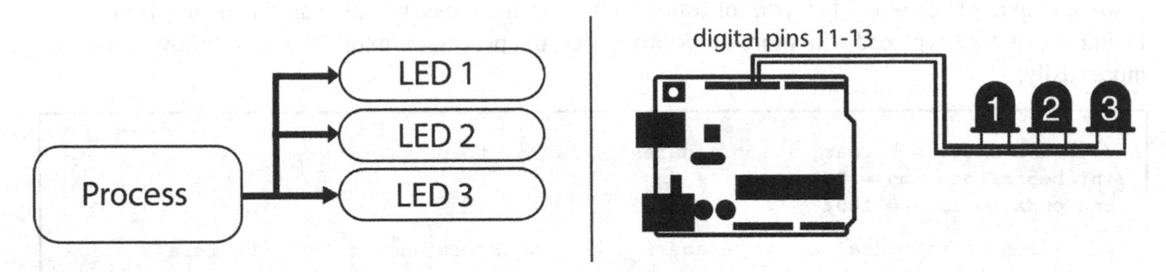

Figure 4.16 Multiple digital output system. ***In the diagram, the system from example 1 has now been extended to include three actuator outputs (LEDs). The digital pins 11–13 will be use to drive three LEDs, which the Arduino will light in sequence by using the delay() function.***

The schematic for this system is again similar to the previous example, where the resistor/LED combination is replicated across three separate output loops that are driven by digital pins 11–13 – all connected to a common ground (Figure 4.17):

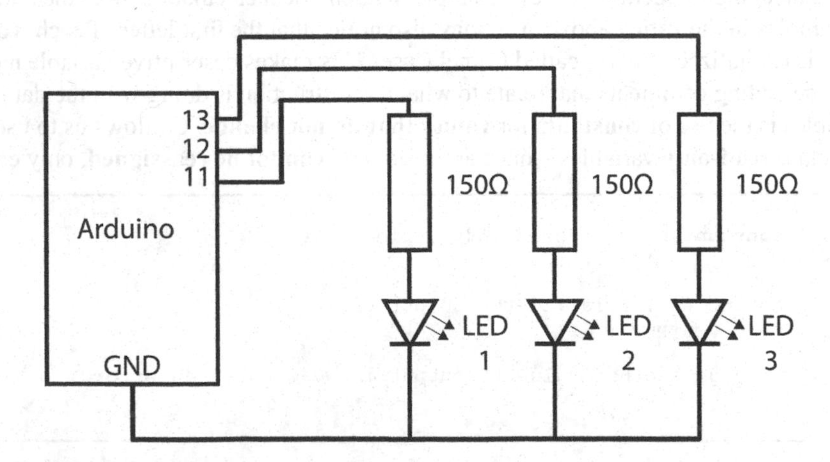

Figure 4.17 Multiple digital output schematic. ***The schematic shows three identical resistor/LED loops connected to the Arduino, which will drive them using digital pins 11–13. All three loops are connected to a common ground (GND) also provided by the Arduino.***

This schematic can be used to create a breadboard layout by replicating the resistor/LED combination from the previous example (the same combination that has also been used in previous chapter projects). A staged approach will be taken to the code in this project, to test the Arduino code by switching an LED on/off with a single loop before extending it to control the other LEDs in the circuit (Figure 4.18).

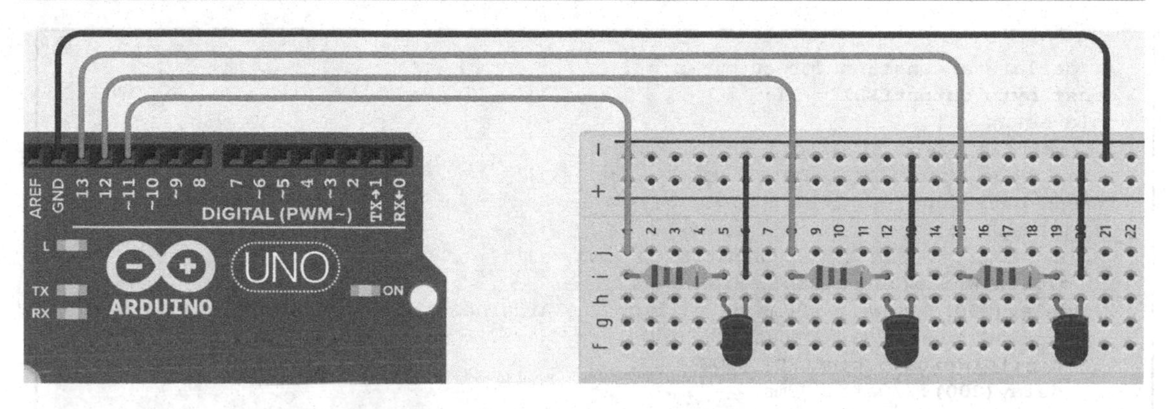

Figure 4.18 Multiple output breadboard layout. ***The layout consists of three resistor/LED loops connected to a common ground provided by the Arduino (GND). Each loop is identical to the previous, allowing the code for a single loop to be tested prior to extending it to the other loops.***

The connection instructions for stage 1 of this circuit are as follows:

Stage 1 (LED 1):

1. Add a 150Ω resistor between pins [i1–i5].
2. Add an LED – anode pin (h5) to cathode (h6).
3. Add a connector to the negative (ground) rail [i6–GND].
4. Add a connector between the negative (ground) rail (near column 21) and the Arduino top left GND pin.
5. Add a connector between Arduino digital pin 11 and breadboard pin j1 [AD11–j1].

As in the previous example, the system diagram from Figure 4.16 is extended to include the processes that the Arduino will execute to cycle a digital pin from HIGH/LOW with a delay to flash the LED (Figure 4.19):

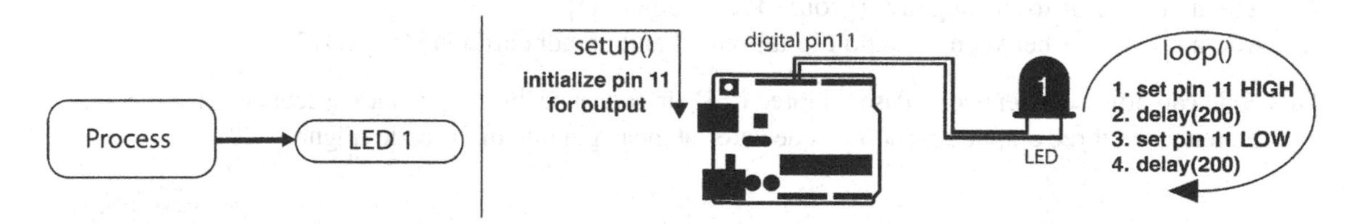

Figure 4.19 Stage 1 system code process diagram. ***In this diagram, the setup() function will be used to configure digital pin 11 for output, whilst loop() will be used to flash the LED with a delay of 200 milliseconds. The const declaration of outputPin1 is omitted for brevity.***

Now the first stage of the circuit is connected, we can take code from the previous example as a basis for adding delay() function calls to flash the LED on/off:

```
// declare a constant for output pin 11
const byte outputLED1 = 11;
void setup() {
  //initilalize digital pin 11 for output
    pinMode(outputLED1, OUTPUT);
}//end of setup()
void loop() {
  //set digital pin 11 to output HIGH (5V)
    digitalWrite(outputLED1, HIGH);
    delay(200); //wait 200ms by calling the Arduino delay() function
  //now set pin 11 to output LOW (0V)
     digitalWrite(outputLED1, LOW);
     delay(200);//wait 200ms
}//end of loop()
```

If you have entered the code correctly, when you run the Tinkercad simulator the first LED should flash on and off at 200ms intervals. If it does not flash on/off, then you may want to check the output pin declaration and also the output pin connected to stage 1 of your breadboard circuit. Once this stage of the simulation is working, you can proceed to add the second stage of two resistor/LED loops. To save time these loops will not be tested individually, but in practice it makes sense to connect each one in turn and run the code on your Arduino to avoid simple layout errors. The connection instructions for stage 2 of this circuit are as follows:

Stage 2 (LED 2 and 3):

1. Add a 150Ω resistor between pins [i8–i12].
2. Add an LED – anode pin (h12) to cathode (h13).
3. Add a connector to the negative (ground) rail [i13–GND].
4. Add a connector between Arduino digital pin 12 and breadboard pin j8 [AD12–j8].
5. Add a 150Ω resistor between pins [i15–i19].
6. Add an LED – anode pin (h19) to cathode (h20).
7. Add a connector to the negative (ground) rail (i20-GND].
8. Add a connector between Arduino digital pin 13 and breadboard pin j15 [AD13–j15].

This code can now be extended to flash all three LEDs in sequence, by simply adding sequence instructions that reference the three output pin constants declared at the beginning of the code (Figure 4.20):

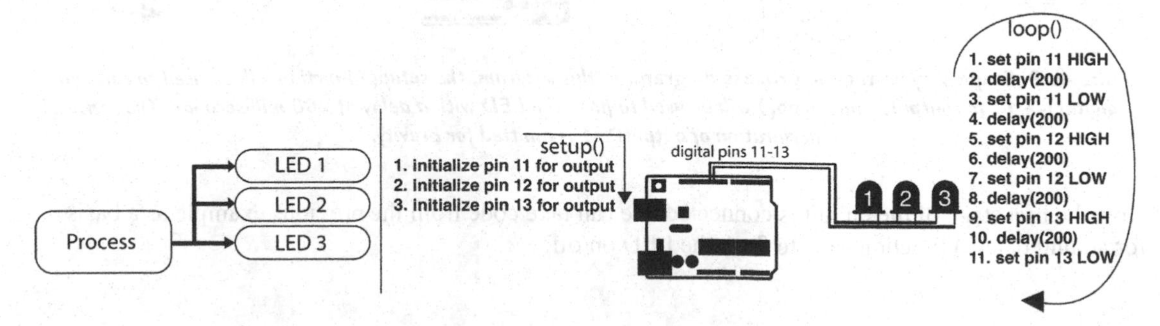

Figure 4.20 Full system code process diagram. ***In the full system process, three digital output pins are configured in setup() for output using const values. The loop() function follows the stage 1 structure of High/delay/Low/delay for all three output pins in sequence.***

The code for this process diagram is essentially a replication of the stage 1 code for three output pins, so it is a good idea to copy/paste the existing code in each section and change the names for each stage involved:

```
// declare constants to hold our output pins 11-13
const byte outputLED1 = 11;
const byte outputLED2 = 12;
const byte outputLED3 = 13;

void setup() {
  //initialize all 3 digital pins for output
  pinMode(outputLED1, OUTPUT);
  pinMode(outputLED2, OUTPUT);
  pinMode(outputLED3, OUTPUT);
}//end of setup()
void loop() {
  //set LED 1 on digital pin 11 to output HIGH (5V)
  digitalWrite(outputLED1, HIGH);
  delay(200);//wait 200ms
  digitalWrite(outputLED1, LOW);
  delay(200);//wait 200ms
  //now LED 2
  digitalWrite(outputLED2, HIGH);
  delay(200);//wait 200ms
  digitalWrite(outputLED2, LOW);
  delay(200);//wait 200ms
  //now LED 3
  digitalWrite(outputLED3, HIGH);
  delay(200);//wait 200ms
  digitalWrite(outputLED3, LOW);
  delay(200);//wait 200ms
}//end of loop()
```

If you have entered the code correctly, all three LEDs should flash on/off in sequence. This sequence should then repeat indefinitely because the code is executed within loop(), which runs for as long as the Arduino has power (Figure 4.21).

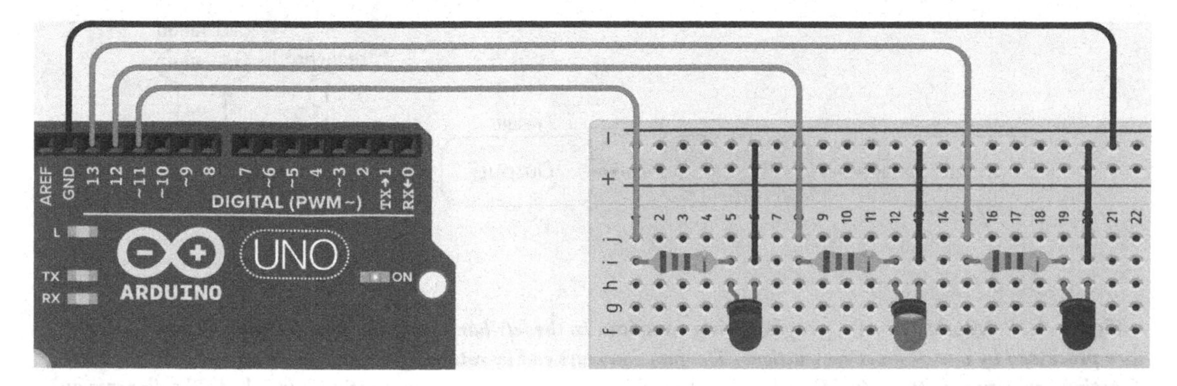

Figure 4.21 Multiple digital output simulation in Tinkercad. ***Once the code is entered you can press the Start Simulation button. If the code is entered correctly, the LEDs should flash on/off in sequence and then repeat the entire sequence indefinitely.***

This example extends the previous single LED example, showing how quickly the Arduino can provide digital control of electronic circuit components. With just three resistor/LEDs and a few lines of code we have created a sequence of flashing lights, something that would have been much more complex and taken significantly longer to build with purely analogue components. Although you are still building simple circuits, the point is to extend your knowledge with each example to work towards more interesting audio milestones.

You may already have noticed that the code listing in this example is significantly longer than that of the previous, and as the code grows so does the difficulty in following its structure (and also the potential for errors). One means of reducing complexity is to place instructions inside functions – named blocks of code that can be executed by calling them (hence function calls). The next section will look at functions as a means of reusing code instructions. You have used functions like digitalWrite() and delay() already, now you will learn more about how they work (and how to build your own).

4.6 Functions – encapsulating code

Returning to the example from section 4.3, where the setup() and loop() functions were first encountered:

```
void setup() {
  // put your setup code here, to run once:
}//end of setup()
void loop() {
  // put your main code here, to run repeatedly:
}//end of loop()
```

Looking at this code, the first thing to notice is the snippet `void setup()` – this is the function **declaration**. In the same way as a variable or constant, **you must declare a function before you use it**. A function behaves very much like the input/process/output system from Figure 4.2, where variables (known as parameters) can be inputs to a function, instructions in the function process the data and the function can then return a result as an output (Figure 4.22).

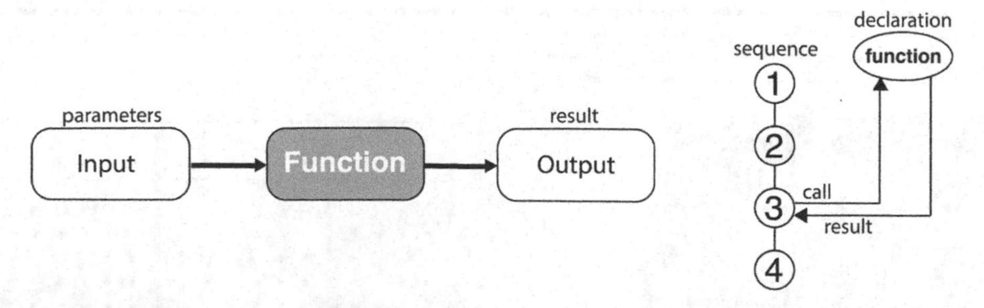

Figure 4.22 Workflow of a programming function. *In the left-hand diagram, function inputs (parameters) are processed by one or more instructions. The function must end by returning a result as an output. Although the function can return nothing (void) and does not need inputs, it must be declared before it is called. The diagram on the right shows how a function must be declared before use, where the function call is a sequence instruction that receives the returned result.*

Functions process parameters to output results, they behave like systems within code. A function declaration begins with the return type (which can be void if nothing is returned) followed by the unique name of the function. A function also uses braces ({}) to encapsulate the instructions it will execute:

```
void anExampleFunction() {

  // any instructions between the opening and closing braces are considered part
 //of the function
  // 1. this function has no input parameters (declared inside the brackets)
  // 2. this function has a void return type so it does not return a result

}//end of anExampleFunction()
```

The **return type void** is very common in programming – you may have heard the phrase '*null and void*' that comes from financial transactions in banking, and it may help to remember that a **void function has a null return**. Other return types can also be used to return a variable as a function result (discussed in the next chapter). The function structure in the code listing above is similar to that of the setup() and loop() functions we have already used. These functions do not have any inputs, primarily because setup() is only called by the ATMega328P when power is provided to the Arduino, and loop() is called repeatedly until power is disconnected – thus they do not need any other information before they execute.

Figure 4.9 introduced functions that take **input parameters**, such as pinMode(pinNumber, mode) and digitalWrite(pinNumber, value). These parameters provide data that the function needs in order to execute its instructions. In the case of pinMode() the function requires both the specific pin and whether it is INPUT/OUTPUT in order to set up that pin correctly, for digitalWrite() the function again needs to know the specific pin for output and also whether it should set a 5V(HIGH) or 0V(LOW) value to that pin. In general terms, **input parameters are declared inside the brackets** that come after the function name (and before the braces that encapsulate the instructions of the function):

```
void parameterFunction(byte param1, int param2) {

  // input parameters can be accessed by using the names param1 and param2
  // notice that the type of the parameter must be specified in the declaration
  // you may recognize the following function call - it now uses param1
  pinMode(param1, OUTPUT);
  // for brevity param2 isn't used in this short example
}//end of anExampleFunction()
```

The listing above may initially seem confusing, as a function is being called from inside another function! In practice, this is no different than calling pinMode() from the loop() function in our previous examples. This helps to highlight the power of functions – a set of instructions can be encapsulated and executed as often as required simply by **calling their unique function name**. The additional benefit of encapsulation is **obfuscation** – we do not need to know how pinMode() works in order to use it. This is probably the most powerful aspect of functions, where other programmers can provide us with code **libraries** containing **functions** that we do not necessarily need to understand or even read – we only need to know how to use them based on their input parameters and return values.[2]

Now we have looked at the three core elements of a function, we can return to our previous example to consider how a function could be used to simplify our code. Functions are often used to reduce repetition in code, where repeating a set of similar instructions is not only more difficult to read but can also introduce errors when writing them. For these reasons, we can look at the stage 1 code from the previous example to see how it is repeated in the other stages of the flashing LED code:

```
//…truncating the loop() function to show relevant code
  digitalWrite(XX, HIGH); //set digital pin XX to output HIGH (5V)
  delay(200); //wait 200ms
  digitalWrite(XX, LOW); //set digital pin XX to output LOW (0V)
  delay(200); //wait 200ms
```

From this code listing, we can see the only thing that actually changes in these four lines of code is the digital output pin number (denoted by **XX** in the listing). Thus, we can now construct a function to execute these lines of code using an input parameter to specify the pin number required:

```
void flashOutputPin(byte outputPinNumber) {

  digitalWrite(outputPinNumber, HIGH); //set pin outputPinNumber to HIGH (5V)
  delay(200);//wait 200ms
  digitalWrite(outputPinNumber, LOW); //set pin outputPinNumber to LOW (0V)
  delay(200);//wait 200ms
}//end of flashOutputPin()
```

This function carries out the same processes as our previous code, but the structure now makes it significantly easier to use. We can now call this function whenever we need it, allowing us to work with multiple output pins by using a single sequence instruction to call flashOutputPin():

```
//we can call flashOutputPin() by passing an output pin number
// as an input parameter to the function - this would flash pin 12
flashOutputPin(12);
// this would flash pin 10
flashOutputPin(10);
```

This section began by stating that **you must declare a function before you use it**, so the process of building a function must happen first in the code before it is called. By encapsulating code, we can significantly reduce our programs in terms of writing time and also potential errors. In the next example, the previous code from example 2 will be replaced with function calls to `flashOutputPin()` to reduce the code involved. In so doing, we are performing the same process as calling functions like pinMode() or digitalWrite() in previous examples, only now we have constructed the function ourselves. Before you do this, a short tutorial is provided on how to write basic code where some of the fundamental elements you need to practise will be discussed to help you avoid making simple mistakes. To help you get the most out of the tutorial, it is recommended that you **read over this section again**, as it has introduced a fundamental programming concept in a very short space of time. The goal here is not to make you expert programmers, but rather to introduce the fundamental concepts needed to program an Arduino for electronic circuit control. Learning programming (like electronics) take time and practice, but being able to read code is the first step in learning to write your own.

4.7 Tutorial – how to write code part I

So far, this chapter has covered the basics of variables and functions and how they operate. Code listings are provided for all examples so that you can simply copy/paste (recommended!) into the editor to minimize errors, but this only helps you to learn how to **read** code. The next step is to learn how to **write** your own code, and there are several simple principles that can help avoid significant time spent in errors and debugging later in the development process (some debugging strategies will be introduced in the example project). When you are coding, the tendency is often to focus on the outcome of the code and in so doing details can be omitted or skipped to get 'closer to the goal'. This can lead to significant errors in your code, and also require a large amount of debugging to track them down. The first thing to remember about computer programming is that it is procedural and structural – **it is not linear text** as you are familiar with from natural language. One of the biggest barriers to learning how to write code effectively is trying to write code linearly from beginning to end (as you would natural language) rather than **building the structures** first, before populating them with instructions:

```
void anExampleFunction () {

  // if you begin by setting up the structure of the function then you avoid
  // the risk of missing a parameter bracket or encapsulating brace
  // in the declaration

}//end of anExampleFunction ()
```

In this listing, a **function stub** has been built – though there are no instructions to execute it has a valid structure, which means it can now be called to test the function itself (as shown in Figure 4.19). This is the first recommended step to take when working with functions – declare the structure of the function first, and then test it by compiling and calling the function to make sure it works ok. Comments have been added to explain what the code is doing, and this is very important for longer code to avoid forgetting what you originally intended a function or code block to do. **Adding a comment after the closing brace** is also recommended to indicate where the function ends. This can often seem like extra effort (particularly in larger code examples) but it is designed to avoid one of the most common errors in programming – deleting a brace or adding an extra one that changes the entire structure of the code. To illustrate this, look at the following example:

```
void setup() {
  // put your setup code here, to run once:
  pinMode(13, OUTPUT);
}//end of setup()

//now with an extra brace!
void setup() {
  // put your setup code here, to run once:
  pinMode(13, OUTPUT);
}//actual end of setup!
}//end of setup()
```

This may seem quite obvious (partly because the line is commented!) but stray braces are often the result of copying and pasting incorrectly, where a brace is either omitted from the copy or added to the paste. It is very easy to make this mistake, even as an experienced coder, and particularly when you are tired after working on a project for a long time (or learning to code as a new skill). Another simple technique that is often overlooked is to copy/paste wherever possible to avoid syntax errors. Although you must always double check to ensure that you have not inadvertently selected the wrong code (e.g. omitting/including an extra brace in the selection), by using copy/paste you avoid typing the wrong names for variables or constants later in your code:

```
// declare a constants to set up an output pin
const byte outputLED1 = 11;

//if you copy/paste this for use in the code, then you will get:
pinMode(outputLED1, OUTPUT);

//if you type it yourself, it is very easy to write something like:
pinMode(outputLed1, OUTPUT);
```

These basic pointers are designed to help you avoid some of the most common coding errors, which can hinder your learning (and often impact your confidence). A second coding tutorial is provided in the next chapter, but for now you can use what you have learned on functions (and how to write them) in the next example to reduce and reuse the code involved.

4.8 Example 3 – reusing code with functions

The previous code listing declared a function that encapsulated the instructions needed to flash an LED – now we can call this function `flashOutputPin()` for any output pin whenever we need it (Figure 4.23):

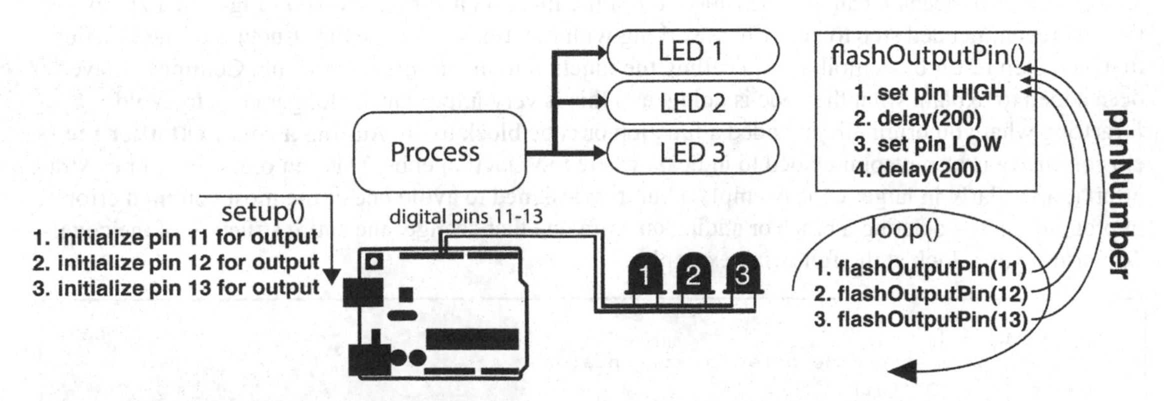

Figure 4.23 System function code process diagram. ***In this system, the loop() function makes three calls to the flashOutputPin() function, passing the input parameter pinNumber to define the relevant output pin for each LED. This function carries out the instructions to flash an LED on a specific output pin, as defined by the function call.***

In this example, we have a total of three LEDs so we will call the `flashOutputPin()` function three times. To do this, go to your previous Tinkercad example project and **create a duplicate** of it (click the

options rollover for your existing circuit in the **circuits browser window** and select duplicate). In the new duplicate project, **delete all existing code** in the editor window and copy/paste the following code to replace it:

```
// declare constants to hold our output pins 11-13
const byte outputLED1 = 11;
const byte outputLED2 = 12;
const byte outputLED3 = 13;

void setup() {

  //initialize all 3 digital pins for output
  pinMode(outputLED1, OUTPUT);
  pinMode(outputLED2, OUTPUT);
  pinMode(outputLED3, OUTPUT);

}//end of setup()

//we must declare our flashOutputPin() function before we can call it in our code
void flashOutputPin(byte outputPinNumber){

digitalWrite(outputPinNumber, HIGH); //set outputPinNumber to 5V (HIGH)
delay(200); //wait 200ms
digitalWrite(outputPinNumber, LOW); //set outputPinNumber to 0V (LOW)
delay(200); //wait 200ms

}//end of flashOutputPin()

void loop() {

  //now we can call flashOutputPin() passing each output pin number in turn
  // as an input parameter to the function, starting with the LED 1 pin
  flashOutputPin(outputLED1);
  //now LED 2
  flashOutputPin(outputLED2);
  //now LED 3
  flashOutputPin(outputLED3);

}//end of loop()
```

If you have copied and pasted the code correctly, the simulation should run in the same way as the previous example, where all three LEDs light flash on/off in sequence. The new code listing replaces the 12 lines of instructions in the loop() function with three sequence instructions that make function calls to flashOutputPin(). This makes our loop() function much easier to read, and also means we are less likely to make errors typing when typing out similar code repeatedly. In addition, if we wanted to add more LEDs to our circuit then each new resistor/LED branch on our breadboard could be controlled by a single additional line of code – this allows us to scale our systems much more effectively. Now we have looked at using functions to reduce our code, we can consider how other forms of output could be added to the current LED circuit. As this is an audio electronics book, it is important that we begin to work with audio projects. Whilst LED circuits are very useful for rapid prototyping, they do not teach us the practicalities of working with audio transducers. To do so, we will now turn our attention to using the Arduino for simple audio output.

4.9 Analogue output – pulse width modulation

So far, our circuits have used digital output pins with delay() timer calls to flash LEDs. This notion of switching an output between digital HIGH(5V) and LOW(0V) is very useful, as we can extend it to employ a technique called pulse width modulation (PWM). PWM exploits the proportion of time that an output signal is either on or off to change the actual output voltage of that signal. Although this may seem a little confusing at first, this section will work through both the concept and its electronics proof to show how the Arduino can produce different output voltage levels. Examples of PWM can be found in many areas of audio, such as subtractive synthesis (you may have heard this in older video games) and also in Class D amplifiers that are used in devices like mobile phones, hearing aids and other portable speakers due to their linear response and low cost.

Using PWM is a straightforward way of obtaining varying output voltage from a digital system that works on specific levels – for the Arduino, this is +5V and 0V (GND). In practice, a modulated wave is not the same as a steady-state voltage level and so extra operations (or alternative methods) must be used. For now, the concept is introduced as a means of explaining how the Arduino tone() function works, which also leads into our study of AC circuits in later chapters. Pulse width modulation works by varying the amount of time a digital output signal is on or off – this change between on/off is known as the duty cycle (Figure 4.24):

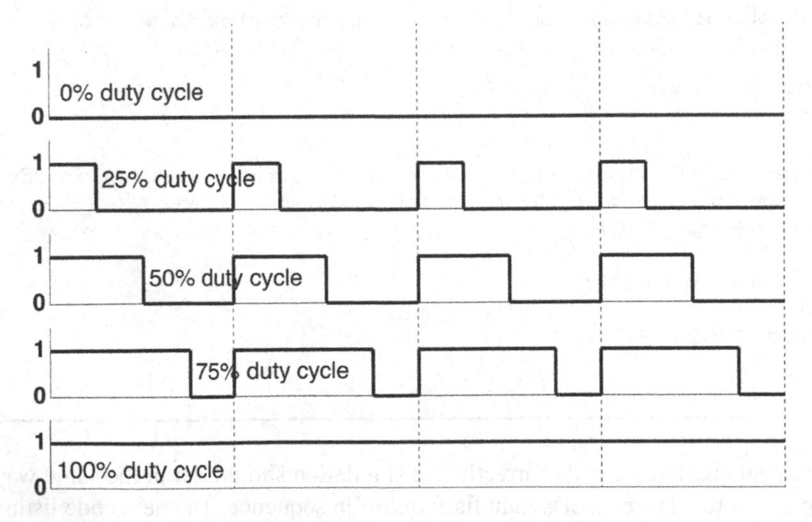

Figure 4.24 PWM duty cycle. ***In the diagram, the duty cycle of a digital output pin is shown as a series of percentage values. Each percentage holds the pin at 5V (HIGH) for longer, which increases the average voltage output as a result.***

In the diagram, increasing the duty cycle means that the output voltage stays at 5V(HIGH) for longer, which changes the overall electrical power in the circuit – this is a new concept. In chapter 3, we learned about Ohm's Law as the fundamental equation used in DC circuit analysis:

$$\textit{Voltage}, V = IR \tag{4.1}$$

V is the potential difference in volts (V)
I *is the current in amperes (A)*
R *is the resistance in ohms (Ω)*

The principle of this equation lies in the relationship between voltage and current, where an increase in current will increase the voltage in a circuit (assuming the overall resistance does not change). Thus, Ohm's Law shows that voltage and current are proportional to one another, which leads to another fundamental equation, for electrical power:

$$\textbf{\textit{Power}}, \boldsymbol{P = IV} \tag{4.2}$$

P is the power in watts (W)
V is the potential difference in volts (V)
I is the current in amperes (A)

Electrical power defines the total **work done** to move a charge past a specific point over time, where we recall that current is already defined in amperes (charge in coulombs per second) – now we can combine this with the potential for charge to move (voltage). Thus, the total work done is defined as the potential difference to move a charge multiplied by the charge flowing past a specific point – this is the power required by a circuit. Although power electronics are not used in this book (as the circuits all involve small signals) it is nevertheless highly important when working with power amplifiers and loudspeakers. You may have encountered an audio amplifier for guitar or keyboard that is rated in watts, which effectively denotes how much work is done by the circuit over a specific time.

This equation can be used to explain how varying the duty cycle of a digital pin changes the average power of a signal, and hence reduces the output voltage it creates. The key to this relationship lies in understanding that the current flowing through an Arduino output pin is constant (maximum 40mA) and so the **only thing that can vary is voltage**. If the voltage is 0V(LOW) for a certain amount of time, then the power at that time is also 0W (because Power, $P = IV = \mathbf{I} \times 0$). Thus, the **average** voltage varies with the average power because current is constant. We can use this simple relationship to determine the average voltage for specific duty cycles:

For an output voltage of 5V (duty cycle of **100%**,) and a maximum pin current of 40mA the power is:

$$V_{HIGH} = 5\text{V}, \quad I = 0.04\text{A}$$

$$\text{High Power, } P_{HIGH} = IV_{HIGH} = 5 \times 0.04 = \mathbf{0.2W}$$

For an output voltage of 0V (duty cycle of **0%**,) the power is:

$$V_{LOW} = 0\text{V}, \quad I = 0.04\text{A}$$

$$\text{Low Power, } P_{LOW} = IV_{LOW} = 0 \times 0.04 = \mathbf{0W}$$

Now for a duty cycle of **75%**, the average power will be an equal combination of the P_{HIGH} and P_{LOW} values:

$$\text{Average Power, } P_{AVG} = 0.75P_{HIGH} + 0.25P_{LOW} = (0.75 \times 0.2) + (0.25 \times 0) = 0.1 + 0 = 0.15\text{W}$$

From this average power value, we can now calculate the **average voltage** for a 75% duty cycle:

$$P_{AVG} = 0.15W, \quad I = 0.04A$$

$$P_{AVG} = IV_{AVG}$$

$$\therefore V_{AVG} = \frac{P_{AVG}}{I} = \frac{0.15}{0.04} = \mathbf{3.75V}$$

Note: the same value can be arrived at by simply multiplying V_{HIGH} by the duty cycle.

For a duty cycle of **50%**:

$$\text{Average Power, } P_{AVG} = 0.5P_{HIGH} + 0.5P_{LOW} = (0.5 \times 0.2) + (0.5 \times 0) = 0.1 + 0 = 0.1W$$

$$V_{AVG} = \frac{P_{AVG}}{I} = \frac{0.1}{0.04} = \mathbf{2.5V}$$

For a duty cycle of **25%**:

$$\text{Average Power, } P_{AVG} = 0.25P_{HIGH} + 0.75P_{LOW} = (0.25 \times 0.2) + (0.75 \times 0) = 0.1 + 0 = 0.05W$$

$$V_{AVG} = \frac{P_{AVG}}{I} = \frac{0.05}{0.04} = \mathbf{1.25V}$$

Although we could have simply multiplied the maximum voltage by the duty cycle, the proof is included both to introduce the concept of electrical power and also to introduce the concept of **varying voltage** within a circuit. PWM can vary an output voltage, but this technique can also be used to change the **frequency** of the output signal. The final example project uses a simple form of PWM based on the Arduino tone() function to drive a piezo loudspeaker as output. In later chapters, we will look at alternating current (AC) signals – signals that can change over time. AC circuit analysis forms the basis of all audio electronics, allowing us to work with more interesting circuits like filters. For now, we can say that the Arduino can approximate different output voltages using PWM, where the **output voltage varies over time**.

4.10 Example project – automatic tone player

The Arduino **tone()** function provides a simple form of PWM, where the duty cycle is varied to approximate specific output frequencies. In this instance, we are not focussing on the average output voltage of the pin, but rather the **frequency** that the signal varies at. We will discuss audio frequency in more detail in later chapters when we learn about AC circuits, but you may already be familiar with the fundamental concept of the frequency of a sine wave being perceived as a tuned pitch by the human ear (Figure 4.25).

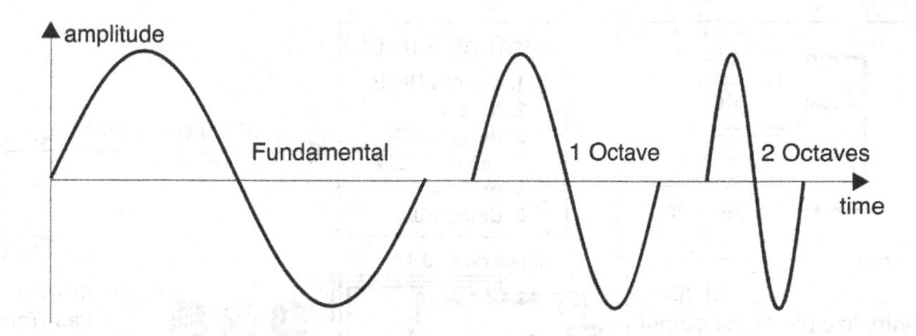

Figure 4.25 Pitch perception by the human ear. ***The human hearing mechanism is incredibly complex, but at its simplest level changes in sound pressure are converted to sound by our ears. When these changes in pressure are periodic they have a specific frequency, they are perceived as a sound of a certain pitch. As the frequency of a sound increases, so does its perceived pitch.***

Frequency is measured in cycles per second (hertz), where each complete sine wave in the diagram in Figure 4.25 represents a single cycle of that wave – periodic waves repeat over time. Higher frequencies like octaves occur when the number of cycles per second doubles, and the diagram shows the relationship between a fundamental frequency and its two nearest upper octaves. Western music defines musical pitches based around a concert pitch of A4 (the A note above middle C) of 440Hz, and this can be used to derive a small set of frequencies that relate to specific pitches in this example. To do this, we can use the Arduino tone() function that can vary the duty cycle on a digital output pin to create periodic waves at specific frequencies. The **tone()** function call specifies **parameters for pin number and frequency** (also optional duration):

```
//example Arduino function calls for tone(pin, frequency)
//play a tone at pitch A4 - concert pitch, A above middle C
tone(13, 440);
//play a tone at pitch C4 - middle C
tone(13, 262);
//play a tone at pitch C5 - 1 octave above middle C
tone(13, 523);
```

We can combine the tone() function to output musical pitches with our existing flashing LED circuit from example 3 by adding a piezo loudspeaker to the circuit. In so doing, we can emit sounds on one output pin (connected to the piezo speaker) whilst also flashing the LEDs on other output pins. To do this, we must also modify the existing `flashOutputPin(byte outputPinNumber)` function to accept **another input parameter** – in this case a note frequency for output to the piezo. The system diagram for this project is also an extended version of the previous system diagram for example 3, where we now add a piezo loudspeaker as an additional output to our system (Figure 4.26).

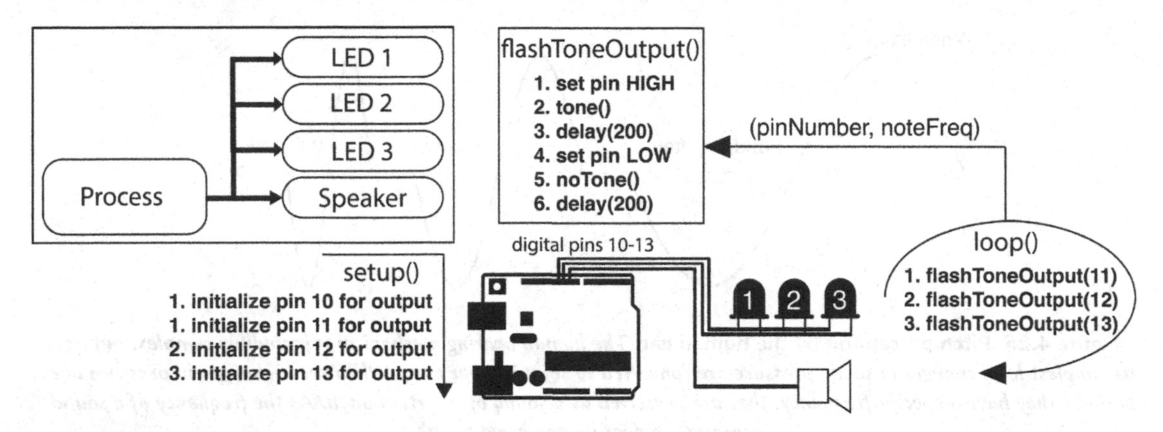

Figure 4.26 Example project system code process diagram. ***In this system, the loop() function makes three calls to the flashOutputPin() function that takes pinNumber and noteFreq parameters, which must be declared first before use. This function now also executes tone() and noTone() calls to send specific frequencies to the output pin connected to a piezo loudspeaker.***

This diagram shows how our system can now be expanded to drive an additional loudspeaker output. This also means we need to update our circuit schematic to reflect the additional loudspeaker component to be added on pin 13 (Figure 4.27).

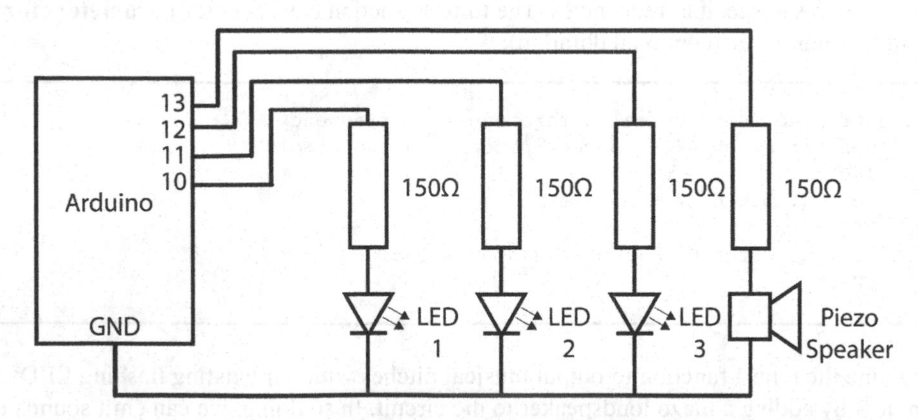

Figure 4.27 Final project schematic. ***The schematic shows four loops, including the three resistor/LED loops connected to digital pins 10–12 from the previous example circuit. The fourth loop connects a piezo loudspeaker to pin 13, which also has a 150Ω resistor connected to protect the loudspeaker.***

The schematic in Figure 4.27 extends the previous project (Figure 4.17) to include a piezoelectric loudspeaker. As with the LEDs, a 150Ω resistor is added in series with the loudspeaker to limit the current that can flow through it. Although the Arduino can output a maximum of 40mA on a single pin (and ~200mA) overall, we do not want to reach this limit at any point – to avoid burning out the Arduino and/or the loudspeaker. With a 40mA limit, we could have used a value of 125Ω

($I = \frac{V}{R} = \frac{5}{125} = 40mA$) to reach the required value, but for practicality 150Ω ($I = \frac{V}{R} = \frac{5}{150} \approx 33mA$) is used to allow you to keep using the same resistors as before (the introduction notes that the 220Ω resistors can also be used, as these are required for the MIDI circuits in chapter 5).

The output pins must be reconfigured to accommodate the loudspeaker on our breadboard, but this is a relatively simple matter because we declared constant values for each pin at the beginning of our previous code example. We will also need to redefine the `flashOutputPin()` function to accept more than one input parameter so we can pass in the frequency needed to call the tone() function. This requires one important piece of information – what type of variable does tone() use to define frequency? The answer to this question is found in the Arduino language reference,[3] which is more commonly known as an application programmer interface (API):

> *... from the Arduino API reference:*
>
> **Syntax**
>
> `tone(pin, frequency)`
>
> **Parameters**
>
> `pin`: the pin on which to generate the tone
>
> `frequency`: the frequency of the tone in hertz - `unsigned int`

The API tells us that the tone() function requires an unsigned integer for frequency, which means we will need to use this type of value as the input parameter to the updated function. This is a very common task in programming – finding out what other functions require and configuring your own code to work with those requirements. As you progress your programming work, you will become used to scanning various APIs (each library of code should have one!) to find an explanation of how to use a particular function or data structure – the concept is introduced now to highlight the correct procedure for working with other functions quickly and accurately. The behaviour of the function has changed, so we should also rename the `flashOutputPin()` function to better describe what it now does – this is a common problem in larger code projects, where a function gets repurposed but does not now have a name that matches its behaviour:

```
//we must declare our flashOutputPin() function before we can call it in our code
void flashToneOutput(byte outputPinNumber, unsigned int noteFreq){
  digitalWrite(outputPinNumber, HIGH); //set outputPinNumber to 5V (HIGH)
  tone(outputSpeaker, noteFreq);
  delay(200); //wait 200ms
  digitalWrite(outputPinNumber, LOW); //set outputPinNumber to 0V (LOW)
  noTone(outputSpeaker);
  delay(200); //wait 200ms
}//end of flashToneOutput()
```

In the listing, we have now added an unsigned integer called noteFreq to our input parameters, and also renamed the function to `flashToneOutput()` to update the description of its behaviour. The noteFreq parameter is used in the tone() function call to specify the output pitch that the Arduino will send on the pin defined as outputSpeaker (13). This means we also need to update the three function calls to use the new function name and structure, but first we must redefine the output pin numbers and also define the frequencies we intend to use in our example:

```
// declare constants to hold our output pins 10-13
//10-12 are LEDs, 13 is a piezo loudspeaker
const byte outputLED1 = 10;
const byte outputLED2 = 11;
const byte outputLED3 = 12;
const byte outputSpeaker = 13;

//also declare constants for the note frequencies used
const unsigned int noteC4 = 262;
const unsigned int noteD4 = 294;
const unsigned int noteE4 = 330;
```

This listing shows how we have reconfigured the output pins to connect a piezo loudspeaker on pin 13 – this is a simple task because we only need to redefine them once, which highlights the power of variables and constants within our code. We have also defined unsigned integer constants for the pitches C4 (262Hz), D4 (294Hz) and E4 (330Hz) that we will pass into the tone() function calls later in the code. Next, we must reconfigure the output pins in the setup() function, to reflect the changes we have made to the system:

```
void setup() {
  //initilalize all 4 digital pins for output
pinMode(outputLED1, OUTPUT);
pinMode(outputLED2, OUTPUT);
pinMode(outputLED3, OUTPUT);
pinMode(outputSpeaker, OUTPUT);
}//end of setup()
```

Now we have redefined our pin numbers, declared unsigned integer constants for our frequencies and initialized our output pins, we can edit the loop() function to update our calls to the `flashToneOutput()` function:

```
void loop() {
  //now we can call flashOutputPin() with both an LED pin
  //and also a note frequency for tone() to play
  flashToneOutput(outputLED1, noteC4);
  //now LED 2
  flashToneOutput(outputLED2, noteD4);
  //now LED 3
  flashToneOutput(outputLED3, noteE4);
}//end of loop()
```

The function call now takes a pin number and a note frequency as inputs, using them to light LEDs and call tone() to emit pitches on the piezo loudspeaker that we will now connect to pin 13. The updated breadboard layout rearranges the output pins to make it easier to set out the components, but in practice most piezo loudspeakers that come with Arduino hobby kits are not as big as the simulator object in Tinkercad (Figure 4.28).

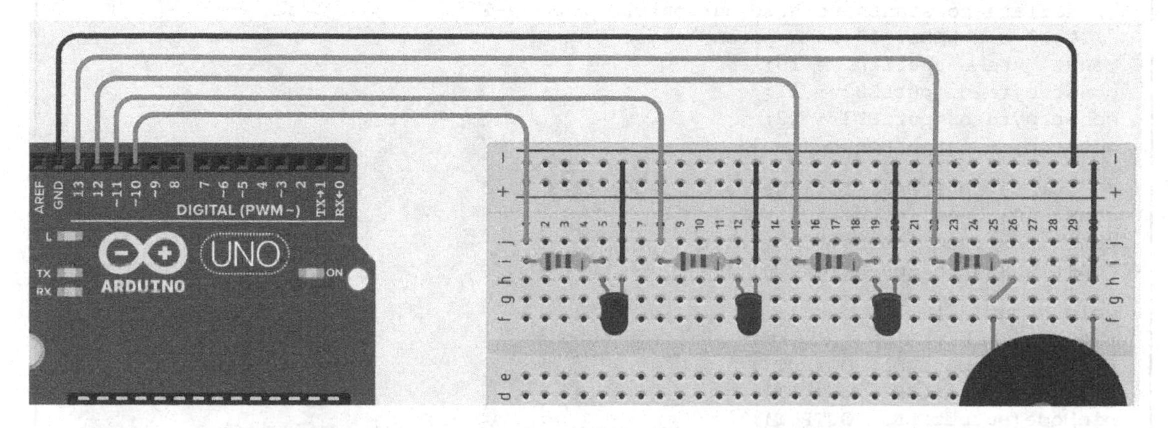

Figure 4.28 Final project breadboard layout. ***The output pins have been rearranged from the previous circuit, to allow signal connections to be aligned for a neater visual layout. The LEDs are connected on pins 10–12, while the piezo loudspeaker is connected on pin 13 (the piezo image is truncated due to size). Note the bridging wire from [g25–h26] to allow the resistor/piezo series combination to be fitted into the top half of the board. The final circuit implementation will differ slightly, as most piezo loudspeaker components are much smaller than the Tinkercad object.***

Now we have a full breadboard layout, we can simulate the project in Tinkercad before transferring it to an actual breadboard. To make this process more straightforward, a full list of pin connections is provided alongside a full code listing – allowing you to copy/paste the code into your own Tinkercad example.

Tinkercad circuit connections

The connection instructions for this project circuit are as follows:

1. Add a 150Ω resistor between pins [i1–i15].
2. Add an LED – Anode pin (h5) to cathode (h6).
3. Add a connector to the negative (ground) rail [i6–GND].
4. Add a connector between Arduino digital pin 10 and breadboard pin j1 [AD10–j1].
5. Add a 150Ω resistor between pins [i8–i12].
6. Add an LED – anode pin (h12) to cathode (h13).
7. Add a connector to the negative (ground) rail [i13–GND].
8. Add a connector between Arduino digital pin 11 and breadboard pin j8 [AD11–j8].
9. Add a 150Ω resistor between pins [i15–i19].
10. Add an LED – anode pin (h19) to cathode (h20).
11. Add a connector to the negative (ground) rail [i20–GND].
12. Add a connector between Arduino digital pin 12 and breadboard pin j15 [AD12–j15].
13. Add a 150Ω resistor between pins [i122–i126].
14. Add a bridging wire (resistor and speaker) between pins [g25–h26].
15. Add a piezo loudspeaker between pins [g25–g30].
16. Add a connector to the negative (ground) rail [h30–GND].
17. Add a connector between Arduino digital pin 13 and breadboard pin j22 [AD13–j22].
18. Add a connector between the negative (ground) rail (near column 30) and the Arduino top left GND pin.

Full code listing

```
// declare constants to hold our output pins 10-13
//10-12 are LEDs, 13 is a piezo loudspeaker
const byte outputLED1 = 10;
const byte outputLED2 = 11;
const byte outputLED3 = 12;
const byte outputSpeaker = 13;

//declare constants for the note frequencies used
const unsigned int noteC4 = 262;
const unsigned int noteD4 = 294;
const unsigned int noteE4 = 330;

void setup() {
  //initilalize all 4 digital pins for output
pinMode(outputLED1, OUTPUT);
pinMode(outputLED2, OUTPUT);
pinMode(outputLED3, OUTPUT);
pinMode(outputSpeaker, OUTPUT);
}//end of setup()

//we must declare our flashOutputPin() function before we can call it in our code
void flashToneOutput(byte outputPinNumber, unsigned int noteFreq){
digitalWrite(outputPinNumber, HIGH); //set outputPinNumber to 5V (HIGH)
tone(outputSpeaker, noteFreq);
  delay(200); //wait 200ms
  digitalWrite(outputPinNumber, LOW); //set outputPinNumber to 0V (LOW)
  noTone(outputSpeaker);
  delay(200); //wait 200ms
}//end of flashToneOutput()

void loop() {
  //now we can call flashOutputPin() with both an LED pin
  //and also a note frequency for tone() to play
  flashToneOutput(outputLED1, noteC4);
  //now LED 2
  flashToneOutput(outputLED2, noteD4);
  //now LED 3
  flashToneOutput(outputLED3, noteE4);
}//end of loop()
```

If you have followed the pin setup for the circuit correctly, and copied the entire code listing into the editor, then the Tinkercad simulation should play a simple three-tone melody sequence (C4, D4, E4) whilst also flashing each of the three LEDs in turn. At this point, we can now run the code directly on an Arduino connected to a breadboard circuit to fully implement our system design from Figure 4.26. To do this, we need to set up a new piece of software – the Arduino Integrated Development Environment (IDE).

Arduino Uno setup

Go to the Arduino website download section at www.arduino.cc/en/Main/Software and download the latest Arduino IDE build for your operating system. *The version number is omitted in this image to avoid confusion.*	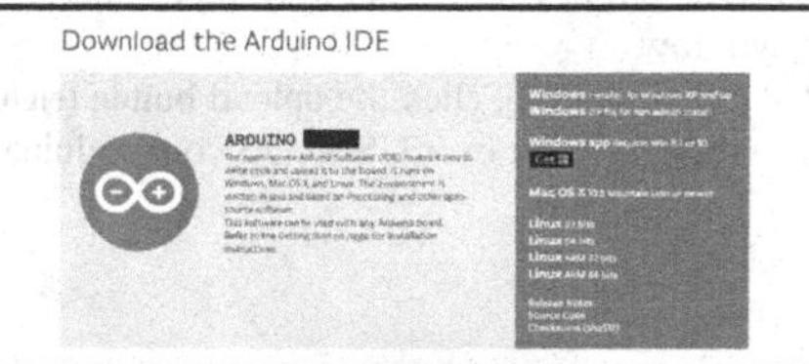
Once the download has completed, go through the setup instructions for your particular operating system (Windows/OSX/Linux). *OSX screenshots are provided for reference.*	
Once the Arduino IDE is installed and you launch the application, the code editor window shows a predefined setup() and loop() function structure where you can compile (and debug) code. The debugging window is shown in black at the bottom of the screenshot – this is where errors and other profiling information will be shown on compile or upload.	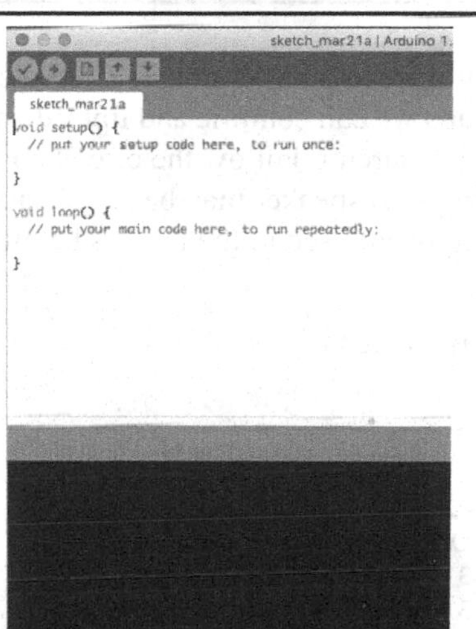
To set up your board, you can follow the instructions at www.arduino.cc/en/Guide/HomePage where we are using an Arduino Uno board. Connect the board using a USB port, and then follow the online instructions (they are for PC, but OSX follows a similar path). With the board connected, specify the board type (Tools>Board) and also the input port for that board (Tools>Port) to configure it for upload from the Arduino IDE.	
Now we have the board configured, launch an example sketch (e.g. File>Examples>01. Basics>Blink) to load the code listing and show it in the editor window. Use this sketch to test that our Arduino board can compile and run code.	

1. Click the tick box in the top left of the editor to compile the sketch; if there are no errors you will get a 'Done Compiling' Message above the debug window.
2. After compiling, click the upload button (right arrow next to the tick box) to load the code over USB onto your Arduino.

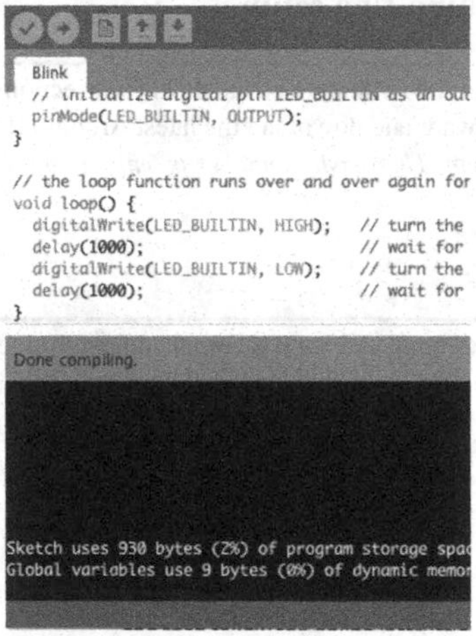

If this process has completed successfully, the onboard LED on the Arduino should now flash. This means we can compile and upload code onto our Arduino, so you can proceed to building the example project circuit. Follow the circuit connection instructions from the code listing above, noting that your own piezo speaker may be significantly different in size to both the Tinkercad object and also to the component used in the images below, and so may be connected to different pins as a result.

Project steps

For this project, you will need:

1. Arduino Uno board
2. One small breadboard (30 rows)
3. 3 × red LEDs
4. 3 × 150Ω resistors
5. 1 × piezo loudspeaker
6. 6 × connector cables and 4 × connector wires

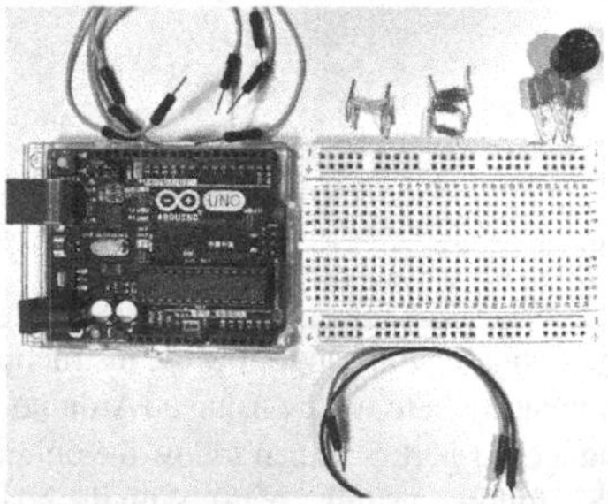

1. Add a 150Ω resistor between pins [i1–i15].
2. Add an LED – anode pin (h5) to cathode (h6).
3. Add a 150Ω resistor between pins [i8–i12].
4. Add an LED – anode pin (h12) to cathode (h13).
5. Add a 150Ω resistor between pins [i15–i19].
6. Add an LED – anode pin (h19) to cathode (h20).
7. Add a 150Ω resistor between pins [i122–i126].

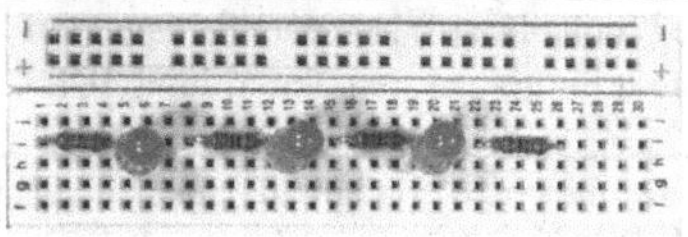

<table>
<tr><td>1. Add a connector to the negative (ground) rail [i6–GND].
2. Add a connector to the negative (ground) rail [i13–GND].
3. Add a connector to the negative (ground) rail [i20–GND].
4. Add a piezo loudspeaker – the image shows a component connected between pins [g26–g28].
5. Add a connector to the negative (ground) rail [h28–GND].</td><td>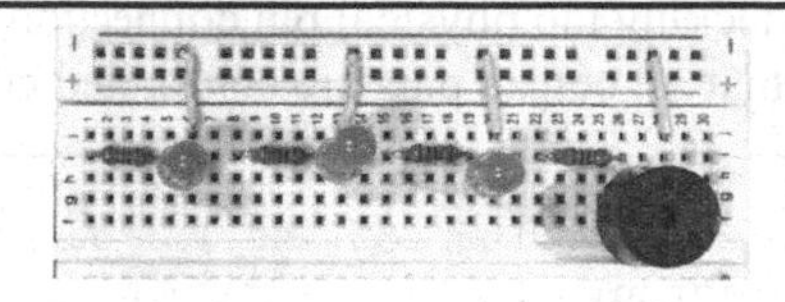</td></tr>
<tr><td>1. Add a connector between Arduino digital pin 10 and breadboard pin j1 [AD10–j1].
2. Add a connector between Arduino digital pin 11 and breadboard pin j8 [AD11–j8].
3. Add a connector between Arduino digital pin 12 and breadboard pin j15 [AD12–j15].
4. Add a connector between Arduino digital pin 13 and breadboard pin j22 [AD13–j22].
5. Add a connector between the negative (ground) rail (near column 30) and the Arduino top left GND pin – not shown in image.</td><td></td></tr>
<tr><td>To test our code, we can copy/paste the listing provided above into a new sketch (third icon in the image) in the Arduino IDE editor. Once pasted we can compile it (first icon, tick mark) then upload it (second icon, right arrow) to the board.</td><td></td></tr>
<tr><td colspan="2">If everything is connected correctly, the three LEDs should flash in time with the speaker emitting three sequential tones (C4, D4, E4).
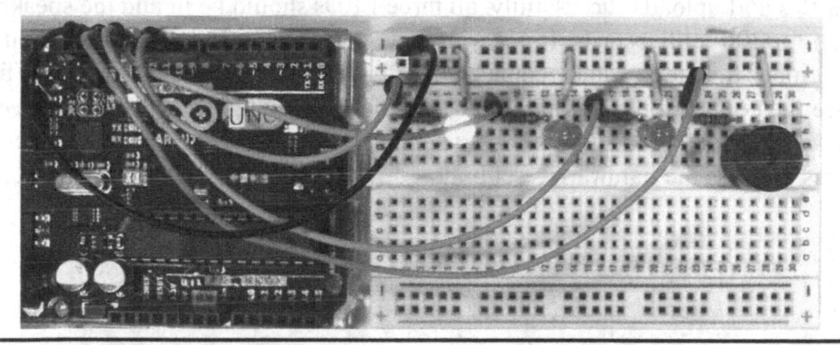</td></tr>
</table>

Debugging

If the circuit does not execute all processes as required, we will need to begin debugging our code and also our circuit – this is where the system diagram helps to break down the processes and outputs involved. The Arduino IDE does not provide extensive debugging support (partly because it is running on a physical electronic circuit where other issues may have been created) and so the following steps are provided for you to follow when uploading any new piece of code to your Arduino:

1. **Check circuit connections** – *you can never assume your components are not an issue.*
2. **Check pin configurations** – *ensure no setup errors have stopped output.*
3. **Check function structure and call** – *ensure your function is built properly and can be called from loop().*
4. **Check function operation** – *if the function is built correctly, the problem may be the instructions it executes.*

After checking the physical pin connections, a stripped-down version of the code can be compiled to ensure that all output pins are configured correctly and are operational:

```
//first debugging routine for our example project circuit
void setup() {
  //initilalize all 4 digital pins for output
pinMode(10, OUTPUT);
pinMode(11, OUTPUT);
pinMode(12, OUTPUT);
pinMode(13, OUTPUT);
}//end of setup()

void loop() {
  //test the digital write output to all LEDs to ensure they light up
  digitalWrite(10, HIGH); //set outputPinNumber to 5V (HIGH)
  digitalWrite(11, HIGH); //set outputPinNumber to 5V (HIGH)
  digitalWrite(12, HIGH); //set outputPinNumber to 5V (HIGH)
  //this line tests the loudspeaker to ensure it is functional
  //we add a duration to the function call to prevent a drone!
  tone(13, 262, 200);
}//end of loop()
```

If this code compiles and uploads successfully, all three LEDs should be lit and the speaker will emit a tone of C4 pitch for 200 milliseconds. If not, you may find that a component is not connected to the correct pin – remember the piezo will be connected differently, so follow the signal path from [j22–h27] to make sure the connection from output pin 13 to GND is complete. When you have checked your breadboard layout, you should now be able to light all three LEDs and sound a tone – a quick way to repeat the test is to press the hardware reset button on the Arduino itself (Figure 4.29).

Figure 4.29 Arduino hardware reset button. ***The reset button is highlighted in white at the top left of the Arduino board. This can be used to quickly reinitialize code and retest the component layout.***

With the LEDs lit and the speaker emitting a tone, we can now focus our attention fully on the code we have written. There are three possible areas where errors may have been made:

1. In the copy/paste of the listing, where a brace may have been omitted from the copy, or the paste was not performed in a blank editor file.
2. The configuration of the output pins is not correct – rather than focussing on how this happened, we should focus on checking it to fix the problem.
3. The function call is not executing properly – it is possible the pins are configured for output, but either the function call or the function itself is not executing properly.

We can address the first issue quickly by selecting everything in the editor and performing a new copy/paste of the full code listing again. The second issue requires us to use another amended code listing, to ensure that all constants have been declared correctly and that the pin configurations are correct (don't forget to copy each page in turn):

```
// declare constants to hold our output pins 10-13
//10-12 are LEDs, 13 is a piezo loudspeaker
const byte outputLED1 = 10;
const byte outputLED2 = 11;
const byte outputLED3 = 12;
const byte outputSpeaker = 13;
//declare constants for the test note output
const unsigned int noteC4 = 262;

void setup() {
  //initilalize all 4 digital pins for output
pinMode(outputLED1, OUTPUT);
pinMode(outputLED2, OUTPUT);
pinMode(outputLED3, OUTPUT);
pinMode(outputSpeaker, OUTPUT);
}//end of setup()

void loop() {
  //first debugging routine for our example project circuit
  //test the digital write output to all LEDs to ensure they light up
  digitalWrite(outputLED1, HIGH); //set outputPinNumber to 5V (HIGH)
  digitalWrite(outputLED2, HIGH); //set outputPinNumber to 5V (HIGH)
  digitalWrite(outputLED3, HIGH); //set outputPinNumber to 5V (HIGH)
  //this line tests the loudspeaker to ensure it is functional
  tone(outputSpeaker, noteC4);
}//end of loop()
```

If this code executes correctly, then the problem is either with the function call or with the function declaration itself. Whilst copy/pasting stripped down code is a simpler way to check for problems (and more useful when you are learning to code) a more common solution involves commenting-out non-essential lines to test the structure of our function declaration:

```
//..edit of previous full code listing for project
//comment every instruction after the first digitalWrite()
void flashToneOutput(byte outputPinNumber, unsigned int noteFreq){
  digitalWrite(outputPinNumber, HIGH); //set outputPinNumber to 5V (HIGH)
```

```
//   tone(outputSpeaker, noteFreq);
//   delay(200); //wait 200ms
//   digitalWrite(outputPinNumber, LOW); //set outputPinNumber to 0V (LOW)
//   noTone(outputSpeaker);
//   delay(200); //wait 200ms
}//end of flashToneOutput()

void loop() {
  //test a single call to the function
  flashToneOutput(outputLED1, noteC4);
  //now LED 2
//   flashToneOutput(outputLED2, noteD4);
  //now LED 3
//   flashToneOutput(outputLED3, noteE4);
}//end of loop()
```

This code listing shows where comments can be added to strip down the function call to `flashToneOutput()`. This allows us to focus on a specific instruction executed from a single function call, to narrow down where potential problems may have occurred. In the listing, if the call to `flashToneOutput()` executes correctly then LED1 will light up, telling us that the function is processing the first parameter (pinNumber) correctly. We can then make two further tests:

1. uncomment the other two digitalWrite() calls in the function to ensure they are not the problem;
2. comment all digitalWrite() calls and focus on the note() function:

```
//..edit of previous full code listing for project
//comment every instruction after the first digitalWrite()
void flashToneOutput(byte outputPinNumber, unsigned int noteFreq){
//   digitalWrite(outputPinNumber, HIGH); //set outputPinNumber to 5V (HIGH)
     tone(outputSpeaker, noteFreq);
//   delay(200); //wait 200ms
//   digitalWrite(outputPinNumber, LOW); //set outputPinNumber to 0V (LOW)
//   noTone(outputSpeaker);
//   delay(200); //wait 200ms
}//end of flashToneOutput()

void loop() {
  //test a single call to the function with values rather than constants
  flashToneOutput(outputLED1, noteC4);
  //now LED 2
//   flashToneOutput(outputLED2, noteD4);
  //now LED 3
//   flashToneOutput(outputLED3, noteE4);
}//end of loop()
```

In this listing, we are now focussing solely on the tone() output of the circuit. If the LEDs are functional, then the only problem in the function is the call to the PWM output to the piezo loudspeaker. At this point, if there is still a problem then rewriting the code section by section is perhaps the only way to chase down any other small errors (e.g. unclosed braces, non-terminated instructions) that may have crept in during the coding process. Whilst this seems pedantic (and undoubtedly tiresome) one of the most common errors in programming is repeatedly trying to fix the same thing and unintentionally changing other elements of the code as a result. Without tracking what is functional in the current build, it is quite easy to add (or remove) elements with the intention of changing them later, only to then get side-tracked in another issue or potential solution. This is particularly common when working on larger projects (or when tired after many hours spent trying to get something working!) and so this recommendation is actually more time-efficient in the long run.

4.11 Conclusions

In this chapter, the three core programming instruction types of sequence, selection and iteration were introduced (the next chapter will discuss selection and iteration in more detail). The chapter then looked at the concept of variables, named references to locations in memory where the ATMega328P microcontroller will store data. Functions were also discussed, named structures that encapsulate code instructions – these are the key to powerful and reusable code. Finally, the Arduino was used for pulse width modulation (PWM) output to generate audio frequencies and drive a piezo loudspeaker as output.

This chapter has covered a lot of ground on computer programming in a short time – it is not expected that you will simply 'pick up' programming as a result of this short introduction, so be patient with yourself when it comes to learning how to read and (particularly) write code. As stated at the beginning of this chapter, a full programming course is not provided in this book but rather an introductory guide to help you use the Arduino to control your electronics circuits. The study of the practical projects in these chapters is designed to consolidate your learning, so take some time to read over this chapter again and complete all the projects – they will significantly improve your understanding.

The next chapter will continue the introductory guide to programming, where you will learn how to include an external code library to send MIDI messages over USB to a digital audio workstation (DAW) running on a computer. This library will then be used in the final project, which takes piezoelectric sensors as inputs to an Arduino MIDI drum trigger – a fully functional project that you can connect to your own DAW to control any software drum machine that will accept MIDI data.

4.12 Self-study questions

Several example projects have been provided in this chapter. Given the significant amount of practical material you have just covered in a short space of time, study questions will not be included at this point. Rather, a short (optional) crossword based on the programming concepts in this chapter is provided as an additional memory aid.

Programming concepts

```
           F
           U
        S  N  C
       SEMICOLON
        L  T  M
 VARIABLE  I  M
        C  O  E
   ITERATION  N
    Y E I     T
    P S O
   SEQUENCE
      L
PARAMETERS
```

Across

(4) Every sequence instruction must be terminated with a ________(9)
(5) A named storage location for data _______(8)
(6) An instruction where you repeat something more than once ________(9)
(9) An instruction that does something once ________(8)
(10) A function can receive input _________(10)

Down

(1) We can encapsulate programming instructions inside a _______(8)
(2) An instruction where you make a choice based on a condition ________(9)
(3) Two forward slashes (//) indicate a ______ in your code (7)
(7) When declaring a function, you must specify the return ____(4)
(8) A function can return a ______(6)

All answers are provided for each chapter at the end of the book.

Chapter 4 summary

The following may help you remember some of the key points in this chapter:

Programming instruction types

Sequence instructions – do something **once**
Iteration instructions – **repeat** something more than once
Selection instructions – make a choice based on a **condition**

Systems process sensors to create actuator outputs

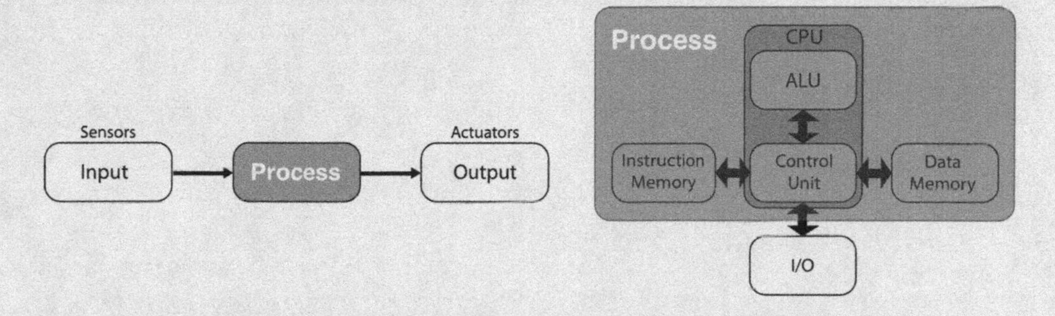

Variables are named locations in memory

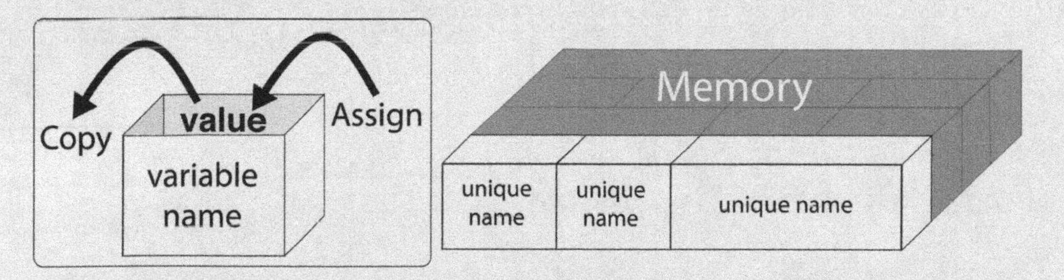

Functions process parameters to output results

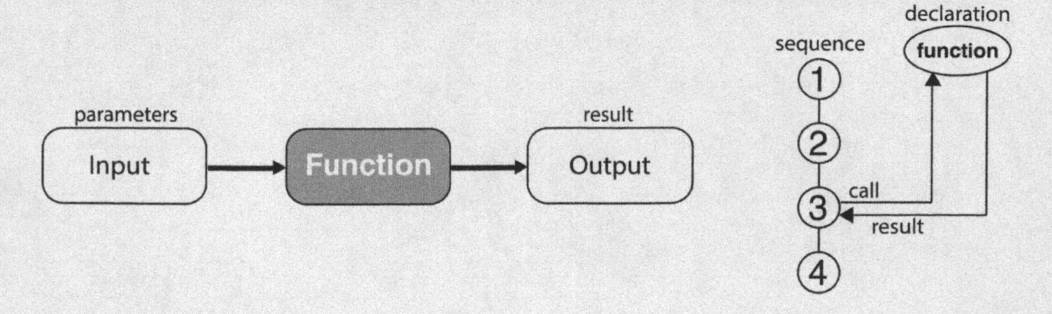

You must declare a function before you use it

Notes

1 FFTW is an open-source Fourier transform analysis library. Although platform-specific libraries may run faster in some cases (such as Apple's Accelerate framework), the FFTW is compatible with any system that can run C code – this is good example of how advanced coding is performed using the C language. For more information, visit http://fftw.org.
2 You can view the complete list of Arduino functions at www.arduino.cc/reference/en/#functions.
3 For more information, see www.arduino.cc/reference/en/language/functions/advanced-io/tone/.

CHAPTER 5

Digital systems 2 – Arduino input

This chapter continues the introductory guide to programming, where you will learn about the core instruction types of iteration and selection (for system input). Several practical examples are provided that build towards the final MIDI Drum Trigger project as a learning milestone within the book, so it is recommended to spend some time working through them carefully to become familiar with the concepts involved and how to apply them. As chapters 4 and 5 are not a substitute for a full programming course, the chapter begins with a brief recap on the previous chapter to reinforce the core concepts involved. The chapter will discuss arrays (groups of variables) and iteration (repeating instructions) as a means of working through large amounts of data quickly – the ability to process data is the most powerful aspect of any computer. Arrays are used as a means of extending data, allowing note sequence information to be stored in a single logical structure. The chapter also looks at the second instruction type (iteration) as a for loop to process each element of an array in sequence and output sound to a piezo loudspeaker using the Arduino tone() function. This process is then extended by including an external code library to send MIDI messages over USB to a digital audio workstation (DAW) running on a computer. By replacing the previous tone() function with MIDI messages, the Arduino can be used as a MIDI controller within a larger MIDI system.

At this point, you have learned how to use programming instructions to generate output from the Arduino, but the real value for audio electronics systems lies in using selection instructions to control outputs based on system inputs. This chapter will look at how the Arduino can receive and analyse input using selection instructions, which are introduced by example (they are arguably the most complex aspect of programming to learn). Digital inputs to the Arduino from push-button switches are then used as the conditions in a selection statement, to determine what MIDI information to output. In so doing, you will also learn about the three stages of testing needed when working with a full Arduino system:

1. **Test the code** – if the Arduino code is wrong, then the system cannot process information.
2. **Test the board** – if the circuit is wrong, then the system cannot receive inputs and create outputs.
3. **Test the system** – if the code and the board are functional, we must then test the full system.

Finally, you will learn how to analyse analogue input from a piezo sensor, as part of the final project to construct a MIDI drum trigger. The chapter builds towards a final project that uses piezoelectric sensors as inputs to an Arduino MIDI drum trigger – a fully functional project that you can connect to your own DAW to control a software drum machine that will accept MIDI data. At that point, you will have built your first full electronics system that takes input from a piezo sensor, processes it using the Arduino and outputs MIDI Note messages. This is a learning milestone in this book, which combines audio systems design, DC circuit analysis and Arduino C programming to produce a fully functional MIDI drum trigger system. If you have followed the text and examples correctly, you will complete this chapter with your first functional electronics system that has numerous possibilities for extension and augmentation into a more robust and flexible MIDI controller. More importantly, you will understand the concepts involved in designing and building such a system, which arguably equips you with the knowledge required to progress your learning in areas such as C programming – beyond the introduction provided in this book.

What you will learn

How to store data in arrays, and process them using iteration instructions
How to work with external code libraries like SoftwareSerial to generate serial MIDI data
How to build a standard MIDI OUT interface
How to use selection instructions to process Arduino digital input
How to use the Arduino to sample (and analyse) analogue input
How to analyse analogue piezo input to build a MIDI drum trigger system

5.1 Programming recap

The previous chapter introduced the three core programming instruction types:

Programming instruction types

Sequence instructions – do something **once**
Iteration instructions – **repeat** something more than once
Selection instructions – make a choice based on a **condition**

We learned that all **sequence** instructions must be **terminated with a semicolon** (;) to prevent the compiler from seeing the next instruction as part of the previous – the semicolon is the fundamental means of punctuation in C/C++ programming. The chapter also covered functions – structures that encapsulate code (Figure 5.1).

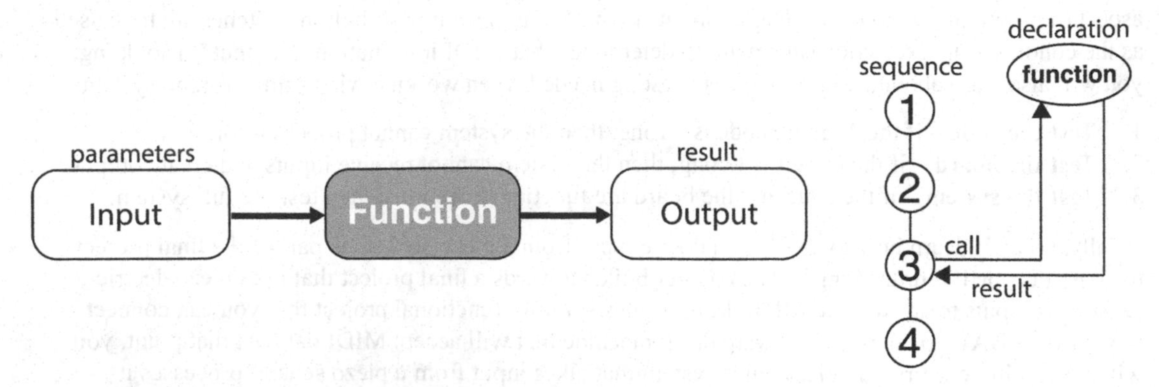

Figure 5.1 Workflow of a programming function. ***A function processes parameters to output results, and we must declare a function before we can call it in our code.***

Functions were used in the examples to reduce and reuse code, allowing us to add PWM output to a piezo loudspeaker relatively quickly to our existing blinking LED code. In so doing, we also made use of constants and variable – named storage locations in memory (Figure 5.2).

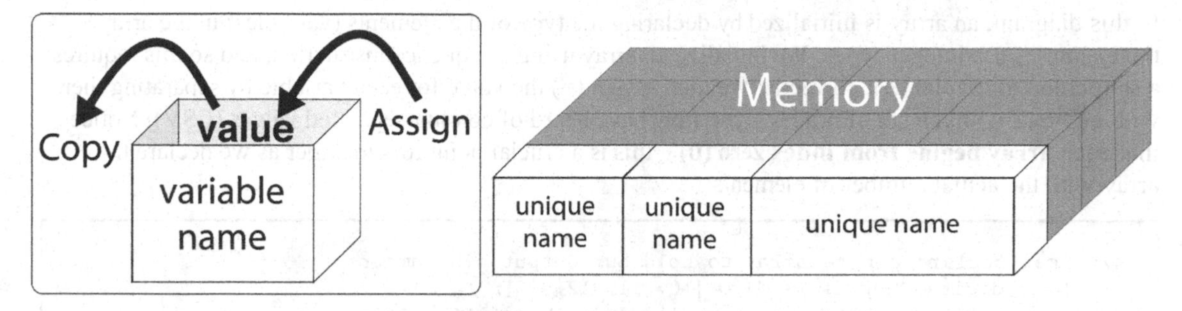

Figure 5.2 Variable storage example. ***In the diagram the unique name of the variable references a specific location in memory. Note the different sizes of boxes, defined by the variable type, and also that a value is assigned using the equals (=) sign.***

In the previous chapter, we learned that functions group code instructions in a single uniquely named structure. Now we can extend our learning to include arrays – uniquely named data structures that group variables in a single referenced location in memory.

5.2 Data structures and iteration – arrays and loops

The previous chapter introduced variables, which are named memory locations, that can be used to hold a reference to a value that may change. The natural-language comparison was made with variables as nouns, that instructions (verbs) can process – you cannot have a verb without a noun. This thinking can now be extended to group related variables together, making them much more easily accessible within a single structure known as an array (Figure 5.3).

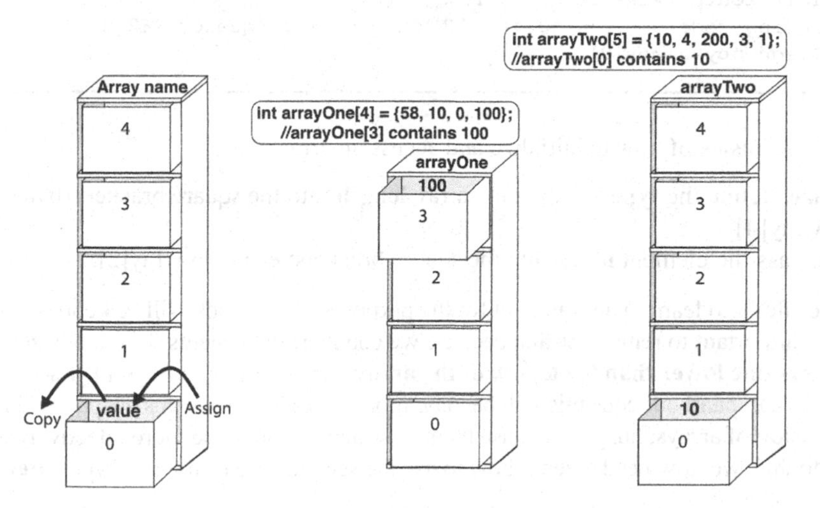

Figure 5.3 Conceptual array storage example. ***In the diagram, multiple variables can be grouped together in a uniquely named array. Each element (variable) in the array is assigned an index starting from zero, and each element value can be accessed by combining the array name with this specific index.***

In this diagram, an array is initialized by declaring the type of the elements (variables) in the array – the examples use integer types. We initialize an array using a sequence instruction, and so this requires a semicolon to terminate it correctly. We then assign (=) the value for each variable by separating them with commas within a set of braces – you may have heard of comma-separated values (CSV). Notice that each **array begins from index zero (0)** – this is a crucial point to remember as we declare the array with the actual number of elements:

```
//we can declare a byte array to hold our output pin numbers
    byte digitalOutputPins[4] = {10, 11, 12, 13};
//we can access an array value by its index - counting from zero
    byte currentOutputPin = digitalOutputPins[3];
// the variable currentOutputPin now holds the value 13
//we can also declare an array to hold 3 tone frequencies (e.g. last chapter)
    int toneFrequencyValues[3] = {262, 294, 330};
//we can declare an empty array to hold 2 input values from our sensors
    int analogueInputValues[2];
//we can assign a value to an array index using (=)
    analogueInputValues[1] = 200;
```

When declaring a new array, we specify the number of elements in the array inside the square brackets. We can then access a **specific array element by passing its index** into the square brackets:

```
//we can access an array by passing the index of the
//element we want into the square brackets
//we can then use this with functions we know like pinMode() and tone()
//note: we are using the arrays declared in the previous listing
//the following code would set pin 12 for output
pinMode(digitalOutputPins[2], OUTPUT);
//the following code would generate PWM output of frequency 262Hz
tone(toneFrequencyValues[0]);
```

Now we know the basics of how to initialize and access an array:

1. To **initialize**, define the **type** and pass the array **length** into the square brackets: byte exampleArray[4].
2. To **access**, pass the element **index** into the square brackets: exampleArray[2].

Arrays can be difficult to learn in abstract, but for the purposes of this book will be kept as straightforward as possible. It is important to remember that because we count array elements from index zero, the **last element is always one lower than the length of the array**. Arrays allow us to group associated data together, and as we expand our code this will become more necessary. Thus this chapter will not go into extensive discussion of arrays, but you will use them to structure your code more effectively in later examples. To do this, we now need to learn how to use the second core instruction type – **iteration**.

We can repeat certain instructions in our code by defining them within a **for loop** that **iterates a specific number** of times. For loops require start and end values, and also the increment (step size) value used to count between them (Figure 5.4).

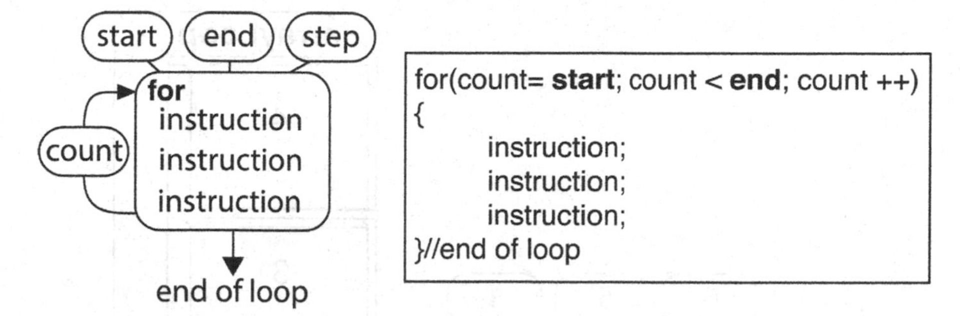

Figure 5.4 For loop workflow. ***The diagram shows a for loop structure containing programming instructions. To iterate the loop, a count variable is incremented between two values – start and end. The step value defines how the count variable is incremented (++ means a step of 1) until the end value is reached and the loop exits.***

The diagram shows how a for loop uses a count variable to determine how often to iterate the instructions within the loop. The count requires a start and end value, alongside a step value that defines how much to increment the count by each time (this is usually 1):

```
//example for loop - between a beginning and end value
// we begin by declaring a variable to hold the count (byte loopCounter)

// this loop will count between 0 and 3 (loopCounter = 0; loopCounter <4;)
// in increments of 1 (loopCounter ++)
    for (byte loopCounter = 0; loopCounter < 4; loopCounter ++) {
   // the instructions inside the braces will execute each time the loop iterates

}//end of for loop
```

In the listing, a variable called loopCounter is declared to control a for loop (you will encounter many examples that use single letters, but descriptive names will be used in this book). When the loop is set up inside the brackets, three sequence statements (terminated by a semicolon) are declared to do the following:

1. Assign the value 0 to loopCounter – this defines the **start** of the loop.
2. Evaluate the result of the expression loopCounter < 4 – this defines the **end** of the loop.
3. Increment the loopCounter variable by 1 (loopCounter ++) – this defines the **step** of the loop.

Expressions (e.g. x<y, a>b) will be discussed later in the chapter when we learn about conditions, but for now we can see that a loop starting at 0 and stopping at less than 4 (i.e. 3) will iterate 4 times. Any instructions inside the braces will execute each time the loop iterates, and we can specifically control how often this happens when we set the loop up. Loops are very powerful coding instructions that are used frequently, particularly in conjunction with arrays to allow all elements to be processed in the same way. To use a for loop with an array, we need to know the array length so we can iterate through all of its elements (Figure 5.5).

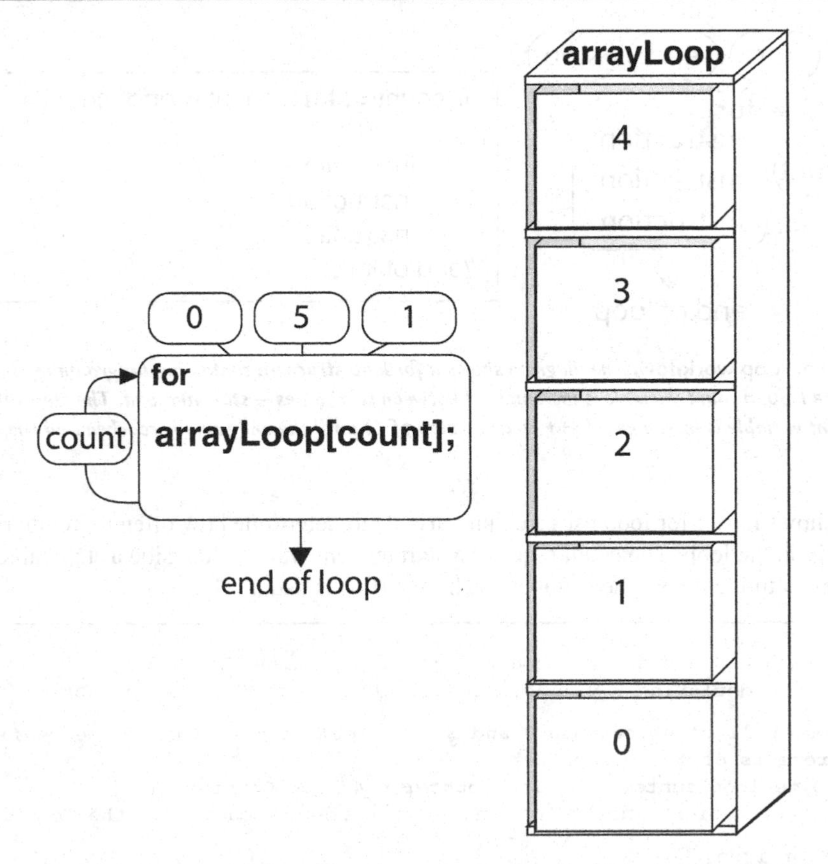

Figure 5.5 Array iteration workflow example. ***In the diagram, the for loop can access each element in the array by using the current count variable as the array index. The count variable is incremented by 1 each time the loop iterates. The length of the array is used to set the end of the loop (in this case 5), to ensure that all elements are processed.***

This leads to a code structure that requires both the array and its length to be defined in order to iterate through it using a for loop:

```
//define the length of our array
const byte arrayLength = 5;
//we can use this value to define an example array of 5 elements
byte ArrayLoop[arrayLength] = {9, 10, 11, 12, 13};
//now we can use a for loop to iterate through the array
for(byte count = 0; count < arrayLength; count ++){
   //this example uses the array values to
   //set up pins (9-13) for output
   pinMode(ArrayLoop[count], OUTPUT);
}//end of for loop
```

The listing above defines a constant called arrayLength that is used to create an array of five elements – don't forget **arrays begin at index 0**. We can then initialize our array by stating its type (`byte`) and passing `arrayLength` inside the square brackets. Next, we use a for loop that counts between 0 and `arrayLength` in increments of 1 (++) – don't forget **arrays begin at index 0**, so we will actually be **counting from 0 to arrayLength−1** (4). Inside the loop, we access each array element by passing the loop counter into the array square brackets as the index required. We can then use the value stored at this index to setup a digital pin for output using the pinMode() function.

Take some time to read over the last code listing, as the combination of data and iteration is very useful for common coding tasks like pin configuration and digital output. In the previous chapter example project we learned how to declare constants corresponding to specific Arduino pins, how to initialize those pins for output and then how to use a function to send digital signals (including PWM) to them to light LEDs and emit sound from a piezo loudspeaker. A smaller version of this project can show how iteration can be used with an array to significantly streamline this code – whilst also ensuring that fewer mistakes are made due to manual repetition of the instructions.

5.3 Example 1 – tone array output

In this example, a stripped-down version of the final project from the previous chapter is used that does not require LEDs, to reduce the overall build time involved. In the chapter 4 project example, three tones (C4, D4, E4) were output in a simple sequence using the flashToneOutput() function that combined LED output with tone() function calls to send PWM output to the piezo loudspeaker. In this example, we will extend this tone sequence to output 16 tones in a much longer musical pattern sequence. The instructions for this longer tone sequence could be coded manually, but this would take significantly more time and would also be prone to errors (e.g. due to the fatigue of repetition).

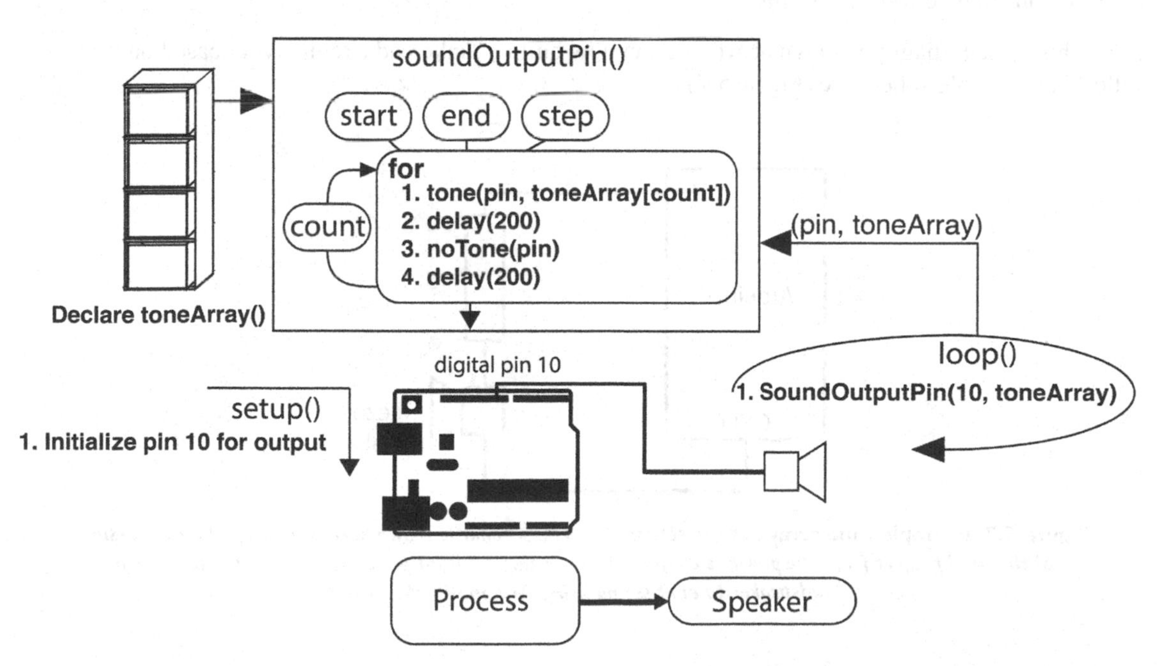

Figure 5.6 Example tone array output system code process diagram. ***This diagram is largely based on amendments to the system used in the chapter 4 example project (the system workflow has been omitted for clarity). In this system, an array (toneArray) holds a sequence of tones. The function soundOutputPin() is used to execute the calls to tone(), passing in each index of toneArray as the frequency value within a for loop of the same length as toneArray().***

Instead, we will use **an array to hold our note frequency** values and then **iterate through them using a for loop** to output this much longer melody to the piezo loudspeaker. This demonstrates the power of arrays when combined with for loops as a means of working with larger amounts of data (Figure 5.6).

The system shown in Figure 5.6 is fairly straightforward in electronic terms, consisting of a single PWM digital output that is used to drive a piezo loudspeaker. The setup() function initializes pin 10 for digital output using pinMode(), whilst loop() makes a call to the soundOutputPin() function that processes the array. The setup() and loop() functions are short, as the majority of the processing work is carried out by soundOutputPin(). At first glance, the function soundOutputPin() can appear a little redundant, as we could simply execute these instructions in loop(). Though this is appears to be the simpler option, we are actually using the function call to soundOutputPin() to keep the loop() function simple and easier to read. In addition, it is good practice to avoid calling a function every time loop() executes to keep our system processes to a minimum (thus saving battery life and reducing needless instruction executions).

In addition, parameters are still being passed into the soundOutputPin() function for both the output pin (10) and also the toneArray(), even though these could simply be stated within the function itself to make the call simpler. This is done to reinforce the concept of passing parameters within function calls, which also has the added effect of making us fully consider the data used by the processes involved when declaring the function itself – we could call a function that uses lots of different data, but we would be less likely to track that data from the call to that function. Although the previous chapter did not go into extensive detail on functions, the aim is still to present examples that show how best to use them to control electronics systems.

From this system diagram shown above, we can now build a Tinkercad circuit layout based on the following example schematic (Figure 5.7).

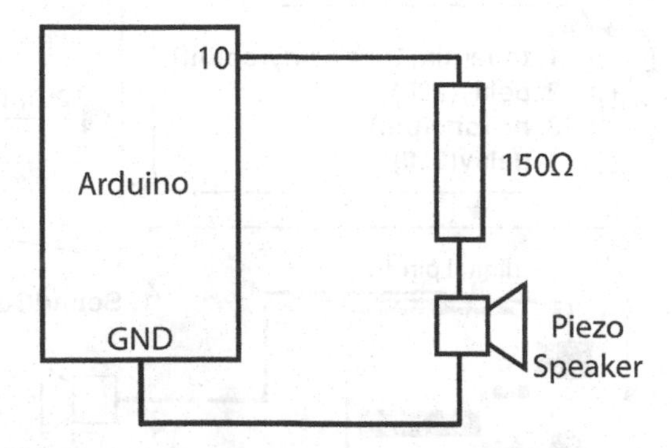

Figure 5.7 Example tone array output schematic. ***The schematic shown here is a stripped-down version of the final project from the previous chapter. Here, a single output pin (10) is connected to a piezo loudspeaker to emit sound using Arduino PWM output.***

This schematic is fairly straightforward, but if you are unsure of how a digital output pin from the Arduino can generate PWM output to drive a piezo loudspeaker then it is recommended to look over the previous chapter again to refresh your memory of Arduino output. It should also be recalled that the 150Ω resistor is added in series with the loudspeaker to limit the current flowing through the circuit. The Arduino can only output a maximum of 40mA on a single pin so we want to avoid burning out either the Arduino and/or the loudspeaker. You should also remember that we have gone slightly higher than needed (125Ω) to 150Ω ($I = \frac{V}{R} = \frac{5}{150} \approx 33mA$) to allow us to use the same resistors as in our other examples to keep our circuit costs down (and that 220Ω resistors can be used instead). This schematic can now be translated into an Arduino breadboard layout (Figure 5.8).

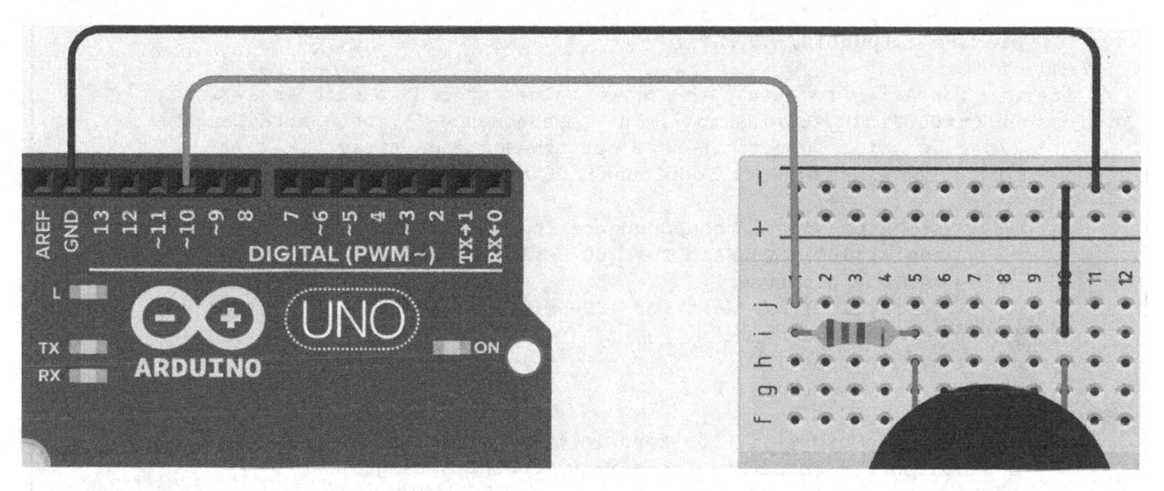

Figure 5.8 Tone array output breadboard layout. ***The layout for this circuit is fairly straightforward, where digital output on pin 10 is routed through a 150Ω resistor that limits current through the piezo loudspeaker.***

This breadboard layout is fairly straightforward, and as previously noted is a stripped-down version of the chapter 4 example project. As with the circuit layout in that chapter, it is again noted that in practice you may need to rearrange the pin layouts based on the specific piezo loudspeaker you have, as the Tinkercad component provided is significantly bigger than many others available. The connection instructions for this circuit are as follows:

1. Add a 150Ω resistor between pins [i1–i5].
2. Add piezo loudspeaker between pins [h5–10].
3. Add a connector to the negative (ground) rail [i10–GND].
4. Add a connector between the negative (ground) rail (near column 11) and the Arduino top left GND pin.
5. Add a connector between Arduino digital pin 10 and breadboard pin j1 [AD10–j1].

From the system code process diagram, we know that we will have to declare an array named toneArray() to hold our note frequencies, a constant to define the output pin (10) alongside three functions in our code. Setup() and loop() will be kept as simple as possible, with the majority of the instructions being contained within the soundOutputPin() function:

```
//declare a constant for the PWM speaker output on pin 10
const byte outputPin = 10;
//now declare a constant for the toneArray length (used by the for loop)
const byte toneArrayLength = 16;
//now declare toneArray to hold our 16-note sequence
//sequence is (C4, D4, E4, F4, G4, A4, B4, C5, B4, A4, G4, F4, E4, D4, C4, C5)
int toneArray[toneArrayLength] = {262, 294, 330, 349, 392, 440, 494, 523, 494,
   440, 392, 349, 330, 294, 262, 523};

void setup()
{
       //configure pin 10 for audio output
       pinMode(outputPin, OUTPUT);
}//end of setup()
// iterate toneArray to output the pitch values stored in each array index
void soundOutputPin(byte pinNumber, int toneSequence [], byte arrayLength){
       //set up a for loop to iterate our toneSequence Array
    for(byte loopCounter = 0; loopCounter < arrayLength; loopCounter ++)
    {
              tone(pinNumber, toneSequence[loopCounter]);
              delay(100); // Wait for 1000 millisecond(s)
              noTone(pinNumber);
              delay(100); // Wait for 1000 millisecond(s)
       }//end of for loop
}//end of soundOutputPin()
void loop()
{
       //call soundOutputPin() to keep loop() simple
       soundOutputPin(outputPin, toneArray, toneArrayLength);
}//end of loop()
```

The listing above shows the complete code for the example – you can copy/paste this into the Tinkercad simulator, or upload it directly onto an Arduino Uno board using the Arduino IDE. An array is declared containing 16 note frequencies with `toneArray[toneArrayLength]`, and then the `soundOutputPin()` function is used to iterate through this array to generate PWM output at the frequencies specified in the array using the `tone(pinNumber, toneSequence[loopCounter])` function. The `toneArrayLength` constant is used to define the length of our array and also the length of our loop, to ensure we iterate through all the elements.

If the circuit has been built correctly, you should now hear a looping sequence of tones from the piezo loudspeaker as output. To save time, an actual build of this circuit will not be provided as the next example on Arduino MIDI output will require a physical circuit to be built for full system testing – this will require you to follow a series of tests that will help translate a simulation to an actual circuit that cannot be fully simulated by Tinkercad. In programming terms, you have now reached an important stage in our understanding of code processes. You can now declare and call a function that iterates through an array of variables to generate Arduino output. In the next section, we will consider how to extend this circuit by replacing the piezo loudspeaker with a simple MIDI hardware interface – allowing us to use our Arduino to generate a MIDI sequence as output.

5.4 Working with external libraries – serial MIDI output

Chapter 2 introduced the Musical Instrument Digital Interface (MIDI) protocol, which uses a 5-pin DIN (Deutsches Institut für Normung) connector to establish communication between devices (Figure 5.9).

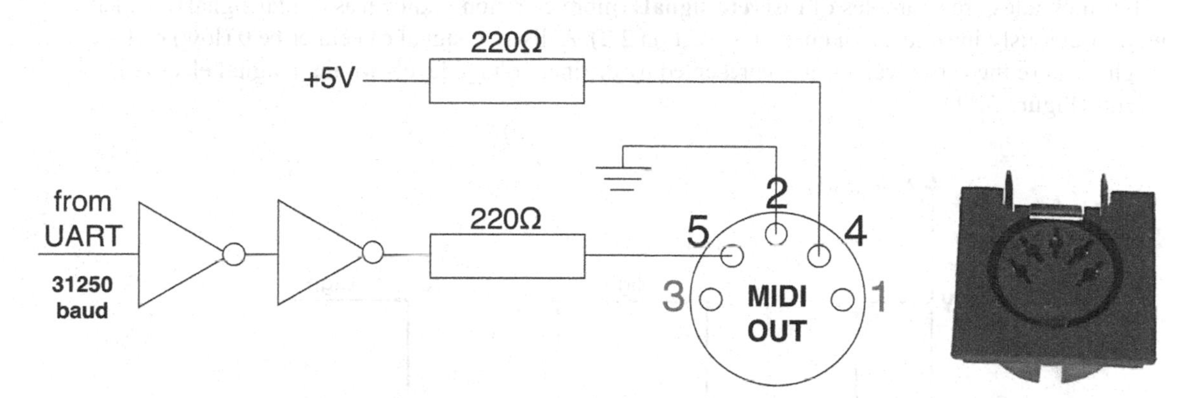

Figure 5.9 Standard MIDI OUT interface (adapted from the MIDI Manufacturers Association, 1985). ***The diagram shows the schematic for a MIDI OUT interface. The MIDI message is sent from a universal asynchronous receiver transmitter (UART) at 31250 baud through two operational amplifiers (triangles) and a current-limiting resistor (220Ω) to pin 5 of a DIN connector. Pin 4 of this connector is set to +5V (also with current limit) and pin 2 is grounded – pins 1 and 3 are not used. The right-hand image shows a 5-pin DIN socket, with output terminals pointing upwards.***

In Figure 5.9, two triangles are shown that represent operational amplifiers (these will be discussed in more detail in chapter 7). The rest of the schematic uses three pins of the 5-pin DIN connector, with pins 4 and 5 having current-limiting resistors (the MIDI specification defines 220Ω) for both signal pin 5 and pin 4 (+5V). The signal from the Universal Asynchronous Receiver Transmitter (UART) is a serial communication of binary bits grouped into bytes, based on the MIDI message format (Figure 5.10).

Status				Channel				Data 1								Data 2							
Note On (9)				1 (binary 0)				Middle C (60)								Velocity (100)							
1	0	0	1	0	0	0	0	0	0	1	1	1	1	0	0	0	1	1	0	0	1	0	0

Figure 5.10 MIDI message example. ***The figure shows the composition of a MIDI Note On message, with its binary equivalent listed underneath. The first byte defines Status, which combines the message type (always starting with a binary 1) with the MIDI channel that the message applies to. The second and third bytes provide Data (which always starts with a binary 0).***

The table details a Note On message, which is a common 3-byte MIDI message sent in a serial (sequential) format. A Note On message is configured as follows:

1. **Status** – four status bits define the message type, where Note On is 1001, or decimal 9 (Note Off is 1000, decimal 8).

2. **Channel** – four bits that define the MIDI channel (1–16) that the message is intended for (counting from binary 0).
3. **First data byte** – defining a 7 bit note number from C-2(0) to G8(127). *On an 88-key piano A0 (21) to C8 is (108).*
4. **Second data byte** – defining the velocity of the note (from 0 to 127).

MIDI messages are examples of **discrete signals** (more commonly known as digital signals), which were previously introduced in chapter 2 (section 2.2). A discrete signal can either be 0 (low) or 1 (high), where these two values are represented by distinct voltage levels within a digital electronic circuit (Figure 5.11).

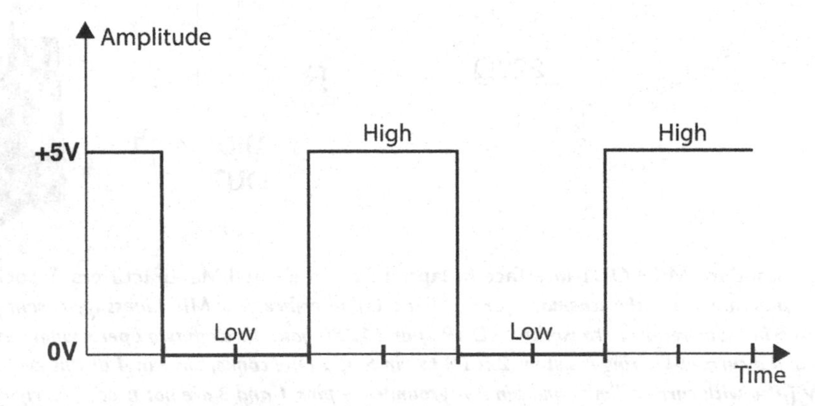

Figure 5.11 A digital signal (taken from chapter 2, section 2.2). ***The diagram shows an example of a discrete signal, more commonly known as a digital signal. The output amplitude levels (y-axis) are defined as either low or high, where specific voltages are used for each of these levels. On the Arduino, these are commonly +5V (HIGH) and 0V (LOW).***

Thus **MIDI is a digital system** that sends data as a combination of binary bits that are represented by voltage levels, where these bits are grouped into 8-bit bytes for processing. An important thing to note is that all MIDI values are in binary, so they are counted starting from 0. This means that MIDI **channel 1 is binary 0**, which is important to know as all MIDI software (and hardware) interfaces count from decimal 1 to 16! In contrast, a Note Off message only requires 2 bytes (to specify the status/channel and note number) but instead most modern devices use a second Note On message with zero velocity as part of a technique known as **running status**. Running status assumes that the first MIDI message contains the status/channel required, and unless another status byte is received then every subsequent first or second data bytes will be assumed as being another Note On message for that channel. This also explains why MIDI data bytes (e.g. notes and velocities) have a 7-bit range (0–127), because the most significant bit (MSB) of the message must always be zero to avoid it being seen as a possible status byte (Figure 5.12).

Note On	Note number	Velocity	Note number	Velocity	Note number	Velocity
10010000	00111100	01100100	00111100	01111101	00111100	00000000

Figure 5.12 MIDI running status example. ***The most significant bit (MSB) of a byte is its highest-value digit, and in MIDI this defines whether it is a status byte (all data bytes have MSB 0). Thus, unless a device receives a byte with MSB 1, it interprets all subsequent bytes as Note On data.***

In this example, the same note (C4) is played three times, once with velocity 100 (1100100), then with a louder velocity of 125 (1111101) and then finally with velocity 0 (which is effectively a Note Off). By streamlining the data sent, MIDI can be much more responsive to note timings because it does not have to send status messages with each one. Now we understand a little more about MIDI messages, we can learn how the Arduino can be configured to send serial data on a digital output pin in the form of MIDI messages. In Figure 5.9, it was shown that the MIDI UART runs at 31250 baud, where **baud** is a unit of transmission speed that defines the number of **bits per second** that can be can transmitted by MIDI. Our Arduino also has a built-in UART, which can be set up to output on a digital pin at this rate (Figure 5.13).

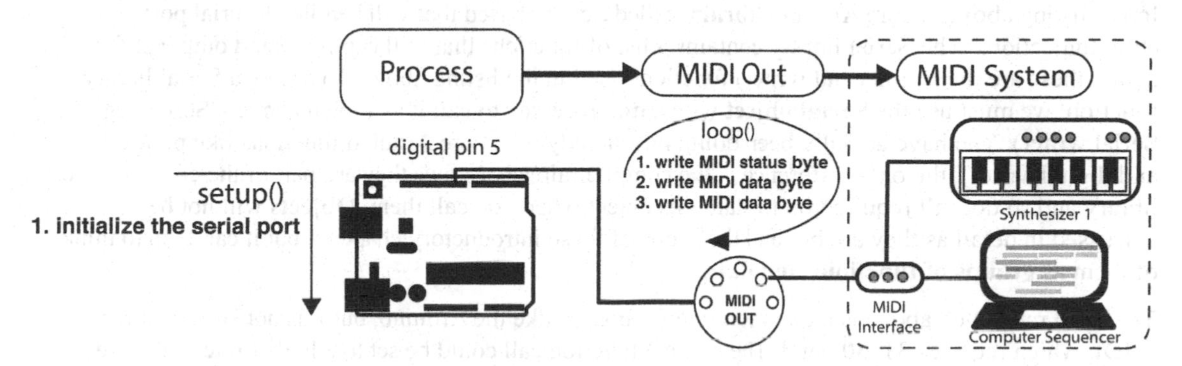

Figure 5.13 Arduino MIDI output code process diagram. ***The diagram shows how the Arduino serial port can be used to write MIDI data to an output connector that may then form part of a wider MIDI system. Serial communication is initialized in setup (), then a MIDI message is written as a sequence of bytes in loop(). This system will be partitioned for coding development, but include the wider MIDI system during testing.***

In this diagram, a multi-stage system has been defined that combines an Arduino MIDI output circuit with a computer sequencer and/or synthesizer that can receive MIDI output. This is the first system that must also model its outputs as inputs to another system, so you may want to revisit chapter 2 (particularly Figures 2.4 and 2.7) to look over system boundaries and partitioning again. In the system above, a boundary is defined between the MIDI output connector and another MIDI system that may comprise any number of MIDI-compatible devices. This is one of the most powerful aspects of MIDI, in that any device that conforms to the MIDI specification can be connected to another. The MIDI sequence player circuit will form part of a larger MIDI system, which is partitioned to focus on the sequence player in terms of code and electronic components. When these elements are tested, they will also need to be combined within the full system to ensure everything is functional.

The Arduino UART can output serial data on digital pin 0 and receive it on pin 1 by default, but this connection must be used carefully on the Arduino Uno (or any ATMega328P board), as these pins are also reserved for serial communication with a computer running the Arduino IDE. This is why the system diagram stipulates pin 5 for output (see below). To understand how to program the Arduino UART to generate serial output, we can begin by using a single call to the Arduino Serial library:

```
void setup() {
  // set the Arduino serial port baud rate:
     Serial.begin(9600);
}//end of setup
void loop() {
  //writes the integer 25 as a binary number (11001) to the serial port
     Serial.write(25);
}//end of loop
```

In the listing above, a core Arduino **library** called Serial is used that will handle all serial port communications. The Serial library contains a list of functions that will configure and output serial data – these are the begin() and write() functions used in the listing above. To access a Serial library function, we must use the **Serial object** with a **dot accessor** to call it as shown above – Serial.begin(), Serial.write(). You have actually been doing this already with core Arduino functions like pinMode() and digitalWrite(), the only difference is the compiler already knows they are part of the core Arduino library and so doesn't require you to state this object when you call them. **Objects** will not be discussed in detail as they are beyond the scope of these introductory chapters, but it can help to think of them as **groups of functions and data**.

The baud rate listed above is a common rate for boards like the Arduino, but it is not sufficient for MIDI, which requires 31250 baud. The begin() function call could be set to a higher rate, but there is also another issue that must be addressed – the pin conflict with USB. As noted in Figure 5.13, the Arduino Uno uses digital pins 0 and 1 for serial communication over USB, so these pins cannot be used for any other purpose while the Arduino is connected to a computer running the Arduino IDE. A workflow could be implemented where the board is unplugged before running any serial output, but this has the potential to confuse things when you are learning and even damage the Arduino by not disconnecting an output pin or running the code with the USB still connected. Instead, another Arduino library called **SoftwareSerial** can be used that can write serial data to other digital pins. To do this, we **must include its library of functions** at the beginning of the code using the following statement:

```
//we use #include<> to include other libraries, which have the extension <lib>.h
#include <SoftwareSerial.h>
//now we declare a new SoftwareSerial object for use with MIDI
// pin 8 for receive(RX) and pin 9 for transmit(TX)
SoftwareSerial SerialMIDI(8, 9);
void setup() {
//first, start the main serial port at 9600 baud
  Serial.begin(9600);
//now we can start our new object to output MIDI on pin 9
  SerialMIDI.begin(31250);
}//end of setup
```

In the listing, the library is included so its functions can be used by calling a **constructor** function to create a **new SoftwareSerial object** called SerialMidi that receives on pin 4, and transmits on pin 5. Calling object functions is a more advanced programming technique, so while the terms used are explained for clarity, it is recommended that when beginning to learn both audio electronics and Arduino programming (at the same time!) it is better to **know how to use objects** rather than fully understand how to build them – there are many object-oriented programming (OOP) resources available should you wish to extend your knowledge after completing this book.

For now, the serial port is initialized (at standard 9600 baud) and then the SerialMidi software object is configured at 31250 baud. With the ports configured, a MIDI message can now be sent by the SerialMIDI object using its write() function:

```
void loop(){
//use the SerialMIDI object to write out a 3 byte MIDI message
    SerialMIDI.write(0x90);//this is a note on status byte
    SerialMIDI.write(60);//this is the note number for middle C (C4)
    SerialMIDI.write(100);//this specifies a velocity of 100 (range 0-127)
}//end of loop
```

MIDI messages are sequential, so by writing each byte in turn a **serial MIDI message** is output to digital pin 9. The first byte is the **status byte (status/channel)**, which for a Note On message (1001) on MIDI channel 1 (0000) gives a binary value 10010000. In the code listing, this is converted into **hexadecimal** (base 16 numbers) to make it easier to program the value than by typing it in binary (where a mistake could easily be made in the digits). Although hex may seem complex, it is very common in computing – if you've ever used a computer graphics program then you will be familiar with hex colour codes. This book will not cover the hexadecimal number system, but as with objects we can make use of it to send a Note On channel 1 status byte (channel 2 would be 0x91, channel 3 0x93 etc).

A note on pacing your learning

The code in the listings above significantly extends your understanding of programming, even though we are still working at an introductory level. In the previous section on arrays, it was recommended to spend some time reading over the code listings to become more familiar with the processes (and concepts) involved. In this section, it is requested that you do **not spend significant time on include files and serial communication objects** at this stage in your learning. The code processes have been described to help you use them, but the remainder of this chapter (after the next example) is dedicated to working with **Arduino input which is much more important** to a systems approach to learning electronics. As a result, you are asked to consider leaving this section until you have worked through more of the content in this book - you can always use MIDI in your circuits, but understanding input sensors (and the selection instructions that process them) will be much more helpful in your understanding of audio electronics systems.

5.5 Example 2 – MIDI sequence player

This example combines the array-processing code from example 1 with MIDI output to create a MIDI sequence player. The main addition to the previous system is the use of the SoftwareSerial library to create an object that can output serial data to our MIDI connector. An array can then be used to iterate through a sequence of MIDI note numbers and output them on digital pin 9 (Figure 5.14).

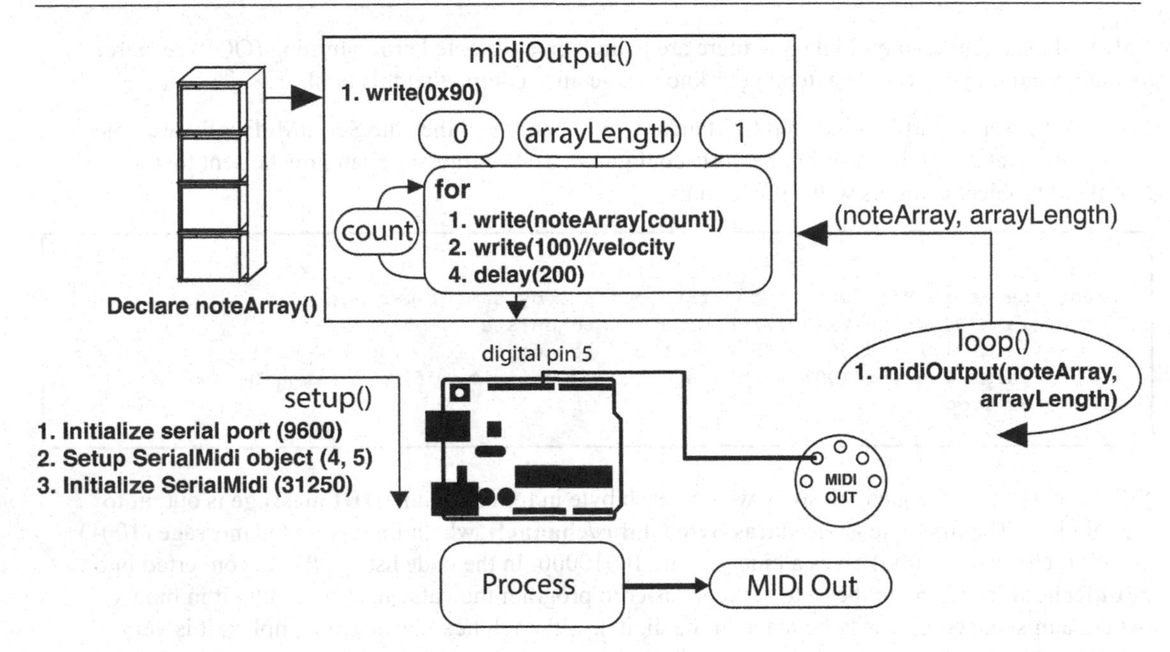

Figure 5.14 Updated Arduino MIDI output code process diagram. ***The diagram shows how we can combine serial communication with an array of MIDI notes to output a sequence of Note On messages to a MIDI connector. In the updated function midiOutput(), we can declare an initial status byte and then iterate through noteArray to output MIDI data bytes in running status.***

The system in Figure 5.14 is very similar to example 1, with the main difference being the replacement of tone() function calls with write() function calls to a SoftwareSerial object. You will also note that the system from Figure 5.13 is partitioned to exclude the overall MIDI system that could be connected to the output of our Arduino circuit. This was done to **focus coding development on the MIDI sequence player circuit**, but the example will revert to the larger system again later for testing. The schematic for this system uses an Arduino digital output pin connected to the standard MIDI out schematic from Figure 5.9 (Figure 5.15).

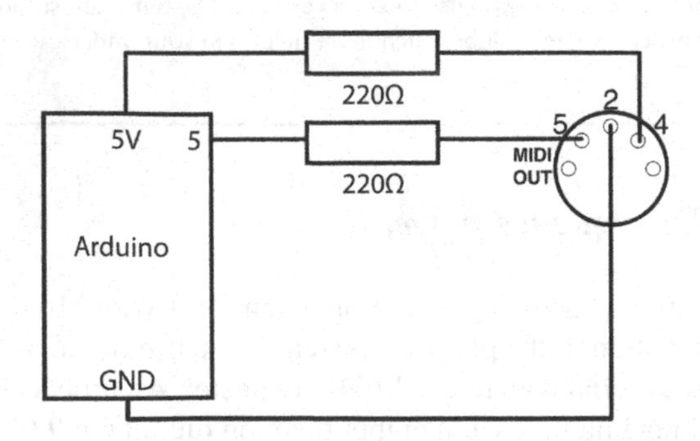

Figure 5.15 Arduino MIDI output schematic. ***In the schematic, digital output pin (9) is routed through a 220Ω resistor to pin 5 of the MIDI DIN connector. A +5V signal from the Arduino is connected to pin 4 (using another 220Ω resistor) and pin 3 is connected to GND.***

The schematic shows digital pin 9 on the Arduino connected to pin 5 on a 5-pin DIN connector, with a current-limiting resistor being added to conform to the MIDI hardware interface specification shown in Figure 5.9. In addition, the 5V Arduino power pin is connected to pin 4 (also with current-limiting resistor) and pin 2 is connected to ground (GND) – this is a **standard MIDI Out interface** that is now interoperable with other MIDI equipment.

This circuit cannot be fully simulated in Tinkercad, as MIDI communication is beyond the scope of the tool – it does not extend to a communication technology like MIDI, nor provide any DIN connections. Tinkercad does provide a useful list of standard electronics components that are aligned to the most common uses of the Arduino, and we have used the LED and piezo loudspeaker actuator outputs provided until this point. Now we are beginning to move outside the scope of this tool and so will have to plan simulation and testing in different ways, to ensure that the simulator can still be used where possible. In this project, the code can still be checked to ensure it is compiling and executing correctly by tracking the output it generates. To do this, we can use the Serial Monitor window provided in Tinkercad (at the bottom of the coding editor) to print out values generated by the code. Whilst **the SoftwareSerial object cannot be accessed** for MIDI output, the print() and println() functions of the core Serial object can be used to output the MIDI status and data bytes to the Serial Monitor window (Figure 5.16).

```
1  void setup() {
2  //start the main serial port at 9600 baud
3      Serial.begin(9600);
4  }//end of setup
5
6  void loop(){
7  //use the Serial object to write out a 3 byte MIDI message
8    Serial.print(0x90);//this is a note on status byte
9    Serial.print(60);//this is the note number for middle C (C4)
10   Serial.print(100);//this specifies a velocity of 100 (range 0-127)
11   Serial.println("");// adds a carriage return to each message.
12 }//end of loop
13
```

Serial Monitor

```
14460100
14460100
14460100
14460100
14460100
14460100
14460100
14460100
```

Figure 5.16 Using the Tinkercad Serial Monitor. ***In the diagram, a short code listing provides an example of how output can be written to the Serial object using the print() function. The data being written is then displayed in the Serial Monitor window below the editor in decimal format (the status byte 0x90 is 144 decimal). Println() is used to add a carriage return at the end of each message, to make it easier to read.***

This diagram shows how Tinkercad can be used to trace a simulation of the data output, which allows the project code to be tested to ensure it will output the correct values before we connect our MIDI interface circuit. You can look at the Serial Monitor window to see if the data has been generated correctly, but you must remember to **click the Clear button** at the bottom of the Serial Monitor to ensure the data is tracing from the beginning of the current simulation (otherwise the monitor will simply add it to the previous output).

With code output now traceable, testing can be planned to reduce the scope of any errors we may encounter. This is a useful process when working with microcontrollers, as we will always be combining code debugging, electronic schematic and breadboard layout issues and then finally checking that everything is physically built correctly! Sanity checks have been added during previous project builds (e.g. build stages) but now we will make a more focussed delineation between what can be tested by simulation and what must be physically tested:

1. **Test the code** – if the data is not correct, then the circuit will not function. Tinkercad can be used for this stage of the project.
2. **Test the board** – the circuit layout should reflect the schematic, to ensure all components are connected as required. There will be no output if we have not configured our layout properly, even if the code is functional.
3. **Test the system** – the outputs of the circuit should behave as required within the broader design of the system involved. When streaming binary data to a MIDI interface we will need to have a wider MIDI system to test this data.

These three testing stages can be used to check the entire project, to avoid making an error in one area that compounds others. For example, if there is a coding error that outputs the wrong data (or none at all) then there is no point in connecting the breadboard circuit. Similarly, if the code is tracing out as required, we know that either the serial communication is not configured properly or the breadboard layout is wrong – this can be tested with a simple output (like an LED) to quickly check for connection issues. If both of these stages are functional, the wider MIDI system can be checked with another known device to ensure it is capable of receiving and responding to messages on the specified channel.

The first testing stage involves the code needed to generate MIDI output, which will iterate through an array of MIDI note values to output them as serial MIDI data through a SoftwareSerial object on digital pin 9. The code from example 1 will be used as the basis for the MIDI output project, where the array-processing code has been moved into a dedicated function both to allow it to be tested separately and also to keep loop() simple. Now, this function can be reworked to output MIDI data to the SoftwareSerial object using the note values stored in the array. We can **also use our Serial object to print the same data** to the Serial Monitor, to ensure that we are building the structure of each MIDI note (and the sequence) correctly:

```
//use #include<> to include other libraries, which have the extension <lib>.h
#include <SoftwareSerial.h>
//declare SoftwareSerial object pin 4 for receive(RX) and pin 5 for transmit(TX)
SoftwareSerial SerialMIDI(4, 5);
//now declare a constant for the noteArray length (used by the for loop)
const byte noteArrayLength = 16;
//now declare noteArray to hold our 16-note sequence - we can use bytes
//sequence is (C4, D4, E4, F4, G4, A4, B4, C5, B4, A4, G4, F4, E4, D4, C4, C5)
byte noteArray[noteArrayLength] = {60, 62, 64, 65, 67, 69, 71,
                                   72, 71, 69, 67, 65, 64, 62, 62, 72};
void setup()
{
//first, start the main serial port at 9600 baud
   Serial.begin(9600);
//now start the new object to output MIDI on pin 5
   SerialMIDI.begin(31250);
}//end of setup()
```

```
// iterate noteArray to output the note values stored in each array index
void midiOutput(byte midiSequence [], byte arrayLength){
        //set up a for loop to iterate our midiSequence Array
    for(byte loopCounter = 0; loopCounter < arrayLength; loopCounter ++)
    {
    //use the SerialMIDI object to write out a 3 byte MIDI message
    SerialMIDI.write(0x90);//note on status byte
    SerialMIDI.write(midiSequence[loopCounter]);//note number from array
    SerialMIDI.write(100);//velocity of 100 (range 0-127)
    delay(100);
      //also write the data to Serial Monitor to check the structure is correct
    // Serial.print(0x90);//note on status byte
     Serial.print(midiSequence[loopCounter]);//note number for middle C (C4)
    // Serial.print(100);//velocity of 100 (range 0-127)
     Serial.println("");//add a carriage return to each message
    }//end of for loop
}//end of midiOutput()
void loop()
{
        //call midiOutput() to keep loop() simple
        midiOutput(noteArray, noteArrayLength);
}//end of loop()}
```

In the listing, the SoftwareSerial library is included using `#include <SoftwareSerial.h>` to allow serial communication to be sent on digital pins other than 0 and 1 (which are used by the Arduino IDE to upload the code over USB). The SoftwareSerial object `SoftwareSerial SerialMIDI(4, 5)` was then set up to handle communications on pins 4 (Rx) and 5 (Tx). An array of integers was declared to hold the MIDI note values `int noteArray[noteArrayLength]`, which uses a constant `noteArrayLength` for the length of the array. Notice that as long as they are separated by commas, the array values can be declared across lines in the code – languages like C will continue to read the text after an array comma as being part of the array, so the declaration can be split to make it easier to read.

In setup(), the pinMode() calls from the previous examples are replaced to configure output pins with two begin() calls to start the Serial and SoftwareSerial `SerialMIDI` objects – you cannot run SoftwareSerial without the Serial object, and the Serial object can also be used to print data to the Serial Monitor for checking in the code tests below. The Serial object is configured to run at 9600 baud (which is a common value), while the SoftwareSerial object is set to 31250 baud for MIDI communication.

A call is then made in loop() to the `midiOutput` function – passing in the `noteArray` of MIDI note values and its length. The function iterates through `midiSequence []` to access each value in noteArray (this array is passed in through the function call to `midiOutput()`). These values are then written to the SoftwareSerial object using `SerialMIDI.write`, which writes a status byte `(0x90)` and then a velocity data byte `(100)` after the note value to complete the Note On message (see Figure 5.10 for a Note On message example of middle C4 at velocity 100). The Serial object is then used to output the same data to the Serial Monitor, allowing us to check everything is correct. The function can then be tested in two steps: the first step checks the notes from the array by commenting out the print() calls for status and velocity, the second step checks the entire MIDI Note On messages by removing the comments and running all of the code:

```
//comment the serial output to check note values only (code test step 1)
   // Serial.print(0x90);//note on status byte
Serial.print(midiSequence[loopCounter]);//note number for middle C (C4)
   // Serial.print(100);//velocity of 100 (range 0-127)
Serial.println("");//add a carriage return to each message
```

If both code tests run correctly, you should see a list of note numbers that contain the array values for step 1 and then a list of full MIDI Note On messages for step 2 (Figure 5.17).

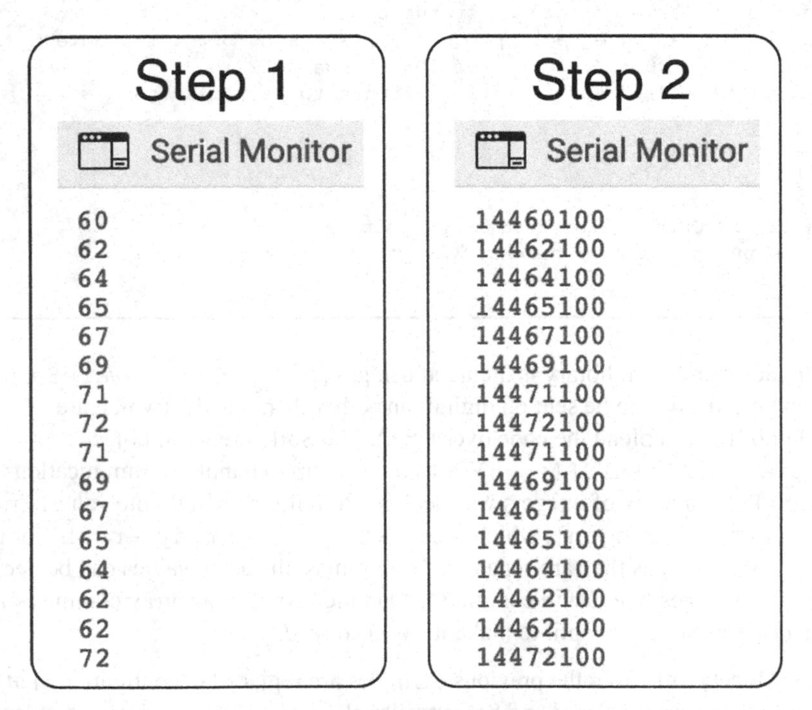

Figure 5.17 MIDI output code testing steps. ***The diagram shows a listing of notes only (to check against the noteArray data) and then a second test listing with the complete MIDI message for each note. If these listings display correctly, we can move on to test our SoftwareSerial object.***

If code testing does not create both of these outputs, you must debug the code listing before continuing on to testing the board. Some simple debugging steps were introduced in the example project at the end of the previous chapter, but the first step is to check that you have performed the copy/paste of the previous code listing correctly – particularly because the listing now spans two pages in the book, it may be possible that a character was omitted.

With the code tested, we can move on to stage 2 of testing – **test the board**. Full MIDI output cannot be simulated on a breadboard circuit within Tinkercad, so full testing of the board is not possible without a physical circuit. The output of the SoftwareSerial object can be tested by using a resistor/LED loop as a visual validator of digital output on pin 5. In section 4.9 of the last chapter, the basics of pulse width modulation (PWM) were discussed, where a square wave can be output from a digital pin with a varying-length duty cycle to create an average voltage as a result. In this case, we can exploit the fact that serial binary communication is effectively no different than PWM output – the only difference is the shape of the wave (Figure 5.18).

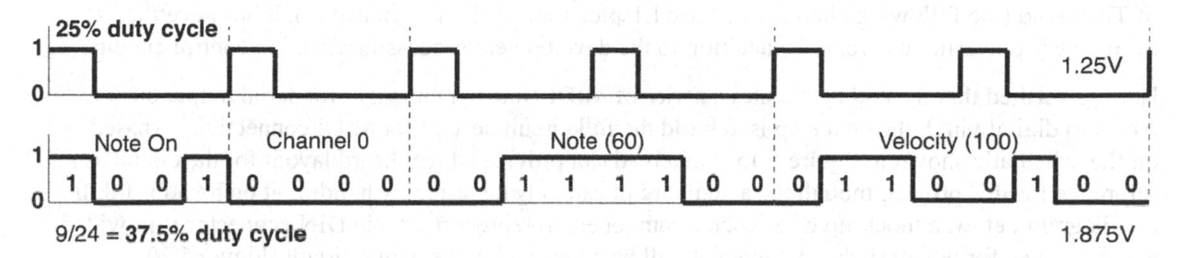

Figure 5.18 Serial binary communication signal output. ***In the diagram, each binary 1 is 5V, while a binary 0 is 0V (GND). Thus, the average output voltage for any binary data signal will be somewhere between 5V (all binary 1's) and 0V (all binary 0's). For serial communication, the average of this output signal can be used to light an LED as a simple test of output connectivity.***

In this diagram, the data for a typical MIDI Note On message (C4, velocity 100) is shown as a series of binary 1 (5V) and 0 (GND) output values. If considered as a PWM output, this data equates to a 37.5% duty cycle as there are a total of 9 binary 1 bits in a 24 bit message ($\frac{9}{24} = 37.5\%$). For a 5V digital output from the Arduino, this gives an average output voltage of 5 × 0.375 = 1.875V, which provides enough current ($I = \frac{V}{R} = \frac{1.875}{150} = 12.5mA$) to power the resistor/LED combination used in the circuits thus far (though the LED will be dimmer than in those circuits). This means the SoftwareSerial object output can be connected via digital pin 5 to a simple resistor/LED series loop to test whether it lights up (Figure 5.19).

Figure 5.19 Serial communication test circuit layout. ***Although not a full test of the MIDI output circuit, connecting an LED to the digital output of the SoftwareSerial object allows it to be tested to see if the data is being serialized and output to the correct pin. If the LED lights up, then some combination of 0's and 1's is being sent to digital pin 9 to provide this current.***

The circuit in the diagram in Figure 5.19 can be used as a (crude) digital signal tester – if the LED lights up then binary data is being sent by the SoftwareSerial object to digital output pin 9. By quickly testing the Arduino output connection to the board in this way, we can check the final piece of the code is functional without leaving the Tinkercad simulator. Although this project has now reached the limits of Tinkercad (the following chapters will use LTspice for AC circuit simulation), it has nevertheless been a very powerful and versatile addition to the development process for Arduino control circuits.

Having verified that the code generates a series of MIDI Note On messages for serial output on Arduino digital pin 9, the next step is to build the full circuit needed for MIDI connectivity – based on the schematic shown in Figure 5.15. Tinkercad can provide a breadboard layout for the circuit as a reference for this project, though the actual circuit can only be tested by building it on breadboard. In the diagram below, a mock-up of a generic component to represent a 5-pin DIN connector is provided as a reference for where such a component will be connected in the actual circuit (Figure 5.20).

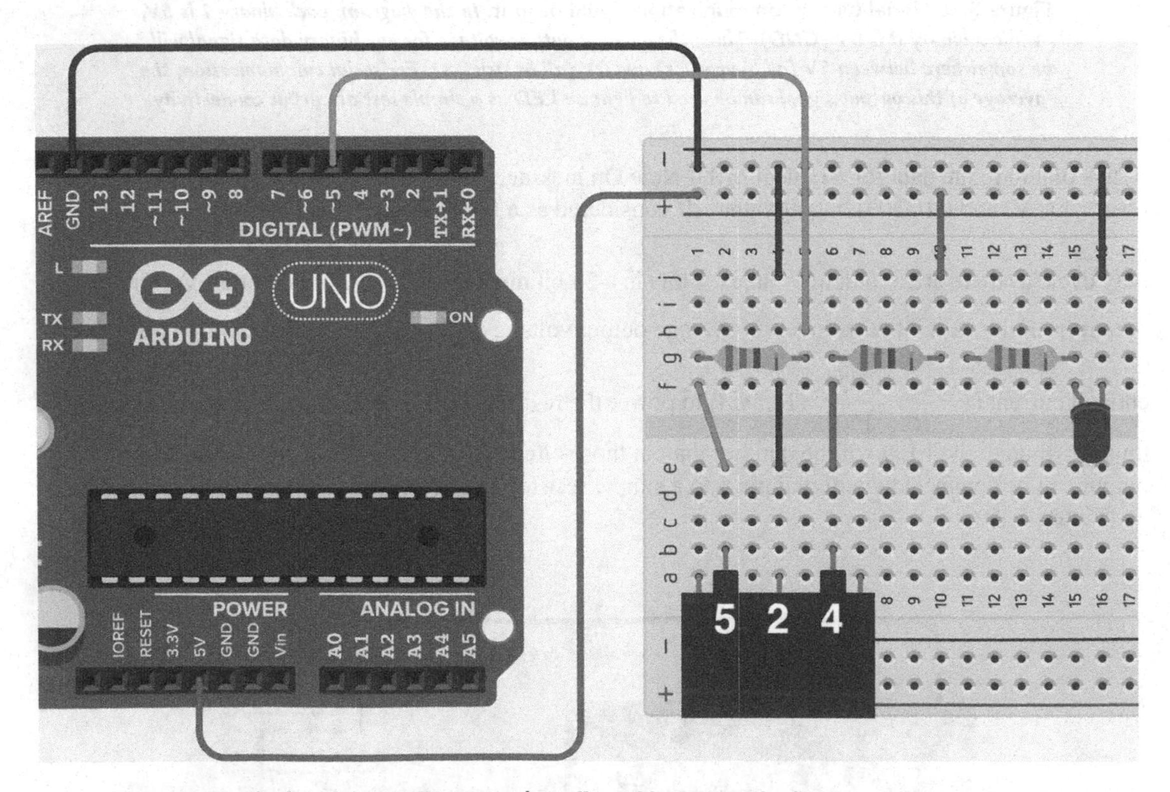

Figure 5.20 Final Arduino MIDI output breadboard layout. *In the diagram, +5V power from the Arduino runs through a 220Ω resistor to pin 4 of the DIN connector. Pin 2 of the DIN is connected to GND, whilst digital output from our SoftwareSerial object on Arduino pin 9 is connected through another 220Ω resistor to pin 5 of the DIN. An LED has also been added on pins [f15–f16], which can be used to test for serial output with digital pin 5 serial output connected to pin h11 (dashed line).*

In the diagram, the layout for a standard MIDI output interface based on the Arduino is shown. The Arduino provides +5V and GND connections for pins 4 and 2 (respectively) of the MIDI DIN connector, alongside a serial connection from digital pin 9 to pin 5 of the DIN that carries MIDI messages output by our code. This circuit can also be quickly modified for connectivity testing by placing an LED on pins [f15–f16] and connecting it to GND (right-hand diagram). If pin 5 of the MIDI connector is then disconnected (by removing breadboard connection j5) this effectively becomes the same serial communication test circuit as shown in Figure 5.19. This circuit can be used to test the SoftwareSerial code and then also test the board to ensure that MIDI messages are being sent to the MIDI DIN connector. Having said this, it cannot be verified that these MIDI messages are correctly structured until they are tested within the wider MIDI system introduced in Figure 5.13. To test this system, we must now build the actual circuit and connect it to a suitable MIDI interface that is capable of receiving (and responding to) MIDI data.

Project steps

For this project, you will need: 1. Arduino Uno board 2. One small breadboard (30 rows) 3. 1 × LED (for the test circuit) 4. 3 × 220Ω resistors 5. 1 × MIDI OUT connector (and MIDI cable to connect to a MIDI system) 6. 3 × connector cables and 7 × connector wires	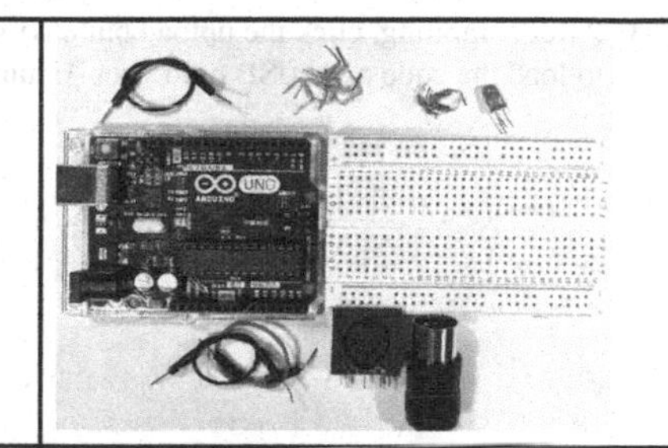
Add a 5-pin DIN MIDI Connector between pins [a1–a7]. The connector has a total of five pins (left-hand image), where the outer pins [a1] and [a7] are MIDI pins 1 and 3 (not used). MIDI pin 5 and MIDI pin 4 are extruded from the MIDI connector, this means that MIDI pin 5 is connected to [b2] and MIDI pin 4 is connected to pin [b6]. MIDI pin 2 is connected to pin [a4].	
1. Add a 220Ω resistor between pins [g1–g5]. 2. Add a 220Ω resistor between pins [g6–g10]. 3. Add a connecting wire from breadboard pin j4 to the negative (ground) rail [j4–GND]. 4. Add a connecting wire from breadboard pin j10 to the negative (ground) rail [j10–GND]. 5. Add a connecting wire between pins [f1–e2] – this is the MIDI DIN pin 5 connection. 6. Add a connecting wire between pins [f4–e4] – this is the MIDI DIN pin 2 connection. 7. Add a connecting wire between pins [f6–e6] – this is the MIDI DIN pin 4 connection.	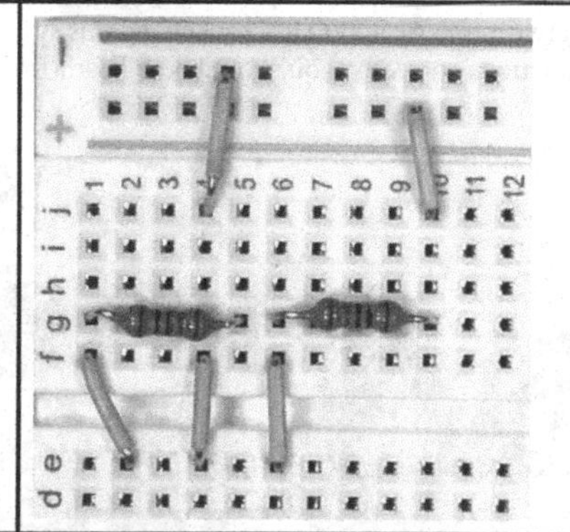
1. Add a connector from Arduino digital pin 5 to breadboard pin h5 [AD5-h5] – *this is the MIDI OUT connection.* 2. Add a connector from the negative (ground) rail to Arduino GND pin [ADGND–GND]. 3. Add a connector from the positive rail to Arduino +5V pin [AD5V–V+].	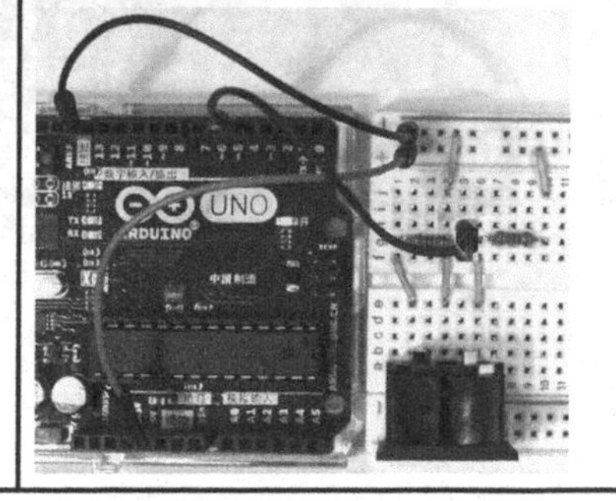

For the LED test circuit:

1. Add a 220Ω resistor between pins [g11–g15].
2. Add an LED – anode pin (f15) to cathode (f16).
3. Add a connector to the negative (ground) rail [j16–GND].
4. Connect the Arduino digital pin 5 connector to pin f11 (not shown).

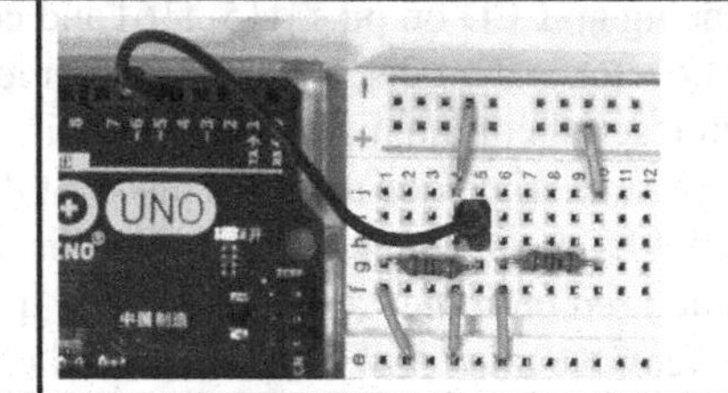

1. Launch the Arduino IDE and create a new sketch called chp5_Example2. Copy/paste the full code listing for this project into the editor window.
2. Click the tick box in the top left of the editor to compile the sketch; if there are no errors you will get a 'Done Compiling' Message above the debug window.
3. After compiling, click the upload button (right arrow next to the tick box) to load the code over USB onto your Arduino.

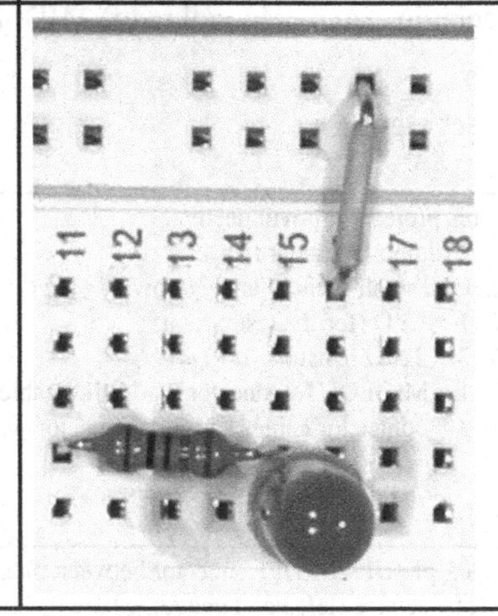

If this process has completed successfully, we can use the LED test circuit to check for serial MIDI output on digital pin 5.

Using whatever MIDI equipment you have available (either synthesizer or computer MIDI interface) connect a MIDI cable between the Arduino DIN socket and this system. Configure the system to receive MIDI data on channel 1. If the system functions correctly, you should now hear a tone sequence being output by the synthesizer in the MIDI system.

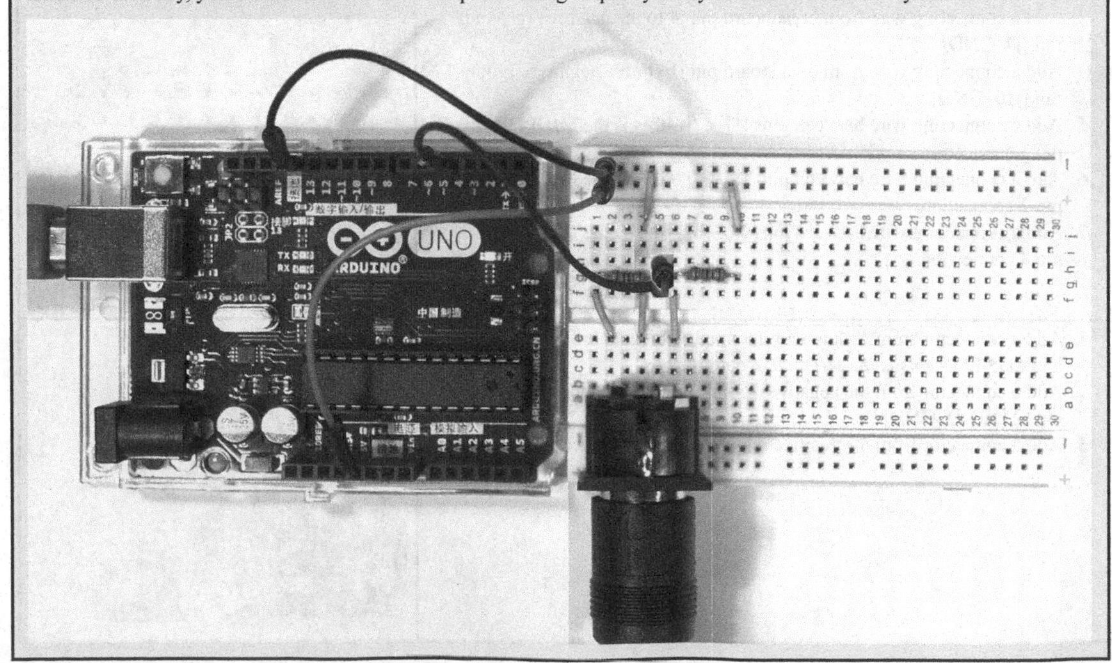

If everything has been built and tested correctly, you should now have a MIDI sequence player running on your Arduino that controls a MIDI-enabled synthesis device configured to respond on MIDI channel 1. Although the tone sequence will quickly become tedious to listen to (!), this represents a significant milestone in your learning, as you have now constructed a full MIDI output interface that can control another MIDI system to create sound output. In the following sections, this interface will be extended to respond to sensor input, first from switches (to learn about selection instructions) and then with a piezo sensor that acts as a drum trigger.

5.6 Conditions and digital input

At this point, a lot of programming concepts have been covered in a short space of time. The introductory programming chapters have discussed variables and sequence instructions, then looked at arrays (groups of variables) and functions (groups of instructions). This chapter introduced iteration as a means of processing the values in an array, leading to the MIDI sequence player example in the previous section. Now, we will look at the third (and arguably most complex) instruction type – selection. Selection instructions are essential to many areas of computing, but for this book they are primarily used to **evaluate sensor input** to a system. A **selection instruction makes a choice based on a condition**, where the condition is evaluated to be either **true or false** (Figure 5.21).

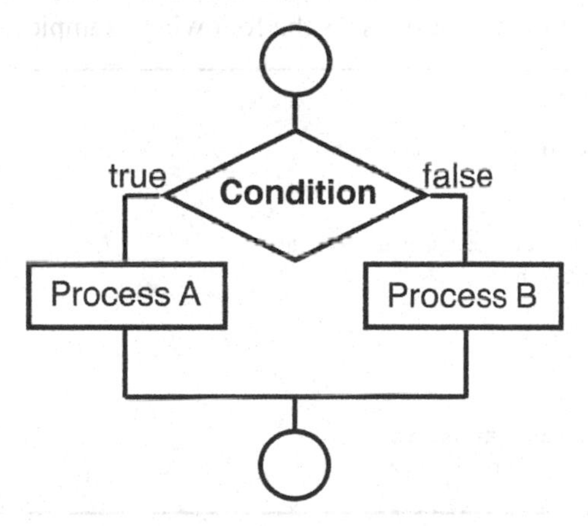

Figure 5.21 Selection instruction workflow. ***The diagram shows how a selection instruction will execute different instructions based on the evaluation of a condition. If the condition is true, then one process will execute, if it is false then another process will be executed instead. After completing the selection, the code continues with the next instruction in the sequence.***

In computer programming, **Boolean algebra** is used to evaluate the logic of a condition to arrive at an outcome that is either **true or false**. Boolean algebra was first proposed by George Boole in 1854, and it has since been extended to become the foundation of all computer logic. The basic structure of a selection statement reflects this logic:

```
if(condition){
    //condition is true - process A
}
else{
    //condition is false - process B
}//end of selection
```

The code listing above shows the core structure that must be used in order to execute a selection statement – if/true else/false. The conditional true/false is used because they are the logical opposite of one another, but in practice it is more likely to use the distinction of digital HIGH/LOW on the Arduino as these can easily be mapped to the different voltage levels of +5V (HIGH) and 0V (LOW) in a circuit. Note that a pair of braces is used for each outcome of the condition, with a total of four (alongside two brackets to hold the condition) being required. This is an important point to remember before studying examples of selection instructions, as Boolean algebra can often become very confusing. When logical operators are combined with these extra syntactic elements of brackets and braces it becomes very **easy to make mistakes with selection** instructions, so it is recommended to return to the listing above if you become confused or frustrated with conditional logic.

Working from the basic structure of a selection instruction, the next step is to learn how to **define the condition** it will evaluate. This book provides an introduction to programming (it does not go into logical operators in depth), and it can be argued that understanding conditions is one of the most difficult aspects of programming to learn. The approach will be to **keep the condition simple**, to avoid some of the more confusing arithmetic and logical operators that are often used in coding tutorials and examples. This is done to provide an introduction, which can be extended over time to include more complex logic to model system input for human interaction with sensors.

A condition is any arithmetic or logical statement that can be evaluated as being either true or false. Conditions are thus constrained in their options, as the following examples show:

```
if(raining){
    //it is raining, so use an umbrella
}
else{
    //it is not raining, no umbrella needed
}//end of raining selection
if(tired){
    //time to rest!
}
else{
    //wide awake, continue studying!
}//end of tiredness selection
```

The above examples are deliberately simple, but already highlight the limitations of simple conditional logic. The definition of rain is seasonal, geographical and is rarely as simple as on/off. Similarly, tiredness is a complex combination of mental and physical fatigue alongside other emotional and motivational factors. These examples highlight the complexity of modelling human interactions – what initially appears to be a simple coding task can rapidly become a philosophical discussion – complexity cannot easily be modelled with simple logical outcomes. For this reason, the conditions in this book use switches as system inputs that function as on/off values. Although this makes teaching system inputs much simpler, real-world circuits will often be much more nuanced. Chapter 2 (section 2.5) introduced the simplest form of sensor input – the push-button switch. Now this switch operation can be used in a selection statement to define its condition (Figure 5.22).

The diagram defines the digital states HIGH (+5V) and LOW (0V) as being the two possible outcomes of the condition of a push-button switch (Arduino digital output was discussed in chapter 4 sections 4.2 and 4.3). This is where conditions are both simple and powerful; by working with digital signals (HIGH/LOW) as inputs they can easily be used in coding processes. For this condition, a +5V source

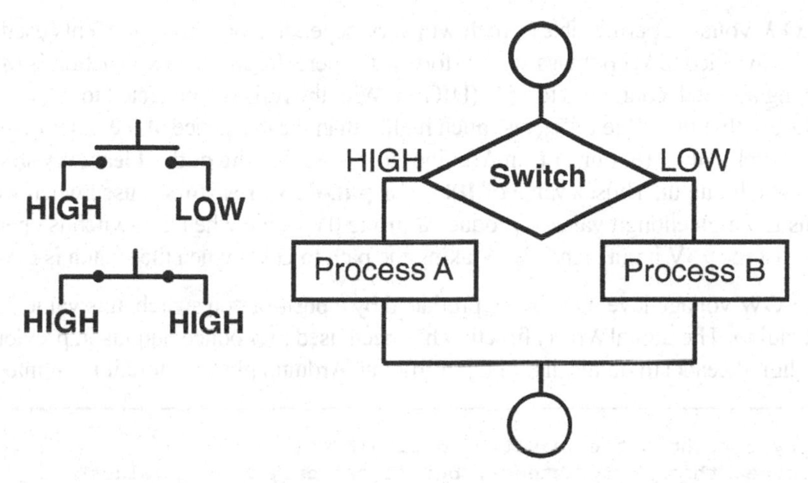

Figure 5.22 Push-button switch selection example. ***The diagram shows the schematic representation for a push-button switch with a HIGH input. Either a HIGH (+5V) or GND (0V) signal can be output by opening or closing the switch. The right-hand selection statement workflow shows how this HIGH/LOW output value can be used to define the two outcomes of the condition of the switch.***

can be connected to the switch and then the Arduino digitalRead() function can be used to determine the output value of the switch. If the switch is closed, then +5V (HIGH) will be present on both pins and the condition will be evaluated as true – when the switch is open, the pin will be seen as 0V (GND). In practice, an additional step is needed to ensure that the signal on the output pin will be 0V (LOW) when the switch is open, by adding a pull-down resistor to ensure the pin will always go to GND (Figure 5.23).

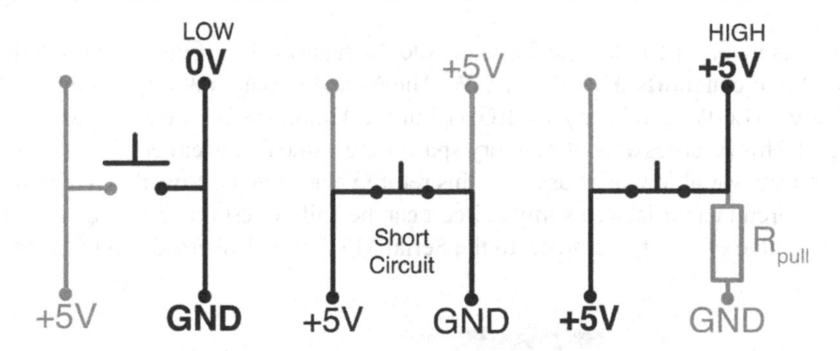

Figure 5.23 Pull-down resistor example. ***In the right-hand schematic, a resistor (Rpull) is added to tie the bottom right pin of the switch to 0V (GND). When the switch is closed, the path to the output pin has less resistance than the path to GND, so the output will be +5V. Without the resistor, when the switch is open (left-hand schematic) the actual voltage on the right-hand pin may 'float' to somewhere other than 0V due to stray currents or noise. In addition, simply connecting the pin straight to GND (middle schematic) would create a short circuit from the +5V input to GND when the switch is closed.***

The explanation for pull-down resistors can become complex, as the concept of defining LOW (0V) and HIGH (+5V) voltage levels has practical constraints relating to what actual levels will be acceptable – i.e. how much less than 5V is still HIGH and how much greater than 0V is still LOW. This primarily affects the calculation of the resistance value needed for R_{pull}, which should take into account both the minimum HIGH and maximum LOW voltages permissible (which will vary depending on the components used). These calculations will be avoided to keep things straightforward, where the simpler explanation is to avoid a short circuit when closing a switch connected to +5V (HIGH). With the resistor connected to GND, the primary focus is then ensuring that the value of R_{pull} is much higher than the resistance of the other component connected to the switch output (in our case an Arduino) to ensure that **the path of least resistance** will always be to the switch output. Thus, a value of **10kΩ for pull-down resistors** is used when working with the Arduino – this is a high enough value to produce a strong 0V signal when the switch is open, and also high enough to resist the +5V input signal from taking the path to GND when the switch is closed.

With a HIGH or LOW voltage level now being produced by a push-button switch, this value can be used as an input to the Arduino. The digitalWrite() function has been used to produce outputs in previous circuits, now we can use digitalRead() to obtain digital input from an Arduino pin and store it in an integer variable:

```
// define a byte to hold the result of digitalRead()
//Arduino defines this as an integer, but it can only have 2 values!!
byte buttonInputState = 0;//we must always initialize a variable with a value
//define an input pin to read digital data from
byte dataInputPin = 5;

void setup() {
pinMode(dataInputPin, INPUT);     // sets digital pin 5 as input
}//end of setup()
void loop(){

// read the state of the input pin by calling digitalRead()
//and store the result in the buttonInputState variable
 buttonInputState = digitalRead(dataInputPin);
}//end of loop()
```

In the listing, a variable of type byte is declared to hold the return value of the function digitalRead() – this will be one of two **constants**: **HIGH** or **LOW**. The Arduino library (Wiring.h) defines these constants as binary 0 (LOW) and binary 1 (HIGH), but the Arduino reference example defines the return type as an integer! This reserves a lot of memory space for a value that is either a binary 0 (LOW) or 1 (HIGH), and so a **byte** variable will be used for this return value instead. Now that a digital pin value can be input and stored in a variable, a simple circuit can be built to test this by using the core Serial object to trace the value of `buttonInput` to the Serial Monitor in Tinkercad (Figure 5.24).

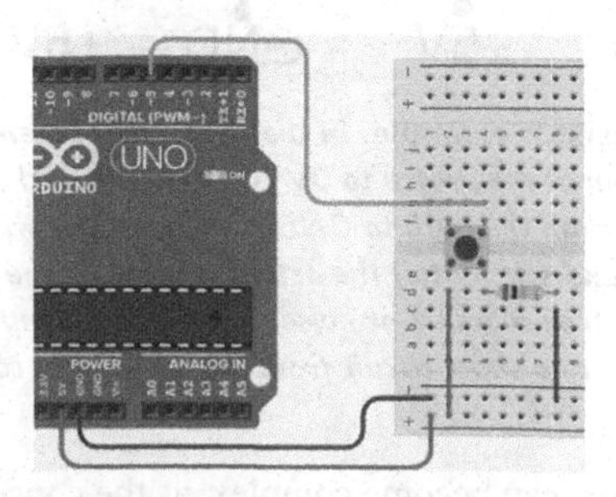

Figure 5.24 Button input test circuit. ***This circuit connects a push-button input to the +5V pin on the Arduino, with the open switch output being held to 0V (LOW) using a pull-down resistor. The output of the switch is connected to Arduino digital pin 5, where it can be read using the digitalRead() function call. If the switch is pressed, a +5V (HIGH) signal will be sent to digital pin 5.***

This circuit will be extended in the next two examples to build the MIDI drum trigger project, so a full build process is not provided at this point, but to follow the development process through Tinkercad the connection instructions for this circuit are as follows:

1. Add a push-button switch between f2 (top left) and e4 (bottom right) [f2–e4].
2. Add a 10kΩ resistor between pins [d4–d8].
3. Add a connector to the negative (ground) rail [c8–GND].
4. Add a connector to the positive (+5V) rail [d2–V+].
5. Add a connector between the bottom ground rail (near column 1) and the Arduino bottom left GND pin.
6. Add a connector between Arduino digital pin 5 and breadboard pin g4 [AD5–g4].
7. Add a connector between the bottom positive rail (also column 1) and the Arduino bottom left +5V pin.

The previous code listing can be extended to output the value of `buttonInput` to the Serial Monitor:

```
// define a byte to hold the result of digitalRead()
byte buttonInputState = 0;
//define an input pin to read digital data from
byte dataInputPin = 5;
void setup() {
  // initialize dataInputPin for push button input
  pinMode(dataInputPin, INPUT);
  Serial.begin(9600);
}//end of setup()
void loop() {
       // read the push-button value:
       buttonInputState = digitalRead(dataInputPin);
       //now trace this value to the Serial Monitor
       Serial.print(buttonInputState);
       //add the carriage return to make the output
       Serial.println("");
}//end of loop()
```

If this code executes correctly, when the push button is pressed you should see a change from 0 to 1 in the Serial Monitor output trace in Tinkercad (Figure 5.25).

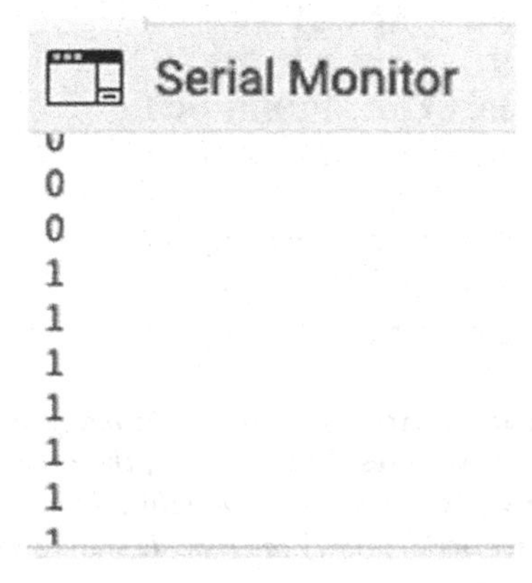

Figure 5.25 Push-button test circuit output. ***The image shows how the Serial Monitor output changes from 0 (LOW) to 1 (HIGH) when the push button is pressed. The baud rate (9600) means the repeated output of the value can often lag behind the actual button contact, but for our purposes this data is sufficient for testing digital input.***

If the code has compiled correctly and the circuit is configured properly, the previous listing can be extended to include a **selection statement** that tests the value of the digital input pin to execute different instructions in each case:

```
// define a byte to hold the result of digitalRead()
byte buttonInputState = 0;
//define an input pin to read digital data from
byte dataInputPin = 5;
void setup() {
  // initialize the push-button pin as an input
  pinMode(dataInputPin, INPUT);
  Serial.begin(9600);
}//end of setup()
void loop() {
       // read the dataInputPin value
       buttonInputState = digitalRead(dataInputPin);
  // selection statement that writes out a message when the button is pressed
  if (buttonInputState == HIGH) {
    Serial.print("Button Pressed!");
    Serial.println("");//add a carriage return for legibility
  }
  else {
    Serial.print("No Input!");
    Serial.println("");//add a carriage return for legibility
  }
}//end of loop()
```

If the code compiles and executes correctly, you should now see the output trace shown in Figure 5.26 in the Tinkercad Serial Monitor window.

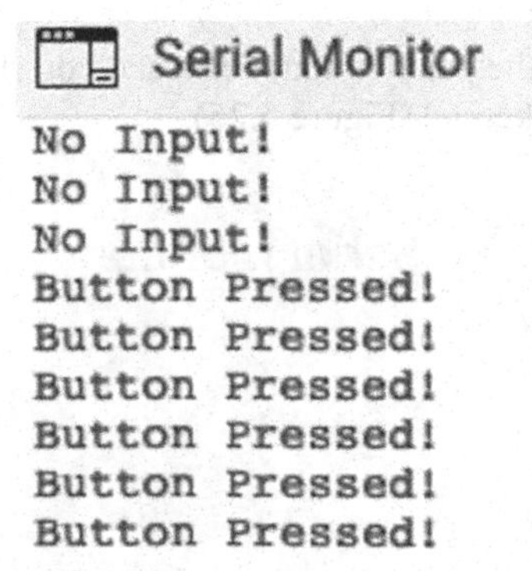

Figure 5.26 Push-button selection statement output. *The image shows the Serial Monitor output based on the selection statement that evaluates the condition of the buttonInputState variable. If the push button is pressed then a Serial print() call is made to state this, if the push button is not pressed (else) then another Serial print() call is made stating no input.*

At this point, digital input can be read from a switch as the basis of a selection statement that executes different processes as a result. In the next example. this workflow will be used to control MIDI Note messages for output. Before doing so, there is one final aspect of selection instructions that can be included to further extend the flexibility of the system code – the **else if** statement (Figure 5.27).

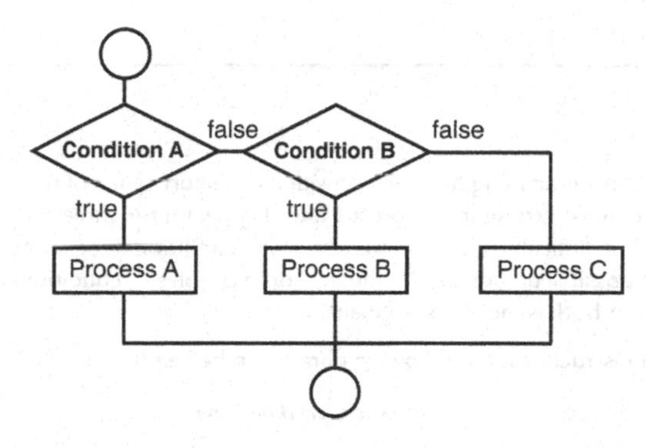

Figure 5.27 Else if statement workflow. ***The diagram shows how a selection instruction will execute different instructions based on the evaluation of multiple conditions. If the first condition is true, then one process will execute, if it is false then a second condition is evaluated. If this is true a second process will execute, if it is false then a third process will be executed instead.***

In the diagram, the 1st condition (Condition A) is evaluated in the if statement. If the **1st** condition is **true** then the selection executes **Process A** and the statement ends. If the **1st** condition is **false** then the selection evaluates a 2nd condition (Condition B). Again, if the **2nd** condition is **true** then the selection executes **Process B** and the statement ends. However, if the **2nd** condition is also **false** then the selection executes the code inside the else statement (**Process C**):

```
if(conditionA){
    //condition A is true - Process A
}//end of if
else if(conditionB){
    //condition B is true - Process B
  }//end of else if
else{
    //neither condition is true - Process C
}//end of selection
```

In the listing, the additional else if statement allows multiple conditions to be tested at once. This means the number of inputs that can be processed can be increased, allowing the Arduino to respond to multiple sensors at the same time. There is one limitation of this statement, however: if condition A is true, then condition B is never evaluated. This is a crucial point in terms of the logic of the statement, as a priority has been created between the switches where switch A will always override switch B. This is not an issue for this example, but if this code were to be extended to select between more switches (i.e. a rudimentary MIDI keyboard) then a more flexible selection statement would be needed where **all conditions are mutually exclusive** of each other.

An else if statement is used in the next example, where two push-button switches are used to generate different MIDI Note messages for output through the Arduino MIDI interface. It is worth noticing that there are now six braces and four brackets (for the two conditions) in the code – it is becoming more difficult to keep track of the structure and processes involved. These issues will be discussed in the following short tutorial on how to code, to allow the more effective implementation of the subsequent MIDI switch example.

The importance of else

A lot of information on Arduino programming has been provided in a short space of time and examples have deliberately been kept simple to avoid getting into more advanced code. Of the three core instruction types, selection instructions are arguably the most difficult to learn as they combine conditional logic with system (and user) modelling to produce multiple possible outcomes. Although short examples of conditional statements have been used, the fundamental element in both is the else statement.

When learning about selection instructions, the following phrase can be helpful:

Else means *everything* else

This means that the else statement is where every other possible outcome that the conditions did not test for must be dealt with. You may have used a web browser and got the **404: page not found** error – this is effectively an example of an everything else condition – it has little to do with the actual error involved. Similarly, the previous example based on if(raining) effectively covers everything from sunshine through clouds to snow (which is not rain!). For this reason, it is important to note the importance of the else option – it is often the one that contains most of the problems the system must deal with. It is important to be aware of the scope of else when developing robust and useful audio systems – at the very least, the code should not ignore potential problems with a default response like '404: page not found'.

5.7 Tutorial – how to write code part II

The previous chapter introduced the concept of coding structure as a means of highlighting how code is not linear text. This chapter has covered a lot of conceptual ground, and it is important to take stock and revisit the structural and procedural elements of the code involved. The structural elements of both iteration and selection instructions are significant, and the use of brackets and braces (as with functions) requires a dedicated approach to writing them. The structures for iteration and selection are the best place to begin:

```
for (start; end; increment) {
  //instructions execute each time the loop iterates
}//end of for loop
```

```
if(condition){
    //condition is true - process A
}
else{
    //condition is false - process B
}//end of selection
```

Before thinking about how often the loop has to iterate, or how best to define the condition, it is crucial to **set up the structure of the statement**. As the previous if/else if/else statement shows, six braces and four brackets leaves a lot of room for error. Comments have been provided on structures throughout the examples, and whilst this is not a method seen in most coding tutorials it is nevertheless a useful way of keeping track of iteration, selection and function stubs before they manifest an error due to a missing brace – often with a compiler error that points somewhere else.

The Tinkercad and Arduino IDE debuggers have not been discussed in any detail, as their proper integration into a coding workflow is outwith the scope of this book. Having said this, debuggers can often flag problems and errors with meaningful explanations if the coding structure is sound (particularly in shorter code examples), so it is recommended to do the following when code does not compile:

1. **Count the brackets and braces** – they should both be (separate) even numbers.
2. **Check the semicolons** – every sequence instruction must have one.
3. **Check for missing loop elements** – e.g. declaring the count variable type and that the end condition is correct.
4. **Check for badly formed conditions** – can the condition evaluate to true/false or LOW/HIGH?
5. **Arrays and loops count from 0** – also, array elements must be initialized before use.

This chapter has covered most of the basic elements of computer programming that relate to the Arduino, but is not a substitute for coding experience – this will take time. Reading code written by others is a great way to learn, and unlike written work a lot of code is freely accessible for use by others. This can often be a little surprising if you have never encountered computer programming before now:

> **A good programmer knows how to use the code of others**

Whenever possible, applying and amending code from other sources is arguably the best way to become proficient at programming. It is very rare, even after many years of coding development experience, that you will start writing any computer program from scratch. It is much more common to take code from other sources and adapt it to solve a problem. Whilst on some occasions this can cause problems (e.g. trying to understand third-party code that works, but doesn't explain how!), in most instances the computer programming community are very helpful and supportive of those trying to learn new skills.

5.8 Example 3 – MIDI switch controller output

In this example, the previous MIDI output circuit is adapted to include two push-button switches. The code will also be reduced in some areas, by removing the array iteration previously used to generate a sequence of MIDI notes for output. In this example, a selection instruction responds to switch inputs by sending different MIDI Note data as serial output based on the condition of each switch (Figure 5.28).

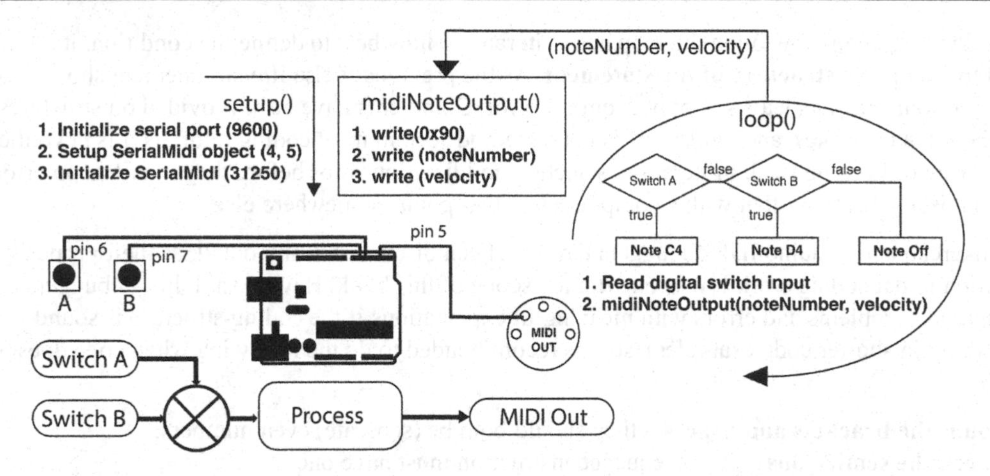

Figure 5.28 Arduino MIDI switch controller code process diagram. ***The diagram shows how input from two switches is used by a selection statement to output different MIDI Note messages. The selection statement calls the function midiNoteOutput() to serialize the three data bytes in each MIDI message, to keep the code in loop() to a minimum.***

This example contains the **first full system** that processes sensor inputs to create actuator outputs. A function called midiNoteOutput() has been added to handle the serialization of the note and velocity data for each message, partly to keep loop() from becoming too cluttered as it will now have an if/else if/else selection statement (with lots of brackets and braces to lose track of!). In addition, sending the three MIDI Note On bytes is a good example of when a function is needed – to reduce and reuse the code involved. The MIDI Note velocity byte can be used to effectively create a Note Off message by writing velocity 0 for the note – the code listing will show that this logic is deliberately simplified but not without some drawbacks as a result. An updated schematic can now be derived for the example, which combines switch inputs with serial output (Figure 5.29).

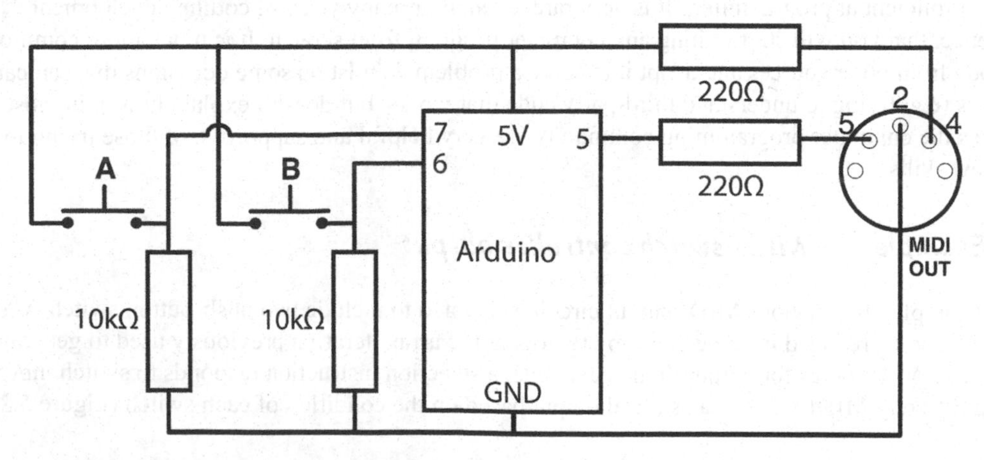

Figure 5.29 Arduino MIDI switch controller schematic. ***The diagram shows two push-button switches (A and B) with 10kΩ pull-down resistors connected as inputs to digital pins 6 and 7 on the Arduino. Pin 4 is configured for serial output to a standard MIDI OUT interface that uses 220Ω resistors for current limiting. Note the connection between the switch A output and Arduino pin 7 has a small bump in the wire, to show that it is not connected to the 5V input for switch B.***

In the schematic, digital input and output are combined to produce a full MIDI OUT controller circuit. The switches provide +5V signal inputs to pins 6 and 7 when they are connected, with pull-down resistors being added to ensure no short circuits occur. The Arduino then processes these signals using a selection statement with two conditions (if/else if/else) to generate serial output to the standard MIDI OUT interface on the right of the schematic. This schematic can now be translated to a breadboard layout that can be prototyped (though not implemented) in Tinkercad (Figure 5.30).

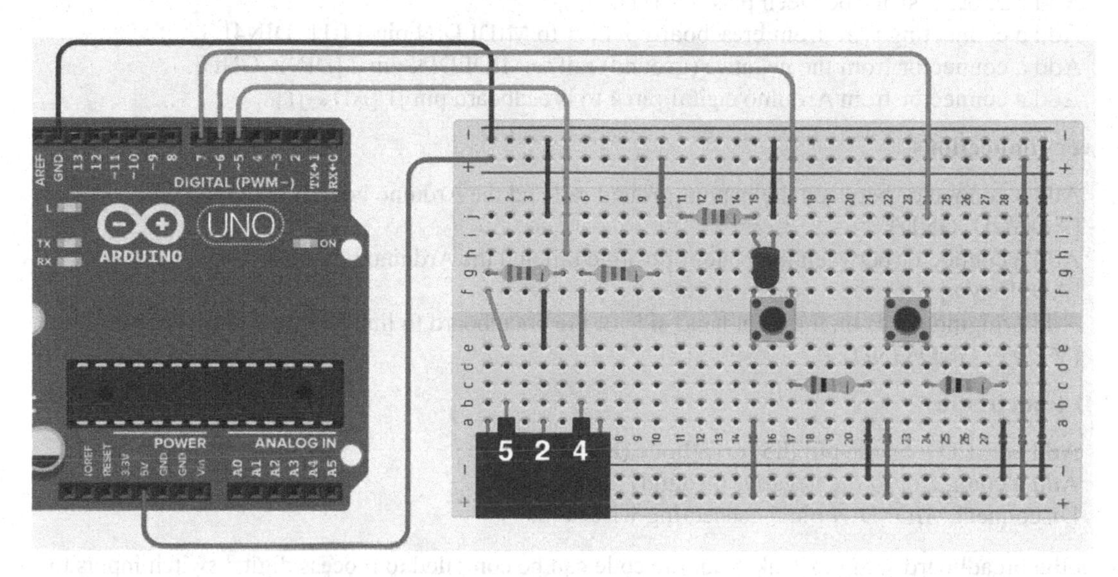

Figure 5.30 Arduino MIDI switch controller output example. ***The diagram shows the breadboard layout for two push-button input switches (with pull-down resistors) that change the MIDI Note information sent to the MIDI OUT hardware output interface. The LED serial test circuit shown in the previous example has been omitted for clarity.***

The breadboard layout shown in the figure combines two digital switch inputs on pins 6 and 7 with a MIDI OUT interface (bottom right of the board). A similar circuit was used in the previous example, and this circuit extends it by bridging the power rails to provide access to both the top and bottom of the breadboard. The switches have 10kΩ pull-down resistors connected to GND, but the layout omits the LED serial tester circuit used in the previous example to keep things simple (these components can be added very quickly when needed for testing). As with the previous example, this circuit will be adapted again to build the MIDI drum trigger project, so a full build process is not provided at this point. For Tinkercad simulation and testing, the connection instructions for this circuit are as follows:

Switch inputs:

1. Add a push-button switch between f12 (top left) and e14 (bottom right) [f12–e14].
2. Add a 10kΩ resistor between pins [d14–d18].
3. Add a connector to the negative (ground) rail [c18–GND].
4. Add a connector to the positive (+5V) rail [d12–V+].
5. Add a connector between Arduino digital pin 6 and breadboard pin g14 [AD6–g14].
6. Add a push-button switch between f20 (top left) and e22 (bottom right) [f20–e22].
7. Add a 10kΩ resistor between pins [d22–d26].
8. Add a connector to the negative (ground) rail [c26–GND].

9. Add a connector to the positive (+5V) rail [d20–V+].
10. Add a connector between Arduino digital pin 7 and breadboard pin g22 [AD7–g22].

MIDI OUT interface:

1. Add a 220Ω resistor between pins [i1–i15].
2. Add a connecting wire from breadboard pin j5 to MIDI DIN pin 5 [j5–DIN5].
3. Add a 220Ω resistor between pins [i7–i11].
4. Add a connecting wire from breadboard pin j11 to MIDI DIN pin 4 [j11–DIN4].
5. Add a connector from the negative (ground) rail to MIDI DIN pin 2 [DIN2–GND].
6. Add a connector from Arduino digital pin 4 to breadboard pin j1 [AD4–j1].

Power connections:

1. Add a connector between the bottom ground rail and the Arduino bottom left GND pin [ADGND–GND].
2. Add a connector between the bottom positive rail and the Arduino bottom left +5V pin [AD5V–V+].
3. Add bridging wires on the right-hand side of the breadboard to link both sets of power rails (V+/V+, GND/GND).

LED test circuit:

1. Add an LED – anode pin (h5) to cathode (h6).
2. Add a connector to the negative (ground) rail [j6–GND].
3. Disconnect MIDI DIN pin 5 connecting wire at j5.

With the breadboard setup in Tinkercad, the code can be compiled to process digital switch inputs to generate MIDI serial data outputs. As the code for a complete digital system combines more complex processes, the listing will be covered in three sections to discuss each one individually:

```
#include <SoftwareSerial.h>//include SoftwareSerial for MIDI data output
SoftwareSerial SerialMIDI(3, 4);//SoftwareSerial object with pin 4(TX) for
  MIDI output
byte buttonOneState = 0;//setup 2 variables for push-button state
byte buttonTwoState = 0;
const byte buttonOneInputPin = 6;//input pins to read the button output
const byte buttonTwoInputPin = 7;
//setup starts the serial port/SoftwareSerial MIDI and configures switch
  input pins
void setup()
{
    Serial.begin(9600);
    SerialMIDI.begin(31250);
    pinMode(buttonOneInputPin, INPUT);
    pinMode(buttonTwoInputPin, INPUT);
}//end of setup()
```

The initial setup in this example involves both switch input and serial output. An input switch component has been added that requires additional code elements to use it. First, two byte variables (`buttonOne, buttonTwo`) are defined for the buttons that can either be digital 0 (LOW) or 1 (HIGH) depending on whether the switch has been pressed. Byte constants are then defined for the digital input pins involved (pins 6 and 7) – these pins must not change during the operation of our circuit, so it makes sense to declare them as constants. In the setup() function, the Serial and SoftwareSerial (`SerialMIDI`) objects are started so serial MIDI data can be output to the MIDI OUT interface circuit connected to digital pin 5. The pinMode() function is also used to configure pins 6 and 7 for digital input, so the switch voltage levels can be read from these pins. With the setup completed, the `midiNoteOutput()` function can be declared to output MIDI data bytes:

```
// midiNoteOutput is used to output the 3 serial data bytes in a Note On message
void midiNoteOutput(byte noteNumber, byte velocity){
    SerialMIDI.write(0x90);//note on status byte
    SerialMIDI.write(noteNumber);//note number - either C4 or D4
    SerialMIDI.write(velocity);//velocity - either 100 or 0 (Note Off)
    //use Serial Monitor to check structure - commented unless testing
    //Serial.print(0x90);//status
    //Serial.print(noteNumber);//note number
    //Serial.print(velocity);//velocity
    //Serial.println("");//add a carriage return to each message
}//end of midiOutput()
```

The function takes two input parameters – the MIDI note number and the note velocity. The status byte of a MIDI Note On message is hexadecimal 0x90, so with this information all three bytes of the message can be written to the SoftwareSerial output object that will output to the MIDI OUT interface on pin 5. Calls to the serial object are also included that write data to the Serial Monitor in Tinkercad, though these are commented in the listing as they are not needed when running the actual circuit. Recall that in example 2 three testing steps were defined, which this section of the `midiNoteOutput()` function focusses on:

1. **Test the code** – ensure that the function structure is correct before testing that the call to `midiNoteOutput()` from loop() is working (listing below).
2. **Test the board** – before connecting the LED serial test circuit, use the Serial Monitor calls to check data is being built and transmitted by the ATMega328P.
3. **Test the system** – the full MIDI controller system cannot be tested until the code and board are running correctly.

With the `midiNoteOutput` function declared, it can now be called from the selection statement in loop():

```
//loop contains the selection that calls midiNoteOutput for each switch input
void loop()
{
  buttonOneState = digitalRead(buttonOneInputPin);//read in switch values
  buttonTwoState = digitalRead(buttonTwoInputPin);
  //now the selection based on these values
  if(buttonOneState){
      midiNoteOutput(64, 100);//switch one pressed, call midiNoteOutput() for C4
  }//end of if
  else if(buttonTwoState){
      midiNoteOutput(66, 100);//switch two pressed, call midiNoteOutput() for D4
  }//end of else if
```

```
    else{
        midiNoteOutput(64, 0);//this is our default condition - all notes off
        midiNoteOutput(66, 0);
    }//end of else
}//end of loop()
```

In loop(), the selection statement if/else if/else structure must be coded before adding the instructions that will be executed in each case of the condition. This is important to avoid introducing simple errors into the code due to a missed bracket or brace, alongside checking that the logical if/else if/else structure is built correctly. Each stage of the selection statement is commented to avoid this problem – there are now eight braces in loop(), and they can easily get confusing! The digital state of the buttons can then be obtained by calling the core Arduino function digitalRead(), which will return either a LOW (0V) or HIGH (+5V) based on the voltage level measured on pins 6 and 7. The result of these calls is then stored in the `buttonOne` and `buttonTwo` variables, so they can be evaluated in the selection statement.

The logic of the selection statement aligns with the code process diagram shown in Figure 5.28, where each condition relates to one of the input push-button states. The end of section 5.6 noted the important of else: **else means *everything* else.** This point is demonstrated here, where all MIDI notes must be off to prevent a switch triggering a note of indefinite length (anyone who has worked with MIDI for a long time will have experienced occasions when the system must be disconnected to turn off rogue note messages!). This is handled in the code by setting the default else condition to all notes off, but this has a limitation – what happens if switch A is held down and switch B is pressed at the same time?

The selection structure will prioritize switch A over switch B, so in practice switch A will always trigger in this case. Thus the system will not respond adequately to this condition, creating a monophonic (single signal) MIDI controller as a result. This is not a significant issue for this example as the focus is on building a first complete system from input to output. Having said this, in a practical MIDI controller the limitation of monophonic operation can have significant practical implications for chords and other concurrent note patterns (such as faster note sequences). The Tinkercad simulation should run this code correctly and trace out MIDI Note On messages to the Serial Monitor (Figure 5.31 overleaf).

At this point in the example, the Tinkercad simulation will not be able to work with MIDI data and connectivity (other than running the LED serial test circuit). The code and (to some extent) the board have been tested, but not the overall MIDI system that will be controlled by the MIDI data. Whilst the aim of this book is to provide full step by step examples throughout this book to avoid any ambiguity or confusion, the focus in this section is to introduce the processing of digital input within the system – not build a full circuit. This is the first full Arduino system in this book, and although this circuit will not be built step by step it still represents a significant advance in your learning. Sensor inputs can be evaluated using selection statements, and the outcome of a condition can execute different processes contained within a dedicated function. In the next section, this will be extended by learning how to measure analogue input signals with the Arduino. This will lead onto the final project of a MIDI drum trigger system that will output through a standard MIDI OUT hardware interface – a fully functional MIDI control system.

```
Serial Monitor
640
660
640
660
640
660
64100
64100
64100
64100
64100
640
660
640
660
640
660
640
660
```

Figure 5.31 MIDI switch controller serial output. ***The image shows the Serial Monitor output when switch A has been pressed. The default condition is Note Off (Note On with 0 velocity) for note numbers 64 (C4) and 66 (D4), but when the switch contact provides +5V to digital pin 6 the selection statement makes function calls to midiNoteOutput() to send a MIDI Note On message for note C4 at velocity 100.***

5.9 Analogue input – percussion sampling

In the previous section, switch input worked on the premise of a discrete digital signal being present – the switch output would either be LOW (OV) or HIGH (+5V). Chapter 2 (section 2.2) discussed **continuous signals**, where the voltage level of the signal continuously varies over time (Figure 5.32).

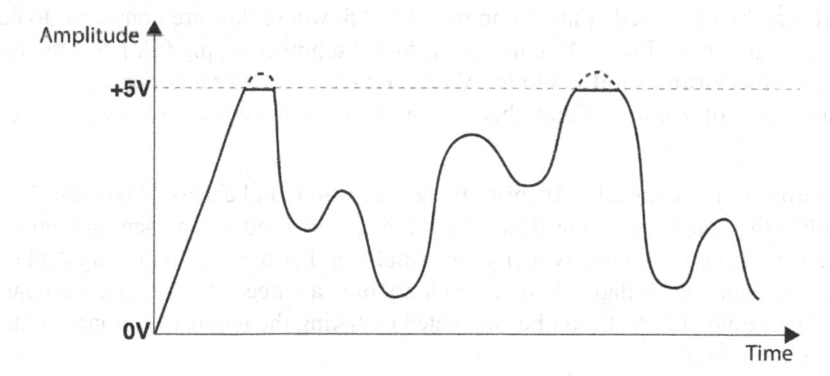

Figure 5.32 An analogue signal (taken from chapter 2, section 2.2). ***The diagram shows an example of an analogue signal, also known as a continuous signal. The output amplitude level (y-axis) of the signal continuously changes over time (x-axis). The Arduino can sample (measure) analogue signals based on a +5V reference, where +5V is the maximum input value the Arduino can accurately define. Notice that two parts of the signal are above the maximum (dashed lines) – their values will become clipped to the maximum input level of +5V when sampled.***

In the diagram, a continuous signal constantly varies in level over time – this distinguishes it from the digital signals in the previous section, which were always at one of two specific voltage levels.

The diagram also shows where the signal exceeds the measurement range for illustrative purposes, as in reality the problem of measurement is compounded by the sample rate and bit depth available (as briefly discussed in chapter 2). Thus, when sampling a continuous signal accuracy is lost due to lack of sampling resolution (Figure 5.33).

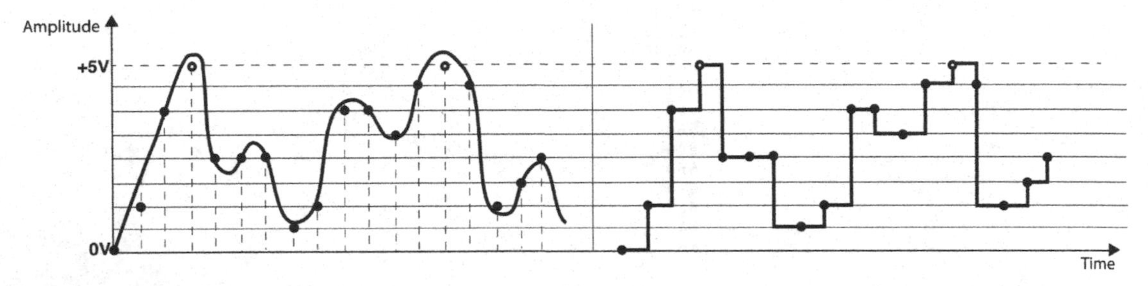

Figure 5.33 Continuous signal sampling example. *In the diagram, a continuous signal (clipped peaks shown as white circles) is shown on the left. A sampled version of the signal is shown on the right, where much of the shape of the original signal has been lost. Although this book will not work with sampled audio signals, it is still important to understand the impact of sampling resolution on the data it obtains to learn how a threshold for an input sensor can be set in code.*

In the diagram, the effects of sampling a continuous waveform are shown in the right-hand signal, where simple nearest-neighbour interpolation has been used to highlight the gaps between the data points. Although in practice more advanced interpolation methods would be used for audio signals, they require more processing power than that available on the ATMega328P. On the Arduino, analogue signals can be measured as inputs on pins A0–A5, where they are converted to data values by analogue to digital conversion (ADC), using the Arduino power supply (5V) as a reference for measurement. The **maximum input value is 5V** because the Arduino cannot operate above 5V, so it cannot measure any voltage higher than this (it would not have a voltage to compare the input signal with).

The Uno can sample an input signal at **10-bit/10kHz resolution**, and chapter 2 (section 2.3) showed that the bit depth is the number of values (the range) that can be used to represent the input signal. Section 2.2 also (briefly) covered binary, where the small number of values in binary (either 0 or 1) means that a large number of digits (known as binary bits) are needed to represent a quantity. The range of values for a given bit depth can be calculated by taking the number of binary values (2) and squaring it by the number of bits:

$$\begin{aligned} Binary\,range &= 2^n \\ \therefore Binary\,range(10\,bits) &= 2^{10} = \mathbf{1024} \end{aligned} \tag{5.1}$$

n is the number of bits

Note: 2 is a scalar representing the number of binary values (either 0 or 1).

Thus, a **10-bit binary** number provides a range of 1024 values, which will equate to an **integer** type in the code. This means that the Arduino can represent an input voltage between 0V and 5V (Arduino supply) using a range of 1024 values, giving an input resolution of:

$$Arduino\ Input\ Resolution = \frac{5}{1024} = 0.0049 = \mathbf{4.9mV} \quad (5.2)$$

Note: this assumes a supply of 5V (the Arduino can also run on 3.3V).

Whilst a 10-bit depth is more than adequate for many types of input sensors (e.g. switches, potentiometers) you will be familiar with 16-bit (e.g. compact disc) and even 24-bit audio formats that provide a much greater range of input signal measurement (24-bit gives a range of 16,777,216 values). With a resolution of 4.9mV, the Arduino should be able to read our sensor input with a sufficient level of accuracy. Although in audio-processing terms this would not be acceptable, for this use of sampling to detect input sensor changes the definition of a threshold (see below) based on a much lower resolution allows a simpler approach to be taken when processing the data obtained.

Chapter 2 (section 2.3) discussed the effect of reducing/increasing the sample rate, where just as with bit depth a lower sample rate also reduces the resolution of the digital representation of the signal. The Arduino also does not have enough memory or a high enough clock speed for audio sampling, and whilst this book will not provide an extended discussion on Nyquist's theorem, it is sufficient to remember that the minimum frequency being sampled must be **at least** half of the sampling rate. For sensor input however, with a sampling rate of 10kHz the Arduino is again more than adequate for the input sensors used in this book:

$$Arduino\ Sample\ Rate = 10kHz, Period, T = \frac{1}{Frequency}$$

$$\therefore Arduino\ Sample\ Resolution = \frac{1}{10,000} = 0.001 = \mathbf{1ms}(\text{milliseconds}) \quad (5.3)$$

Note: in practice, processing each sample will take much longer.

To provide some perspective, a sample resolution of 1 millisecond means it can theoretically detect all musical note durations within a wide range of tempos (Figure 5.34).

Beats per minute (BPM)	Note duration (milliseconds)					
	Whole	1/2	1/4	1/8	1/16	1/32
60	4000	2000	1000	500	250	125
80	3000	1500	750	375	188	94
100	2400	1200	600	300	150	75
120	2000	1000	500	250	125	62.5
140	1716	858	429	215	107	53.5
160	1500	750	375	188	94	47
180	1332	666	333	167	83	41.5
200	1200	600	300	150	75	37.5

Figure 5.34 Musical note durations. ***The table shows the durations (in milliseconds) of various standard musical notes (semibreve, minim, crotchet, quaver, semiquaver and demisemiquaver) for a range of tempos (60–200 bpm). This shows how the Arduino can sample an input sensor at 10kHz (1 millisecond resolution) and be capable of detecting a 1/32 note at over 200bpm.***

The figure shows common note durations for a range of tempos, to illustrate the difference in time resolution between a 37.5 millisecond demisemiquaver note at 200 bpm and the much higher sampling resolution of the Arduino at 1 millisecond. This table aims to clarify both the detectable time resolution of a sensor input (which is much smaller than most musical notes!) and also the response time of a processor like the ATMega328P (which runs at 16MHz cycles per second). With such high resolution in both sampling and clock speed, it would seem feasible that timing will never be an issue – the ATMega328P can easily generate serial MIDI Note On messages at the output rate required. This is noted in advance of building the MIDI drum trigger project, where the effective responsiveness of the trigger is not based solely on the resolution and execution speed of the Arduino. In practice, the input signal must also be measured relative to a *threshold* to define when the input sensor data is either digital HIGH or LOW (Figure 5.35).

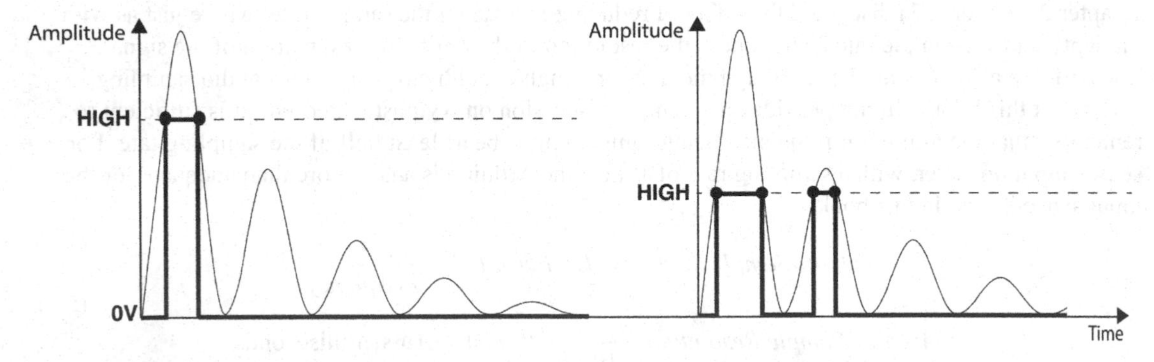

Figure 5.35 Thresholding analogue sensor input. ***In the diagrams, a critically damped sine wave mimics the idealized high initial amplitude and short duration of a percussive sound (called a transient sound). The threshold voltage (dotted line) must be set at the right level to detect the first oscillation of the input wave and thus the beginning of the percussive sound. The left-hand diagram shows an appropriate level, but if the threshold is not set correctly (e.g. too low) as in the right-hand diagram it can potentially detect multiple instances of the same sound.***

In the diagram, the fast onset time of the signal (known as the attack) is typical of the striking collision that generates a percussive sound (often called a drum stroke). The decay of the sound (known as the release) is also quite rapid, and the critically damped sine wave shown is close to that of a bass drum sound (particularly for electronic drums). The next chapter will discuss analogue audio signals (and the crucial role of sine waves) in more detail in relation to AC circuit theory. For now, the detection of the first onset threshold of a piezo input sensor can be used as a MIDI drum trigger. To do this, the Arduino must sample the continuous analogue input signal from the piezo and then use a condition to determine when this sampled value goes above the defined threshold. The threshold must be greater than 0, but the circuit must be tested to determine where the best value will be to help avoid multiple onset detections from the same sound (as shown in the right-hand diagram in Figure 5.35). The threshold must also be used to avoid detecting other vibrations that are not related to the actual drum stroke that produces the percussive sound (see Figure 5.36 overleaf).

The diagram gives some basic examples of the multiple sources that can occur within a percussion signal – the sensor does not discriminate on their source, it simply detects all vibrations. These additional vibrations manifest as noise in the input signal, and thus the threshold is used to avoid detecting them.

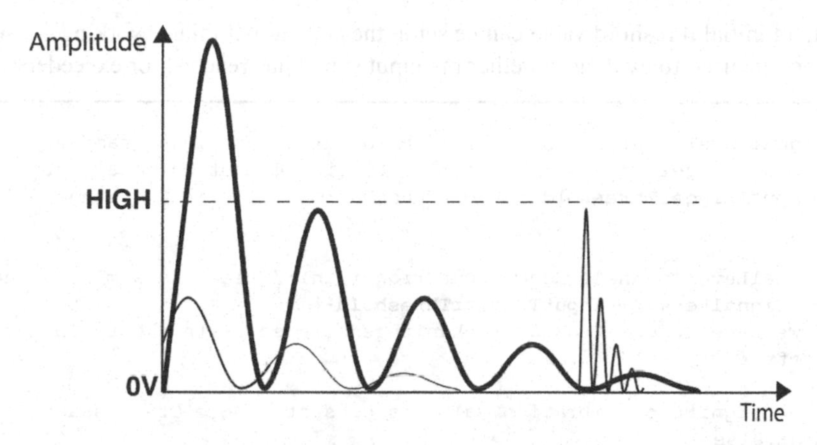

Figure 5.36 Analogue sensor vibration noise. ***The diagram shows the effect of other unwanted percussive signals that are also detected by a piezo sensor, which is very sensitive to all movement. It is common for multiple percussive instruments to be placed in close proximity (the right-hand transient mimics another drum head), or for more than one drumstick or beater to be used in tandem (the bottom left transient mimics the decay from a previous stroke) – all of which can generate vibrations as noise.***

This creates another problem if the threshold is set too high to avoid false triggering due to noise – the potential range of input signals is now reduced to detect only the very loudest sounds (Figure 5.37).

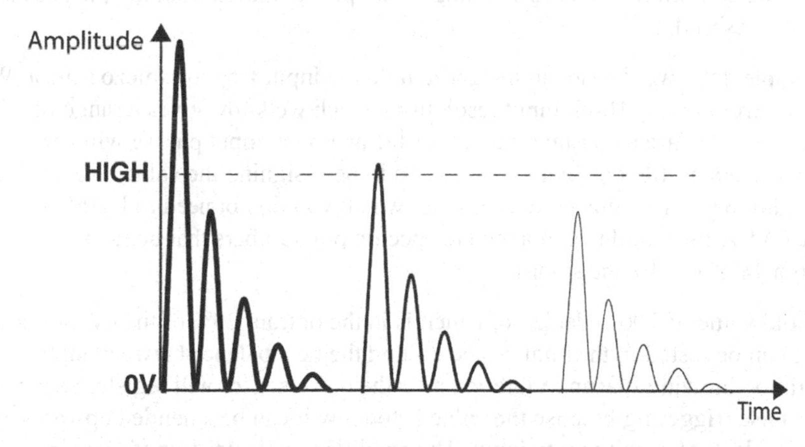

Figure 5.37 Reduced vibration sensitivity. ***The diagram shows the impact of setting a higher onset detection threshold to avoid false triggers due to noise. Of the three percussive sounds shown, only two have an amplitude greater than the defined threshold (the middle sound is just over the threshold). This means the detector will miss the quieter sound on the right, which may significantly reduce the expressiveness of the MIDI drum trigger.***

Too low a threshold can lead to the detection of multiple onsets, but the threshold level must also not be set too high – thus reducing the sensitivity of the trigger. These issues are discussed to highlight the complexity involved in building a fully functional MIDI drum trigger interface – the difference between this simple Arduino example and a commercial interface is a significant amount of design, modelling and testing to develop accurate onset detection across a wide range of drums and percussion.

For this project, an initial threshold value can be set in the code to get things working – a selection statement can then be used to evaluate whether the input signal has reached (or exceeded) this value:

```
int inputSignalLevel = 0;//a variable to hold our input signal sample
const int inputTriggerPin = A0;//this maps to pin 14, but A0 makes more sense!
const int inputTriggerThreshold = 200;// use a low value in the range 0-1023
void loop()
{
    inputSignalLevel = analogRead(inputTriggerPin);//read the sample value
    if(inputSignalLevel > inputTriggerThreshold){
        //we have an onset greater than threshold so write a Note On message
    }//end of if
    else{
        //no significant vibration value is present - Note Off message
    }//end of else
}//end of loop
```

In the loop() function, the value of the analogue input pin is obtained using analogueRead() and then stored in the variable `inputSignalLevel`. This variable is then used to evaluate the result of the expression `inputSignalLevel > inputTriggerThreshold` in the selection statement. The mathematical symbols for greater than (>), less than (<) and equivalent to (==) are widely used in selection statements to evaluate a quantity, where the result of the expression is either a Boolean true or false. This means the condition is only responding to samples obtained from the piezo sensor that are above the defined threshold.

To obtain this sample data, we declare an integer to hold the input from our piezo sensor. We use an integer type as the Arduino has 10-bit input resolution, which we know gives a range of 1024 possible input values. Next, we declare a constant integer for the analogue input pin we will connect the piezo sensor to. This can seem a little confusing as we seem to be assigning the value 'A0' to an integer, but actually in the Arduino pin assignment header code (which you do not need to learn!) all analogue pins are defined as A0/A1/A2/etc. and then mapped to specific pin numbers that begin after the digital pins (A0 is actually pin 14, A1 is 15 and so on).

An initial threshold value of 200 is declared, which is in the bottom 20% of the overall range of 0–1023. This value can be tested in the final project to find the best balance between input sensitivity and signal-to-noise ratio, knowing that any vibration near the piezo sensor will register at some level. If testing produces false triggering because the value is too low it can be amended upwards, or conversely reduced if there is a lack of sensitivity to input. The condition evaluates true if `inputSignalLevel > inputTriggerThreshold` and if this is the case a MIDI Note On message can be sent using a function call (as in the previous example). This may seem reasonably straightforward, but there is also a second condition that must be taken into account – what happens after the first onset value, when the sampled values that follow it are **also** still above the threshold (Figure 5.38).

In the diagram, there are two specific samples of interest – the first sample above threshold (onset), and the first sample after the signal crosses back down below threshold again (offset). These events equate to MIDI Note On (onset) and MIDI Note Off (offset) messages that will define a drum stroke, but they must be separated out from all the other samples that are higher than the defined threshold. The MIDI Note On message must only be sent **once** when the input signal reaches threshold – no other onset should be registered until the signal goes below the threshold again.

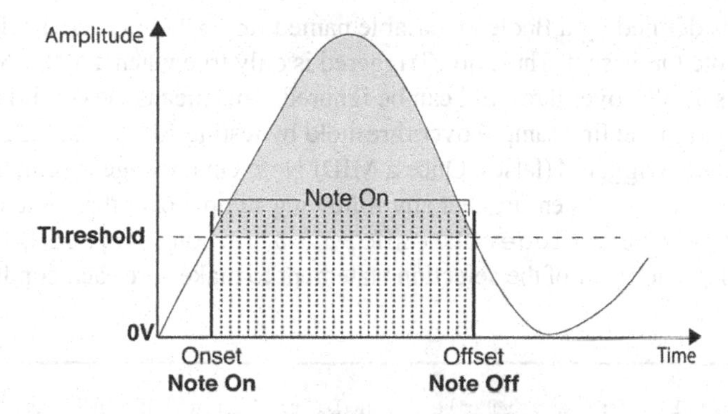

Figure 5.38 Analogue signal onset and offset detection. ***The diagram shows the sampled area of the input signal that will fulfil the thresholding criteria, but in so doing will also create multiple MIDI Note On messages if the logic in the previous code listing is used. To avoid this, a second condition must be added that specifically defines the first sample above threshold, to distinguish between all the other possible values that fulfil this condition.***

This requires **a second condition** for the input signal that checks to see if a previous MIDI Note On has occurred, which can be implemented in the selection statement using **Boolean algebra**:

```
A || B //disjunction - either A OR B can be true for a true result
A && B //conjunction - both A AND B must be true for a true result
A && !B //negation - A must be true AND B must be false (NOT true)
```

The concepts of conjunction and negation will be introduced using code examples, as a full discussion of this topic is beyond the introductory programming concepts discussed in this book. Having said this, if you wish to progress into digital electronics then you will inevitably encounter **truth tables** as a useful way of defining all the logical outcomes for a specific set of conditions (Figure 5.39).

Condition A	Condition B	Output	Action
>threshold	noteTriggered		
TRUE	FALSE	**TRUE**	**Note On**
FALSE	FALSE	FALSE	Do nothing
TRUE	TRUE	FALSE	Do nothing
FALSE	TRUE	**TRUE**	**Note Off**

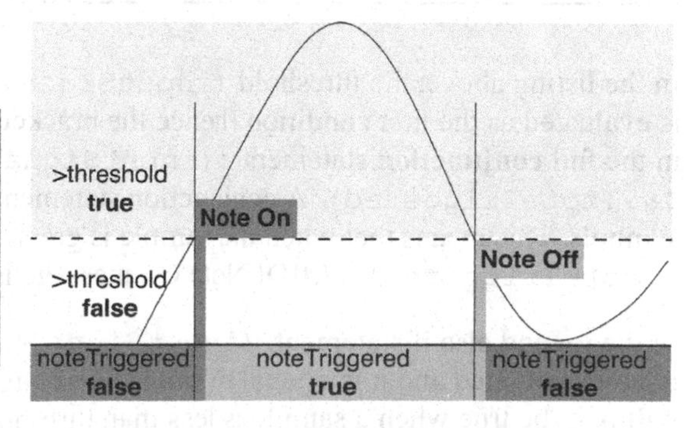

Figure 5.39 Onset/offset detection truth table. ***The table lists all possible logical inputs for two conditions and the specific outputs that must be true. The right-hand graph shows how the onset condition requires that the sample >threshold be true, and also that no previous sample has triggered a MIDI Note On (noteTriggered = false). The offset condition requires the sample >threshold be false and also that noteTriggered = true. Notice that noteTriggered does not change the selection at the same time as the onset/offset, but rather in the next call to loop().***

This second condition is defined by a Boolean variable named noteTriggered, which is set to true when a MIDI Note On is sent. Thus, noteTriggered is only true when a MIDI Note On has happened and so other samples over threshold can be ignored. This means the conditions can be combined to only respond to the first sample over threshold by testing for !noteTriggered – which equates to NOT noteTriggered (false). Once a MIDI Note On message is sent, this can be set to noteTriggered = true to ensure that any following sample over threshold does not meet the combined criteria of >inputTriggerThreshold &&!noteTriggered. To do this, extra brackets must be added to each part of the selection statement to make sure each condition is evaluated separately:

```
int inputSignalLevel = 0;//a variable to hold the input signal sample
const int inputTriggerPin = A0;//this maps to pin 14, but A0 makes more sense!
const int inputTriggerThreshold = 200;// use a low value in the range 0-1023
boolean noteTriggered = false; //use this Boolean to stop multiple onsets
//setup() has been omitted for brevity….
void loop(){
  inputSignalLevel = analogRead(inputTriggerPin);//read the sample value
    if((inputSignalLevel > inputTriggerThreshold) && !noteTriggered){
        //first sample so send a Note On
        midiNoteOutput(38, 100);//General MIDI Note number for snare drum sound
        noteTriggered = true;//make sure no other onset can trigger a message
    }//end of if (>inputTriggerThreshold)

    else if((inputSignalLevel < inputTriggerThreshold) && noteTriggered){
        //last sample so send a Note Off
        midiNoteOutput(38, 0);//MIDI Note Off message for snare drum
        noteTriggered = false;//reset the Boolean so another note can trigger
    }//end of else if (<inputTriggerThreshold)

    else{
        //no significant vibration value is present - do nothing
    }//end of else
}//end of loop
```

In the listing above, the threshold (inputSignalLevel > inputTriggerThreshold) is evaluated as the first condition (hence the brackets). Then both conditions are evaluated together in the full **conjunction** statement ((inputSignalLevel > inputTriggerThreshold) && !noteTriggered). A conjunction statement means **A AND B** – as denoted by the && symbols. This means that when the sample is greater than threshold **AND** noteTriggered is false (!noteTriggered) a MIDI Note On message is sent.

In the second else if statement, (inputSignalLevel < inputTriggerThreshold) is now evaluated and it is crucial to **note the change to less than (<)**. This condition will only be true when a sample is less than threshold, which is the same as saying >inputTriggerThreshold = false. Now this result can be evaluated within a second conjunction statement ((inputSignalLevel < inputTriggerThreshold) && noteTriggered) to find the first sample after a MIDI Note On that has crossed below the defined threshold. When this condition is true, a MIDI Note Off message can be sent to terminate the previous MIDI Note On – thus avoiding sending a note that does not switch off.

As previously noted, an in-depth discussion of conjunction and negation to evaluate combined logic conditions is beyond the scope of this book, but examples are provided to allow a full MIDI drum trigger project to be built in the following section. It is recommended to consult more comprehensive programming tutorials and texts on this topic to help disambiguate both these concepts and their application. Combinatorial logic takes a significant amount of time to understand and can prove to be a major source of programming problems. For now, the final project will build the first full practical system in this book, in part to define a milestone in your learning and also to provide you with the building blocks needed to extend this project in the future if you so wish.

5.10 Final project: MIDI drum trigger

The MIDI drum trigger project combines both the electronics and programming topics covered in the book so far, where Arduino code is used to read an input sensor (piezo) value and select the specific circumstances for which this input will generate a MIDI Note On (or Off) message as output to a serial MIDI OUT interface. Before beginning, take a moment to remind yourself **how much you have learnt in a short space of time** – this project combines DC circuit theory with Arduino C programming to build a full MIDI control system that is compatible with any standard MIDI interface, and represents a significant learning milestone in this book. As stated, the MIDI Drum trigger system uses the Arduino to sample and analyse piezo sensor input to create serial data output in the form of MIDI messages (Figure 5.40).

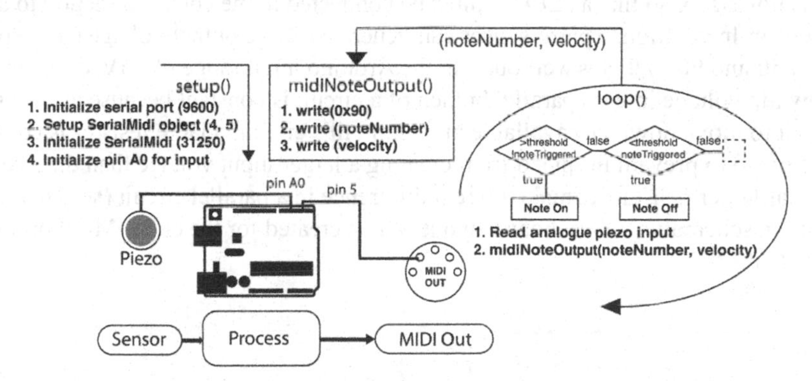

Figure 5.40 MIDI drum trigger code process diagram. ***In the diagram, a piezo sensor that will act as a drum trigger is connected to analogue input pin A0. The sensor input is read by the Arduino and used in a selection statement that determines the first onset (Note On) and offset (Note Off) of a drum stroke, and outputs a MIDI Note message in each case.***

The main element of the Arduino code process is the if/else if/ else statement that evaluates the combined threshold and noteTriggered conditions to determine the first onset and offset of the input drum stroke. Notice that both conditions in the selection are reversed in each case relative to the MIDI Note message they relate to, while the else condition has no output:

1. **Note On** Condition: **>threshold && !noteTriggered**
2. **Note Off** Condition: **<threshold && noteTriggered**

The schematic for this system is fairly similar to the previous example systems in this chapter, where a sensor input is connected to the Arduino, which then sends serial data as MIDI Note messages to the 5-pin DIN MIDI interface as output (Figure 5.41).

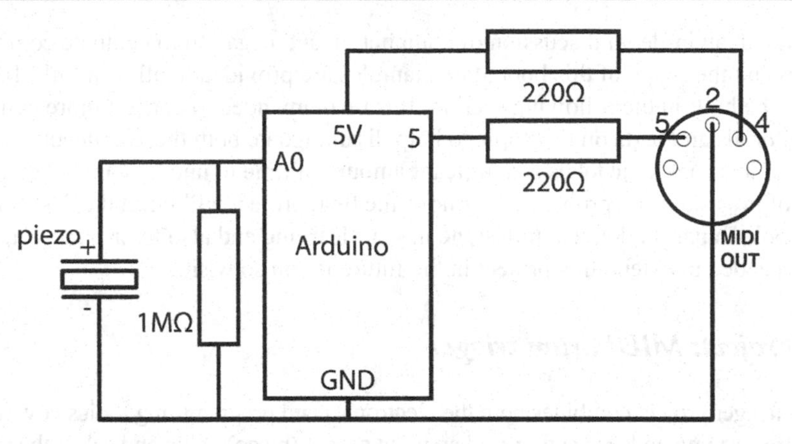

Figure 5.41 MIDI drum trigger schematic. *In the diagram, a piezo sensor is connected in parallel with a 1MΩ resistor, which limits the voltage that the sensor can output to protect the Arduino. The piezo sensor is polarized, where the positive lead of the sensor is connected to the analogue input pin A0 on the Arduino (negative to GND). The Arduino digital output pin 4 is used to transmit serial data (in the form of MIDI Note messages) to the standard 5-pin DIN socket MIDI OUT interface.*

A piezo sensor is polarized, so like an LED it must be connected in the correct direction to avoid damaging the sensor. In addition, a piezo sensor can generate a large output voltage (some piezos can produce between 50 and 90V) that is well outwith the Arduino input range of +5V. Chapter 3 (section 3.4) showed how the voltage in each parallel branch of a circuit is constant because electrons have the same potential to move down any available branch. In this circuit, a large resistor is connected in parallel to the piezo to prevent the piezo from creating a larger input voltage than the Arduino can handle – recall that larger resistances make no real difference in a parallel circuit (see chapter 3 section 3.1.5 Q9). From the schematic, a breadboard layout can be created for the entire MIDI drum trigger circuit (Figure 5.42).

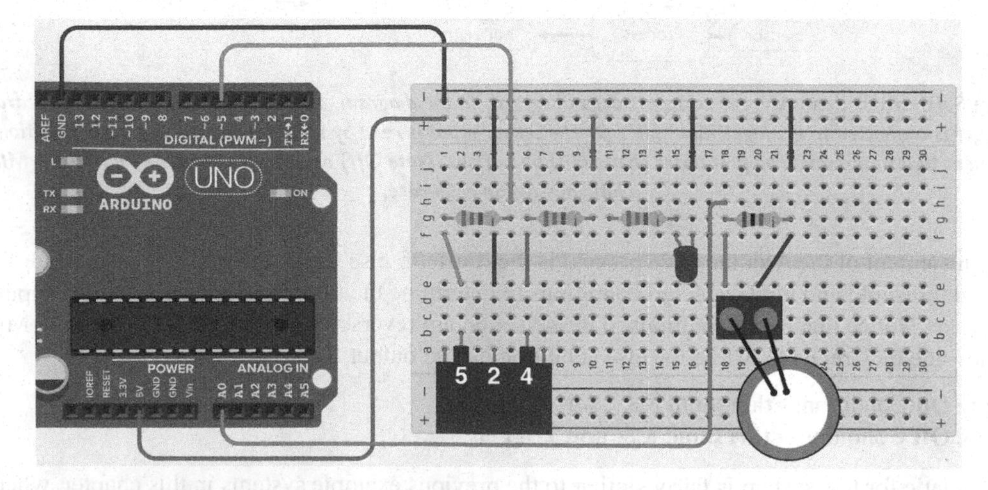

Figure 5.42 MIDI drum trigger breadboard layout. *The diagram shows the connections for a piezo sensor input (on Arduino pin A0) and the MIDI OUT interface connected to Arduino digital pin 5 (the serial test LED circuit is also shown between rows 11 and 16). This system will sample and analyse piezo drum trigger input to create MIDI Note messages as output through the MIDI connector component.*

A PCB mount screw terminal block is used to connect the piezo sensor to the breadboard (Figure 5.43).

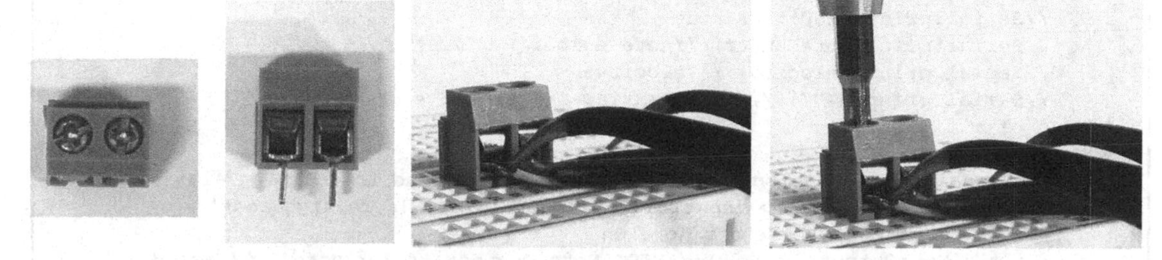

Figure 5.43 PCB mount screw terminal blocks. ***The left-hand images show a top (left) and side (right) view of a screw mount terminal block, which is useful in providing a stable breadboard connection for audio input and output cables. The middle image shows how the terminal block can be mounted on breadboard to allow wires to be pushed into each terminal more easily. The right-hand image shows how each terminal can then be screwed closed to hold the wire in place – thus preventing it being pulled out.***

Screw terminal blocks help to prevent the audio wires from moving and becoming detached from the breadboard (which can easily happen with a direct wire inserted into the breadboard). When connecting a wire to the block, it can help to pre-mount the terminal block to the breadboard to provide a solid surface to work with. To insert the wires, it can be easier to unscrew one terminal on the block, and with a wire in position (inside the terminal clamp) screw the terminal to clamp down on this wire while holding it in position (the figure shows how tweezers can be used for this).

In the case of two wires (either for a piezo sensor or for an audio connector as used in chapters 7–9), one common issue can be too much exposed wire outside the terminal block. This can potentially short out the signal by allowing both wires to make contact with one another outside the terminals – so the terminal effectively becomes a short circuit. It is recommended to take a little time when stripping the wire for connection to ensure that a relatively short (<1cm) length of wire is exposed. With the breadboard layout completed, we can now enter the code needed to process piezo sensor switch input to generate MIDI serial data output.

Full code listing

```
//use #include<> to include other libraries, which have the extension <lib>.h
#include <SoftwareSerial.h>
//declare SoftwareSerial object pin 4 for receive(RX) and pin 5 for transmit(TX)
SoftwareSerial SerialMIDI(4, 5);
int inputSignalLevel = 0;//a variable to hold our input signal sample
const int inputTriggerPin = A0;//this maps to pin 14, but A0 makes more sense!
const int inputTriggerThreshold = 200;// use a low value in the range 0-1023
boolean noteTriggered = false; //use this Boolean to stop multiple onsets
void setup() {
     Serial.begin(9600);
     SerialMIDI.begin(31250);
     pinMode(inputTriggerPin, INPUT);
}
// midiNoteOutput is used to output the 3 serial data bytes in a Note On message
void midiNoteOutput(byte noteNumber, byte velocity){
      SerialMIDI.write(0x90);//note on status byte
      SerialMIDI.write(noteNumber);//note number - either C4 or D4
```

```
    SerialMIDI.write(velocity);//velocity - either 100 or 0 (Note Off)
    //use Serial Monitor to check structure - commented unless testing
    //Serial.print(0x90);//status
    //Serial.print(noteNumber);//note number
    //Serial.print(velocity);//velocity
    //Serial.println("");//add a carriage return to each message
}//end of midiOutput()
void loop(){
  inputSignalLevel = analogRead(inputTriggerPin);//read the sample value
    if((inputSignalLevel > inputTriggerThreshold) && !noteTriggered){
      //first sample so send a Note On
      midiNoteOutput(24, 100);//MIDI Note On message for note C1 (bass drum)
      noteTriggered = true;//make sure no other onset can trigger a message
    }//end of if (>inputTriggerThreshold)
    else if((inputSignalLevel < inputTriggerThreshold) && noteTriggered){
      //last sample so send a Note Off
      midiNoteOutput(24, 0);//MIDI Note Off message for note C1 (bass drum)
      noteTriggered = false;//reset the Boolean so another note can trigger
    }//end of else if (<inputTriggerThreshold)
    else{
      //no significant vibration value is present - do nothing
    }//end of else
}//end of loop
```

In the listing above, you should already be familiar with much of the selection statement code used to capture the first sample above threshold (onset) and the first sample after this that crosses below threshold (offset). The same `midiNoteOutput()` function from the previous example is used to send MIDI Note messages to the SoftwareSerial object for output to a MIDI OUT interface. If you copy/paste this code directly into the Arduino IDE, it should compile and upload onto your Arduino board. From here, the first full system prototype can be built.

Project steps

For this project, you will need:

1. Completed circuit from previous build (sequence player)
2. 1 × 2 pin screw terminal block connector
3. 1 × piezo sensor
4. 1 × 1MΩ resistor
5. 1 × connector cable and 3 × connector wires

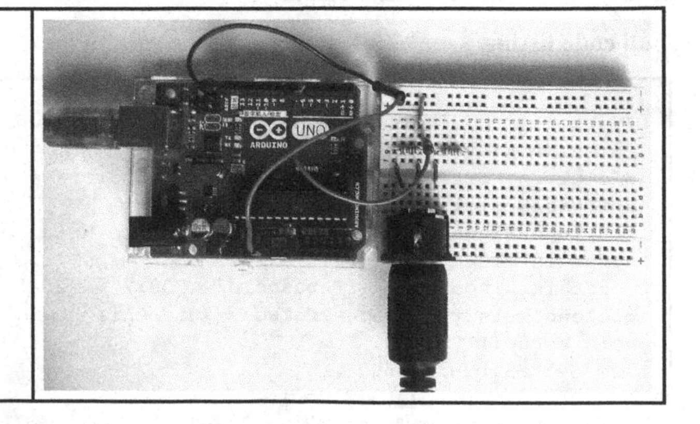

Piezo Sensor Input: 1. Add a 1MΩ resistor between pins [g18–g22]. 2. Add a connector wire before the resistor to breadboard pin e18 [f18–e18] – *this is the parallel path for the piezo sensor.* 3. Add a connector wire after the resistor to breadboard pin e20 [f22–e20] – *this completes the parallel path for the piezo sensor.* 4. Add a connector wire after the resistor to the negative (ground) rail [j22–GND].	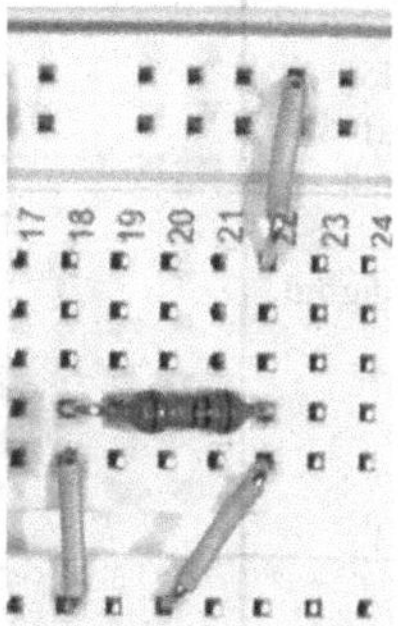
1. Connect the Arduino analogue input pin A0 to pin h18 (Arduino connection not shown) [ADA0–h18].	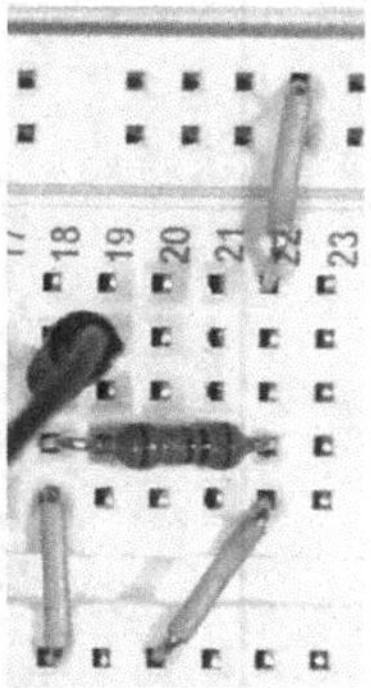
1. Connect the piezo sensor wire to a screw terminal block – if the wires are too long, they can be looped and twisted until they are shorter (only looping will allow them to move around inside the screw terminal block).	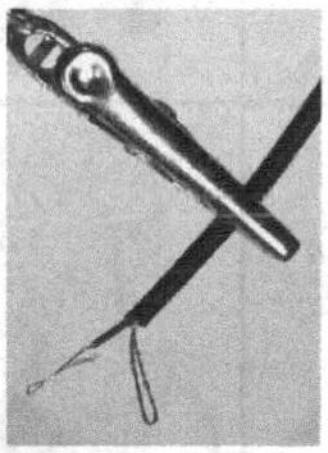 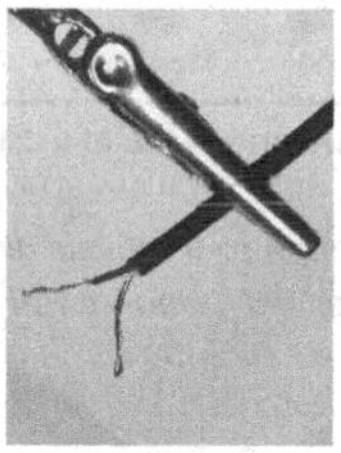
1. Connect the screw terminal block between pins c18 (positive) and c20 (negative) to leave space for the connector wires.	

To add the LED test circuit: 1. Add a 220Ω resistor between pins [g11–g15]. 2. Add an LED – anode pin (f15) to cathode (f16). 3. Add a connector to the negative (ground) rail (j16-GND]. 4. To test serial output, connect the Arduino digital pin 5 connector to pin f11 (not shown).	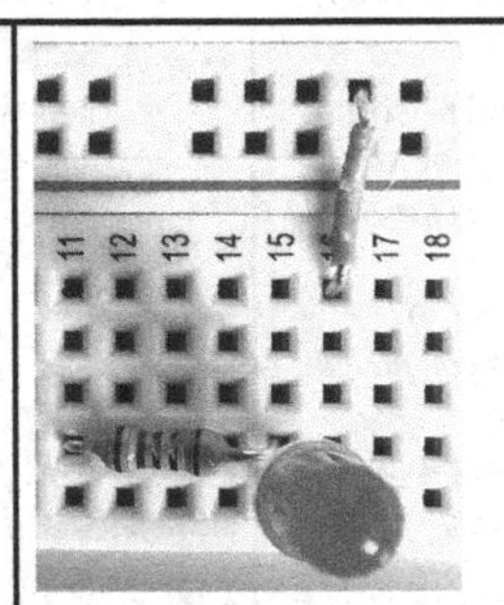

Connect the piezo sensor to the underside of a drum practice pad. *Any surface that is robust enough to withstand drum strokes can be used, but sensor performance will vary.*	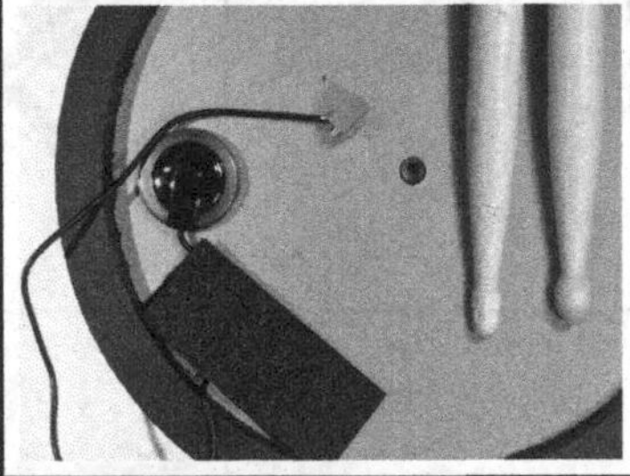

Using whatever MIDI equipment you have available (either synthesizer or computer MIDI interface) connect a MIDI cable between your Arduino breadboard DIN socket and this system. Configure your system to receive MIDI data on channel 1. If the system functions correctly, every time you strike the drum pad a MIDI Note On message should be sent on channel 1 to your MIDI system. With some tweaking of the input threshold, a usable drum trigger can be obtained.

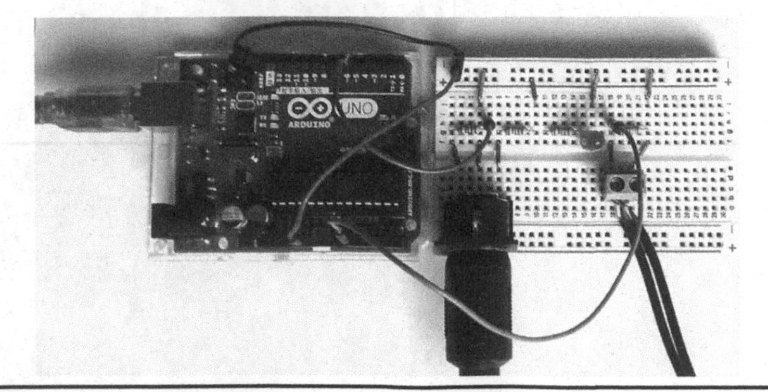

If you are not getting a usable output from your system:

1. **Test the code** – the `inputTriggerThreshold` value may be too high (or low).
2. **Test the board** – recheck pin connections – are the piezo wires properly connected to the screw terminal block.
3. **Test the system** – connect another MIDI interface to ensure it is responding to messages on channel 1.

5.11 Conclusions

As with the previous chapter, this chapter has provided an introductory overview to C programming for the Arduino and so has covered a lot of ground in a short space of time. The chapter began by discussing arrays as logical structures for grouping variables and showed how a for loop can be used as an iteration instruction to process each element of an array in sequence. An array was used to hold a sequence of note frequencies for output to a piezo loudspeaker using pulse wave modulation (PWM) on an Arduino digital pin, where PWM can also be used to output a voltage range between +5V (HIGH) and 0V (LOW). The chapter then covered serial output on the Arduino, and how the SoftwareSerial object allows other digital pins to be used to send digital data (as pins 0/1 are also used by the USB interface that connects the Arduino IDE to our board). The SoftwareSerial object was then used as part of a standard MIDI OUT interface that sent a sequence of MIDI Note On messages (in running status) using note numbers held in an array. This MIDI OUT interface is built to the standard MMA 1.0 specification, and thus should be interoperable with all other MIDI interfaces that use a 5-pin DIN connector.

The chapter looked at how the Arduino can receive and analyse input using selection instructions – the last of the three core programming instruction types. Selection instructions can become complex (and take time to learn), so were introduced by example only. An if/else if/else selection was used to process two push-button switches for MIDI output, to show how digital input can be analysed by the Arduino to create a full electronics system that combines input/process/output. In so doing, the three stages of testing needed when working with a full Arduino system were introduced:

1. **Test the code** – check the code compiles, run software tests in Tinkercad (also with the LED serial test circuit).
2. **Test the board** – component layout must align with the schematic, and all circuit paths must be checked.
3. **Test the system** – once software output is confirmed (Serial Monitor) we must test with a full system.

Selection statements were then extended to work with analogue input from a piezo sensor, which required two conditions – one for threshold to indicate onset/offset and a second condition to prevent multiple MIDI Note messages being sent. This logic structure was used to build a full electronics system that takes input from a piezo sensor, processes this input using the Arduino and then outputs MIDI Note messages as a result. This system represents a learning milestone in this book, combining audio systems design, DC circuit analysis and Arduino C programming to produce a fully functional MIDI drum trigger system.

The next chapter will move on to a completely new topic – alternating current (AC) circuits. Audio signals are continuous, they change over time. Working with continuous signals requires an understanding of the relationship between sine waves and audio signals in terms of both their magnitude and phase. This leads to the definition of impedance as the AC form of resistance, where the varying nature of the signal means a magnitude value cannot be calculated in the same way as DC circuits in chapter 3. This leads on to time-varying components like the capacitor, which are a core building block of all electronic circuits. Capacitors will be used in various parts of the amplifier and filter circuits later in the book, where the frequency-dependent nature of the component provides control over input and output audio signals. The mathematics involved in AC circuits are more complex than in DC circuit analysis, but ultimately the same basic principles apply.

Changing learning topics

The next chapter will have a significant shift in focus from chapters 4 and 5, moving from Arduino programming and practical system examples to theoretical concepts. Be prepared to take a little time to adapt to this shift in topic – **C programming has been covered very quickly** and you are now moving on to a completely different area in electronics. It is not unusual for this to be confusing in itself, so allow some extra time to make the transition in your thinking as you progress through the second half of the book.

5.12 Self-study questions

As this chapter represents a learning milestone in the book, no self-study questions are provided at this point. Instead, consider the following:

1. Reflect on **how much you have learned** about audio systems, DC circuits and Arduino programming in a short space of time – we often focus on what we don't know and forget what we have learned! Also spend some time re-reading the chapter summary sheets – do you know how the concepts involved relate to the final project in this chapter?
2. Think of **how the current project could be improved or extended** to make it a more practically useful system. For example, you could add multiple drum triggers to output on different MIDI Note numbers. You could also perform input velocity triggering where you map the sample value range (0–1023) to the MIDI velocity range (0–127). You could even add MIDI channel assignment where you increment a channel number variable using a push-button switch
3. Take a **short break from this book** to let what you have learned fully sink in – the desire to progress can often slow your learning down if you do not take a break in your studies. The next chapter will focus on AC circuit theory and thus does not provide many practical examples, so you will need a break to help refocus your thinking from the very practical coding work covered in both chapters 4 and 5.

Chapter 5 summary

The following images may help you remember some of the key points in this chapter:

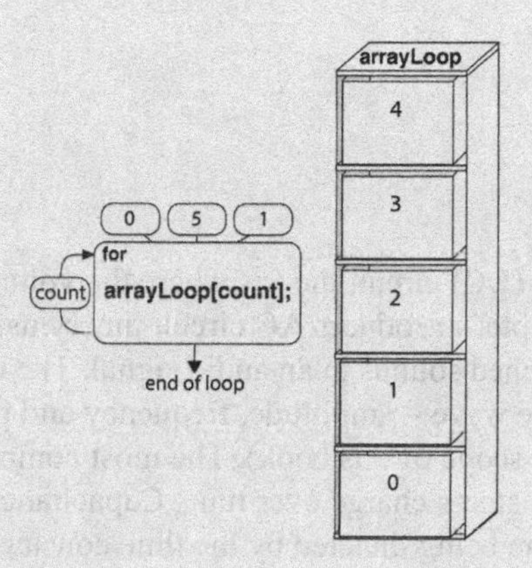

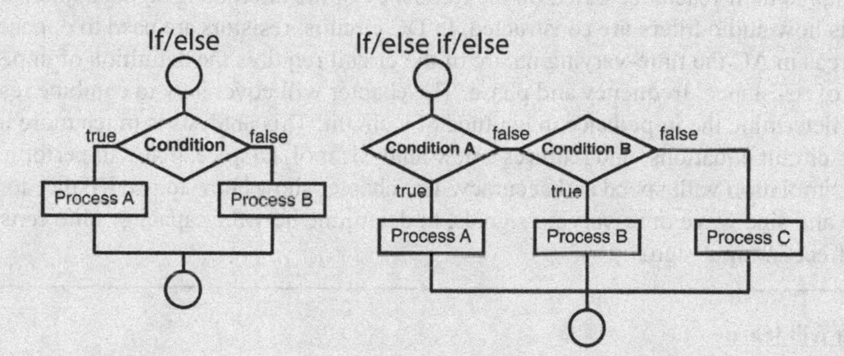

MIDI Note On message

Status	Channel	Data 1	Data 2																				
Note On (9)	1 (binary 0)	Middle C (60)	Velocity (100)																				
1	0	0	1	0	0	0	0	0	0	1	1	1	1	0	0	0	1	1	0	0	1	0	0

CHAPTER 6

AC circuits

Chapter 3 looked at direct current (DC) circuit theory, where the voltage and current levels in a circuit do not change over time. This chapter introduces AC circuit theory using the sine wave, which is the fundamental component of all pitched sounds in an audio signal. The chapter begins by introducing the fundamental elements of a sine wave – amplitude, frequency and phase (though an in-depth discussion of phase is outwith the scope of this book). The most common time-varying component in electronics is the capacitor, which stores charge over time. Capacitance is defined in terms of charge and voltage, with the charging time being dictated by the time constant within an RC series circuit. Capacitors change their reactance based on the frequency of the alternating signal applied across their plates – this is how audio filters are constructed. In DC circuits, resistors are used to oppose the flow of current, whereas in AC the time-varying nature of the circuit requires the definition of impedance – the combination of resistance, frequency and phase. The chapter will cover how to combine resistors and capacitors to determine the impedance magnitude of a circuit. This analysis is much more involved than with DC circuit equations, and requires a new analysis tool, LTspice, that will perform AC circuit analysis and simulation with speed and accuracy. The chapter shows how to use LTspice to analyse both impulse and sine wave time-varying signals, to determine how the capacitor time constant and phase shift affect an input signal.

What you will learn

The fundamental components of sound
The fundamental elements of a time-varying (AC) signal
How capacitors store charge over time
How to combine resistors and capacitors in RC circuits
How to use LTspice as a circuit analysis tool
How to simulate and measure AC circuits elements like time constants and phase shifts

6.1 Audio signal fundamentals – sine waves

This book focusses on audio signals, to learn how to detect and process sounds for output in audio systems. As discussed in chapter 2, all audio systems use some form of sensor (e.g. microphone) and transducer (loudspeaker) to convert sound to and from electricity for processing by audio electronics circuits. This chapter begins by covering the basic building block of all sounds that are detectable by the human ear – the sinusoidal waveform. Sine waves are periodic in nature, they continuously vary their amplitude in a repeating pattern over time (Figure 6.1).

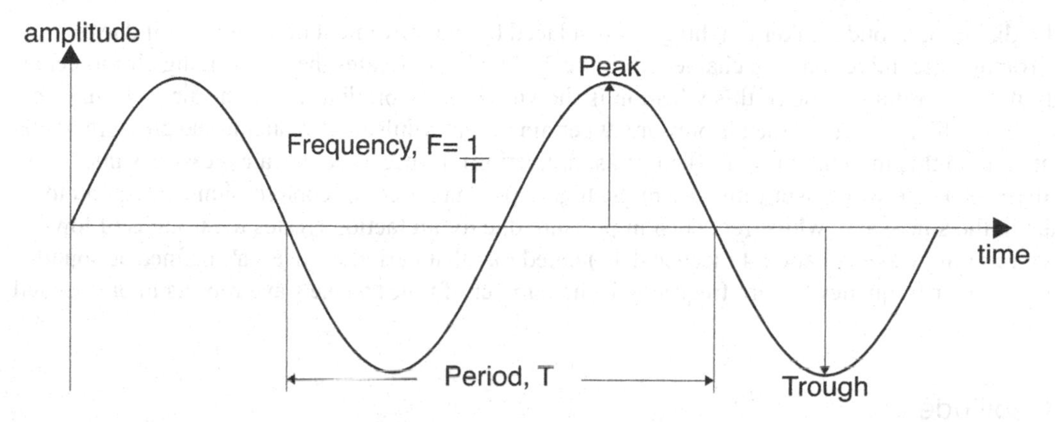

Figure 6.1 A sine wave. ***The diagram shows a sinusoidal waveform that is a continuous signal – it is constantly varying its output. A sine wave has a peak amplitude when the output is highest, and its periodicity (T) is the amount of time before the sine wave begins to repeat itself. As shown in the diagram, the frequency of the sine wave is the reciprocal of the wavelength.***

The diagram shows a sine wave that repeats periodically over time (defined by the Period, T). The amplitude of the sine wave (also called the magnitude in electronics) varies over the period of the sine wave, beginning at 0 and then steadily increasing to the highest peak value. The wave then decreases in amplitude until it reaches the trough value, which can often be a negative value depending on the scale used. The period of the wave is measured between two points of the same amplitude (the diagram defines the period between two downward zero-crossing points). The sine wave is a fundamental concept in both mathematics and science and is used in many different ways by different domains and disciplines that work with continuous signals of some form (you will probably have encountered sine waves at some point in your learning). In audio terms, a sine wave is used to describe the changes in air pressure created by vibrations that are detectable by the human ear as sound (Figure 6.2).

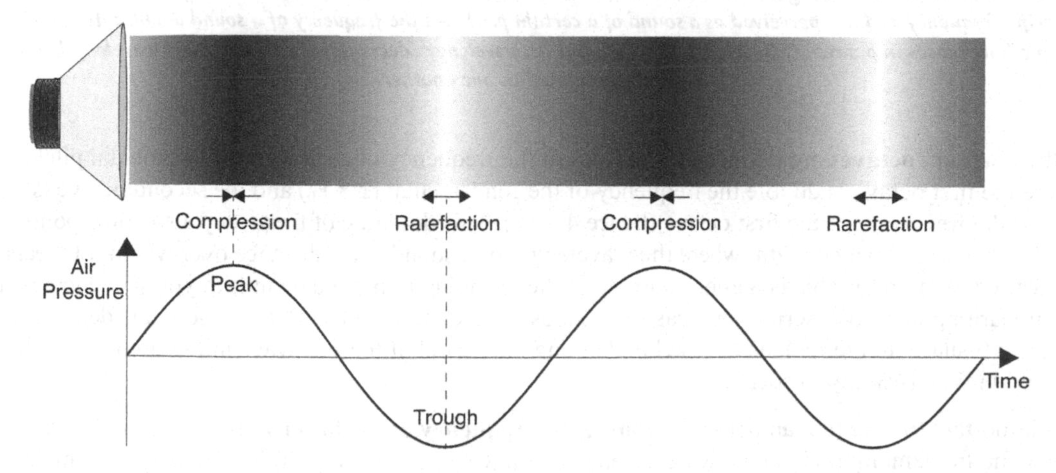

Figure 6.2 Air pressure changes due to sound vibration. ***In the diagram, a loudspeaker diaphragm vibrates air molecules, causing them to compress. The dark bands show where compression occurs, alongside the lighter areas that experience a corresponding reduction in density (rarefaction) as a result of the movement. These points correspond to the peaks and troughs of a sine wave, which can be used to describe the changes in air pressure over time.***

In the diagram, a loudspeaker diaphragm is displaced by the movement of its voice coil due to electromagnetic induction (see chapter 2, Figure 2.32) which vibrates the surrounding air molecules (only the forward direction of this vibration is shown). These vibrations cause the air molecules to compress, which increases the air pressure at certain points, whilst also reducing the pressure at other points due to this movement of air (known as rarefaction). If measured, the air pressure values would ideally plot a sine wave over time, where the higher pressure areas of compression correspond to peaks in the sine wave, whilst reduction in pressure due to rarefaction creates the troughs of lowest pressure in the wave. Chapter 4 (section 4.10) noted that that periodic waves are defined as sounds of a particular frequency, where frequency is the number of times a sine wave repeats in one second (Figure 6.3).

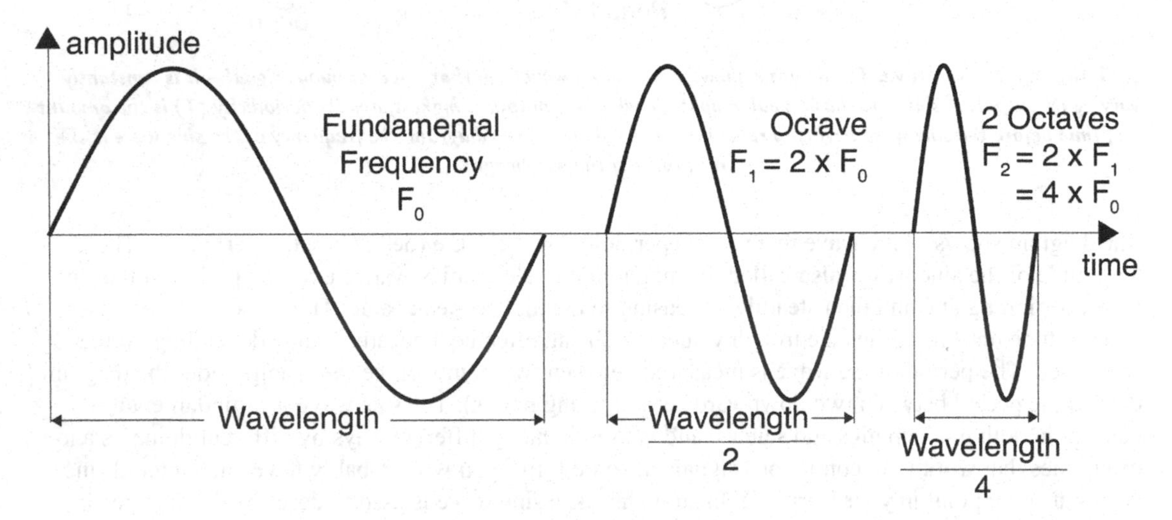

Figure 6.3 Pitch perception by the human ear (adapted from chapter 4, Figure 4.22). ***At its simplest level, the human ear detects changes in air pressure as sound. When these changes in pressure are periodic they have a specific frequency and are perceived as a sound of a certain pitch. As the frequency of a sound doubles its perceived pitch increases in octaves (F0, 2 × F0, 4 × F0), but its wavelength decreases – frequency and wavelength are inversely related to one another.***

In the diagram, **octaves** are defined as multiples of the frequency of an arbitrary fundamental pitch, where the first octave is double the frequency of the fundamental ($2 \times F_0$) and the second octave is double the frequency of the first octave (hence $4 \times F_0$). This doubling of frequency also corresponds to a halving of the wavelength, where the wavelength of a sound is the distance over which it repeats (measured in metres). This is a good example of the changing terms and quantities you will encounter when learning how to describe (or measure) sounds, where the wavelength of the sound is defined as a distance (usually in metres), as opposed to defining the period of the sine wave in Figure 6.1, which is defined in time (usually in seconds).

It is important to note that an **octave is double the frequency of the lower note**, hence three octaves above the fundamental will actually be a frequency of $8 \times F_0$ (or $4 \times F_1$) – this distinction will make more sense when discussing harmonics later in the section. Frequency can be related to wavelength by using the speed of sound in air:

$$\textit{Frequency}, F = \frac{\textit{Speed of sound in air}, c_{air}}{\textit{Wavelength}, \lambda} \tag{6.1}$$

F is the frequency (or number of cycles per second) of the sound in hertz (Hz)

c_{air} *is the speed of sound in air in metres per second* (ms^{-1})

λ *is the wavelength of the sound in metres (m)*

The speed of sound is normally defined as 343 ms^{-1} (for an air temperature of 20°C) and is therefore used as a constant to relate the other two quantities of frequency (number of cycles per second) and wavelength. In practice, the speed of sound varies between ~331 ms^{-1} at 0°C and ~355 ms^{-1} at 40°C and thus for live sound (and indeed some high-end recordings) temperature can be an important factor. To understand this, consider how an instrument (or voice) will emit a sound of a specific wavelength based on its construction (e.g. length of vibrating string, distance to air hole). If the instrument plays the same note then this wavelength does not change, but if the speed of sound has increased due to temperature then the frequency of the sound will also increase, as the following worked example will illustrate:

6.1.1 Worked example – varying the speed of sound

These calculations are illustrative only, and thus approximate the actual value for the speed of sound in air at various temperatures. Also note that a piano tuner would never tune an instrument at such a low temperature!

Q1: *A piano tuner tunes a note to A4 (concert pitch 440Hz) in a concert hall that is very cold – the thermometer reads 0 °C. If the hall is heated by the presence of the audience (who also generate heat) to 20°C before the performance, how much will the frequency of the note A4 on the piano be increased when played?*

For a temperature of 0°C:

$\textit{Speed of sound in air}, c_{air} = 331\ ms^{-1}$, $\textit{Note frequency} = 440\text{Hz}$

$$\textit{Frequency}, F = \frac{c_{air}}{\textit{Wavelength}, \lambda}$$

$$\therefore \textit{Wavelength}, l = \frac{c_{air}}{\textit{Frequency}, F} = \frac{331}{440} = \mathbf{0.75m}$$

For a temperature of 20°C:

$\textit{Speed of sound in air}, c_{air} = 343ms^{-1}$, $\textit{Wavelength}, \lambda = 0.75\text{m}$

$$\textbf{Frequency}, F = \frac{c_{air}}{\textit{Wavelength}, \lambda} = \frac{343}{0.75} = \mathbf{457.33Hz}$$

Answer: A note tuned to A4 (440Hz) at 0°C, when played at 20°C, will have increased in frequency by **17.33Hz** to 457.33Hz.

Notes:

This example aims to show the relationship between the speed of sound (which can change) and frequency if all other factors are considered to remain constant. Although it may seem somewhat spurious, it is not uncommon to encounter tuning problems with instruments in live settings as a performance progresses, particularly if the audience are moving around (and thus generating even more heat than when seated). In chapter 2 (section 2.1) live sound experience was recommended, to better inform your understanding of audio electronics systems and their requirements. If you do get such an opportunity, do not be surprised to encounter a hot performance space that creates unexpected tuning problems (and thus potentially arguments!) as a result of the change in temperature - for this reason (and others), it is important to know that heat and humidity have a significant impact on live sound.

In Western music, musical pitches are defined around **concert pitch**, where the note A4 (fourth octave on a piano keyboard) equates to 440Hz – though opinions on this value differ, it is the standard value defined in ISO 16. The human ear can detect pitches between 20 and 20kHz, though in practice the high range of human hearing degrades with time and exposure to high-intensity sound sources. Chapter 8 will discuss the human hearing range in more detail when considering how audio filters must map onto the non-linear human hearing response, but for reference this range can be considered relative to some of the frequency ranges of common musical instruments (Figure 6.4).

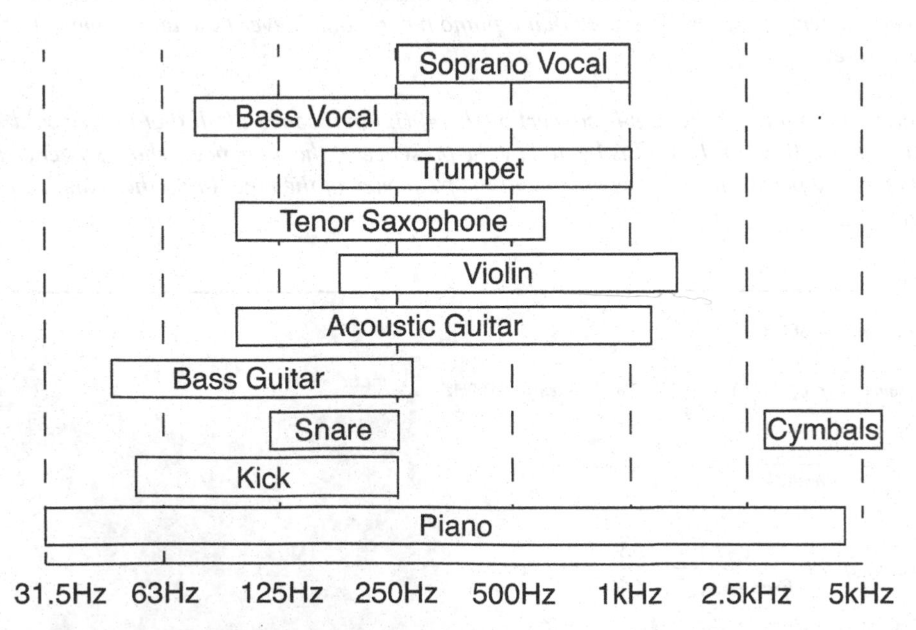

Figure 6.4 Musical instrument frequency ranges. *The diagram shows the indicative frequency range (from low to high) of some common instrument types, including piano, violin, guitar and bass. The lowest (bass) and highest (soprano) human vocal ranges are included for indicative purposes – these are the ranges which composers have traditionally used. Synthesizers are not included as they are not limited in the pitch they can output (i.e. 20–20kHz) – the only limitation is in the mapping to specific MIDI Note Numbers on the MIDI controller used to trigger them in that system.*

The diagram shows a broadly indicative range of pitches for some common instrument types, though in practice these can vary widely for a variety of design and construction reasons (e.g. 5-string bass, 7-string guitar, R&B snare drum). Notice that the frequency scale (in hertz) of the chart is **defined in octave bands** rather than a linear range of frequency – because the human ear has such a wide range, a logarithmic (order of magnitude) axis is normally used to keep data graphs and charts manageable in size. The frequency bands used are taken from ISO 266:1997, which provides a standard set of frequency values across the audible range – though you will encounter different interpretations of these values in practice). Common instrument ranges are provided as a broad reference for later work in chapter 8 on audio filters, where the range of each instrument in a production will dictate not only the filter used but also how that filter may overlap with the other instruments involved. It is not uncommon to boost a snare drum and find that the low end of an acoustic guitar (or vocals) is now less defined – that range of frequencies is now more amplified in general.

This problem would seem manageable if the overlaps in range were used to adjust filters accordingly, but this definition of frequency range only relates to the fundamental frequency of the note involved. In reality, it is important to understand that most instruments (other than very basic synthesizers) do not produce sounds that are simple sinusoids, but rather emit much more complex combinations of multiple periodic waveforms (Figure 6.5).

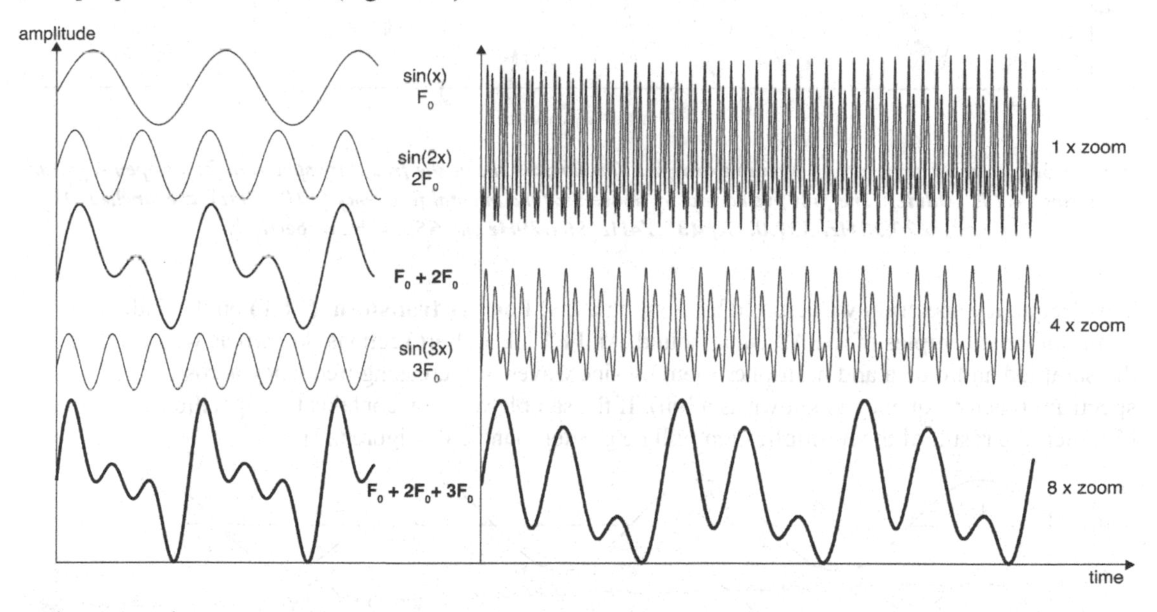

Figure 6.5 Sine wave combination. ***The diagram shows how combining three sine waves of different frequency multiples (F0, 2F0 and 3F0) creates a more complex output waveform as a result. The first wave addition equates to sin(x) + sin(2x), whilst the second wave addition is sin(x) + sin(2x) + sin(3x). The result of adding these waves is a very broad approximation of the audio waveform shown on the right, which is a recording of a guitar string at 108Hz (A2) zoomed to 8 times magnification for comparison.***

The diagram shows how sine waves can be combined to create a much more complex waveform as the output, where three multiples of the fundamental frequency F_0 (defined as sin(x)) are added together. Although the resulting waveform is not the same as the actual audio waveform of a 108Hz (A2) guitar string being plucked, there are enough similarities in the number of peaks present in both waveforms to show that each wave is created by a combination of different frequencies being added together. These sine waves are all linear multiples of the first frequency F_0 (unlike octaves) and are known as the **harmonics** of a sine wave – where the fundamental F_0 is the first harmonic, the second harmonic would be $2F_0$, the third harmonic would be $3F_0$ (and so on).

Musical harmonics are a complex subject that is outwith the scope of this book, but they are presented here to show that we are dealing not only with the pitched frequency range (i.e. the fundamental) of an instrument, but also its additional harmonic components. To highlight this more clearly, let us look at the frequency spectrum of the 108Hz (A2) guitar string waveform shown in the previous diagram (Figure 6.6).

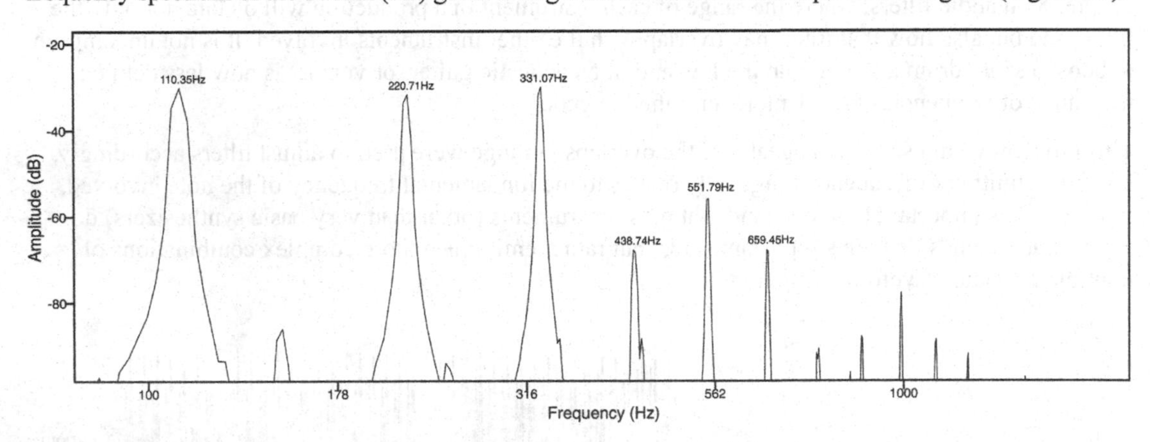

Figure 6.6 Audio frequency spectrum example. ***The diagram shows the frequency spectrum for an open A guitar string (A2 is 108Hz). The first five harmonics above the fundamental frequency (110.35Hz), are labelled at 220.71Hz, 331.07Hz, 438.74Hz, 561.79Hz and 659.45Hz respectively.***

This frequency spectrum was created by running a **fast Fourier transform (FFT)** on the audio file containing the samples of the A2 guitar sound. An FFT takes short sections (known as windows) of the sampled audio data and multiplies them by sine waves of increasing frequency across the audio spectrum (each frequency is known as a bin). If the sample window contains that specific frequency bin, then the result of the multiplication will be greater than zero (Figure 6.7).

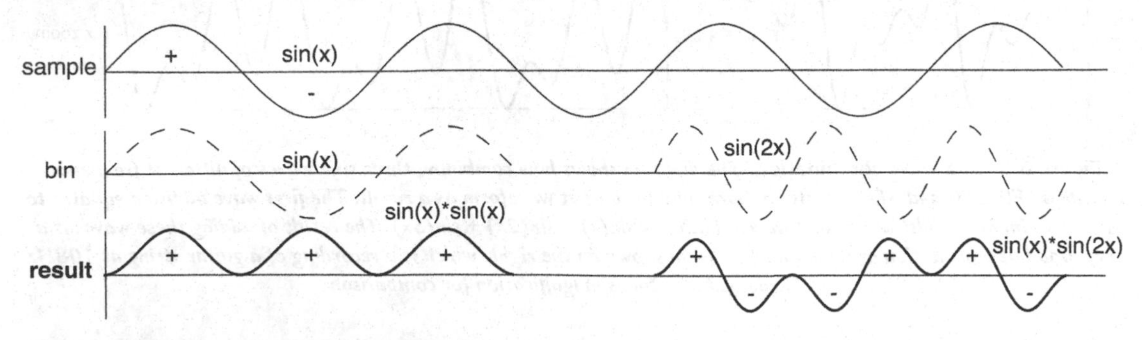

Figure 6.7 Sine wave multiplication examples. ***In the diagram, the left-hand graph shows the result of multiplying sin(x) by itself, where the average value of the resulting waveform is greater than zero. The right-hand graph shows the result of multiplying sin(x) with sin(2x). The overall average value of this waveform is 0, which shows that sin(2x) is not present in the sample.***

The diagram in Figure 6.7 shows a very simplified demonstration of Fourier analysis, where the presence of a specific frequency is detected by multiplying the input sample window by that frequency. The result of each multiplication is analysed based on the average of the entire waveform, where any single frequency sine wave centred on zero will average zero because exactly half of the wave is negative (+/− signs are included in the diagram to show where the cancellation occurs). If the result of multiplication is greater than zero, it indicates the presence of that frequency in the sample. By multiplying each window by every bin, a frequency profile of the entire sample is created – this is how the frequency spectrum in Figure 6.6 was produced. Although digital signal processing is not covered in this book, it is important to have a basic understanding of how a frequency spectrum is produced for your future learning. This section aims to provide some indication of the importance of sine waves in audio – thus the following discussion of AC circuit theory will have a more concrete application, rather than a purely mathematical explanation of the variation of an electrical signal over time.

Understanding harmonics

It can be difficult to remember that a doubling in frequency is an increase of one octave, but also represents the second harmonic of the sound! Chapter 1 discussed the impact of scales, symbols and equations - now this is extended to show that different aspects of sound and audio may use different terms that are measured by the same quantity. This book aims to be as clear as possible, and disambiguate terms whenever they are used, so note that frequency is a measurement scale used to measure the upper **harmonics** of a **specific sound** wave, but it is also used to define the relationship between **two musical pitches** relative to the **octave**.

Harmonic analysis can be very complex and is often proposed as part of the explanation for the tonal characteristics of a particularly favoured instrument - the Stradivarius violin is famed for its unique tone, and the comparison of its harmonic patterns with other violin makers is a popular debate! The discussion here has been kept to integer multiples for simplicity, but even at this level the variance of harmonic content for each fundamental pitch an instrument can produce is significant - no musical instrument has a single harmonic footprint that is consistent across its entire range.

The important point for your understanding is that harmonics are frequency multiples of the fundamental frequency of a sound that augment the tonal characteristics of that sound. Without harmonics, pure sine tones would be very dull to listen to for any length of time and would not have the rich variety of timbres provided by musical instruments. This is an important distinction for audio electronics, as most electronic signal analysis and processing aims to avoid harmonics whenever possible as they degrade the information being sent.

This can also be true for some areas of audio like hi-fi signal amplification, where the aim is to avoid introducing new signal components during the reproduction of a sound. Having said this, in other areas of music technology the preservation and augmentation of the additional harmonic content in a signal is often a significant part of the design. A vacuum tube amplifier is a good example of a non-linear amplifier that can introduce harmonics into a signal under certain conditions (notably odd-order harmonics, which are more musically pleasing to the human ear). Most rock music from the 1960s and 1970s utilizes some form of tube overdrive or distortion - definitely the antithesis of other electronics signal processing that aims to avoid distortion whenever possible!

6.2 AC signals – amplitude, frequency and phase

The sine wave is of fundamental importance in audio, and is similarly fundamental in the analysis of alternating current (AC) circuits. Alternating current was first introduced by Hippolyte Pixii in 1832 as part of an induction generator (which used magnets rotating in a coil to generate current), but is more famously documented in the battle for control of electrical power transmission that occurred in the United States between Thomas Edison (proposing DC) and Nikola Tesla – who had produced a huge number of significant innovations like the AC motor. In modern times, alternating current is used throughout electronics for the transmission of both electrical power and communication signals, where protocols such as WiFi and Bluetooth send wireless AC signals that are then decoded into data.

The previous chapter used time-varying signals to send MIDI Note messages as serial binary data where the signal varied between HIGH (+5V) and LOW (0V) over time. Time-varying signals were also used with pulse width modulation to output tones to a piezo loudspeaker, which again are a form of AC signal. Unlike direct current circuits that remain constant, the common factor in all AC circuits is that the signal changes over time. To analyse such circuits, a sine wave can be used as the mathematical model for a continuous signal input (Figure 6.8).

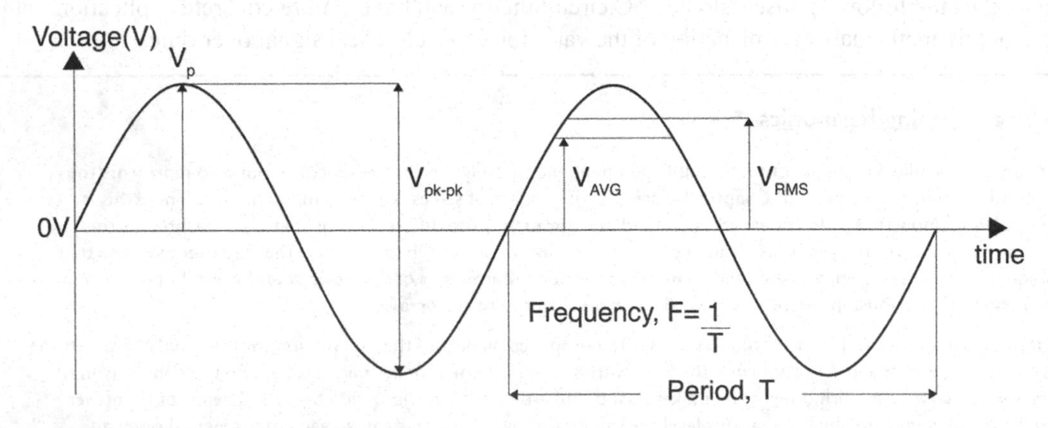

Figure 6.8 An AC signal. ***The diagram shows how a sine wave can be used as a simple model of a time-varying signal, allowing it to be used in electronic circuit analysis. The peak voltage Vp (measured from 0V) is shown, alongside the peak-to-peak voltage Vpk-pk, the average voltage V_{AVG} and the root mean square voltage V_{RMS} – all representing different ways to measure an AC signal.***

The diagram shows how the **peak voltage** (V_p) of an AC signal is measured relative to a reference value of 0V, and this is an important point to remember – **the reference level must be known**. It is common for a signal to alternate between two voltages that are both above the 0V ground reference, and this needs to be taken into account when working with peak voltage values from other parts of an audio system to ensure that levels are consistent. A more general level is the **peak-to-peak voltage** (V_{pk-pk}), which describes the difference between the lowest and highest voltage levels in the signal. The V_{pk-pk} level is often used to define the peak voltage (i.e. divide by 2) but, as noted, this does not take the ground reference level required for measuring V_p. Instead, the **amplitude** of the signal is better defined as **being half of the peak-to-peak voltage** ($Amplitude = \frac{V_{pk-pk}}{2}$)). The next element to consider is the frequency of the sine wave – how often it repeats over a given time. This can be calculated using the period of the wave (in seconds):

$$\textbf{\textit{Frequency, F}} = \frac{1}{\textbf{\textit{Period, T}}} \tag{6.2}$$

F is the frequency of the wave in hertz (Hz)
T is the period of the wave in seconds (s)

As discussed in the previous section, frequency is crucial to any audio signal. Analysing voltage and current in a circuit effectively extends from prior knowledge of DC circuits, but when frequency is included this analysis now relates to how the circuit will respond over time. Note that the relationship between frequency and time shown in the equation above is distinct from the previous relationship defined in equation 6.1 for acoustic measurement of frequency (which used speed and distance). Time-based measurements of electrical signals will be used for the remainder of this book, so it important to remember that **wavelength is not an electrical quantity**.

When describing a sine wave, the **phase angle defines the wave position within its periodic cycle**. This aspect of trigonometry can be confusing when learning how to analyse AC circuits, as the relationship between signal amplitude and sine angle values is often stated in both radians and degrees – many new students see the symbol π in an electronics equation and think of triangles or circles, rather than the position of a sine wave. There are many ways to explain this relationship between amplitude and angles in a sine wave, including the rotating-wheel example (Figure 6.9).

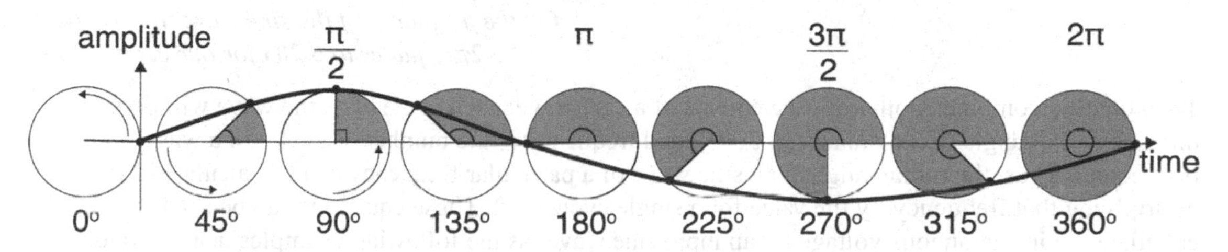

Figure 6.9 Rotating wheel sine angles. ***In the diagram, a point on a wheel rotates anti-clockwise. If this point is measured over time it will trace out a sinusoidal waveform, where the angle of the sine wave is defined by the angle of the rotating wheel. The sine angle is shown in degrees at the bottom of the diagram, and the major radian measurements (π, 2π, etc) are marked at the top for reference.***

The angle values are also given in radians at the top of the diagram as these are commonly used in AC circuit calculations, where a radian is a portion of the radius of a circle (with a value of ~57.3°). Radians will not be used (other than 2π) for the calculations in this book, but it is important to know about them if you wish to progress your studies into more complex AC circuit theory at a later time (many textbooks will provide radian calculations). For now, the diagram shows how the amplitude of a sine wave at any point in time can be referenced to a specific angle between 0 and 360°. As the wheel rotates, the point marked **begins at time 0 with an angle of 0°** on our amplitude/time graph (where the x-axis is time). As time increases, so does the angle of rotation and also the amplitude of the sine wave until it reaches its **peak value at 90°** (or $\frac{\pi}{2}$ radians). From this point, the amplitude of the **sine wave decreases back to 0, but the angle of rotation increases to 180°** (π radians). The amplitude decreases again until it reaches the **trough value at 270°** degrees ($\frac{3\pi}{2}$ radians), before increasing again until it **returns to 0 at 360°** (2π radians).

Any point on a sine wave can be described by combining the peak amplitude of the wave and the specific time at which the point occurs. This leads to the standard equation for the instantaneous (at a specific time) voltage and current of a sinusoidal input signal:

$$\begin{aligned} \textit{Instantaneous Voltage}, V &= V_p \sin(\omega t) \\ \textit{Instantaneous Current}, I &= I_p \sin(\omega t) \end{aligned} \quad (6.3)$$

V is the instantaneous voltage in volts (V)
V_p is the peak voltage in volts (V)
I_p is the peak current in amps (A)
ω is the angle of the sine wave in radians (rad)
t is the time in seconds(s)

The equation above uses the symbol omega (ω) to represent the angle of the sine wave at a given point (in radians). This value for ω can be derived using the frequency of the sine wave in question:

$$\textit{Sine Wave Angle}, \omega = 2\pi f \quad (6.4)$$

ω is the angle of the sine wave in radians per second (rad/s)
f is the frequency of the sine wave in hertz (Hz)
*2π equates to **6.283** for our calculations*

The equation combines both known elements of a sine wave: each cycle of a sine wave will pass through a total angle of 2π radians (0–360°), and frequency is the number of cycles of a wave per second. Thus, the radian angle for a sine wave of a particular frequency can be calculated by multiplying that frequency by the value for a single cycle (2π). These equations can be used to calculate the instantaneous voltage for an input sine wave, as the following examples demonstrate.

6.2.1 Worked examples – finding the instantaneous voltage of a sine wave input signal

The first example will use a known frequency, while the second will use the period of the wave. In both cases, the relationship between the period, T (in seconds), and the measurement time, t (in seconds), is a good indicator of the value of the sine wave – note that both symbols are the same, so be sure to check for capitals!

Q2: *An AC electronic circuit has an input signal that is a sine wave of peak voltage +5V and frequency 200Hz. If the input is measured 50 milliseconds after the signal is connected, what would be the amplitude of the signal?*

$V_p = 5V$, *Frequency*, $f = 200Hz$, *Time*, $t = 50ms = \mathbf{0.05}$ seconds

First, calculate the radian angle for a wave of frequency 200Hz:

$\textit{Sine Wave Angle}, \omega = 2\pi f = 6.283 \times 200 = 1256.64$

Now calculate the instantaneous voltage at 50ms (starting with the value for ωt):

$\omega t = 1256.64 \times 0.05 = 62.832$

$\textit{Instantaneous Voltage}, V = V_p \sin(\omega t)$

$= 5 \times \sin(62.832)$

$= 5 \times 0.00015 = 0.00075V \approx \mathbf{0V}$

Answer: The voltage level of a sine wave signal at 200Hz measured at a time of 50 milliseconds is **~0V**.

> **Notes:**
>
> This example shows how an input sine wave signal of 200Hz will have an amplitude of ~0V at 50 ms. This is not unexpected, as 200Hz has a period of 5 ms (or 0.005 seconds) and 50 ms is exactly divisible by this. Thus, after 50 ms the wave will have completed 10 cycles and so should be at a zero crossing (assuming it started at 0V at time 0). It is important to note that the value for (ωt) is calculated **before** finding the sine of the result, so this step was performed before calculating the instantaneous voltage. Also note that the final result is not exactly 0, but it is a value of 0 to three decimal places – this is due to rounding errors created by taking a shorter value for 2π. The absolute value of π is not known, so there will always be a slight deviation in the calculation as a result.

Q3: *An AC electronic circuit has a sine wave input signal with a peak voltage +5V and a period of 10ms. If the input is measured 52.5 milliseconds after the signal is connected, what would be the amplitude of the signal?*

> $V_p = 5V$, $\quad Period, T = 10ms = \mathbf{0.01}seconds$, $\quad Time, t = 52.5ms = \mathbf{0.0525}seconds$
>
> First, calculate the frequency of a wave with period 10ms:
>
> $$Frequency, F = \frac{1}{T} = \frac{1}{0.01} = \mathbf{100Hz}$$
>
> Now calculate the radian angle for a wave of frequency 100Hz:
>
> $$Sine Wave Angle, \omega = 2\pi f = 6.283 \times 100 = 628.3$$
>
> Finally, calculate the instantaneous voltage at 52.5ms (starting with the value for ωt):
>
> $$\omega t = 628.3 \times 0.0525 = 32.98575$$
>
> $$Instantaneous Voltage, V = V_p \sin(\omega t)$$
>
> $$= 5 \times \sin(32.98575)$$
>
> $$= 5 \times 0.999 = 4.999V \approx \mathbf{5V}$$

Answer: The voltage level of a sine wave signal of period 10ms (100Hz) measured at a time of 52.5 milliseconds is **~5V**.

> **Notes:**
>
> In this example, we have to find the frequency of the input sine wave based on a period of 10ms. This gives a value of 100Hz, and specifies a time value of 52.5ms, which should occur one quarter through the period of the wave. Thus, the calculated value is effectively 5V as the input signal amplitude is at its peak. Note that Period T and Time t use the same symbol, so it is important to double check capital letters when working with sine waves.

Using the instantaneous voltage calculation for an input sine wave, some practical measurements can be made that are used in audio circuits. In Figure 6.8, both the average (V_{AVG}) and the root mean square (V_{RMS}) voltage levels were marked, which are less than the peak value. At first glance, the average voltage would seem straightforward to calculate, but it must be remembered that a sine wave is half positive and negative parts in a single cycle – as shown in Figure 6.7, a normal sinusoid will cancel itself out to give an average value of zero. Instead, the first half of a sine wave cycle is measured as the average of the sum of a number of points within it, which are known as mid-ordinates (Figure 6.10).

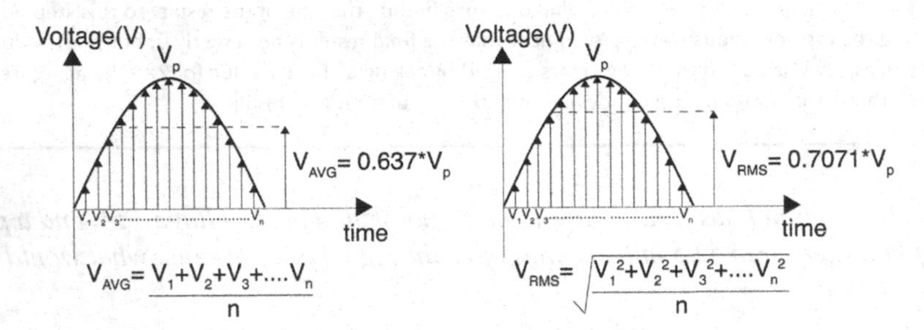

Figure 6.10 Measuring mid-ordinate points on a sine wave. ***The diagram shows how measuring a set of points in the positive half cycle of a sine wave can be used to find the average value (V_{AVG}) of that wave. For a pure sinusoid, this value equates to 0.636*Vp. We can also calculate the root mean square (V_{RMS}) voltage, which at 0.7071*Vp is the equivalent DC voltage level that would be applied to the circuit.***

The diagram shows two methods of defining the average voltage for a sine wave, each using a set of mid-ordinate points that represent the voltage level measured at regular intervals along the positive half of the cycle. In the case of average voltage, the values are summed and divided by the number of points involved. More commonly, the values are squared, summed, divided by the number of points and then the square root of this result is taken – the root mean square (RMS) voltage:

$$\textit{Root Mean Square Voltage}, V_{RMS} = \sqrt{\frac{V_1^2 + V_2^2 + V_3^2 + \ldots V_n^2}{n}} = 0.7071 \times V_p \qquad (6.5)$$

V_{RMS} *is the equivalent DC voltage signal in volts (V)*

V_n *is the voltage at a specific mid-ordinate point n in volts (V)*

n *is the number of mid-ordinate points*

The root of the mean of the sum of squares (RMS) equates to a value of 0 $7071 \times V_p$ for any pure sinusoidal circuit – the value comes from $\frac{1}{\sqrt{2}}$ as a factor of ω (the derivation of the full equation is not required in this book). RMS voltage measurements are used throughout electronics to represent the **equivalent DC** signal voltage that the AC input would generate. This allows the **power in a circuit** or system to be calculated by substituting the equivalent DC value into the electrical power equation introduced in chapter 4 (section 4.9, equation 4.2) – *Power, P = Current × Voltage = I.V*$_{RMS}$ measurements are often included as part of the specification for a piece of electronics equipment, particularly in the case of audio amplifiers and loudspeakers. By using equivalent DC voltage, the power output to a loudspeaker can quickly be calculated to ensure that an amplifier will not damage it.

The final element used to describe an AC signal is its **phase**, which can be difficult to learn because the calculations require more detailed mathematical models that involve complex numbers (combining two quantities as a real and imaginary pair). Having said this, the conceptual aspect of phase is much more straightforward, as it relates to the relative position of one sine wave to another (Figure 6.11).

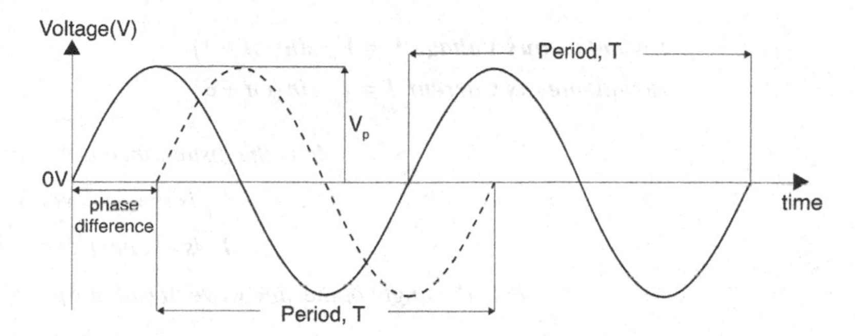

Figure 6.11 Sine wave phase example. ***In the diagram, two sine waves of equal amplitude and frequency are shown. The first wave (solid line) begins with amplitude 0 at time 0, but the second wave (dashed line) does not cross the amplitude axis until later in time, therefore it lags the first wave and there is a phase difference between them – only one cycle of the second wave is shown for clarity.***

The diagram shows two sine waves, where both have equal amplitude (V_p) and frequency (Period, T, is shown). The first wave crosses the origin at time 0, so it begins with an amplitude of 0. The second wave (shown as a dashed line) crosses the 0 amplitude point much later in time, and thus the **second wave lags the first wave**. This can be confusing when you first look at the diagram, as the second sine wave appears visually to be 'ahead of' the first, but it must be remembered that the x axis denotes time, and **the greater the time value the later the wave** will be – it is lagging behind a wave that started its cycle earlier in time. Conversely, it can be said that the **first wave leads the second wave**, as it begins its cycle first. The phase difference between these two waves is defined in terms of the rotational angle between them, where the angles can be compared at a specific point in time to determine the relative lead or lag (Figure 6.12).

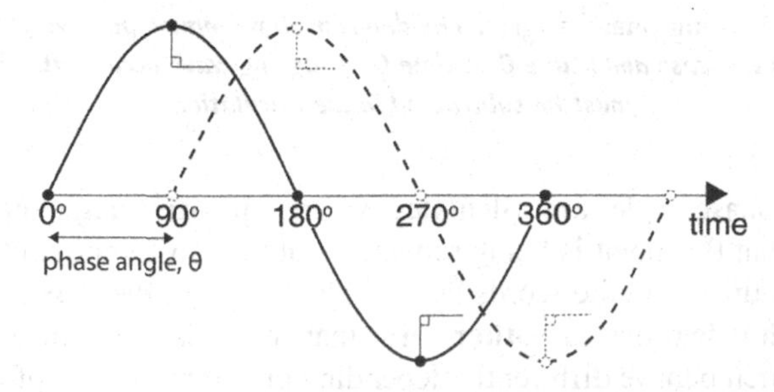

Figure 6.12 Phase angle difference between two sine waves. ***The diagram shows the same sine waves as Figure 6.11, but as both waves have the same amplitude the y-axis is omitted for clarity. The x-axis indicates the angle of the first wave (in degrees), where it can be seen that the second wave has a 0° angle when the first wave has an angle of 90°. The second wave lags the first wave by 90°, which is defined by the phase angle θ.***

The diagram shows the main phase angles for each sine wave over time, where by inspection it can be seen that the second wave is at 0° when the first wave is at 90°. This means the phase relationship can be stated as the angle difference between the two waves in either direction as required, where the **second wave lags the first** wave by 90°. The equations for instantaneous voltage and current can thus be updated, to take into account the phase angle θ of the signal relative to a known reference:

$$\begin{aligned} \textit{Instantaneous Voltage}, V &= V_p \sin(\omega t + \theta) \\ \textit{Instantaneous Current}, I &= I_p \sin(\omega t + \theta) \end{aligned} \tag{6.6}$$

V is the instantaneous voltage in volts (V)

V_p is the peak voltage in volts (V)

I_p is the peak current in amps (A)

ω is the angle of the sine wave in radians per second (rad/s)

t is the time in seconds(s)

θ is the phase angle difference in radians (rad)

In the equation, the value for θ is calculated in radians relative to a known reference. In practice, either a sine wave that has amplitude 0 at time 0 or another wave is used for reference. It is important to reiterate that a **positive phase angle means the signal leads** the reference – addition means the wave will cycle earlier in time (Figure 6.13).

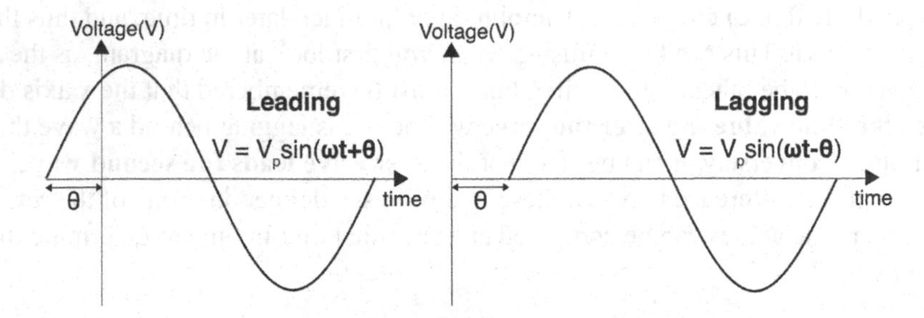

Figure 6.13 Leading and lagging phase angles. ***The diagram shows how a positive phase angle means the wave is leading the reference – in this case amplitude 0 at time 0. A lagging wave occurs later in time, so the phase angle must be subtracted in the calculation.***

The diagram shows how phase angles are calculated, where a **positive** angle **leads** and a **negative** angle **lags**. It may seem that this point is being repeated, but the most important thing for your understanding of phase is the use of the words lead and lag: because the x-axis is time, the **wave furthest left in the graph is leading any other wave** that comes later in time. This chapter also introduces capacitors, which behave differently depending on the frequency of the input signal (only resistors remain constant as frequency changes). Other components (e.g. amplifiers) also change the phase of an input signal, and thus a system containing any of those components (effectively all of them!) has an impact on signal phase. As a final note on phase, note the result of combining (adding) two sine waves of the same amplitude and frequency but with different phase – this illustrates the effect of phase cancellation on signal output (Figure 6.14).

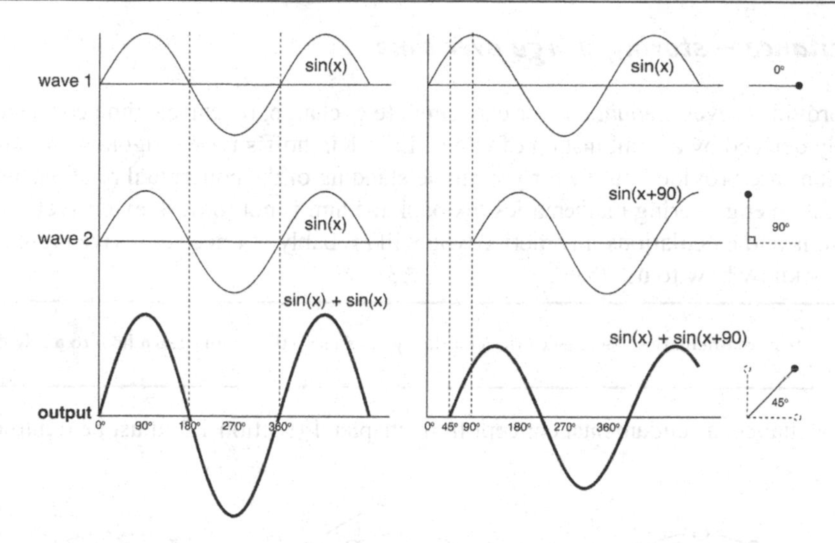

Figure 6.14 Sine wave phase cancellation. ***The diagram shows two sine wave addition examples, where the left-hand example adds two sine waves of the same amplitude, frequency and phase angle – the output is also in phase with these signals. The right-hand example adds two sine waves of the same amplitude and frequency, but the second wave lags the first by 90°. The resulting output wave is lower in amplitude than the first output, and also lags the input signal by 45° (as shown by the phasor diagrams to the right of each wave).***

The diagram shows how phase cancellation can occur when combining signals, where the output signal in the right-hand graph is now lower in amplitude and lagging the original input signal by 45°. A simple example of a **phasor diagram** for each wave is shown to the right of this graph, which uses trigonometry to calculate the resulting phase angle produced when combining multiple signals. At this point, any further discussion of phasors requires mathematical explanation involving complex numbers, which is not needed at an introductory level. Having said this, phase is crucial to working with audio signals and it is something you will learn more about as you progress your audio electronics learning – simulation tools (like LTspice) allow phase angles to be visualized quickly, so their impact can be modelled and evaluated in your own circuits without knowing how to perform the derivations on paper.

Phase cancellation is a potential issue in any instance where different audio signals are combined. Everything from effects processors with multiple paths (e.g. sidechains), buffer amplifiers on a preamp, power amplifiers that provide separate monophonic amplification and indeed any form of stereo signal path may potentially lead to phase alignment issues. The reduction in amplitude seen in the right-hand example will increase as the phase angle increases, until a 180° lag between the inputs will completely cancel out the output signal. This amplitude reduction is compounded by the overall phase lag introduced (in the example, a lag of 90° on one input creates an overall lag of 45°), which will now be carried into the next stage of the audio circuit or system. In this book, the 2-band equalizer used in chapters 8 and 9 is a multiple-path circuit and although each filter stage will introduce phase shifts due to the use of capacitors they are not analysed. This deliberate omission aims to focus instead on the use of simple components to build audio filters, where the frequency-dependent behaviour of the capacitor allows us to build practical audio circuits. In more advanced high-fidelity audio, the issue of phase cancellation cannot be so easily ignored.

6.3 Capacitance – storing charge over time

This section provides several equations for capacitors (e.g. charge, reactance, time constant) that are mathematically derived by a combination of Ohm's Law, Kirchoff's Laws, trigonometry and calculus. These derivations are provided purely for your understanding of the conceptual relationships involved, but as this is not an engineering mathematics textbook the aim is not to work extensively with paper-based derivations and calculations. In practice, you will probably not need to derive these equations very often, only know how to use them:

> if you find aspects of mathematics like calculus difficult then you can skip the proof – **learn how to apply the result!**

To discuss capacitance, a fundamental concept from chapter 1 (section 1.2) must be reintroduced (Figure 6.15).

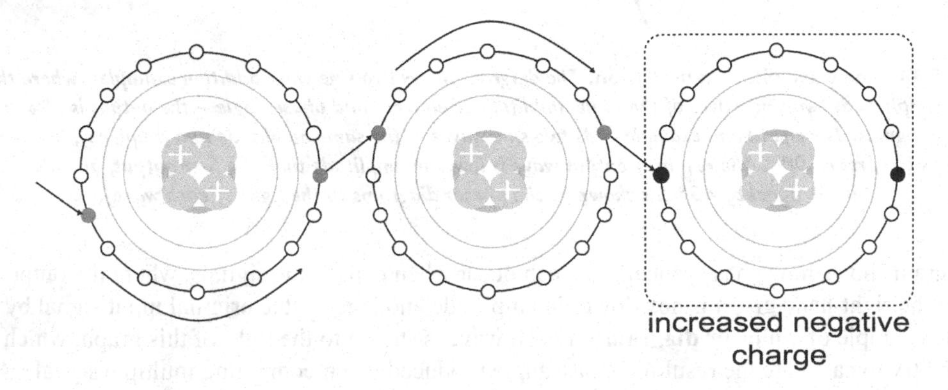

Figure 6.15 Movement of electrical charge (adapted from chapter 1, Figure 1.5). ***Chapter 1 showed that electrical charge is the net sum of negative (electrons) and positive (holes) at a given point in space. The crucial element is that although electrons are displaced from atom to atom, the flow of charge is the relative change in balance between electrons and holes – not the movement of electrons from one point to another.***

This diagram is crucial to understanding capacitors, because it illustrates a fundamental concept that can often be forgotten when learning circuit theory:

> **the net movement of charge is not the net movement of electrons**

What this statement means is that though an electron may move from one atom to another, the **net movement of charge is not dependent on this electron moving the entire length of the circuit.** As one electron moves into a valence shell, other electrons can move out of it – and thus the net charge will change. This is an important distinction to make, as the book has referred to the movement of electrons so far – no distinction was needed when working with DC circuits. Now it is important to note the fundamental distinction between the local movement of an electron (which displaces others in a chain) and the **global movement of charge that passes between atoms**. The charge is carried by electrons, but the specific electrons moving may change. This is crucial to understanding the capacitor, where electrons are prevented from moving across an insulated gap (dielectric), yet the net charge can still move through the capacitor as an AC signal.

Chapter 2 briefly introduced the capacitor. A capacitor stores electrical charge in an **electrostatic field** that is created when a charge is applied between two conducting surfaces (Figure 6.16).

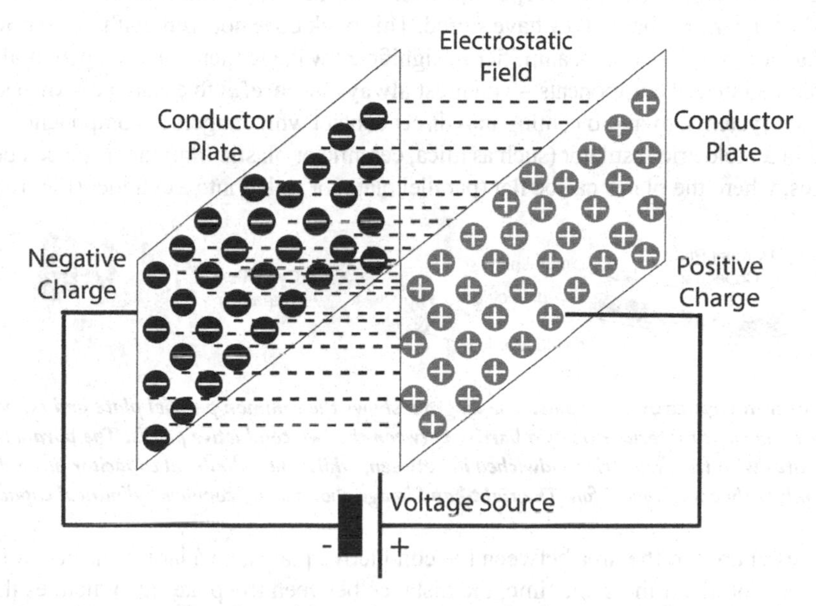

Figure 6.16 Capacitor example, adapted from chapter 2. ***In the diagram, a voltage source is used to supply electrical charge to the conductive plates (one positive, one negative). The build-up of charge between the plates becomes stored electrostatically in the air (or another insulator) between them. Thus, as the charge between the plates builds the capacitor will store more charge over time.***

An electrostatic field occurs between two **stationary charges** (one negative, one positive) that are physically close together. You may have encountered static electricity before if you have rubbed an inflated balloon against a piece of cloth. The rubber in the balloon acts as an insulator that can store charge on its surface and rubbing the balloon against a conductor like cloth (wool is a good conductor) creates static electricity. The movement stimulates electrons to leave the surface of the cloth and travel towards the balloon and this electrostatic charge is then held on the surface of the balloon where the insulating rubber prevents the electrons from moving any further. If the balloon is now placed near a neutral (or positive) surface such as a wall then the charged particles on the surface of the balloon will be attracted to it – hence the balloon 'sticks' to the wall due to the attraction of the charged particles present. This simple example exploits the same principle as lightning strokes, where electrostatic charge is built up under certain cloud conditions and this creates an electric field in those clouds (the formation of a semi-ice water known as graupel is believed to be part of the cause). The Earth's atmosphere acts as an insulator, so a huge electrostatic charge can build up in the clouds that is then attracted towards oppositely charged particles on the Earth's surface. When a suitable conductor is introduced, the charge can flow through it – often with devastating effect given the electrical power in a lightning stroke. This is an important point about electrostatic charge that has direct relevance to capacitors:

Capacitors store electrical charge between their plates

A large capacitor (such as those used in audio power amplifiers) can store a significant amount of charge, and under certain conditions this charge can remain in the capacitor even after the power source that originally charged it disconnects. Thus, **large capacitors can be dangerous** if touched as the human body becomes a conductor for the charge they have stored. This book does not work with audio power circuits, but if you own an audio or instrument amplifier of significant wattage then it has the potential to store this power within its internal components – you must **always be careful** to ensure your own safety when working with such systems, to avoid coming into direct contact with dangerous components. In a capacitor, charge is stored in a dielectric insulator (such as mica, ceramic or plastic film) that is placed between conductive plates, where the plates can be flat (parallel plate) or rolled into a cylinder (Figure 6.17).

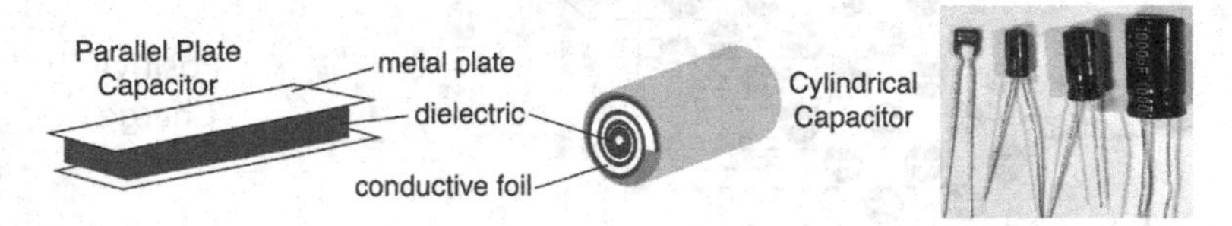

Figure 6.17 Common capacitor examples. ***The diagram shows the common parallel plate and cylindrical capacitor designs, where the dielectric material acts as a barrier between the two conductive plates. The parallel plate capacitor has two metal plates with the dielectric sandwiched in between, whilst the cylindrical capacitor has a dielectric rolled between two sheets of metal foil. The right-hand image shows some common cylindrical capacitors.***

A capacitor stores charge in the area between the conductive plates, so a larger surface area allows more charge to be stored. At the same time, the distance between the plates also dictates the charge that can be stored – **less plate distance allows more charge to be stored** as the electrostatic field is stronger. Physically larger capacitors can usually store more charge because of their larger surface area, and depending on their use can be significant in size. A capacitor stores charge when connected to a voltage source, where the build-up of negative electrons on one plate repels the electrons on the other plate to increase its **overall positive net charge** (Figure 6.18).

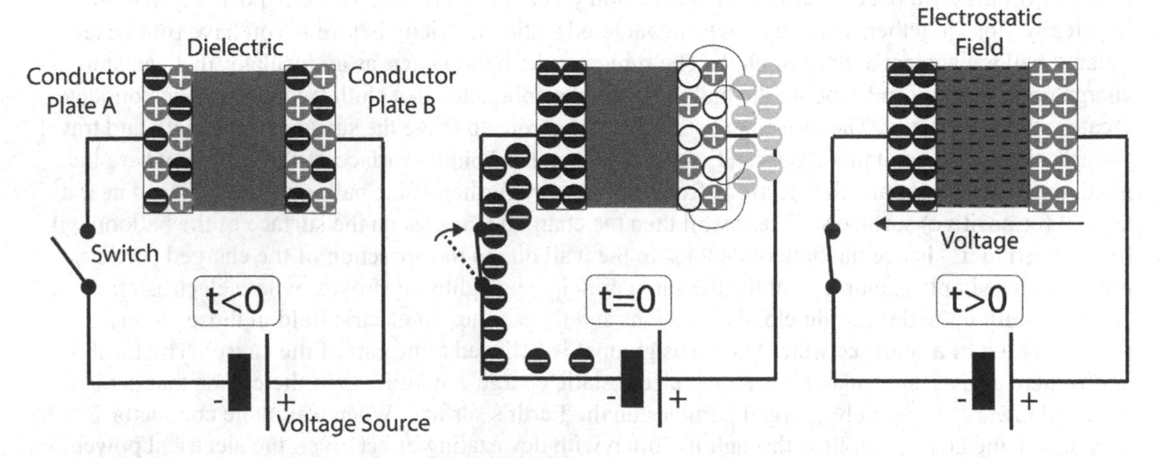

Figure 6.18 Movement of charge in a capacitor. ***In the diagram, when the switch is closed at time 0 the increase in electrons on plate A repels electrons from plate B – which thus becomes more positive. Although the electrons on plate A are now attracted to the holes on plate B, the dielectric prevents them from moving. This creates an electrostatic field between the plates, which stores the charge. The difference in potential between the plates also creates a voltage (potential difference) across them.***

The most important element of this diagram for your understanding is the difference between the net movement of charge and the physical movement of electrons – the electrons cannot cross the dielectric barrier, but **more electrons have moved to plate A so the net charge on plate B becomes more positive**. Thus, charge can flow through a capacitor, even though electrons cannot cross the dielectric. The circuit in Figure 6.18 actually uses a DC power source with a switch to simulate the simplest form of time-varying circuit, where the signal goes from 0 amplitude at time 0 to full amplitude immediately after the switch is closed. Although not an oscillating signal like a sine wave, this signal shows how the build-up of charge between the plates occurs in a capacitor. A capacitor does not fully charge up immediately – it takes **time** for the charge to build up across the plates (Figure 6.19).

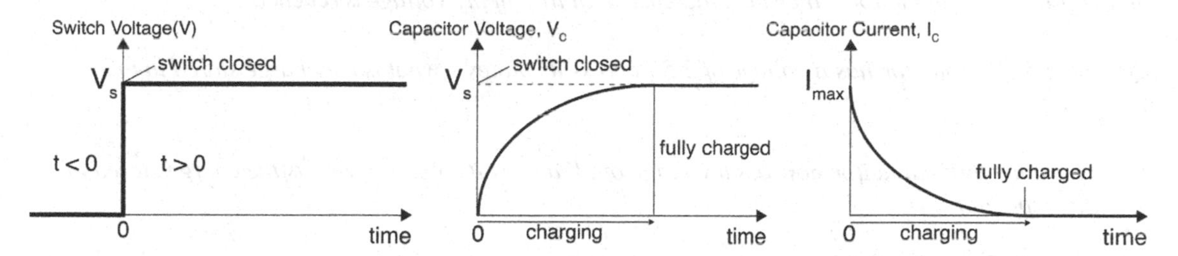

Figure 6.19 Graph of a capacitor charging over time. ***In the diagram, the switch in our circuit from Figure 6.18 is closed at time 0, allowing current to flow. As time increases, the middle graph shows how the capacitor charges up until the voltage across it is equal to that of the supply (Vs). Conversely, the right-hand graph shows how the current flowing through the capacitor reduces until it reaches 0 when the capacitor is fully charged – the capacitor effectively becomes a short circuit.***

The diagram illustrates the importance of capacitors for time-varying signals. Unlike a resistor, the capacitor reacts differently to an input signal over time. A capacitor begins to charge up when a voltage is applied to its plates, but it will take a finite time (known as the **time constant,** τ) for the charge between the plates to build up to full capacity (known as the capacitance, C). When the capacitor is fully charged, the potential difference between its plates is now at the same level as the voltage supplied to them – this is a crucial point, because it also explains why the current in a capacitor falls to 0 when it is fully charged. Once the potential at each capacitor plate is the same as the supply then there is no potential difference between them to move charge – the **capacitor has effectively become a short circuit**.

This can seem confusing at first, as capacitors are used so often in circuits it would seem bizarre for them to completely prevent current from flowing! The reason for this is the example above relates to a **DC circuit**, where the signal does not change. The case of a switch activating the circuit is used to provide a before and after at time 0, to help illustrate how a capacitor charges up. The amount of charge a capacitor can store is defined as its capacitance (in farads), based on the voltage across its plates:

$$\textit{Capacitance, } C = \frac{Q}{V} \tag{6.7}$$

C is the capacitance in farads (F)

Q is the charge in coulombs (C)

V is the potential difference between the capacitor plates in volts (V)

The equation defines capacitance in farads (F), but for audio signals capacitance values in the range μF (10^{-6}) to nF (10^{-9}) and even pF (10^{-12}) are normally used, so it is important to remember that scaling factors will be a significant part of calculations (see chapter 1, Table 1.1). The relationship between charge and voltage determines how much charge is held in the capacitor, as the following example shows:

6.3.1 Worked example – calculating the charge on a capacitor

This example is mathematically straightforward, but it aims to show the relationship between charge and voltage for a capacitor – they both increase until the supply voltage is reached.

Q4: *(a) A 5μF capacitor has a voltage of 2.5V across its plates – what is the charge stored in the capacitor?*

(b) The same 5μF capacitor now has a voltage of 5V across its plates – what is the charge stored in the capacitor?

For a voltage of 2.5V: *Capacitance*, $C = 5F$, *Voltage*, $V = 2.5V$

$$C = \frac{Q}{V}, \quad \therefore Q = CV = \left(5 \times 10^{-6}\right) \times 2.5$$

$$C = 0.000005 \times 2.5 = 0.0000125 = \mathbf{12.5\mu C}$$

For a voltage of 5V: *Capacitance*, $C = 5F$, *Voltage*, $V = 5V$

$$C = \frac{Q}{V}, \quad \therefore Q = CV = \left(5 \times 10^{-6}\right) \times 5$$

$$C = 0.000005 \times 5 = 0.000025 = \mathbf{25\mu C}$$

Answer: (a) The charge stored in the capacitor for 2.5V voltage supply is **12.5μC**.

(b) The charge stored in the capacitor for a 5V voltage supply is **25μC**.

Notes:

This example involves the multiplication of two values, but the scaling factor of the capacitance can cause problems if it is not calculated correctly. It is important to avoid the temptation to miss steps when working with capacitors, as the μF value can easily change the scale of the result if a zero is missed during multiplication. The same capacitance value is used to show the impact of increasing the voltage across the capacitor - the charge it stores increases accordingly. The capacitor is fully charged once the plate voltage reaches the supply voltage level, as there is now no more potential difference to move charge across the plates - thus the current flow drops to 0 and the capacitor becomes a short circuit.

A capacitor takes time to charge, and this is part of what makes them very useful for time-varying AC circuits. In a DC circuit, a capacitor has time to fully charge up and thus become a short circuit, but in an AC circuit the changing polarity of the voltage signals means the capacitor does not have time to do this. Electrical charge is proportional to voltage, so if the supply voltage decreases so will the charge on the capacitor – this circuit can then be combined with a resistor (to give the capacitor a path to discharge through), as an example (Figure 6.20).

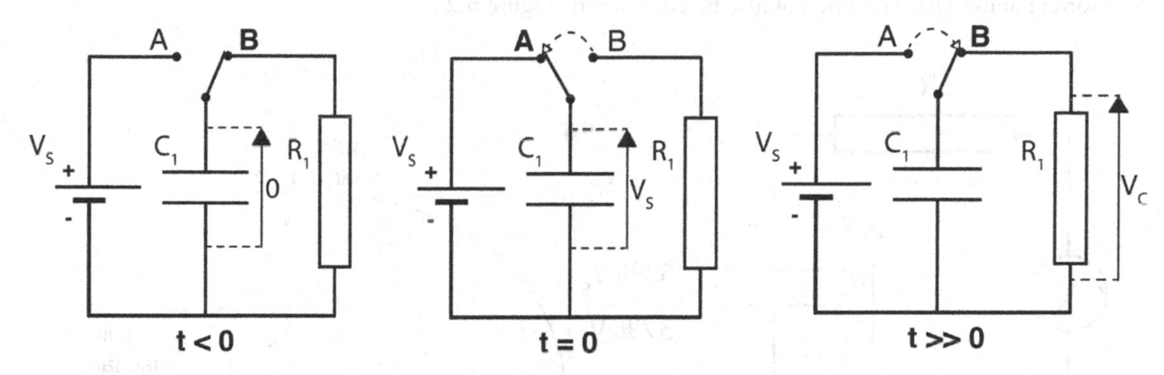

Figure 6.20 Discharging capacitor example. ***In this example, the effect of switching on a power supply to a capacitor that is initially uncharged (switch position A) is shown. The switch changes to position B at time = 0, and shortly after the capacitor will be fully. If at a later time (t >> 0) the supply is disconnected (switch position B), the capacitor will discharge through the resistor R1 until the capacitor voltage returns to 0 – there is then no potential difference to move charge.***

This example circuit has a charging graph as shown in Figure 6.21.

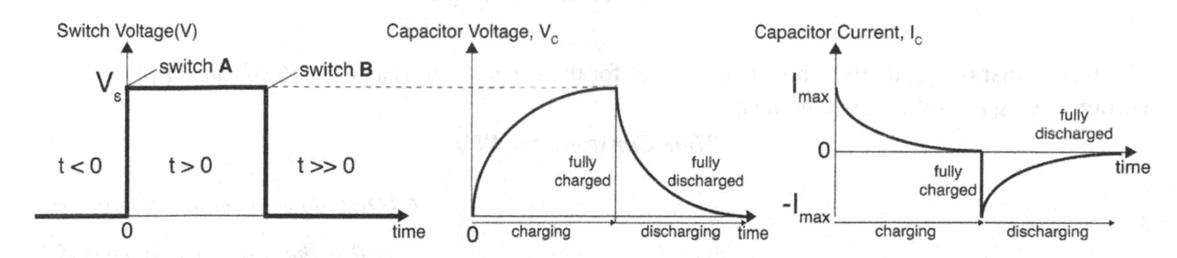

Figure 6.21 Graph of capacitor charging and discharging over time. ***In the diagram, the switch in the circuit from Figure 6.18 is in position A at time 0, allowing current to flow into the capacitor. As time increases, the middle graph shows how the capacitor charges up until the voltage equals the supply ($V_C = V_s$) and no current can now flow (right-hand graph), before the switch is returned to position B at time >> 0. At this point, current can flow in the opposite direction (−Imax) and the capacitor can discharge through the resistor R1.***

The diagram shows how a capacitor takes time to charge by applying a voltage pulse (i.e. switching from position B to A and then back again). The capacitor builds up charge based on the applied input voltage, so when the switch is moved to position A the full supply (V_S) is applied across its plates and it begins to store charge. Once full charge has been reached, the capacitor effectively becomes a short circuit, and so if the switch was not moved back to position B it would oppose any current flow (which drops to 0). Once the capacitor can discharge again (through the resistor R_1) current can flow, but this time it is flowing in the opposite direction to the initial supply voltage – hence the value of $-I_{max}$ shown in the graph.

The idea of a time constant (τ) that governs the charge and discharge time of a capacitor was introduced earlier in this section. Now capacitance has been defined (and also how a capacitor's voltage and current change over time), the time constant (τ) that describes the rate of change in a resistor/capacitor (RC) circuit can also be considered (Figure 6.22).

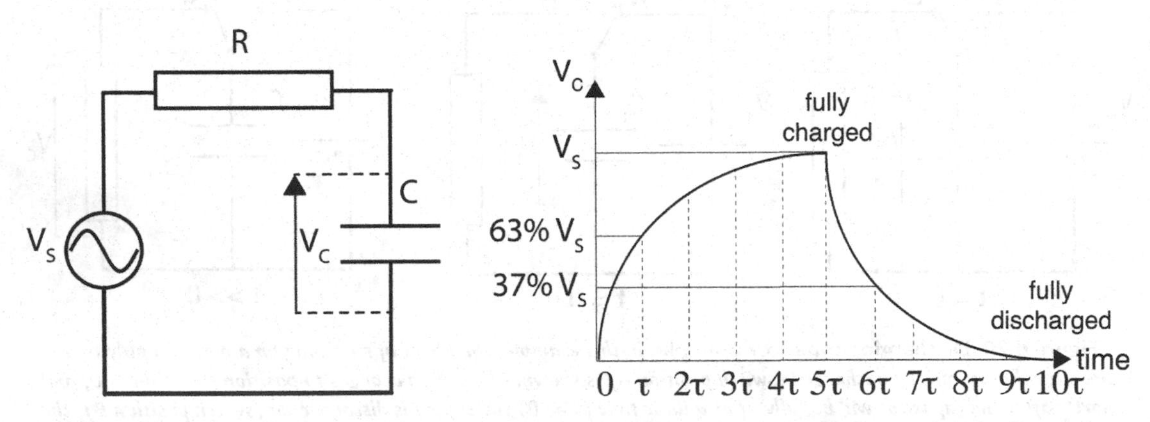

Figure 6.22 Time constant in an RC circuit. ***The diagram shows the charge/discharge graph for the series combination of a resistor and capacitor. The values of both components dictate how long it takes for the capacitor to reach ~63% of its final charge level, which is known as the time constant (τ). The capacitor takes 5 time constants to reach full charge, and from there reduces to ~37% charge after a further time constant (6τ) – decreasing to zero after 5 time constants (10τ in total). This allows the charge graph to be defined by the number of time constants taken to fully charge or discharge.***

The **time constant** (τ) denotes how long it takes for the capacitor to **reach ~63% of the supply voltage**, using the following equation:

$$\textit{Time Constant}, \tau = RC \qquad (6.8)$$

τ is the time constant in seconds (s)

***R** is the resistance in ohms (Ω)*

***C** is the capacitance in farads (F)*

The time constant equates to ~63% because the charging curve is exponential in shape (this equation is derived in the appendix). It will take a total of 5 time constants to reach full charge (when the plate voltage is equivalent to the supply), as each time constant increases the voltage by a further 63% of the previous value. Once fully charged, it will take the capacitor a further 5 time constants to discharge again, where 6τ defines when the discharge cycle reaches ~37% (100 – 63). The circuit above demonstrates a capacitor's ability to store charge, which means they can **become a source of charge** at a later time. This is a common use of capacitors in power-supply circuits, where they are used to provide extra charge at points where the power-supply levels may not be guaranteed.

For example, car audio systems cannot rely on a constant current from the battery, given that the draw from other systems like headlights and windscreen wipers may vary widely over time depending on driving conditions. Even if a separate battery is used for the car audio system, the charge/discharge graphs shown in Figure 6.21 highlight another important use of capacitors – smoothing input signals. In chapter 4 (section 4.9), PWM output was used from the Arduino to light an LED. In that case, the duty cycle of the PWM output was used to approximate the same power (and hence average voltage) required to light the LED, but in practice the binary on/off nature of a digital signal does not provide a smooth enough supply to power electronic systems (particularly audio systems). Smoothing out a time-varying signal (such as PWM) is performed as part of a process known as **rectification** – the negative portion of the AC input voltage is inverted and smoothed to produce a DC output for consistent power supply (Figure 6.23).

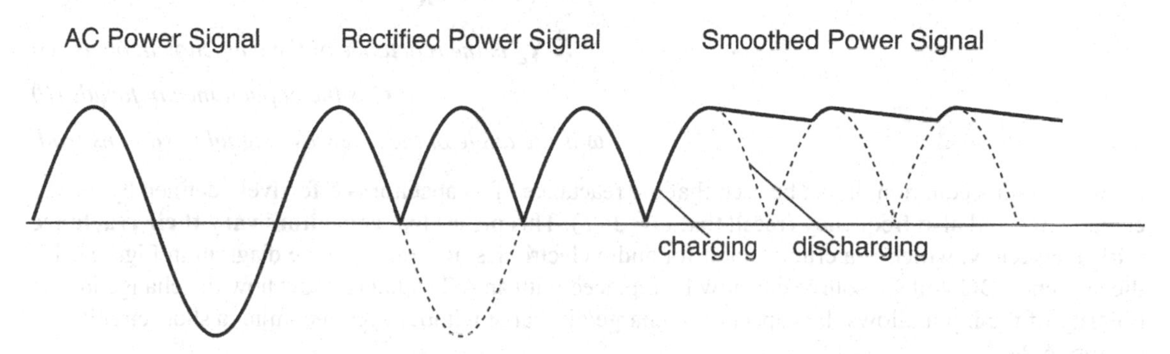

Figure 6.23 Full wave rectification example. ***An input AC power signal (such as a 240V 50Hz mains signal in Europe) must be converted to DC to provide a constant supply for an electronic circuit. To do this, a bridge rectifier circuit is used to invert the negative portion of the sine wave input cycle. Once inverted, the rectified signal must be smoothed using a capacitor to reduce the ripples in the output. In so doing, a rectifier uses a capacitor as a simple form of filter.***

In the diagram, the AC signal is rectified by inverting the negative portion of the wave to produce a fully positive output (diode rectification will be discussed in the next chapter). After inversion, the rectified signal is smoothed by passing it through a parallel capacitor that charges up while the voltage signal is increasing to V_p, and then discharges as the signal decreases from V_p back to 0. Looking at the graphs in Figure 6.21, it can be seen that although the input signal is a square wave, it still rises and falls like a sine wave (just with a gradient of 1 and minus 1). Thus with a sine wave, the current flowing out of the capacitor also increases to peak value (I_{max}) as the input voltage goes to 0 – this is what happens in a rectifier.

Using a capacitor to smooth these types of binary signal is also part of digital to analogue conversion, where the discrete output levels of the digital audio sample data must be smoothed by **filtering** to produce a more suitable output. We will learn more about filters in chapter 8, where we will build circuits for both low- and high-pass filters that use a capacitor as the component that varies with input signal frequency. To do this, we need to know how a capacitor will react at different frequencies, which is defined as its **reactance**. In the appendix, the equation for capacitor voltage in terms of current and capacitance is derived:

$$Capacitor\ Voltage,\ V_C = -\frac{I_p}{\omega C}\cos\omega t \qquad (6.9)$$

I_p is the instantaneous current in amperes (A)

C is the capacitance in farads (F)

ω is the angle of the sine wave signal in radians (rad)

t is the time in seconds (s)

This can then be used to derive the equation for capacitive reactance (X_C):

$$Capacitive\ Reactance,\ X_C = \frac{1}{\omega C} \qquad (6.10)$$

X_C is the reactance of the capacitor in ohms (Ω)

C is the capacitance in farads (F)

ω is the angle of the sine wave signal in radians (rad)

Looking at this equation, it can be seen that the reactance of a capacitor is effectively defined by its capacitance and also frequency (recall that $\omega = 2\pi f$). This means that **capacitors vary their reactance with frequency**, which is a crucial point for audio electronics. Returning to the diagram in Figure 6.18, the switched DC voltage source can now be replaced with an AC signal to show how the change in polarity of the input allows the capacitor to charge/discharge without ever becoming a short circuit (Figure 6.24).

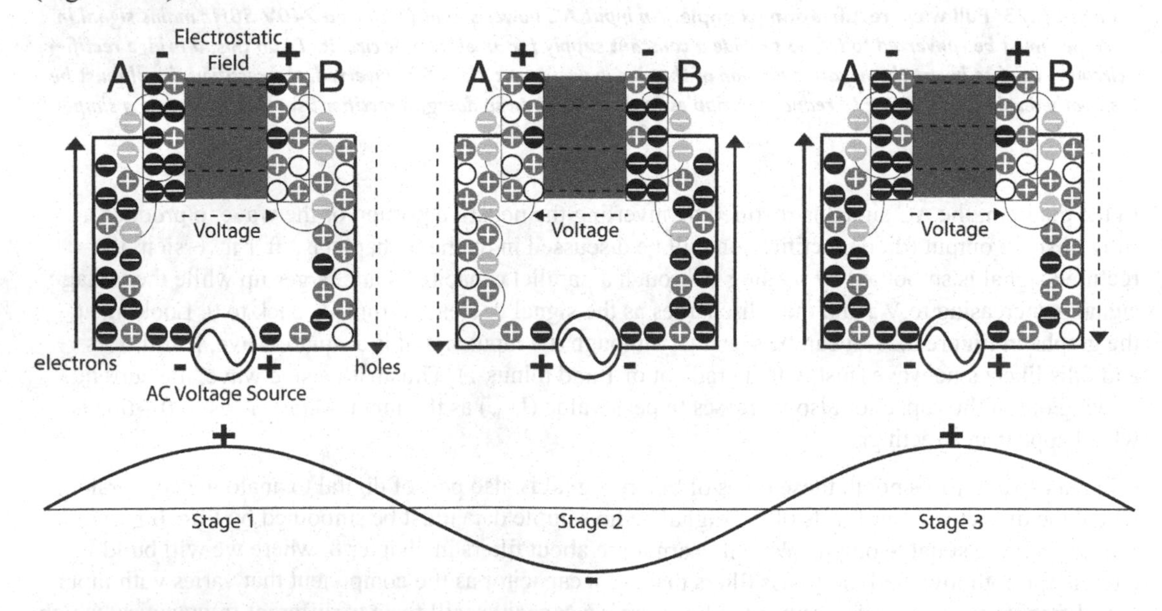

Figure 6.24 Charge movement for an AC signal in a capacitor. ***In the diagram, the alternating signal is constantly changing polarity, and so the voltage at each capacitor plate changes polarity too. This means charge does not have time to build up and so the electrostatic field is not as strong – current can still flow across the plates. If the frequency of the AC signal is high enough, the capacitor will not have time to charge up at all – it effectively becomes an open circuit.***

In the diagram, the constantly varying AC input signal provides a different voltage polarity to each capacitor plate at different points in its wave cycle. When the input signal is highest (**stage 1**), plate A will be negative and plate B positive and an electrostatic charge starts to build up between them. As the input signal now decreases again, the voltage between the plates decreases and current can flow to discharge the capacitor (see Figure 6.21). When the input signal decreases further to its lowest point (**stage 2**) plate A will now be positive and plate B will be negative – an electrostatic charge starts to build up between them again. The input signal will now increase (towards **stage 3**), and as the voltage between the plates decreases current can flow once again and discharge the capacitor. Thus, the capacitor never fully charges up, and so current can flow throughout the wave cycle. If the frequency of the input signal is high enough, the capacitor effectively becomes an open circuit – it does not react to the input signal and current can flow.

It is important to understand the charging ability of capacitors, as they are used extensively in both amplifier and filter circuits. A capacitor can be used to smooth signals, filter input, block DC signals (chapter 9) and provide charge to different parts of a circuit – they are an incredibly versatile component and used throughout electronics. Returning to our equation 6.10 for capacitive reactance, a short example is provided to show how signal frequency changes reactance.

6.3.2 Worked example – calculating capacitive reactance for different input frequencies

This example shows how capacitive reactance varies with the frequency of the input signal – this is used in chapter 8 to build audio filter circuits.

Q4: *(a) A 5μF capacitor has an AC input voltage of 5V at a frequency of 100Hz – what is the reactance of the capacitor?*
(b) The same input voltage now has a frequency of 1kHz – what is the reactance of the capacitor?

For an input frequency of **100Hz**:

$$Capacitance, C = 5\mu F, \quad Voltage, V = 5V, \quad Frequency, f = 100Hz$$

$$\omega = 2\pi f = 6.283 \times 100 = 628.3\,rads$$

$$Capacitive\,Reactance, X_C = \frac{1}{\omega C} = \frac{1}{628.3 \times 0.000005} = \frac{1}{0.0031415} = \mathbf{318.32\Omega}$$

For an input frequency of **1kHz**: $Capacitance, C = 5\mu F, \quad Voltage, V = 5V, \quad Frequency, f = 1kHz$

$$\omega = 2\pi f = 6.283 \times 1000 = 6283\,rads$$

$$Capacitive\,Reactance, X_C = \frac{1}{\omega C} = \frac{1}{6283 \times 0.000005} = \frac{1}{0.031415} = \mathbf{31.83\Omega}$$

Answer: (a) For a 100Hz AC signal input, the capacitor has a reactance of 318.32Ω.
(b) For a 1kHz AC signal input, the capacitor has a reactance of 31.83Ω.

Notes:

This example uses two frequencies that are a multiple of 10 apart, to show how the reactance of a capacitor scales downwards with frequency. As frequency increases, so does the reactance of the capacitor in ohms (Ω). The value of 2π was defined as being 6.283 in equation 6.4, which is a good enough approximation for the frequency calculations in this book. As with all capacitor calculations, the scale of the capacitance value must be correct and so care should be taken to avoid adding/subtracting a zero in the capacitance (which would scale the result by a factor of 10 each way).

This example gives a basic demonstration of how capacitive reactance decreases with frequency. Taking these examples further, scaling to a 10kHz signal would give a reactance of 3.183Ω – effectively negligible in most circuits. Note in the examples above that reactance is measured in ohms (Ω), which is part of the AC definition of resistance – impedance. The next section will show how to combine capacitors in series and parallel, and also how to determine overall circuit impedance when they are combined with resistors. It is recommended that you reread this section again after completing this chapter, to ensure you are comfortable with the behaviour of capacitors – they are arguably the most versatile (and commonly used) component in electronics, and for AC signals (like audio) they are essential.

A note on inductors

This section has looked at capacitors as electronics components that vary their behaviour over time - they react based on the frequency of the voltage signal across them. As this is an introductory text on audio electronics, inductors have been omitted to focus your learning around the most common components used in amplification and filtering. Having said this, induction is core to transducers like loudspeakers and fundamental to AC circuits in general (particularly in communications), so this omission is not based on their lack of importance within electronics - rather on the practicality of focussing on introductory audio circuits. Capacitors can quickly be combined with resistors to form first-order audio filters, and are also used with amplifiers for grounding, coupling and blocking (as shown in the next chapter) - they are much easier to adapt and apply to our practical projects.

By comparison, inductors are now less commonly used in audio electronics, though examples include the famous Crybaby Wah-Wah pedal, which arguably owes its unique sound to the inclusion of an inductor within the filter circuit. In most electronics courses, the combination of a resistor, inductor and capacitor is a core focus of study - inductors are usually given the symbol L, so this type of circuit is called an RLC circuit. This chapter is not intended to be used as a reference text - a more thorough study of topics like AC circuits is recommended, but is largely beyond the scope of this book. The focus on resistors and capacitors is based on getting functional audio circuits working within a practical context.

6.4 Impedance – combining AC components

The previous section reintroduced the chapter 3 concept of resistance to the flow of current as being proportional to voltage over current ($R = \frac{V}{I}$). This relationship was then used to define the reactance of a capacitor, also in ohms (Ω), which varies with frequency. Resistance is a time-invariant quantity, in that a resistor will not change with the frequency of the input signal. When working with components like capacitors that do vary their response based on signal frequency, a term is needed that will encompass both resistive and reactive elements. This term is called **impedance**, which reflects the **time-varying nature of the signals** it is used with.

Capacitors can be combined in both series and parallel – they effectively behave in the **opposite manner to resistors** in DC circuits (Figure 6.25).

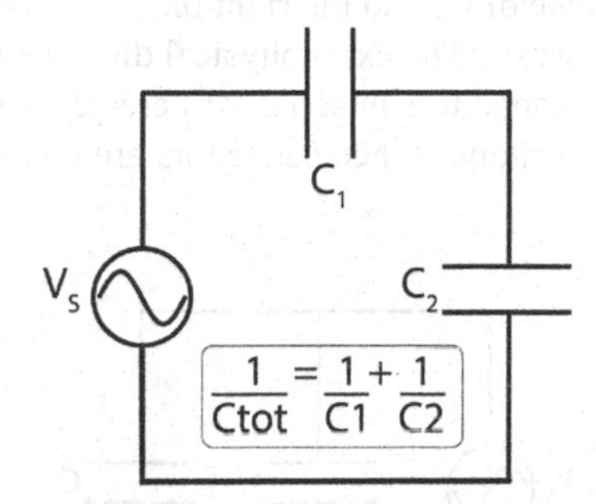

Figure 6.25 Series capacitor circuit. ***The diagram shows two capacitors connected in series, where the sum of the voltages across each capacitor (V_{C1} and V_{C2}) is equivalent to the total voltage (V_{tot}) for the circuit. An AC supply is shown, but time is not included in magnitude calculations.***

In the appendix, the equation for total series circuit capacitance is derived:

$$\textit{Total Series Capacitance}, \frac{1}{C_{tot}} = \frac{1}{C_1} + \frac{1}{C_2} + \ldots \frac{1}{C_n} \quad (6.11)$$

C_{tot} is the total series capacitance in ohms (Ω)

n is the number of capacitors in series

This equation may initially seem slightly counterintuitive – chapter 3 showed that series resistors combine by addition, so why would capacitors behave differently? Although the mathematical proof is reasonably straightforward, it can help to remember that **capacitance is dictated by the distance between the plates** that hold the charge – the greater the distance, the lower the charge that can be held. Thus, if capacitors are added in series **the physical distance** between the total charge from the circuit supply becomes greater than the sum of each individual capacitor distance (Figure 6.26).

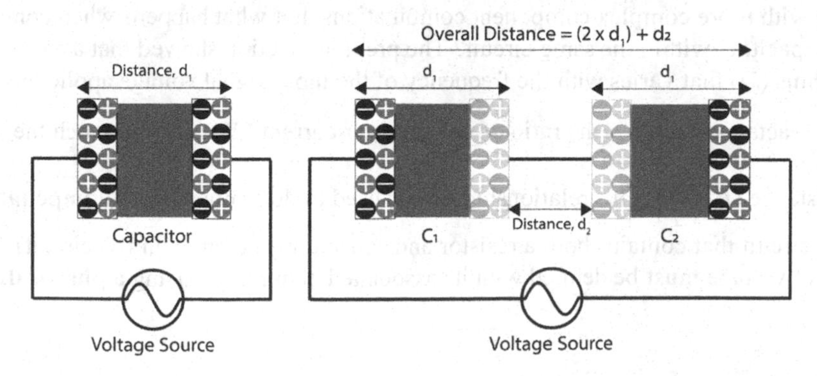

Figure 6.26 Capacitor plate distance in series circuits. ***In the left-hand diagram, a capacitor has a specific plate distance (d_1) that separates the overall charge provided by the supply V_S. When two capacitors are combined in series in the right-hand diagram, the gap between the overall supply charge now includes an additional distance (d_2) due to the physical connection of each capacitor – the total distance of $(2 \times d_1) + d_2$ is greater than $d_1 + d_1$ and so the overall capacitance is reduced.***

The diagram shows how combining two series capacitors includes an additional distance d_2, which reduces the overall capacitance as a result. The charge on each capacitor is less important than the total charge held between the left plate of C_1 and the right plate of C_2 – this is the total capacitance when components are combined in series. The extra physical distance d_2 reduces this total capacitance, which explains why combining two capacitors in series will create a total capacitance that is less than their sum. The equation for total capacitance when capacitors are connected together in parallel can also be stated (Figure 6.27):

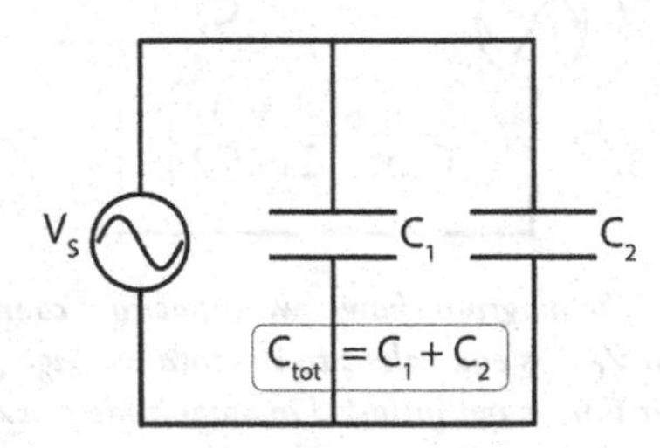

Figure 6.27 Parallel capacitor circuit. ***The diagram shows two capacitors connected in parallel, where the total charge (Q_{tot}) is the sum of all branch charges (Q_1 and Q_2) in the circuit. Again, an AC supply is shown but time is not included in the magnitude calculations.***

$$\textit{Total Parallel Capacitance, } C_{tot} = C_1 + C_2 + \ldots C_n \qquad (6.12)$$

C_{tot} *is the total parallel capacitance in ohms (Ω)*

n *is the number of capacitors in parallel*

In this case, the voltage across each branch in the circuit is equal, and so an equal amount of charge can develop across each capacitor (see the appendix for a full derivation). This means that the total capacitance in a parallel circuit is equal to the sum of all capacitances within it. These two equations show how capacitors can be combined to define an overall capacitance value in the opposite manner to that used for resistors (where the reciprocal value is used for parallel combinations). This is useful when working with more complex component combinations, but what happens when combining resistors and capacitors within the same circuit? The previous section showed that a capacitor has a reactance in ohms (Ω) that varies with the frequency of the input signal voltage applied to it. Ignoring frequency, the reactance becomes the ratio of voltage over current ($X_C = \frac{V}{I}$), in much the same way as it is for resistance ($R = \frac{V}{I}$). This relationship can be used to determine the total **impedance** (symbol **Z**) for a series circuit that contains both a resistor and capacitor (known as an RC circuit). To do this, each component voltage must be defined with its associated phase angle using a **phasor diagram** (Figure 6.28).

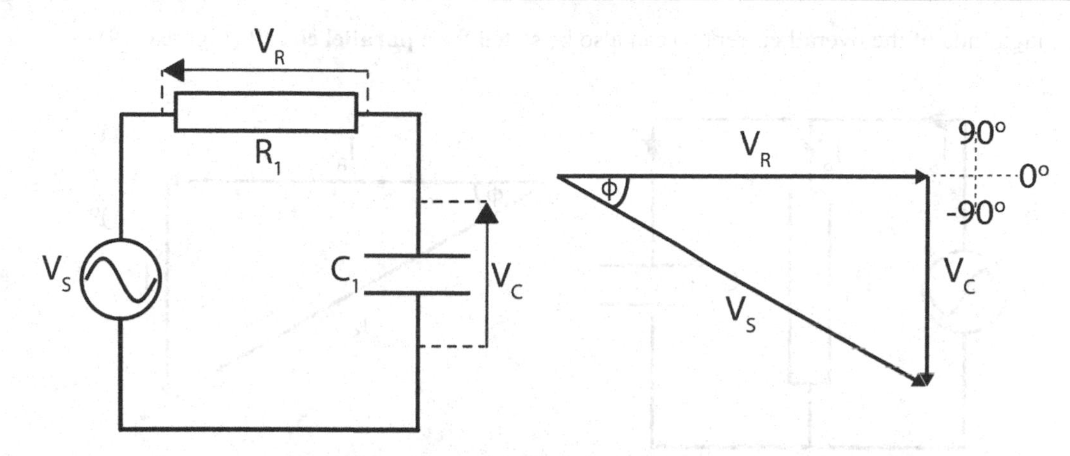

Figure 6.28 RC series circuit phasor diagram. ***The left-hand diagram shows a resistor and capacitor in series, connected to an AC voltage supply (VS). The right-hand diagram shows the resulting phasor diagram for this circuit, where the voltages in the circuit are combined to include the phase angle of each component. As it does not vary with frequency the resistor has phase angle 0°, whilst the capacitor has a phase angle of −90°. The overall phase angle can be found using the arctangent of the reactance divided by resistance:*** $\phi = \tan^{-1}\left(\frac{X_C}{R}\right)$.

The phasor diagram is a common way of analysing an AC circuit by combining component voltages (or currents) with their associated AC phase angles. From trigonometry, $\phi = \tan^{-1}\frac{X_C}{R}$ can be used to calculate the **overall phase** of an RC circuit. Although this book does not go into detail on phase angles, a phasor diagram can be used to show how to derive the **overall impedance (Z)** of the circuit. The resistor voltage (V_R) is shown with a phase angle of 0°, as resistive components do not vary with frequency. The previous section derived the voltage across a capacitor as $V_C = -\frac{I_p}{\omega C}\cos\omega t$ and the negative cosine component of this equation explains why the capacitor voltage in the diagram has a phase angle of −90°. The overall voltage (V_S) of the circuit forms the hypotenuse of this triangle, and trigonometry can be used to define the **magnitude of this voltage**. A similar equation can also be defined for the total circuit impedance (Z) in an RC series circuit:

$$\textit{Voltage Magnitude},\ V_S = \sqrt{V_R^{\,2} + V_C^{\,2}}$$
$$\textit{Series Impedance},\ Z = \sqrt{R^2 + X_C^{\,2}} \tag{6.13}$$

V_S is the supply voltage in volts (V)

V_R is the series resistor voltage in volts (V)

V_C is the series capacitor voltage in volts (V)

***Z** is the total series impedance in ohms (Ω)*

***R** is the series resistance in ohms (Ω)*

X_C is the reactance of the series capacitor in ohms (Ω)

The magnitude of the overall current (I) can also be stated for a **parallel circuit** (Figure 6.29).

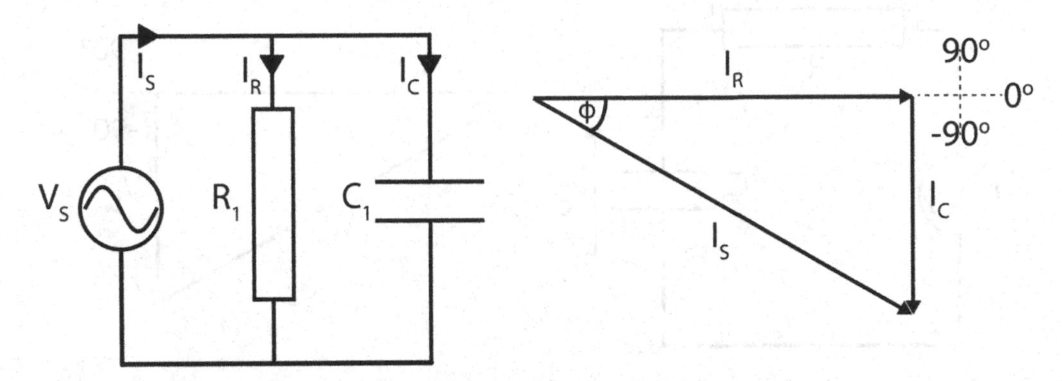

Figure 6.29 RC parallel circuit phasor diagram. ***The left-hand diagram shows a resistor and capacitor in parallel, connected to an AC voltage supply (VS). The right-hand diagram shows the resulting phasor diagram for this circuit, where the currents in the circuit are combined to include the phase angle of each component. As it does not vary with frequency the resistor has phase angle 0°, whilst the capacitor has a phase angle of −90°.***

As the voltage (V_S) in a parallel circuit will be constant across branches, this also allows the total parallel impedance (Z) to be calculated using Ohm's Law:

$$\textit{Current Magnitude}, I_S = \sqrt{I_R^2 + I_C^2}$$
$$\textit{Parallel Impedance}, Z = \frac{V_S}{I_S} \qquad (6.14)$$

I_S is the total parallel current in amps (A)

I_R is the parallel resistor current in amps (A)

I_C is the parallel capacitor current in amps (V)

V_S is the supply voltage in volts (V)

Z is the total parallel impedance in ohms (Ω)

These equations allow a series AC circuit to be analysed in much the same way as DC series circuits in chapter 3. Resistive and capacitive components can be combined by determining their overall impedance as the root sum of squares of the resistance and reactance terms. Thus, to determine the overall current in an AC circuit the total impedance (Z) can be calculated and then Ohm's Law used to solve for a known supply voltage. Extending this thinking, the reactance equations from 6.12 and 6.13 can be combined with their total resistance equivalents (chapter 3, equations 3.3 and 3.5) to analyse more complex component combinations in terms of their total resistance and reactance – as the following examples show.

6.4.1 Worked examples – analysing combined resistive and reactive circuits

These examples show how to analyse combinations of resistors and capacitors within an AC circuit. In the second example, equivalent capacitors and resistors are used to represent the total value for a parallel branch within the circuit – this is a common technique that is worth practising. Phase is also ignored in these calculations for brevity and to reduce complexity, but note that it is crucial to more advanced circuit analysis.

Q5: *The schematic below shows an AC circuit combining a 10kΩ resistor in series with a 5μF capacitor. The circuit has an input voltage of 5V at a frequency of 1kHz:*

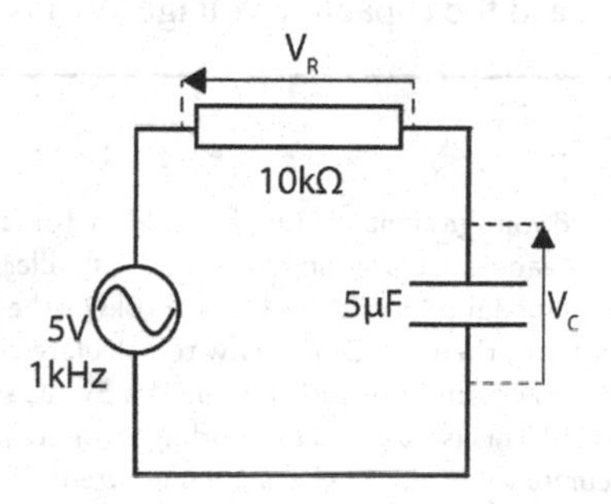

(a) *What is the reactance of the capacitor?*
(b) *What is the magnitude of the total impedance of the circuit?*
(c) *What is the current flowing through the circuit?*
(d) *What are the voltage drops across the resistor (V_R) and the capacitor (V_C)?*

part (a) was already answered in Q4:

For an input frequency of **1kHz**:

$Capacitance, C = 5\mu F, \quad Voltage, V = 5V, \quad Frequency, f = 1kHz$

$\omega = 2\pi f = 6.283 \times 1000 = 6283\, rads$

$$\textbf{Capacitive Reactance}, X_C = \frac{1}{\omega C} = \frac{1}{6283 \times 0.000005} = \frac{1}{0.031415} = \mathbf{31.83\Omega}$$

Knowing the reactance, calculate the magnitude of the total impedance:

$Resistance, R = 10k\Omega, \qquad Reactance, X_C = 31.83\Omega$

$$Impedance, Z = \sqrt{R^2 + X_C^{\,2}} = \sqrt{10{,}000^2 + 31.83^2} = \sqrt{100{,}000{,}000 + 1013.15}$$

$$Z = \sqrt{100001013.15} = 10000.05 \approx 10k\Omega$$

Total impedance (10kΩ) and input voltage (5V), so use Ohm's Law to calculate current (I):

$$\textbf{Current}, I = \frac{V}{R} = \frac{5}{10{,}000} = 0.0005 = \mathbf{0.5mA}$$

Use the current to calculate the voltage drops across the resistor (V_R) and capacitor (V_C):

Resistor Voltage, $V_R = IR = 0.0005 \times 10{,}000 = \mathbf{5V}$

Capacitor Voltage, $V_C = IX_C = 0.0005 \times 31.83 = 0.016 = \mathbf{16mV}$

Answer: (a) For a 5V 1kHz AC signal input, the capacitor has a reactance of 31.83Ω.

(b) The total impedance of the circuit is **~10kΩ.**
(c) The current flowing through the circuit is **0.5mA.**
(d) The resistor voltage (V_R) is **5V**, and the capacitor voltage (V_C) is **16mV.**

Notes:

This example uses the same capacitor and voltage signal as Q4, but adds a 10kΩ series resistor. The first thing to notice is that the capacitive reactance (already calculated in Q4) is much smaller than this resistor, at only 31.83Ω. Thus, the magnitude of the total circuit impedance is effectively still ~10kΩ - the **capacitor has virtually no effect at 1kHz**. With the impedance value determined, then use Ohm's Law to calculate circuit current (0.5mA) and also the voltage drops across each component. This creates a problem: there is a 5V AC supply, but the total of the voltage drops exceeds this - (5 + 0.016) = **5.016V**! This is due to the rounding error created by approximating the total impedance to 10kΩ, where a more accurate value would give a circuit current of ~0.499mA - this would reduce the resistor voltage by around 0.01V, which is effectively the supply. It is interesting to note that the capacitor has no effect, and if the phase angle of the circuit was calculated it would be around 0.18° - effectively a purely resistive circuit.

Q6: *This circuit combines two series 10kΩ resistors in parallel with a 5μF capacitor and another parallel branch containing two series 5μF capacitors. The circuit can be broken into stages to analyse each parallel branch, to then represent the total resistance or reactance as a* ***single equivalent value****. The circuit has an input voltage of 5V at a frequency of 1kHz:*

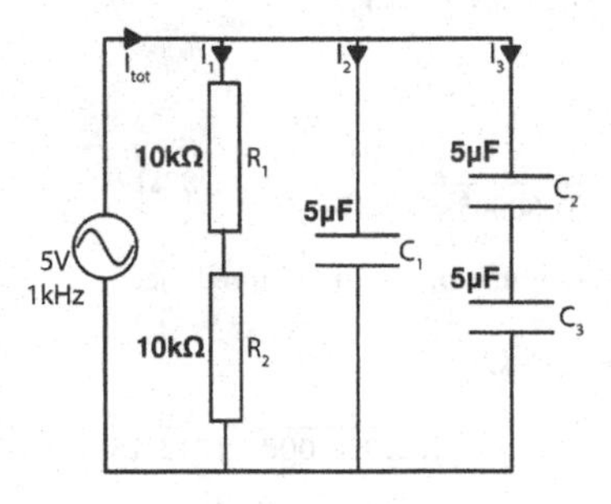

(a) *What is the total reactance of the capacitors?*
(b) *What are the individual branch currents in the circuit?*
(c) *What is the magnitude of the total impedance of the circuit?*
(d) *What is the magnitude of the total current flowing through the circuit?*

The key to answering this question is to break the circuit into stages:

1. Branch 3 (get series capacitance, get branch reactance, get branch current) – replace with equivalent capacitor.
2. Branch 2 (get branch reactance, get branch current).
3. Combine branches 2 and 3 to get total capacitance, and thus total reactance – replace with equivalent capacitor.
4. Branch 1 (get total series resistance, get branch current) – replace with equivalent resistor.
5. Combine both terms to get magnitude of total impedance (square root of sum of squares).
6. Get total current using magnitude of resistive and reactive currents (combine reactive currents).

First, find the total series capacitance of branch 3:

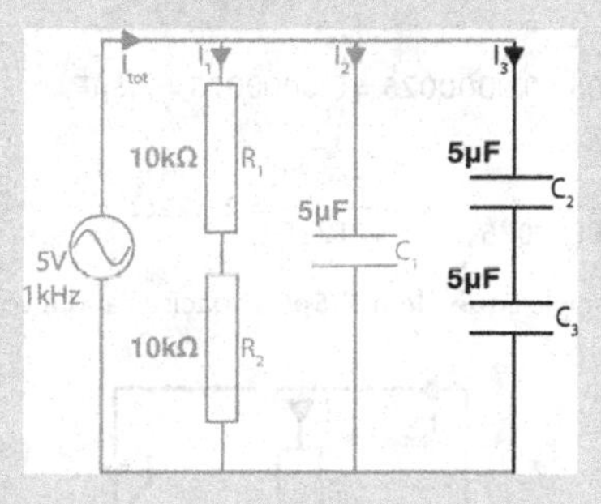

$Capacitance, C_2 = 5\mu F, Capacitance, C_3 = 5\mu F \quad Voltage, V = 5V, Frequency, f = 1kHz$

$$Branch3\,Capacitance, \frac{1}{C_{br3}} = \frac{1}{C_2} + \frac{1}{C_3} = \frac{1}{0.000005} + \frac{1}{0.000005} = 200000 + 200000 = 400000$$

$$\therefore C_{br3} = \frac{1}{400000} = 0.0000025F = 2.5\mu F$$

Now get branch series reactance for an input frequency of 1kHz:

$$\omega = 2\pi f = 6.283 \times 1000 = 6283\,rads$$

$$Branch3\,Reactance, X_3 = \frac{1}{\omega C} = \frac{1}{6283 \times 0.0000025} = \frac{1}{0.0157075} = 63.66\Omega$$

Use Ohm's Law to find **branch current**:

$$\boldsymbol{Branch\ 3\,Current, I_3} = \frac{V}{X_3} = \frac{5}{63.66} = 0.0785 = \boldsymbol{78.5mA}$$

Now replace branch 3 with a single **equivalent 2.5μF** capacitor:

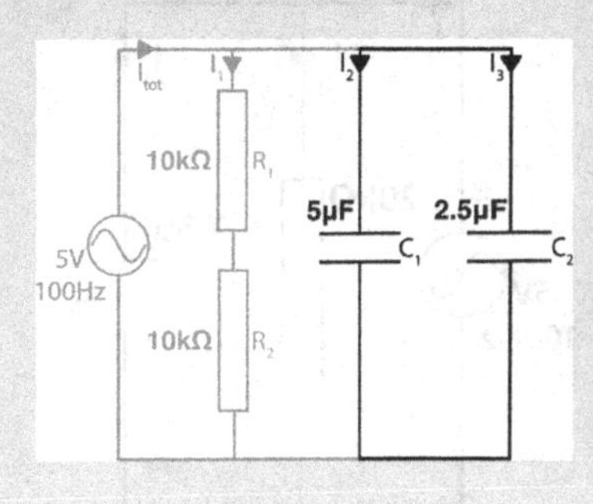

$Capacitance, C_1 = 5\mu F, Capacitance, C_2 = 2.5\mu F \quad Voltage, V = 5V, Frequency, f = 1kHz$

For branch 2, get reactance (see Q4 and Q5) and **current**, then get total parallel capacitance and **total reactance**:

$$Branch\,2\,Reactance, X_2 = \frac{1}{\omega C} = \frac{1}{6283 \times 0.000005} = \frac{1}{0.031415} = \mathbf{31.83\Omega}$$

$$\boldsymbol{Branch\,2\,Current, I_2} = \frac{V}{X_2} = \frac{5}{31.83} = 0.157 = \mathbf{157}\boldsymbol{mA}$$

$$Total\,Capacitance, C_{tot} = C_1 + C_2 = 0.000005 + 0.0000025 = 0.0000075 = 7.5\mu F$$

$$\boldsymbol{Total\,Reactance,\ X_{tot}} = \frac{1}{\omega C} = \frac{1}{6283 \times 0.0000075} = \frac{1}{0.0471225} = \mathbf{21.22\Omega}$$

Now replace branches 2 and 3 with a single **equivalent 7.5μF** capacitor and a total reactance current $\boldsymbol{I_X}$:

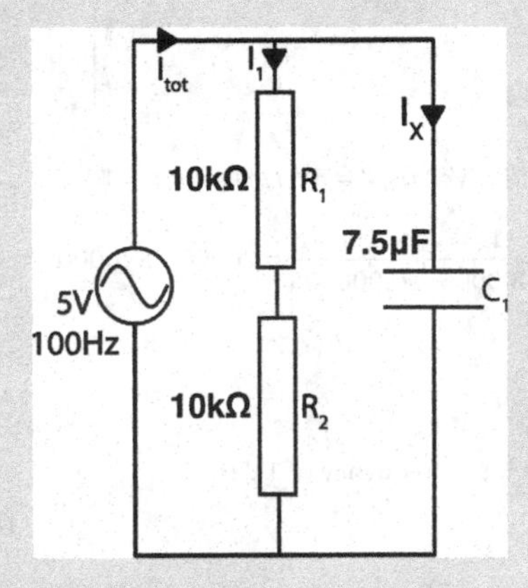

For branch 1, calculate the total series resistance and the branch current:

$$Total\,Resistance, R_{tot} = R_1 + R_2 = 10{,}000 + 10{,}000 = 20{,}000 = 20k\Omega$$

$$\boldsymbol{Branch\,1\,Current,\ I_1} = \frac{V}{R} = \frac{5}{20{,}000} = 0.00025 = \mathbf{0.25}\boldsymbol{mA}$$

Now replace the two series resistors with a **single equivalent resistor** and a total resistance current $\boldsymbol{I_R}$:

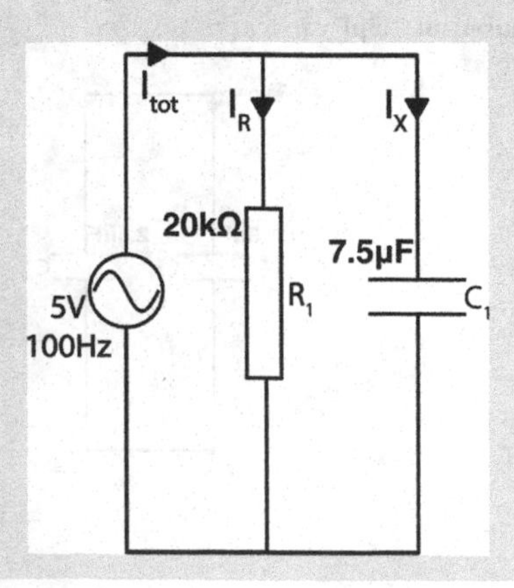

Use the total resistance and reactance to calculate the magnitude of the total impedance:

$$Total\,Resistance, R_{tot} = 20k\Omega, \qquad Total\,Reactance, X_{tot} = 21.22\Omega$$

$$Impedance, Z = \sqrt{R^2 + X_C^{\,2}} = \sqrt{20{,}000^2 + 21.22^2} = \sqrt{400{,}000{,}000 + 450.29}$$

$$\mathbf{Z} = \sqrt{400000450.29} = 20000.01 \approx \mathbf{20k\Omega}$$

All branch currents are known, so calculate I_R and I_X and use the magnitude equation for total current:

$$Reactance\,Current, I_X = I_2 + I_3 = 0.157 + 0.0785 = 235.5mA$$

$$Resistance\,Current, I_R = \frac{V}{R} = \frac{5}{20{,}000} = 0.00025 = 0.25mA$$

$$\mathbf{Total\,Current}, \mathbf{I_{tot}} = \sqrt{I_R^{\,2} + I_X^{\,2}} = \sqrt{0.00025^2 + 0.2355^2} = \sqrt{0.0000000625 + 0.05546025}$$

$$= \sqrt{0.0554603125} = \mathbf{235.5mA}$$

Answer: (a) For a 5V 1kHz AC signal input, the total reactance in the circuit is of 21.22Ω.

(b) The branch currents are $\mathbf{I_1 = 0.25mA}$, $\mathbf{I_2 = 157mA}$ and $\mathbf{I_3 = 78.5mA}$.
(c) The magnitude of the total impedance in the circuit is $\approx$ **20kΩ.**
(d) The magnitude of the total current in the circuit is **235mA.**

Notes:

This example uses the same input signal (5V, 1kHz) with the same resistance (10kΩ) and capacitance (5μF) values, but adds extra components in parallel branches to show how to analyse a more complex circuit. By working in reverse towards the voltage source, the circuit can be reduced to single equivalent resistance and reactance terms, allowing the total impedance and current to be calculated for the circuit. Note the scaling of values, and that the reactive components have little impact on the overall impedance of the circuit (≈ **20kΩ**). Conversely, the resistive branch draws very little current overall (**0.25mA**) and most of the charge in the circuit passes through the capacitive branches (**235.5mA**). This highlights one of the many uses of a capacitor, as a route to ground for excess charge in a circuit that does not significantly alter the resistance (and hence the voltage).

This second circuit example is significantly more complex, as it includes multiple components and uses various circuit analysis techniques to get the required results. It is also important to note that had a parallel branch containing both resistive and reactive elements been included then the calculations would have become significantly more involved, and this is where phasor diagrams become essential. The mathematical calculations required when working with phasors are beyond the scope of an introductory textbook in some ways, as this can often lead to learners losing interest in electronics at an early stage. These examples are included to aid in your understanding, and also demonstrate why simulation tools are so important to practical electronics work. This section has focussed on deriving and demonstrating the basic equations that govern AC circuits, to show why a simulation tool that could automate this process would be preferable. LTspice is capable of calculating these equations both at speed and to a high degree of accuracy, so the next section covers its installation and use.

6.5 Tutorial: installing LTspice

The previous examples help to illustrate the amount of time and effort needed to work through the calculations needed to analyse even simple resistor capacitor circuit combinations. In addition, it is very easy to make mistakes with scales and quantities, and so automated simulation tools become an essential part of designing and prototyping circuits. Tinkercad is a great simulation tool, but it does not carry out some of the more advanced analysis needed for audio signals and circuits. LTspice is a very powerful software tool for electronic circuit analysis and simulation. LTspice is free to use and runs on both the OSX and Windows platforms. This section provides a brief overview of how to use LTspice, prior to analysing two circuits in the example projects that follow. LTspice can be downloaded from the Analog Devices website (www.analog.com) at the following link: www.analog.com/en/design-center/design-tools-and-calculators/LTspice-simulator.html#.

From there, the installation is reasonably straightforward and provides you with a fully functional copy of LTspice – along with a large set of demo circuits. It is important to note that although in many ways a very powerful tool that significantly reduces the time needed to design electronic circuits, LTspice is **perhaps not the easiest user interface** to work with. It is recommended to spend some time learning this interface by practice, as some elements of LTspice can initially be frustrating until you become familiar with the thinking behind them. Screenshots for the OSX platform are provided for information only, so your own interface configuration may be slightly different. One of the first things that can help is to change the interface colour scheme from the control panel – accessed by the hammer icon (Figure 6.30).

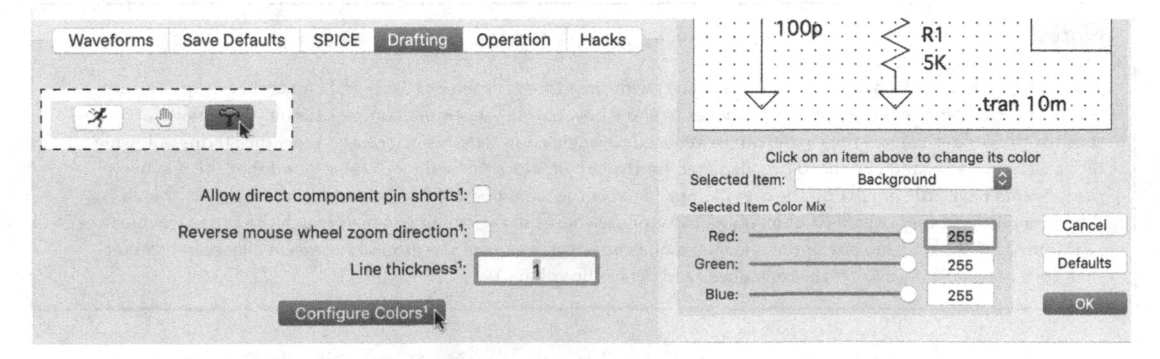

Figure 6.30 LTspice control panel. ***Clicking the hammer icon (dashed line inset) brings up the control panel, where the Configure Colors options can be accessed. From there, setting the background to white (255,255,255) is recommended, alongside setting wires and components to black (0,0,0).***

From this point onwards, the default colour scheme has been changed in the control panel to make it easier to see the schematic in black on a white background – this is also useful if you cannot see certain colours easily. With a new (empty) schematic created, a blank screen is shown; on OSX you can right-click to access most of the main editing commands – on Windows these are provided in a button bar at the top of the screen (Figure 6.31).

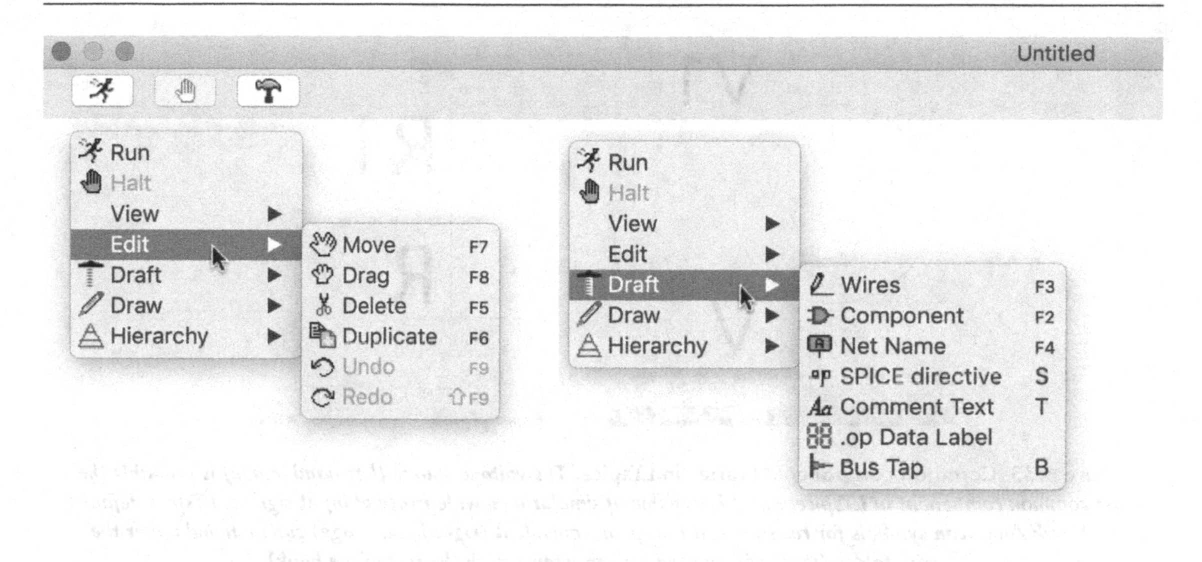

Figure 6.31 LTspice main command set. ***By right-clicking on a blank space within the authoring area (grey window) the main command sets are shown (both Edit> and Draft> are overlaid in this image). These commands cover most of the common tasks in LTspice, but working with them can take a little time to get used to.***

The diagram shows both the Edit and Draft command sets, which cover most common operations within LTspice. The Edit set provides tools for moving, deleting and duplicating components, which can initially confuse as the component will be moved without its connecting wires! Similarly, the delete tool will either delete what is between the scissors in the icon or allow you to drag a bounding box for group deletion (Figure 6.32).

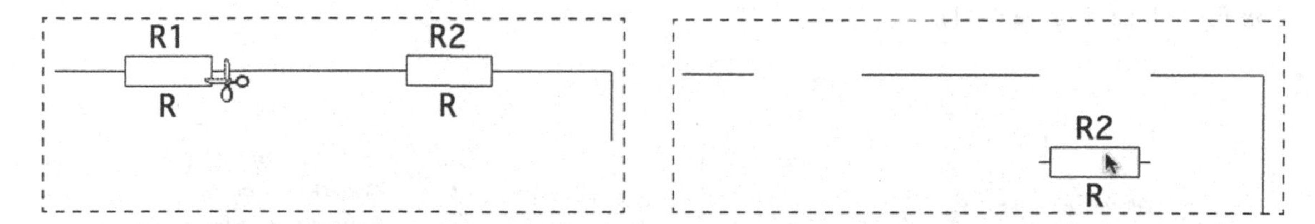

Figure 6.32 Moving and deleting components in LTspice. ***The left-hand image (dashed line box) shows how a single deletion is performed by moving the scissors icon over the component – group deletion (by bounding box) is not recommended until you are familiar with LTspice! The right-hand image shows how the second component is moved without retaining its connecting wires: these broken connections will now cause errors during analysis and simulation.***

Adding components is fairly straightforward (on OSX you can rotate a component using CTRL>R, but it is recommended to spend a little time thinking through the schematic layout (even sketching it quickly on a piece of paper for reference) to help set things out quickly – moving them afterwards can often take more time! When adding a component on OSX, a list of libraries is provided where different options can be chosen. We will be working with some of the most common components, which include voltage sources, resistors and capacitors (Figure 6.33).

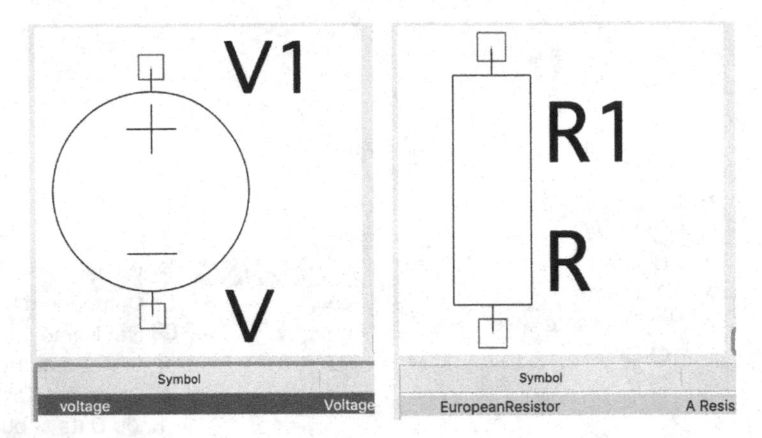

Figure 6.33 Common component libraries in LTspice. ***The voltage source (left-hand image) is probably the most common component in LTspice, and it is capable of simulating a wide range of input signals. LTspice defaults to North American symbols for resistors – a European equivalent (right-hand image) can be found under the misc folder (these will be used for consistency with the rest of the book).***

Voltage sources are flexible and powerful input components, and will be used for all simulations with LTspice in this book. Both DC and AC inputs are possible, alongside sine and pulse waveforms – even loading an audio wave file into LTspice is possible as a voltage source input for simulation! The voltage source is effectively the core of an LTspice circuit, as its configuration determines how the circuit can subsequently be analysed and simulated. The various configuration options will be covered in each example, to introduce them to you in a practical context. The right-hand image shows a European resistor symbol, which is stored in the [Misc]> folder – this symbol is used for consistency with the other schematics in this book, but the North American version (listed as *res* in the main folder) is exactly equivalent so either can be used in your own circuits. Once the components are added, we need to specify values for them. To do this, right-click on a component to bring up the dialog options box for values (Figure 6.34).

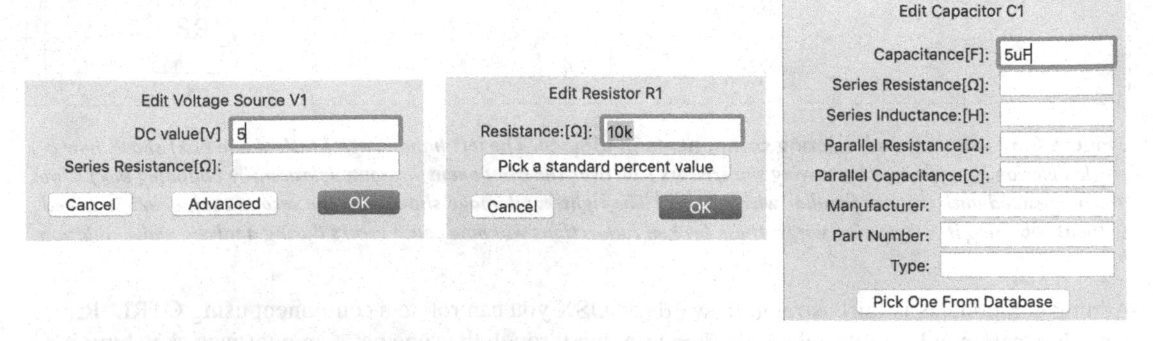

Figure 6.34 LTspice component dialogs. ***The diagram shows dialog boxes for voltage (left-hand image), resistor (middle image) and capacitor (right-hand image) component settings. Notice that capacitors require a capital F to be added after the value – otherwise LTspice will not see it as a capacitance!***

LTspice will accept voltage and resistor values with scaling factors only (i.e. no measurement units), but for capacitors **you must include the farad (F)** units alongside your scaling factor (μ = u, n = n, ρ = p) or LTspice will not factor it into the simulation as a valid capacitance value. There are **exceptions to this rule – 1 MEG for 1MΩ and 1 (no unit) for 1 farad** capacitance. With components added, the next thing to include is a ground (GND) node for the circuit – **no simulation can be performed without a valid GND node** (Figure 6.35).

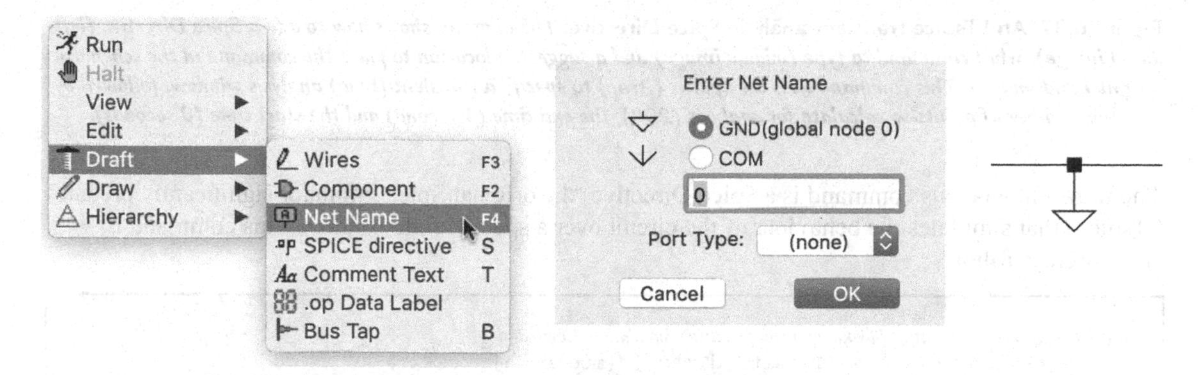

Figure 6.35 LTspice ground node. ***Adding a ground (GND) node is straightforward, where a new Net Name is selected, a GND (node 0) is specified and then the component is moved to a place where it can be connected to the rest of the circuit.***

With a voltage source, components and a ground node (GND) added, wires can now be drawn to connect them into a full circuit (Figure 6.36).

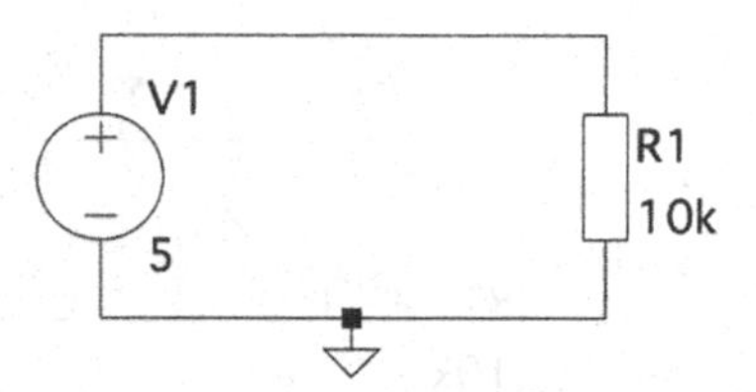

Figure 6.36 LTspice example circuit. ***The diagram shows the simplest possible LTspice circuit for simulation – a 5V DC voltage source connected in series with a 10kΩ resistor. The GND node at the bottom of the circuit is essential, as is a valid Spice Directive to tell the application how to begin analysing the circuit.***

This circuit uses a 5V DC voltage source in series with a 10kΩ resistor, where the DC voltage is the simplest form of input signal that can be provided to the circuit. The other essential component to be added is a valid Spice Directive, which tells LTspice how to analyse the circuit when running a simulation. The circuit in Figure 6.36 shows DC 5V, but the example projects will introduce some of the more useful voltage source functions when the input source is changed from a pulse signal to an AC sine wave. For now, the simplest form of Spice Directive is a transient analysis command (Figure 6.37).

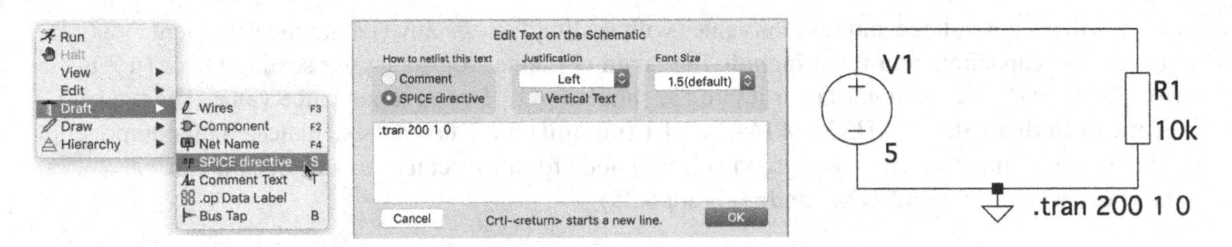

Figure 6.37 An LTspice transient analysis Spice Directive. ***The diagram shows how to add a Spice Directive (left-hand image), what command to type (middle image) and a suggested location to place the command in the schematic (right-hand image). This command uses the syntax (.tran) to specify a transient (time) analysis window, followed by the number of points to calculate for analysis (200), the end time (1 second) and the start time (0 seconds).***

The transient analysis command is a Spice Directive (the original Spice simulator significantly predates LTspice) that simulates the behaviour of the circuit over a specific time window. This command is structured as follows:

From the LTspice wiki: http://ltwiki.org/index.php?title=Simulation_Command

```
.tran <Tstep> <Tstop> [Tstart [dTmax]] [modifiers]
.tran       200      1        0
```

The example above shows a transient command that will start at time 0 and calculate 200 data points before stopping at time 1 second (the command is spaced to align with the structure). LTspice automatically determines the time difference between data points, and if the syntax is correct, a simulation can now be performed by clicking the run icon (Figure 6.38).

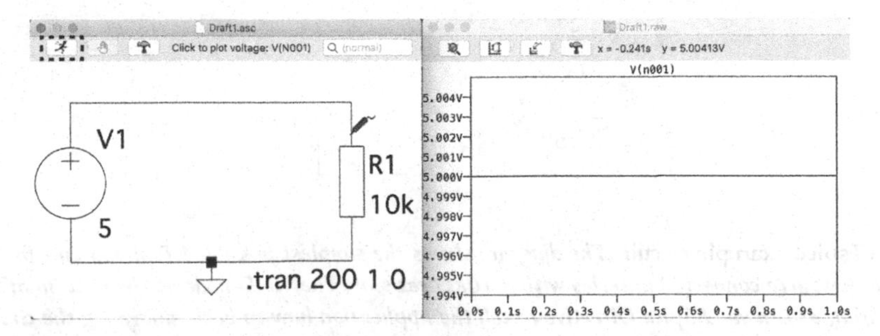

Figure 6.38 LTspice circuit simulation window. ***In the diagram, when the run icon (dashed box, top left) is clicked the simulation begins. The left-hand image shows the full circuit (with transient directive) and the right-hand image shows the resulting simulation window displaying a 5V DC signal trace over 1 second. Note the probe icon above the resistor – this is how simulation values are measured in the circuit.***

LTspice will be used to simulate circuits with a variety of signal inputs (e.g. sine, pulse) and chapter 8 (worked example 8.4.3) will demonstrate how audio wave data can also be loaded for simulation. For now, the tutorial will work through all of the steps discussed above to perform a transient analysis of a 5V DC source connected to a 10kΩ series resistor.

Tutorial steps

Create a new circuit in LTspice and name it chp6_tutorial.	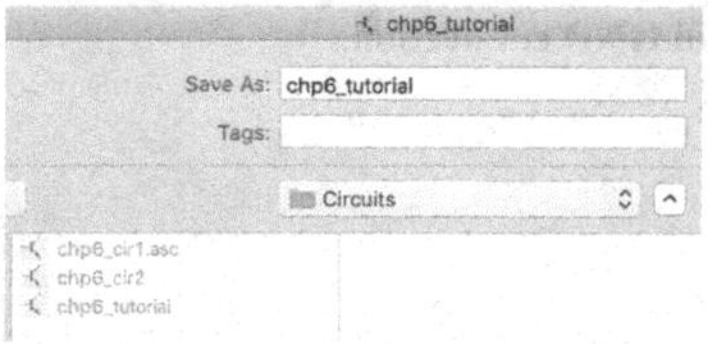
Add a voltage source component (on OSX, right-click and select Add Component, scroll down to <V> in the list).	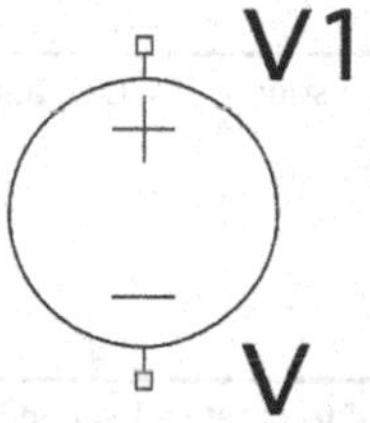
Add a resistor in parallel to the voltage source – a European resistor is shown for consistency, but North American symbols are functionally equivalent.	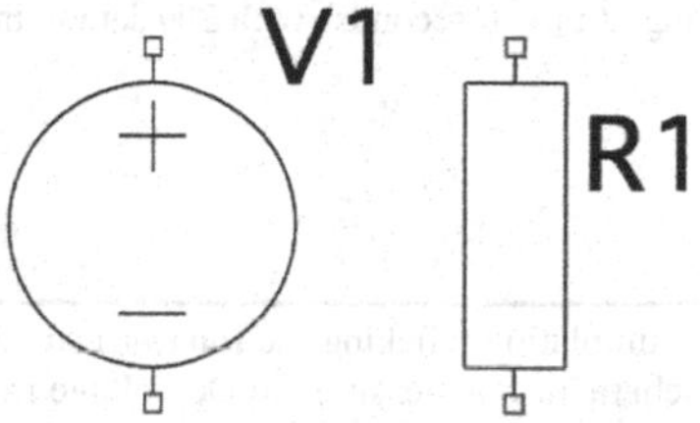
Add connecting wires between the two components, so they are connected in series. LTspice provides a set of coordinate crosshairs when drawing wires to allow accurate measurement of right angles in the circuit.	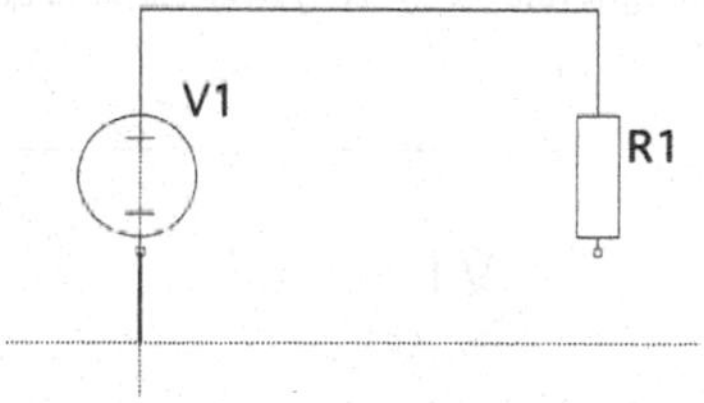
Add a GND connection for the circuit by selecting Draft>Net Name.	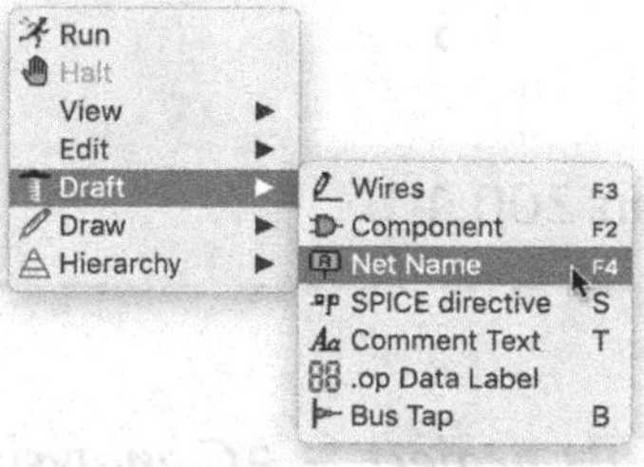
In the Net Name dialog, click GND and it will automatically assign it a Node number.	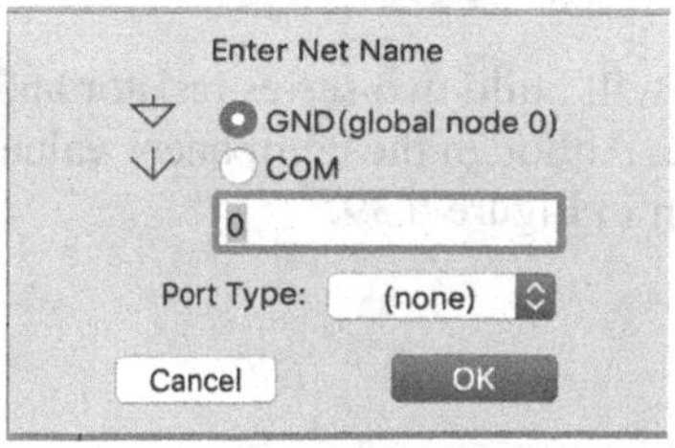

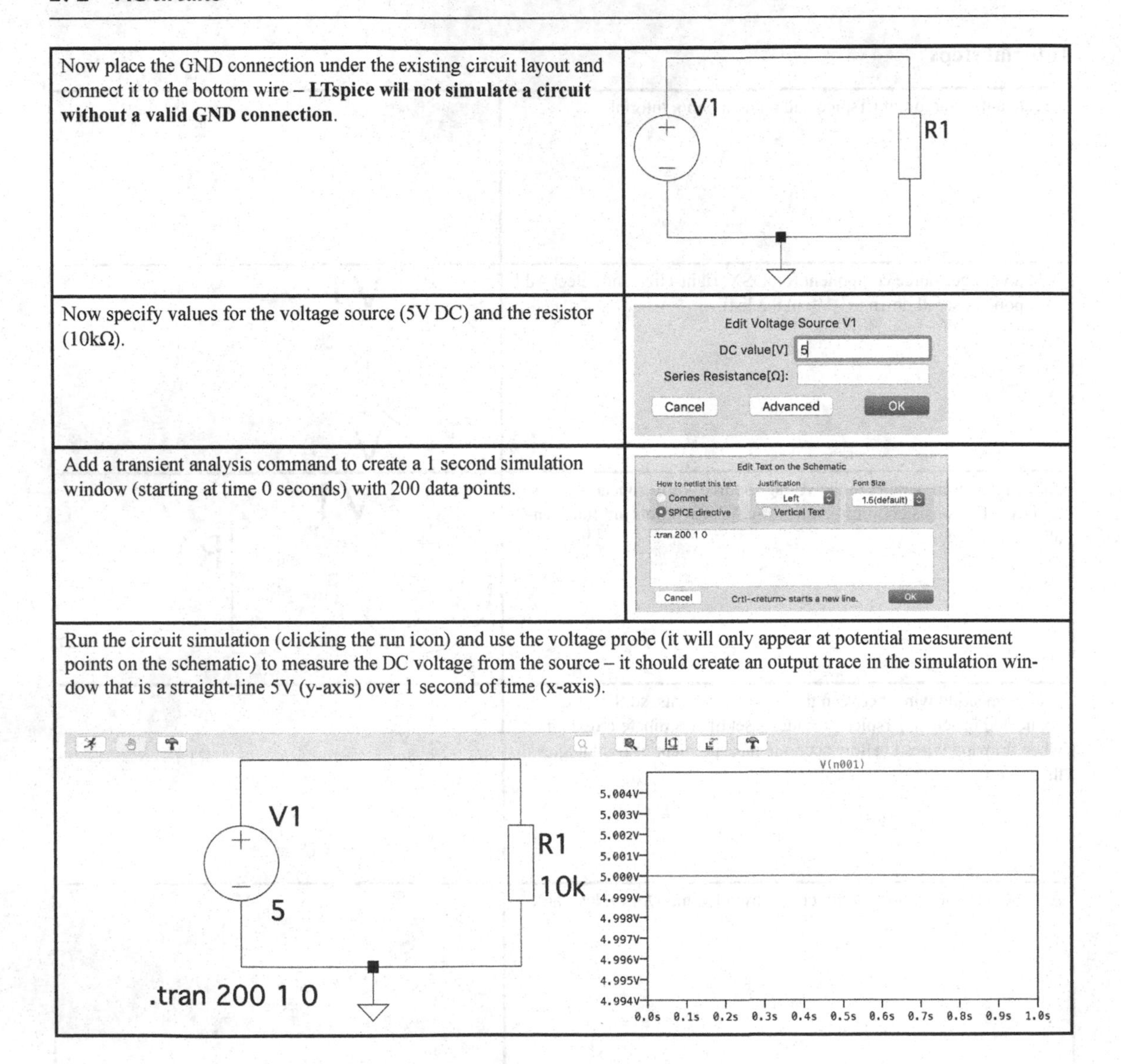

Now place the GND connection under the existing circuit layout and connect it to the bottom wire – **LTspice will not simulate a circuit without a valid GND connection.**	
Now specify values for the voltage source (5V DC) and the resistor (10kΩ).	
Add a transient analysis command to create a 1 second simulation window (starting at time 0 seconds) with 200 data points.	
Run the circuit simulation (clicking the run icon) and use the voltage probe (it will only appear at potential measurement points on the schematic) to measure the DC voltage from the source – it should create an output trace in the simulation window that is a straight-line 5V (y-axis) over 1 second of time (x-axis).	

6.6 Example project – AC analysis with LTspice

This project will build two series resistor and capacitor (RC) circuits to practise simulating circuits with LTspice. Although the component values will differ, both circuits will use the series component layout shown in Figure 6.39.

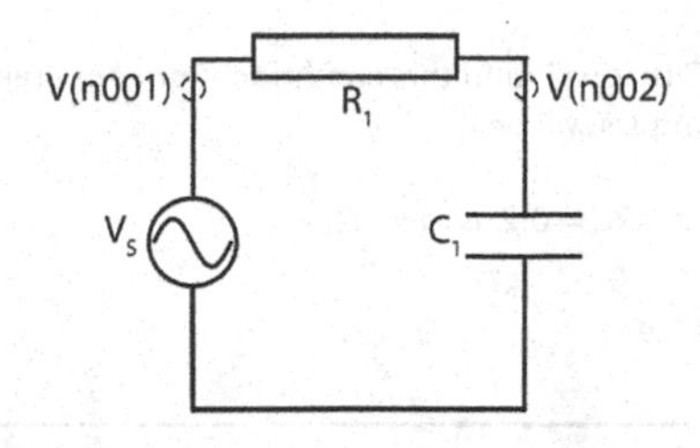

Figure 6.39 Example project schematic. ***The diagram shows an RC circuit, where a resistor and capacitor are connected in series. In each example project, the values of R and C will change (to define a time constant, and to show phase shift at output) but the layout will remain the same. The measurement nodes for the circuit (V(n001) and V(n002)) are also marked – this is where the analysis probe will be placed during simulation (Spice numbers circuit nodes sequentially by measurement).***

6.6.1 Example project – circuit 1

The first example will analyse an RC circuit to show the charge/discharge curve of a capacitor responding to an input pulse waveform (Figure 6.40).

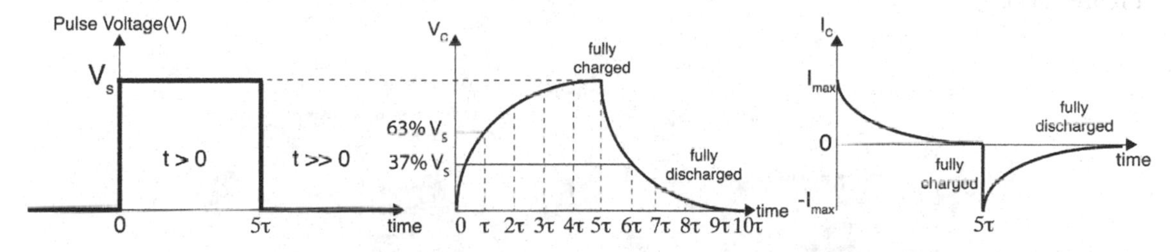

Figure 6.40 Example circuit 1 input waveform, adapted from Figure 6.21. ***In the diagram, a pulse input waveform causes the capacitor to charge and discharge over time based on the time constant (τ) of the circuit. The pulse must remain HIGH (5V) until the capacitor has fully charged, which will take 5 time constants (5τ). Thus, R and C values must be calculated that will align with the duration of the input pulse.***

To set up the LTspice analysis window properly, an input pulse must be defined and then component values calculated for the resistor and capacitor that will produce a manageable time constant (τ) to allow this pulse to be analysed. Equation 6.8 states that the ***Time Constant***, $\tau = RC$ and knowing that the pulse must remain HIGH (5V) for 5 time constants (5τ) to allow the capacitor time to fully charge up, a reasonable length for the input pulse (say 1 second) can be chosen to work backwards to find values for R and C that will produce the time constant (τ) required:

By defining a **pulse length of 1 second** in the LTspice analysis window, this equates to the time needed for the capacitor to fully charge (5τ):

$5\tau = 1\,second, \quad \therefore \textbf{\textit{Time Constant}}, \tau = \frac{1}{5} = \mathbf{0.2}\ \textbf{\textit{seconds}}$

There are numerous values for R and C that will fulfil this condition. For consistency, the capacitance value of **C = 5µF** from the previous worked examples will be used:

$$Capacitance, C = 5\mu F, \qquad Time\,Constant, \tau = RC = 0.2\,seconds$$

$$\therefore R = \frac{\tau}{C} = \frac{0.2}{0.000005} = 40{,}000 = \mathbf{40k\Omega}$$

Now this RC circuit can be built in LTspice, using component values of R = 40kΩ and C = 5µF. Create a new circuit in LTspice and name it chp6_proj1. The process for creating the schematic is as follows:

1. Lay out the components: voltage source, resistor, capacitor and the **GND node**.
2. Connect the wires to form a series circuit between Vs, R, C and GND.
3. Specify resistor and capacitor values of R = 40kΩ and C = 5µF.
4. Configure the voltage source to be a pulse input waveform of 1 second duration (choose Vs = 5V for consistency with previous circuits) – *setting up this command will be discussed below.*
5. Add a Spice Directive to perform a transient analysis of 2 seconds (10τ) to analyse the full charge/discharge curve of the capacitor.

Setting up the voltage source as a pulse input requires the advanced option in the editor dialog (Figure 6.41).

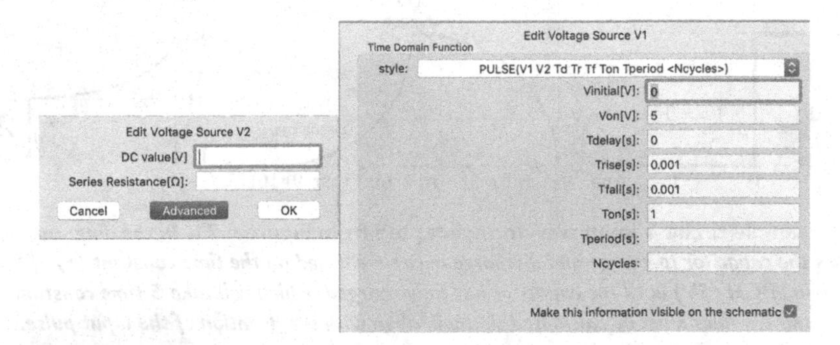

Figure 6.41 Creating a pulse input source in LTspice. ***In the diagram, on OSX clicking on the advanced option in the editor dialog (left-hand image) provides access to a range of settings for various waveforms. Selecting a pulse function from the dropdown options (right-hand image) allows the input to be configured for 0V (LOW) and 5V (HIGH) starting at time, t = 0 (Tdelay = 0) for a duration of 1 second (Ton = 1) with a very short rise and fall time (Trise = Tfall = 0.001 seconds).***

In the configuration window in Figure 6.41, a very short rise and fall time (Trise = Tfall = 0.001 seconds) are used to get the input pulse as close to a square wave as possible – if you continue your studies in digital electronics you will learn more about these kinds of practicality when working with binary signals. For now, a Spice Directive for transient analysis can be used to set up the full circuit for analysis (Figure 6.42).

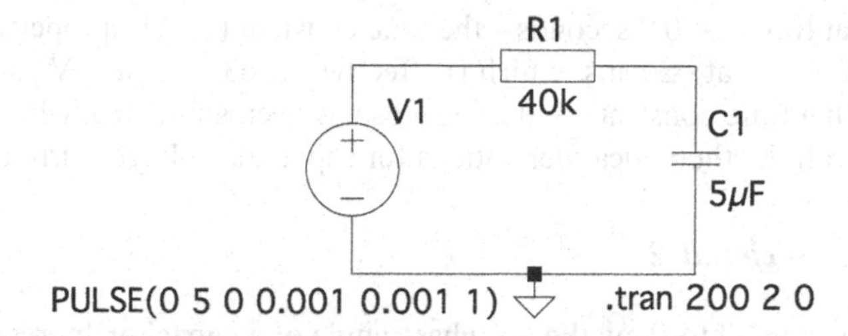

Figure 6.42 Example circuit 1 final LTspice circuit. ***The diagram shows a voltage source (Vs) configured for pulse input, connected in series with a 40kΩ resistor and a 5μF capacitor. A Spice Directive defines a transient analysis command, where the circuit will be analysed for 2 seconds, which should allow the full capacitor charge/discharge cycle (10τ = 2 seconds) to complete. The current probe is also shown near the capacitor C1, which can be combined with the voltage probe to measure all data for the circuit.***

With the simulation running, the voltage and current probes can be used to analyse the V(n001), V(n002) and I(C1) points (see Figure 6.39 schematic) to compare the input pulse with the capacitor charge/discharge response curve (Figure 6.43).

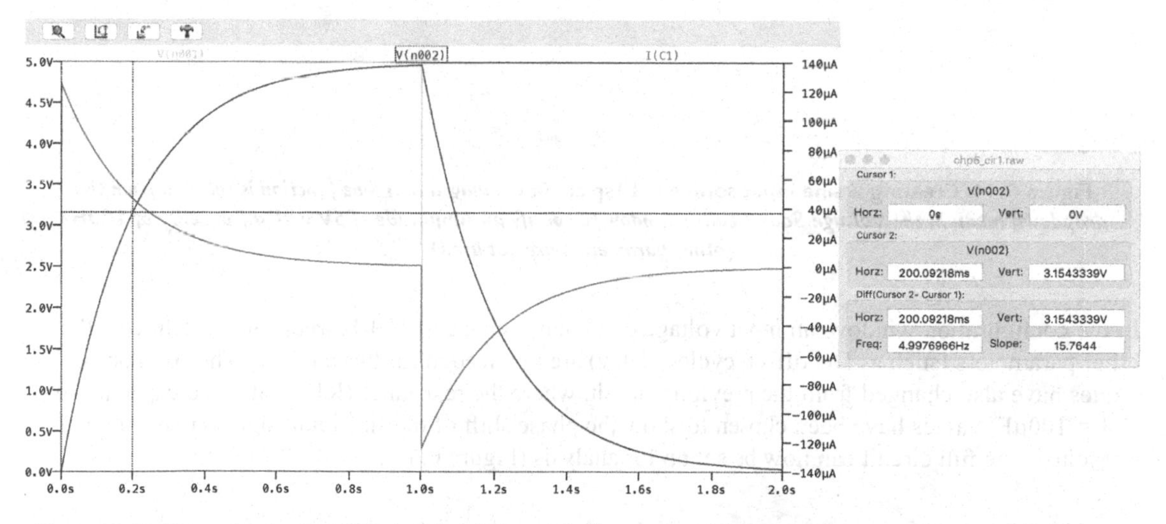

Figure 6.43 Example circuit 1 LTspice simulation. ***The image shows the V(n001) input pulse signal (light grey) that rises to 5V (HIGH) at ˜time = 0 seconds (rise time 0.001 seconds) and then falls to 0V (LOW) at time = 1 second. The other traces show the capacitor charge/discharge curves for both voltage V(n002) and current I(C1) (range provided by second y-axis on the right of the graph), where at t = 1 second (5τ) the capacitor has fully charged (so Vc = Vs = 5V and Ic = 0A). At t = 2 seconds (10τ) the capacitor has fully discharged again. The dotted line shows the added cursor for V(n002), which gives a value of 3.15V (˜63% of 5V) at time t = 0.2 seconds (τ).***

In this simulation window, the traces show capacitor voltage V(n002) and current I(C1) for an input pulse wave of 5V (HIGH) that lasts for 1 second (5τ). The transient analysis window runs for a further 1 second to capture the discharge curve (10τ), where the reversal of current polarity (−Imax = −120μA) and the reduction in capacitor voltage Vc can be seen. Clicking on the V(n002) trace label (at the top of the simulation window) will add a cursor, which can be dragged (using the hand icon) to get close to a value for V(n002) at time, t = 0.2 seconds – the time constant (τ). The properties window for the cursor shows a value of 3.15V at ~200ms, which is effectively ~63% of the 5V supply to the capacitor. Thus, an RC circuit with a time constant τ = 0.2 seconds has been simulated, which produces an output simulation that aligns with the theoretical derivations for capacitor voltage, current and time constant.

6.6.2 Example project – circuit 2

This circuit can now be adapted to show the 90° phase angle of a capacitor, by reconfiguring the input source with a sine wave function. Before doing this, it is recommended to **save a copy of the current circuit and name it chp6_proj2** to allow you to compare the two circuits at a later date. From the new circuit, the voltage source (and transient window) can be reconfigured to work with a sine wave input. The process for editing this schematic is as follows:

1. Specify resistor and capacitor values of R = 16Ω and C = 100μF.
2. Configure the voltage source to be a 100Hz sine input waveform.
3. Add a Spice Directive to perform a transient analysis of 0.05 seconds (5 cycles of a 100Hz wave).

Set up the voltage source as a sine input by choosing the advanced option in the editor dialog (Figure 6.44).

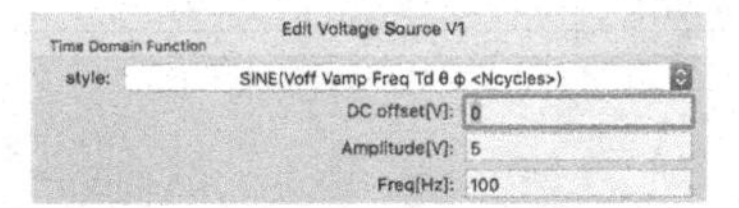

Figure 6.44 Creating a sine input source in LTspice. ***In the diagram, a Sine function is selected from the dropdown menu in the Voltage Source editor window to specify an Amplitude of 5V with a frequency of 100Hz (other parameters are not used).***

In the configuration window, an input voltage of 5V amplitude and 100Hz frequency is defined – the other parameters (such as DC offset, cycles, delay) are not needed for this analysis. The component values have also changed from the previous circuit, where the resistance (R1 = 16Ω) and capacitance (C1 = 100μF) values have been chosen to show the phase shift of the sine input signal created by the capacitor. The full circuit can now be set up for analysis (Figure 6.45).

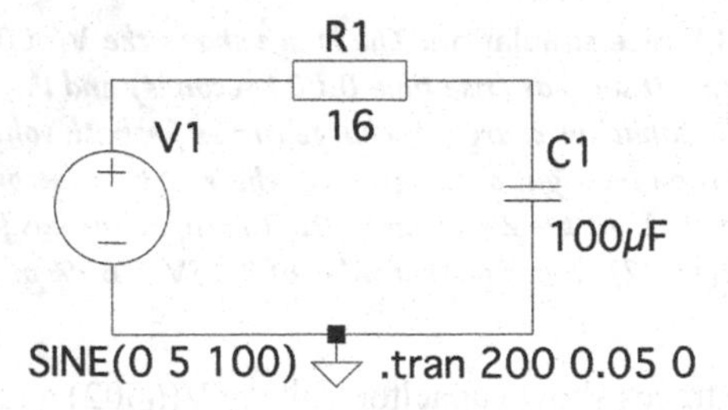

Figure 6.45 Example circuit 2 final LTspice circuit. ***The diagram shows a 5V 100Hz sine wave voltage source in series with a 16Ω resistor (R1) and a 100μF capacitor (C1). A transient analysis window of 0.005 seconds is used to capture 5 cycles of the 100Hz waveform for analysis.***

The LTspice schematic shows the voltage source configuration for a sine wave input, alongside the transient analysis window of 0.005 seconds. Knowing that the input signal will have a frequency of 100Hz, equation 6.2 (from section 6.2) can be used to find the period of one cycle of the wave:

The input signal frequency is 100Hz, so:

$$Frequency, F = \frac{1}{Period, T} = 100Hz$$

Rearranging the terms of the equation gives:

$$\mathbf{Period,\ T = \frac{1}{Frequency, F} = \frac{1}{100} = 0.01 seconds}$$

Using a transient analysis window of 0.05 seconds will show 5 cycles of the input sine wave V(n001), including the phase shift created by the capacitor at the output V(n002). Run the simulation, and use the voltage probe to analyse the V(n001) and V(n002) signals to compare the sine waves at both nodes (Figure 6.46).

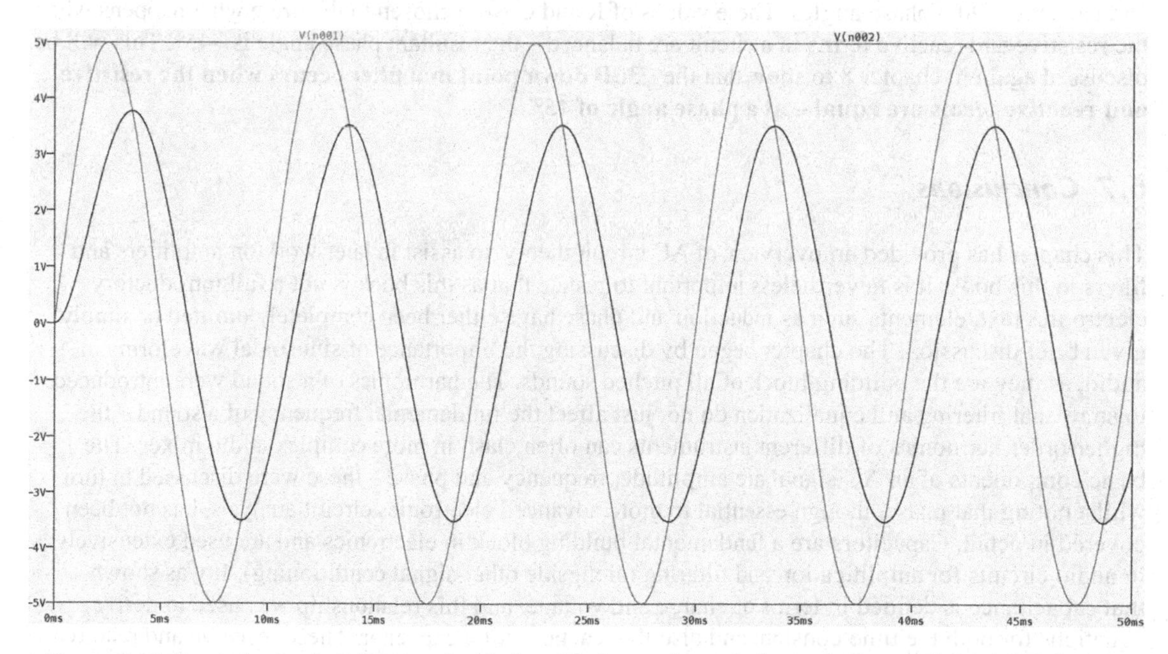

Figure 6.46 Example circuit 2 LTspice simulation. ***The image shows the input sine wave V(n001) in light grey, with the capacitor voltage V(n002) shown in black as the output. The phase angle of the capacitor means the output signal V(n002) lags the input signal – the exact value is calculated below.***

The analysis window shows how the output voltage V(n002) trace lags the input voltage signal V(n001). Recall from Figure 6.28 that the overall phase angle of an RC circuit can be calculated using

$\phi = \tan^{-1}\frac{X_C}{R}$, to verify what is seen in the graph:

First, calculate the capacitive reactance, Xc for an input frequency of 100Hz:

$\omega = 2\pi f = 6.283 \times 100 = 628.3\ rads,\qquad Capacitance,\ C = 100\mu F = 0.0001F$

$$X_C = \frac{1}{\omega C} = \frac{1}{628.3 \times 0.0001} = \frac{1}{0.06283} = \mathbf{15.92\Omega}$$

To get the output phase angle, use the ratio of Xc and R – when these are equal the result will be 45°:

$$\phi = \tan^{-1}\frac{X_C}{R} = \tan^{-1}\frac{15.92}{16} = \tan^{-1}0.995 = 44.85 \approx \mathbf{45^o}$$

The calculation shows that the output phase angle for V(n002) is 45°, which is half of the resistor (0°) and capacitor (90°) phase angles. These values of R and C were chosen to illustrate what happens when the resistive and reactive terms in a circuit are balanced – the resultant phase angle is ~45°. This will be discussed again in chapter 8 to show that the **−3dB down point in a filter occurs when the resistive and reactive terms are equal – at a phase angle of 45°**.

6.7 Conclusions

This chapter has provided an overview of AC circuit theory, to assist in later work on amplifiers and filters in this book. It is nevertheless important to restate that as this book is not a full introductory electronics text, elements such as induction and phase have either been completely omitted or simply given brief discussion. The chapter began by discussing the importance of sinusoidal waveforms in audio, as they are the building block of all pitched sounds. The harmonics of a sound were introduced, to show that filtering and equalization do not just affect the fundamental frequency of a sound – the higher-order harmonics of different instruments can often clash in more complex audio mixes. The basic components of an AC signal are amplitude, frequency and phase – these were discussed in turn whilst noting that phase, though essential to more advanced electronics circuit analysis, has not been covered in detail. Capacitors are a fundamental building block in electronics and are used extensively in audio circuits for amplification and filtering (alongside other signal conditioning). It was shown that capacitance is defined in terms of charge and voltage, and this relationship was used to derive equations for both the time constant and also the reactance of a capacitor. These resistive and reactive elements were then combined as AC impedance, though to focus on an introductory use of RC circuits again no detailed discussion of the role of phase in impedance was provided.

The tutorial covered the use of LTspice, a very powerful cross-platform tool that performs complex circuit analysis tasks without recourse to manual calculations (which can take time and are prone to errors). Although LTspice can take a little time to get used to as an interface, the speed and power of the tool are obvious when compared to the examples in the previous section – it will be used for AC circuit analysis from now on. The final example project showed how to use LTspice with an RC circuit to measure the time constant for a pulse input, and also to show the phase shift introduced by a capacitor for a sine wave input. These two examples can be used as templates for future work, where chapter 8 will show how to configure LTspice for audio wave data input for analysis. This chapter will also show how Bode plots are used to examine a sweep of frequencies as the input to a circuit – they are a very useful tool for working with audio circuits.

The next chapter will look at operational amplifiers as a means of signal amplification. Operational Amplifiers (Op-Amps) are examples of integrated circuits, where many transistors are combined within a single chip that keeps size, cost and noise to a minimum. The chapter will introduce transistor theory both as an explanation of Op-Amps and also to assist in future study, as many well-known audio effects circuits (e.g. pedals) are built around some form of transistor. Transistors can take time to learn in detail, so don't be too concerned if you find this section difficult – it is not essential for practical Op-Amp work in the amplification of audio signals. The chapter contains the first analogue audio project in this book, taking a headphone signal as a sensor input and amplifying it to drive an output loudspeaker as a transducer. A systems approach will be used (as in previous chapters) to help demonstrate how Op-Amps can be used to build a circuit that will be extended in later chapters to include filtering (chapter 8) and Arduino control (chapter 9). In so doing, the chapter moves towards practical audio systems that can inform future study of more advanced topics.

6.8 Self-study questions

For this chapter, the self-study uses LTspice to simulate (and validate) the example circuits in section 6.4.1 (Q5 and Q6). For each circuit, use a sine wave of 5V 1kHz as input, and measure the output voltages and currents using the probe. These exercises will help you to become more familiar with LTspice, which though not very user friendly is powerful enough to warrant the effort! The LTspice schematic is provided below for reference, and the relevant sub-questions listed.

For both questions, use a voltage source with a **5V 1kHz Sine Wave** as input and a **transient analysis window of 0.01 seconds** (this will give 10 cycles of a 1kHz signal, which has a period of $\frac{1}{1000} = 0.001$ seconds

Q1*: Create an LTspice circuit to simulate the following schematic:*

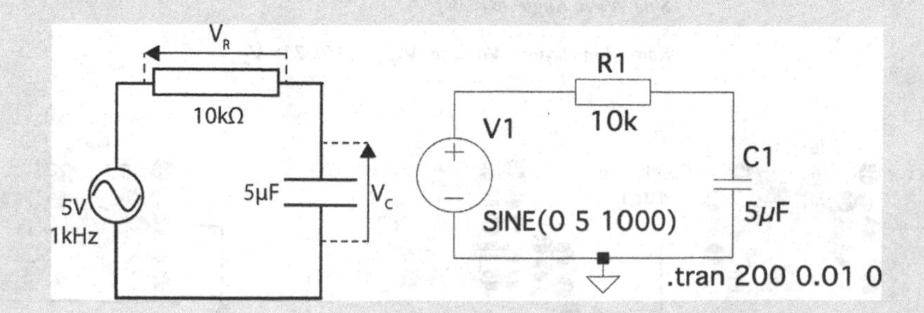

(a) What is the current flowing through the circuit?
(b) What are the voltage drops across the resistor (V_R) and the capacitor (V_C)?

Q2: *Create an LTspice circuit to simulate the following schematic:*

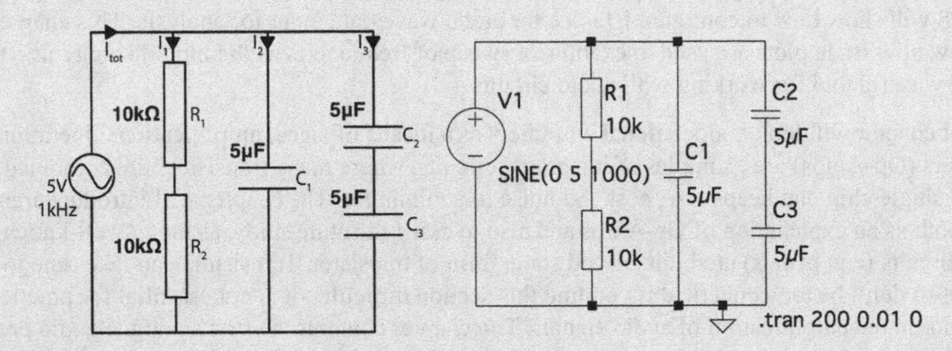

(a) *What are the individual branch currents in the circuit?*
(b) *What is the magnitude of the total current flowing through the circuit?*

All answers are provided for each chapter at the end of the book.

Chapter 6 summary

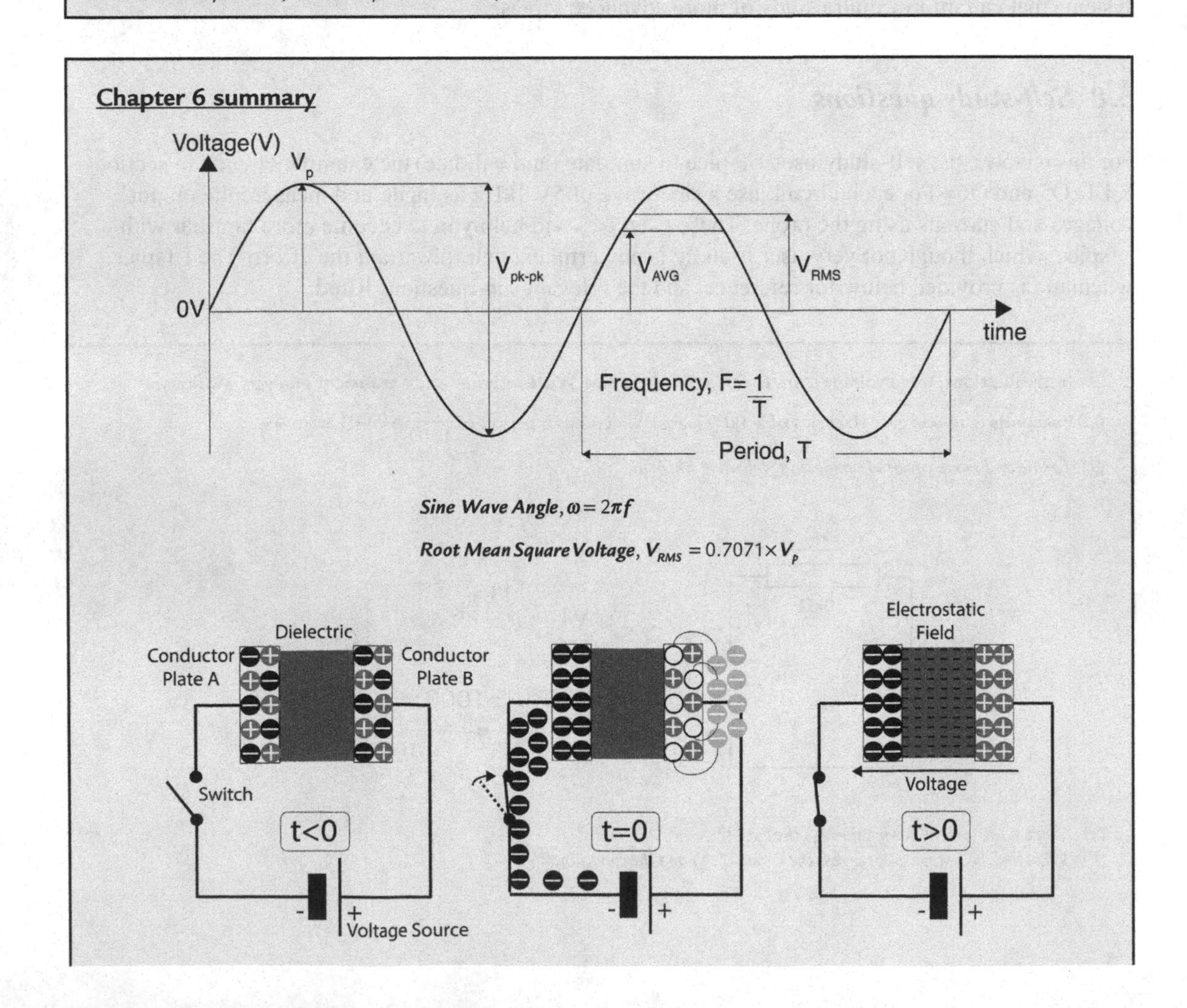

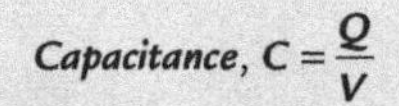

Capacitance, $C = \dfrac{Q}{V}$

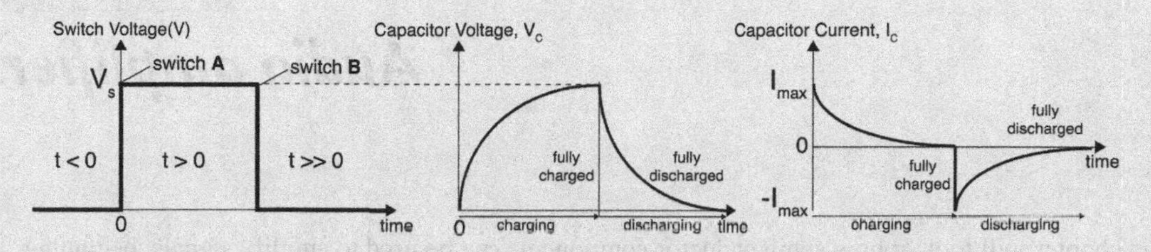

Time Constant, $\tau = RC$

Capacitive Reactance, $X_C = \dfrac{1}{\omega C}$

Total Series Capacitance, $\dfrac{1}{C_{tot}} = \dfrac{1}{C_1} + \dfrac{1}{C_2} + \ldots \dfrac{1}{C_n}$

Total Parallel Capacitance, $C_{tot} = C_1 + C_2 + \ldots C_n$

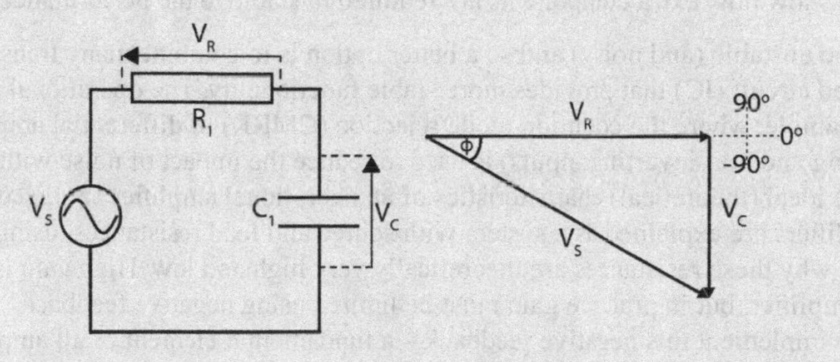

Voltage Magnitude, $V_S = \sqrt{{V_R}^2 + {V_C}^2}$

Series Impedance, $Z = \sqrt{R^2 + {X_C}^2}$

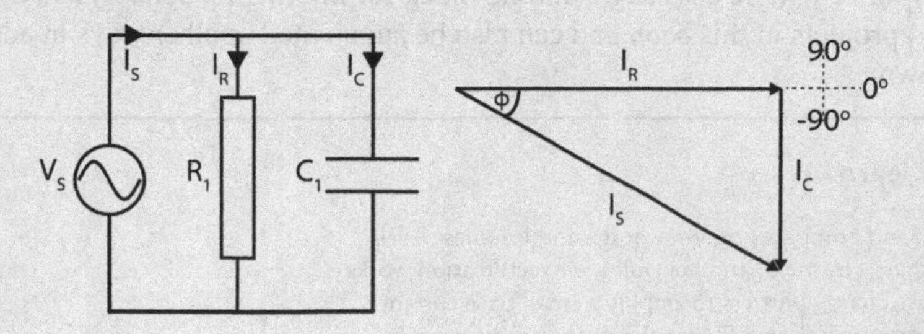

Current Magnitude, $I_S = \sqrt{{I_R}^2 + {I_C}^2}$

Parallel Impedance, $Z = \dfrac{V_S}{I_S}$

CHAPTER 7

Audio amplifiers

This chapter will look at how semiconductor components can be used to amplify signals, beginning with a brief discussion on how to calculate power, voltage and loudness decibel levels and why the relationship between them is important for audio systems. The next topic considers how a charge barrier restricts the flow of charge in a diode to a single direction, which allows it to be used for full wave rectification to convert an AC power signal (from mains supply) to a DC signal for use in an electronic circuit. This principle of the charge barrier can then be used to amplify an input signal within a transistor, where a small current at the base of the transistor will allow a much larger current to flow through the collector of the transistor – the input base current is used to amplify the collector current at output. This book does not build a transistor amplifier, but LTspice is used to simulate both the characteristic output curves (which dictate biasing) and a full bipolar junction transistor (BJT) common emitter circuit to show how extra components are required to stabilize the performance of a BJT.

Transistors can be unstable (and noisy) and so a better option is to combine many transistors within a single integrated circuit (IC) that provides more stable functionality. The operational amplifier is a common IC example, where the common mode rejection (CMRR) of differential amplification (between inverting and non-inverting inputs) is used to reduce the impact of noise within the circuit itself. The ideal (theoretical) characteristics of an operational amplifier are discussed, where operational amplifiers are explained as a system with source and load resistances, using voltage dividers to show why these resistances are theoretically very high and low. High gain is important for an operational amplifier, but in practice gain must be limited using negative feedback. Voltage dividers are again used to implement this negative feedback – a fundamental element of all amplification systems. An overview of DC output blocking, AC power decoupling and Zobel networks to stabilize loudspeaker output is provided, to help explain where some of the 'extra' components in audio circuits come from. The final project uses an LM386 operational amplifier chip to build a minimal audio amplifier. This amplifier will be used as a building block for filtering (chapter 8) and digital circuit control (chapter 9) projects in this book and can also be augmented in other ways in additional research projects of your own.

What you will learn

How to calculate (and compare) power, voltage and loudness levels
How diodes use charge barriers, and how full wave rectification works
How transistors use charge barriers to amplify a small base current
How operational amplifiers use differential inputs to reduce noise
How negative feedback works
How basic noise reduction can be performed (DC blocking, AC decoupling)
How a Zobel network helps balance the varying reactance of a loudspeaker
How to build a minimal audio amplifier

7.1 Amplification

In electronics, amplification defines how much a system will increase (or reduce) the amplitude of an input signal at output (Figure 7.1).

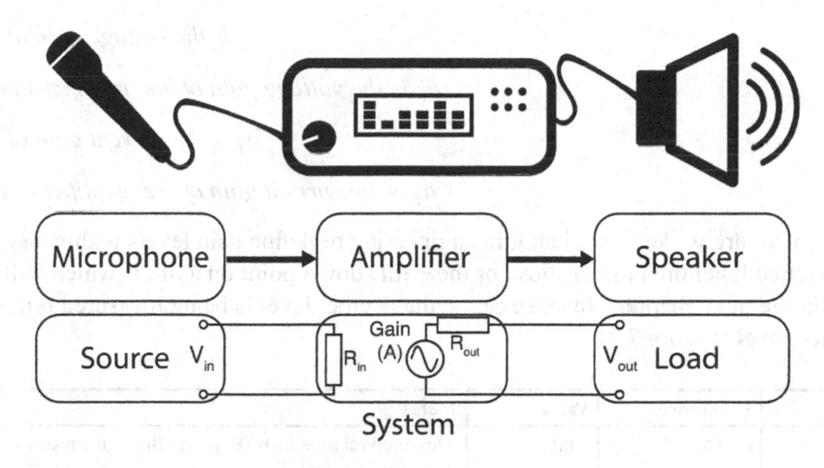

Figure 7.1 An audio amplification system (adapted from chapter 2, Figures 2.3 and 2.6). ***In the diagram, an open-loop audio amplifier system is shown that takes a microphone level input signal and increases it by the gain value (A) for output to a loudspeaker. The source resistance (Rin) and load resistance (Rout) are also shown, as these dictate how the amplifier is connected to other audio systems.***

The diagram shows how an amplifier can be modelled as a system comprising a source and load impedance (as audio signals are AC) – a system must be able to **accept a source and drive a load**. The amplifier provides gain to amplify a signal, where the gain (A) of the system is defined by the ratio of the input and output signals:

$$\textit{Amplifier Voltage Gain}, A_V = \frac{V_{out}}{V_{in}}$$
$$\textit{Amplifier Current Gain}, A_I = \frac{I_{out}}{I_{in}} \quad (7.1)$$

A_V *is the voltage gain of the amplifier*

V_{out} *is the output voltage in volts (V)*

V_{in} *is the input voltage in volts (V)*

A_I *is the current gain of the amplifier*

I_{out} *is the output current in amperes (A)*

I_{in} *is the input current in amperes (A)*

Amplifier gain levels are typically measured using a **decibel** scale, as the ratio of input to output signals can often be too large to represent on a linear scale. Logarithmic scales like the decibel provide an exponential range of values to describe a quantity, and they are used widely in electronics for amplitude and frequency measurements that cover large variations across a scale:

$$\begin{aligned} \textit{Amplifier Voltage Gain}, a_V (dB) &= 20\log_{10} A_V \\ \textit{Amplifier Current Gain}, a_I (dB) &= 20\log_{10} A_I \end{aligned} \quad (7.2)$$

A_V *is the voltage gain of the amplifier*

a_V *is the voltage gain of the amplifier in decibels (dB)*

A_I *is the current gain of the amplifier*

a_I *is the current gain of the amplifier in decibels (dB)*

The equations above are widely used in audio engineering to define gain levels within a system, such as the 0dB reference level on a mixing desk or the −3dB down point on a filter (which will be covered in more detail in the next chapter). In these cases, the decibel level is being measured relative to a known reference level (Figure 7.2)

Quantity	Reference	Value	Notes
Electrical power	0 dBm	1 mW	Measured relative to 1mW power through a resistor – telecommunications standard level
Signal voltage	0 dBu	0.775 Vrms	Measured relative to 1mW power (like dBm but unloaded) – sometimes known as dBv (small v)
Consumer line level	-10 dBV	0.316V	This lower voltage level is used for consumer equipment, and equates to −7.8 dBu
Professional line level	4 dBu	1.2276V	This higher voltage level is used for professional equipment and equates to 1.78 dBV
Signal voltage	0 dBV	1 Vrms	Commonly used for audio electronics, equates to 2.218 dBu
Sound pressure level	0 dB	20 µPa	20 micropascals (air pressure reference level) – lowest audible level
Sound pressure level	130dB	63.2 Pa	Audible pain threshold (where hearing damage will occur)

Figure 7.2 Common decibel reference levels. ***The figure lists some common decibel reference levels, as used in audio electronics systems. Power measures (in dBm) are effectively equivalent to dBu as they are taken from the same 1mW reference value (through a 600Ω resistive load). The dBV standard is widely used for professional audio signals, though dBu is sometimes referred to as dBv (which is confusing!). Example SPL levels for the broad range of human hearing are also provided, though this book does not discuss acoustics in detail.***

Figure 7.2 highlights some of the measurements in audio that use a decibel scale. For audio electronics, the most common is the dBV measurement, which is made relative to a 0dB value for a 1Vrms signal. Vrms is a root mean square voltage (see chapter 6, section 6.2, eqn. 6.5), which equates to the equivalent DC voltage level for an AC signal. When learning about audio, it can be very confusing to work with acoustic equations for sound pressure level (SPL) and sound intensity level (SIL) that also use the decibel scale, so it is important to remember that Bels and decibels are a unit of measurement – **not a quantity.**

Why $20log_{10}$?

The equations listed above use the term $20log_{10}$ to define voltage and current gain, which can initially be confusing when learning about decibel scales. Bels were designed for use with telecommunication systems, where the power gain is calculated as a ratio in Bels. As the Bel scale was so small, the decibel was introduced to provide a greater range when calculating power ratios. The first thing to remember is that the decibel is $\frac{1}{10}$ of a Bel, so any answer provided in decibels must scale the Bel value by a factor of 10. This is similar to working with scales like mV (thousandths of a volt), where 0.001V could be defined in millivolts as 1mV if you scale the value by 1000. The same is true of decibels, where the log_{10} of a ratio would give the answer in Bels, and so it must be multiplied by a factor of 10 to compensate for using decibels. This is not the factor of 20 shown above, however – it requires some further explanation. Power gain can be defined as follows:

$$\text{Power Gain}, a_P(dB) = 10log_{10}(A_I \times A_V) = 10log_{10}\left(\frac{A_{Pout}}{A_{Pin}}\right)$$

where the first equation uses the power relationship $Power, P = IV$ (chapter 4, section 4.9, eqn. 4.2) to determine the **decibel power gain** (a_P) relative to the voltage (A_V) and current (A_I) gain. The second equation is common in electronics, but while power measurements are used for amplifier efficiency, the power gain of an audio system is less widely used when working with audio signals. In most cases, the voltage or current gain of an audio system is used for the very small signal levels (from microphones or other sensors like pickups) amplified for output through a transducer like a loudspeaker. Hence, power can also be defined by using Ohm's Law to substitute the terms:

$$V = IR \rightarrow I = \frac{V}{R}, \qquad Power,\ P = IV = \frac{V}{R} \times V = \frac{V^2}{R}$$

which allows the **power in a circuit to be defined solely in voltage terms** (rearranging also gives current as $P = I^2R$). When there is a squared term in logarithmic values, it can be moved to a scalar outside the logarithm calculation:

$$10log_{10}\left(\frac{A_{Pout}}{A_{Pin}}\right) = 10log_{10}\left(\frac{V_{out}^{\ 2}/R}{V_{in}^{\ 2}/R}\right) = 10log_{10}\left(\frac{V_{out}^{\ 2}}{V_{in}^{\ 2}}\right) = 10log_{10}\left(\frac{V_{out}}{V_{in}}\right) \times 2 = \mathbf{20log_{10}}\left(\frac{V_{out}}{V_{in}}\right)$$

This explanation may appear complex in some ways, as this book does not cover power calculations in detail. Having said this, understanding decibel scales when working with signal levels or amplifier gain is a crucial part of working in any recording or broadcast studio, and so this explanation is provided to disambiguate the factor of 20 that is so often used in audio electronics calculations.

Amplifiers are often rated in terms of their **electrical power** (in watts), but it is important to correlate this measurement to other quantities such as signal voltage that are commonly used when working with audio amplifier circuits. In acoustic terms, sound pressure level (SPL) can be linked to voltage by virtue of the acoustic pressure being used to move a microphone diaphragm to create a signal voltage – these can be considered equal to each other in terms of gain for illustrative purposes. It is also important to consider the psychoacoustic scale of perceived loudness, which will also be discussed in the next chapter on audio filters:

7.1.1 Worked examples – calculating decibel gain values

These examples calculate some basic gain values for electrical power and signal voltage. The final example uses these values in comparison with perceived loudness.

Q1: *Tabulate the power gain A_P and decibel power gain $a_P\,(dB)$ of an audio amplifier with:*

(a) A DC input power of 1W and an output power of 2W.
(b) A DC input power of 1W and an output power of 4W.
(c) A DC input power of 1W and an output power of 10W.

For an input power of 1W and output power of 2W:

$A_{Pin} = 1W, \quad A_{Pout} = 2W$

$$Power\ Gain,\ A_P = \frac{A_{Pout}}{A_{Pin}} = \frac{2}{1} = \mathbf{2}$$

$$Decibel\ Power\ Gain,\ a_P\,(dB) = 10log_{10}\left(\frac{A_{Pout}}{A_{Pin}}\right) = 10log_{10}(2) = 10 \times 0.301 \approx \mathbf{3dB}$$

For an input power of 1W and output power of 4W:

$A_{Pin} = 1W, \quad A_{Pout} = 4W$

$$Power\ Gain,\ A_P = \frac{A_{Pout}}{A_{Pin}} = \frac{4}{1} = \mathbf{4}$$

$$Decibel\ Power\ Gain,\ a_P\,(dB) = 10log_{10}\left(\frac{A_{Pout}}{A_{Pin}}\right) = 10log_{10}(4) = 10 \times 0.602 \approx \mathbf{6dB}$$

For an input power of 1W and output power of 10W:

$A_{Pin} = 1W, \quad A_{Pout} = 10W$

$$Power\ Gain,\ A_P = \frac{A_{Pout}}{A_{Pin}} = \frac{10}{1} = \mathbf{10}$$

$$Decibel\ Power\ Gain,\ a_P\,(dB) = 10log_{10}\left(\frac{A_{Pout}}{A_{Pin}}\right) = 10log_{10}(10) = 10 \times 1 = \mathbf{10dB}$$

Answer:

Power gain, A_P	Decibel power gain, a_P (dB)
2	3
4	6
10	10

Notes:

This example aims to show the relationship between power gain and decibel power gain for a small range of input and output values. A doubling of the power gain equates to an increase of 3dB, whilst a gain of 10 will increase the power output by 10dB. The following examples will add to these results.

Q2: *Tabulate the voltage gain A_V and decibel voltage gain $a_V\,(dB)$ of an audio amplifier with:*

(a) An input signal of 1V and an output signal of 1.41V.
(b) An input signal of 1V and an output signal of 2V.
(c) An input signal of 1V and an output signal of 3.16V.

For an input signal of 1V and output signal of 1.41V:

$$A_{Vin} = 1V, \qquad A_{Vout} = 1.41V$$

$$Voltage\,Gain, A_V = \frac{A_{Vout}}{A_{Vin}} = \frac{1.41}{1} = \mathbf{1.41}$$

$$Decibel\ Voltage\ Gain,\ a_V\,(dB) = 20log_{10}\left(\frac{A_{Vout}}{A_{Vin}}\right) = 20log_{10}\,(1.41) = 20 \times 0.149 \approx \mathbf{3dB}$$

For an input signal of 1V and output signal of 2V:

$$A_{Vin} = 1V, \qquad A_{Vout} = 2V$$

$$Voltage\ Gain,\ A_V = \frac{A_{Vout}}{A_{Vin}} = \frac{2}{1} = \mathbf{2}$$

$$Decibel\ Voltage\ Gain,\ a_V\,(dB) = 20log_{10}\left(\frac{A_{Vout}}{A_{Vin}}\right) = 20log_{10}\,(2) = 20 \times 0.602 \approx \mathbf{6dB}$$

For an input signal of 1V and output signal of 3.16V:

$$A_{Vin} = 1V, \qquad A_{Vout} = 3.16V$$

$$Voltage\ Gain,\ A_V = \frac{A_{Vout}}{A_{Vin}} = \frac{3.16}{1} = \mathbf{3.16}$$

$$Decibel\ Voltage\ Gain,\ a_V\,(dB) = 20log_{10}\left(\frac{A_{Vout}}{A_{Vin}}\right) = 20log_{10}\,(3.16) = 20 \times 0.499 \approx \mathbf{10dB}$$

Answer:

Voltage gain, A_v	Decibel voltage gain, a_p (dB)
1.41	3
2	6
3.16	10

Notes:

This second example shows the relationship between voltage gain and decibel voltage gain for a similar small range of input and output values to the previous power example. The values have been changed to give an indication of how a doubling of the voltage gain equates to an increase of 6dB, whilst for comparison with power gain levels output voltages of 1.41V (~3dB) and 3.16V (~10dB) have been used.

Q3: *Assuming sound pressure level (SPL) to be a correlate for signal voltage in an audio system, draw a graph of decibel power gain $a_P(dB)$, decibel voltage gain/SPL $a_V(dB)$ and perceived loudness (dB).*

(a) *What does this graph show about the increase of electrical power in relation to perceived loudness?*
(b) *What does this graph show about the increase in voltage in relation to both electrical power and perceived loudness?*

The values for power $a_P(dB)$ and voltage $a_V(dB)$ are now calculated, but perceived loudness levels must also be considered, which is outwith the scope of this book (as there are different standards, units and applications). ISO 226:2003 will be discussed in the next chapter to highlight how the human ear does not perceive all audio frequencies at equal levels, but as a workaround the general values for 3, 6 and 10dB values for perceived loudness are provided in the table (which is used to plot the accompanying graph).

Answer:

Decibel gain	Power gain, A_p	Voltage gain, A_v	Perceived loudness (dB)
3	2	1.41	1.23
6	4	2	1.52
10	10	3.16	2

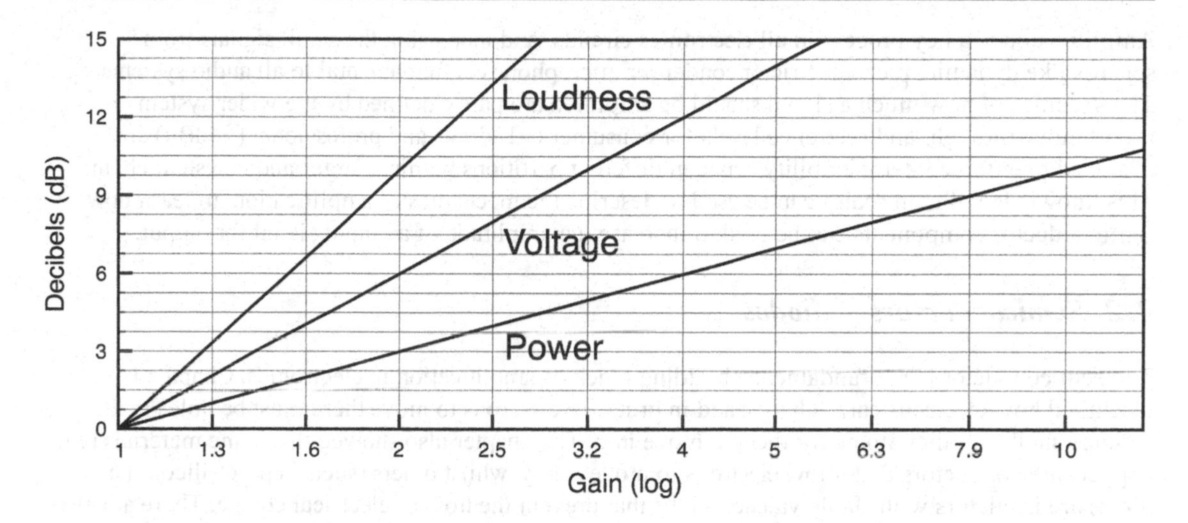

(a) The graph shows that perceived loudness requires a 10dB increase to double the gain, which is significantly higher than either the power or voltage gains. A power gain of 2 will give an increase of 3dB, so doubling the power does not double the loudness. For the perceived loudness to double the power will need to increase by 10dB, which is a gain of 10. Thus, **a 100W amplifier will only sound twice as loud as a 10W amplifier** when heard by the listener. This is often misunderstood in audio, where equipment is defined by electrical power rating rather than SPL output (an acoustic measure) or perceived loudness (how it is heard by the listener).

(b) The graph shows that doubling the voltage gain gives an increase of 6dB, which is twice the power gain. This can often be confusing when working with gain ratios and decibels, particularly with filtering (the next chapter shows how the cutoff of a filter is defined in dB).The voltage would need to increase by a factor of 3.16 (from the table) to double the perceived loudness of the sound, which is much less than the power increase of 10 required.

Notes:

The graph shown above illustrates the relationship between electrical power gain, voltage gain (corresponding to SPL) and perceived loudness. Not all of the data values from the previous table are shown, as the aim is to highlight the general disparity between the three quantities. Increasing perceived loudness requires the greatest decibel increase of the three quantities, and though there is a simple relation between doubling loudness and power (a factor of 10), values for voltage gain and 3/6 dB values are less easily calculated.

The previous examples aim to familiarize you with the use of power, voltage and loudness gain calculations – all of which are measured on a decibel scale. It is important to reiterate that decibels are a unit of measurement, not a quantity. This can often be confusing when working with similar-sounding scales like dBu, dBm, dbv and dBV – there are also others that have been omitted to preserve some clarity in the discussion! When beginning to learn about amplifiers, the main thing to remember is:

Amplification defines how much the output signal amplitude increases relative to the input signal

Amplification is a key process in all electronics circuits, and increasing the small signals from sensors like dynamic, piezoelectric or condenser microphones is fundamental to all audio systems. The specifics of how much a signal should be amplified are partly defined by the wider system it will move through, and reference levels for consumer (−10dBV) and professional (+4dBu) are designed to enforce interoperability between different partitions within a larger audio system chain. This knowledge of gain scales can be used to describe the mechanics of amplification, to learn how semiconductor components can be used to increase the amplitude of an input signal for output.

7.2 Semiconductors – diodes

The semiconductor is the fundamental building block of amplification in electronics. Chapter 1 discussed how electrons carry charge, and in order for electrons to move there must be holes in the valence shells of other atoms for them to move into. The chapter also showed that some materials (e.g. copper) are conductors that allow electrons to move easily, whilst others (such as pure silicon, i.e. glass) are insulators with stable valence shells that prevent the flow of electrical charge. There are also materials that can be chemically altered to behave as **both** conductor and insulator – these are known as **semiconductors**. The materials used in semiconductors are created by chemically **doping** the element **silicon** (Si) to change its atomic structure (Figure 7.3).

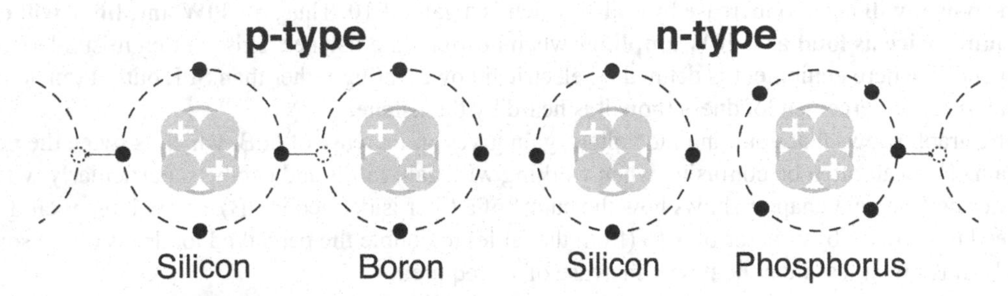

Figure 7.3 Silicon doping. ***The left-hand diagram shows how adding boron atoms to silicon imbalances the existing electron pairs. Silicon electrons can move to create new pairs with the boron atoms – this leaves holes in the existing structure, which now becomes more positive. The right-hand diagram shows how adding phosphorus atoms to silicon creates the opposite imbalance; the extra valence electrons in phosphorus are unpaired and so are capable of moving out of the structure – they can become charge carriers.***

In the example in Figure 7.3, silicon has a balanced atomic structure, where four valence electrons form pairs that will not move easily (electron pairing was briefly discussed as part of electromagnetism in section 2.5). If this silicon is now doped by adding boron atoms (three valence electrons) it creates a positive imbalance by breaking the existing valence pairs, so a silicon electron can now pair with a boron electron – leaving holes in the structure for other electrons to move into. On the other hand, doping silicon with phosphorus atoms (which have five valence electrons) means the unpaired electrons create a negative imbalance in the overall atomic structure – the extra electrons are free to move as charge carriers. These doped materials can be combined to create the **simplest form of semiconductor, the diode** (Figure 7.4).

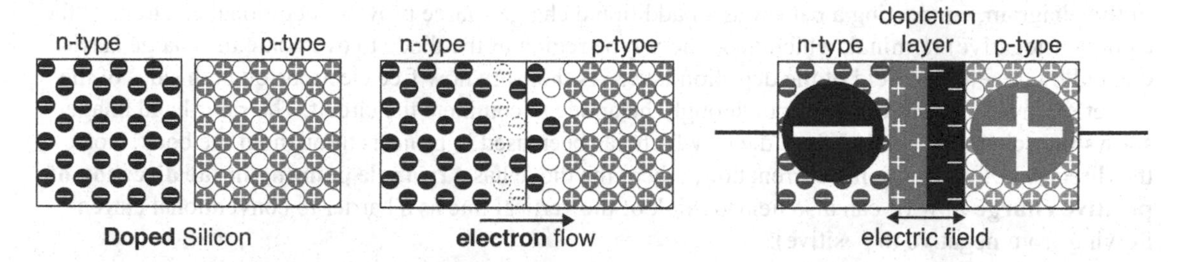

Figure 7.4 A semiconductor diode. ***The left-hand diagram shows negatively (n-type) and positively (p-type) doped Silicon that can be used to create a diode. The middle diagram shows how combining these materials allows free electrons in the n-type region to flow into holes in the p-type material. The right-hand diagram shows how this movement of charge creates an electric field of the opposite polarity between the doped regions – this is called the depletion layer. The opposite polarity of the depletion layer now opposes any further movement of charge between the regions.***

The left-hand image shows how a diode combines an **n-type** (negative) material that contains a surplus of **electrons** with a **p-type** (positive) material that contains extra **holes** for those electrons to move into. These two materials are then joined as either a p-n junction or an n-p junction diode, where the junction between them is known as the **depletion layer**. In the neutral state (without a power source being connected), some electrons from the n-type material move across to fill the holes in the p-type region (middle image). These moving **electrons are now at the edge of the p-type** material, but by moving they have also **left holes behind them at the edge of the n-type** region. This creates an electrical field in the depletion layer that is in the **opposite direction to the flow of charge**, which means no further charge can flow (right-hand image).

Chapter 6 (section 6.3) showed how a capacitor creates an electric field between its plates that reduces the flow of current to zero, and in some ways the depletion layer in a diode performs a similar function – electrons cannot cross the charge barrier created by the small depletion layer to get to the positive holes on the other side, because the potential of the barrier is the opposite of the potential between the two regions. This charge barrier can be overcome by connecting another source of charge that floods the n-type region with even more free electrons, eventually breaking down the polarity of the depletion layer and allowing current to flow (Figure 7.5).

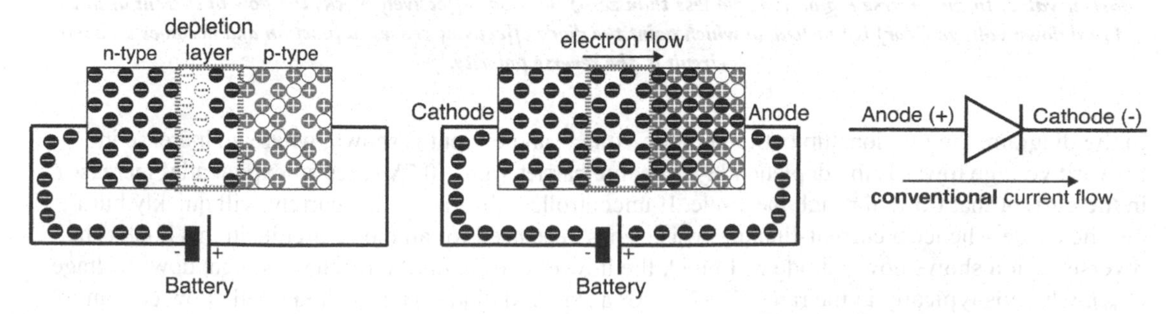

Figure 7.5 Diode current flow. ***The left-hand diagram shows how adding an additional source of charge (a battery) provides more electrons to the n-type region of the diode. The middle diagram shows how these extra free electrons enter at the cathode to flood the n-type region and overcome the depletion layer, which allows current to flow through the anode back to the positive terminal of the battery. The right-hand diagram shows the schematic symbol for a diode, noting that conventional current flow is defined in the opposite direction between the positive anode and negative cathode.***

In this diagram, connecting a battery as an additional charge source provides additional electrons at the cathode (negative) terminal, which flood the n-type region of the diode to overcome the charge barrier created by the electric field of the depletion layer. As there are now free electrons on both sides of the depletion layer, current can flow out through the anode to complete the circuit. The right-hand image shows the schematic symbol for a diode, which has been used in projects throughout the book. Note the direction of conventional current flow, where the diode has a **triangle pointing in the direction of positive charge** flow (it can also help to think of the vertical line as a barrier to conventional current flowing from negative to positive).

Returning to real electron flow, free electrons will only flood the depletion layer if there is a **potential difference between the terminals** of the charge source (battery) that can overcome the charge barrier. This potential difference (known as the forward bias voltage) equates to **0.7V for silicon** diodes (0.3V for germanium), and once it is reached the diode will allow more current to flow without any significant increase in voltage (Figure 7.6).

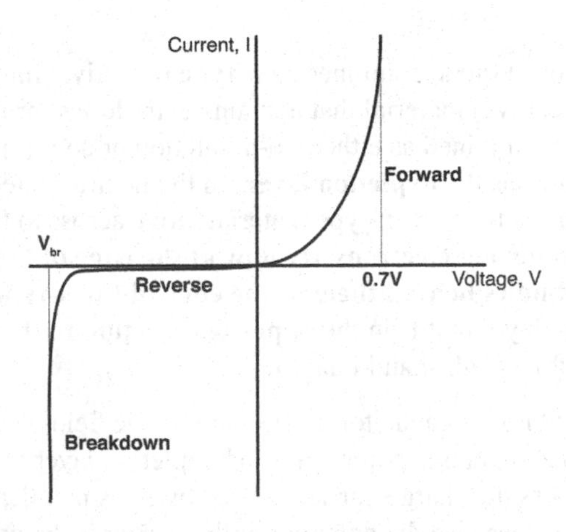

Figure 7.6 Diode voltage vs current graph. ***In the diagram, the current flowing through a typical diode for various input voltages is shown. In the forward region (voltage greater than zero) the voltage and current will increase as expected until the voltage reaches the charge barrier (0.7V). At this point, the diode will allow current to flow as if it were a short circuit (and hence rapidly burn out) but voltage will not significantly increase beyond the charge barrier value. In the reverse region (voltage less than zero) the diode effectively blocks the flow of current until the breakdown voltage (Vbr) is reached, at which point the diode effectively ceases to function and becomes an open circuit in the reverse polarity.***

In the diagram, the relationship between diode voltage and current is shown, where an increase in forward voltage towards the depletion layer charge barrier value (0.7V) sees an exponential increase in the current that flows through the diode. If uncontrolled, this increasing current will quickly burn out the diode – hence a current-limiting resistor has been used for all diode circuits in this book. The reverse region shows how a diode will block the flow of current until it reaches its breakdown voltage (V_{br}), which is typically in the range 50–75V for a standard diode. Thus, a diode will allow current to flow in the forward direction once the applied voltage is greater than the potential of the charge barrier (0.7V), but will oppose the flow of current in the reverse direction until it eventually breaks down at high voltage levels. This relationship between voltage and current means a diode has many uses in electronics – for digital circuits, it only allows signals to pass in one direction, which is the logic equivalent of true/false (if forward true, else reverse false). As chapter 6 (section 6.3) noted, diodes can be used in analogue circuits to rectify a sinusoidal AC voltage to produce a DC signal as output (Figure 7.7).

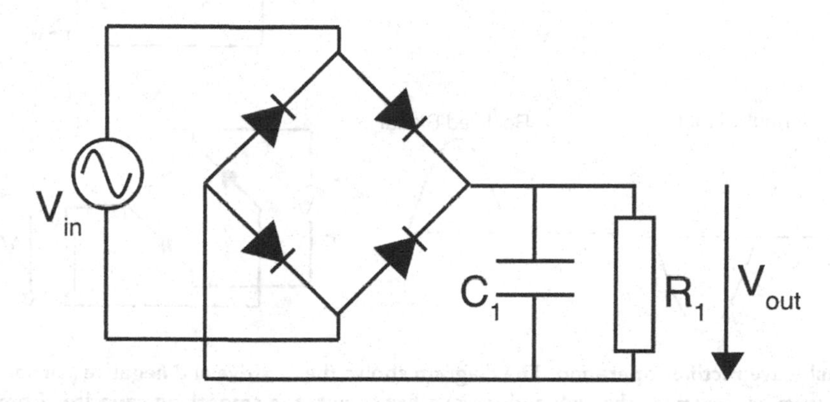

Figure 7.7 Full wave rectifier circuit with capacitive smoothing. ***The diagram shows how diodes can be used to form a full wave rectifier circuit, where an input AC voltage signal (V_{in}) is rectified and smoothed by the capacitor (C_1) to produce a DC output signal across the load resistor (R_1). The arrangement of the diodes allows current to flow in two paths based on the polarity of the AC input signal, where each path presents the same polarity across the load resistor R_1.***

The circuit in Figure 7.7 shows how diodes can be used to route a signal of changing polarity to the same output path, which uses a smoothing capacitor (C1) to shape the rectified output voltage (see section 6.3) that is output across the **load resistor R_1**. Chapter 2 (section 2.1, Figure 2.6) introduced the concepts of source and load within electronic systems, where matching the input and output impedances of different stages of a system is crucial to ensuring they can operate correctly with each other. AC circuits amplify signals either for further processing (e.g. filtering in chapter 8), connection to another system (e.g. within a larger audio recording system such as the one shown in section 2.1, Figure 2.7) or for output to a transducer like a loudspeaker. In all cases, these systems are said to present as sources to drive loads, where impedance matching is an important element of effective power transfer within the circuit. Returning to the diode rectifier above, the operation of the circuit is based on a varying-polarity AC signal input (Figure 7.8).

Rectifier circuits are needed in any electronic system that uses a mains power supply. In audio systems such as effects pedals and preamplifiers, this is often performed by a dedicated DC power supply that steps down the mains voltage (220V in Europe, 110V in North America) with a transformer, and then rectifies and smooths this reduced voltage for connection to an audio circuit. This book does not go into detail on power systems as the Arduino (like many audio preamps effects and pedals) can be powered from a single 9V battery that provides DC at output. In practice, stable power supply (and conditioning) is a significant part of any commercial recording studio or live audio system. Many professional musicians use some form of power conditioner in their systems both to reduce noise and to prevent damage to their equipment – the rectifier circuit shown in Figure 7.8 is an initial stage within this process.

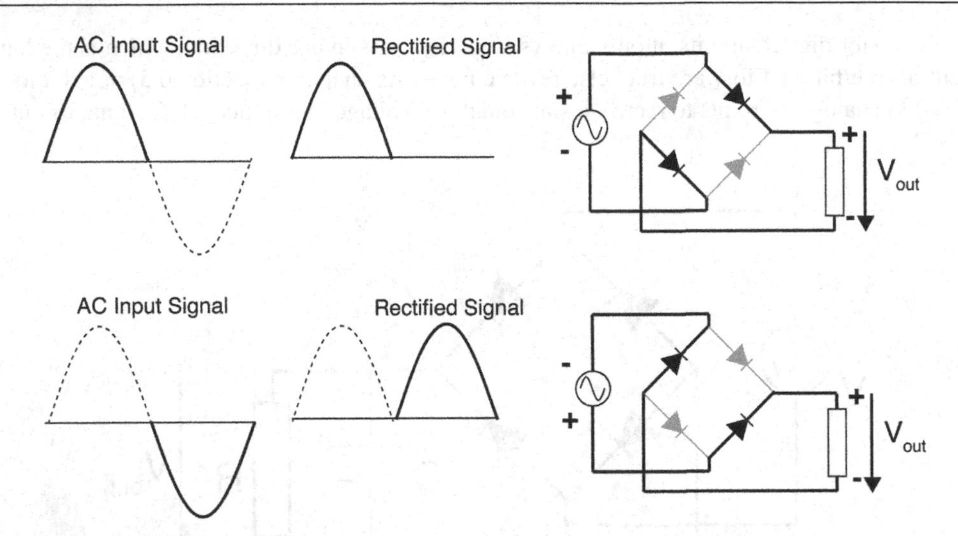

Figure 7.8 Full wave rectifier operation. The diagram shows the positive and negative portions of a single cycle of a sine wave as they pass through a diode rectifier circuit (the smoothing capacitor from the previous diagram has been omitted for clarity). In the positive cycle (top image), only the downward-facing diodes allow current to flow, which routes the AC input signal to the load resistor with a positive (top) to negative output polarity. In the negative cycle (bottom image), only the upward-facing diodes allow current to flow and so even though the positive terminal of the AC signal has now changed, it is still routed to the positive (top) of the output load resistor.

Diodes have many other uses in electronic systems, and many guitar effects pedal circuits exploit the 0.7V forward voltage of a diode to clip an input signal to distort it at output (Figure 7.9).

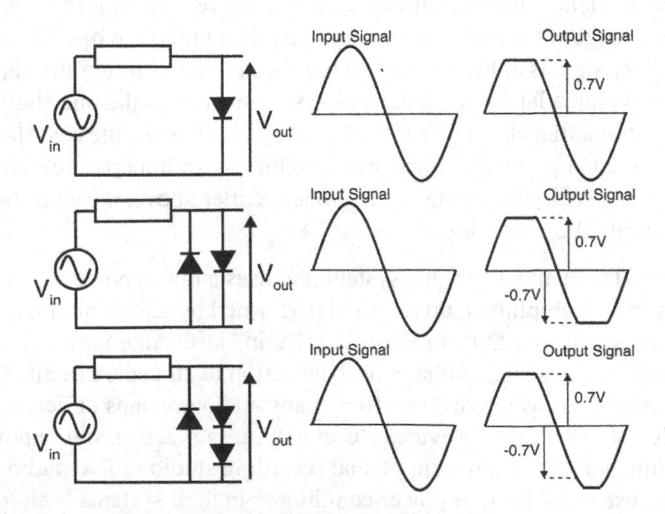

Figure 7.9 Diode clipping of an input signal. ***The top diagram shows how the forward voltage of a diode will peak at 0.7V, which clips the positive half of the input signal at this level. Using a pair of antiparallel diodes (middle diagram) allows both sides of the input AC signal to be clipped, and more exotic versions of this configuration such as asymmetric clipping with a third diode (bottom diagram) can shape the output signal in different ways.***

Distorting an input audio signal is a fundamental process in many genres of music, and thus a staple element of musical production in general. A full discussion of distortion methods and principles is beyond the scope of this book, but combining amplification feedback with diode clipping is a common technique used in many pedal designs. As you extend your studies, you may encounter diodes being used in effect pedal circuits for either soft clipping (in the feedback path) or hard clipping (at the amplifier output) distortion, where different combinations of diodes (and perhaps capacitors) are used to shape the signal.

Many of these effects circuits are based around **transistor** amplification, which extends the principle of a charge barrier to control and **amplify** an input signal. Although single transistor circuits have effectively been superseded by operational amplifiers (which combine them for stability), there are many example schematics for classic effects and amplifiers that will employ transistors (or vacuum tubes). This book will not cover valve amplification (nor build a transistor amplifier), but the following section details how a transistor operates primarily to help introduce operational amplifiers. Operational amplifiers have higher performance and are much easier to use, so this section on **transistor theory can be skipped** if you find it difficult to follow – it is not needed for the practical project circuits in this book. In some ways, the information on transistors is also provided to assist in the analysis of the many classic schematics that employ them – should you wish to extend your learning in this direction. Having said this, the behaviour of a transistor is more complex to follow (and the calculations more involved) than an operational amplifier – it is a topic that cannot be covered fully in an introductory text.

7.3 Semiconductors – transistors

The previous section discussed the principle of a diode as being two oppositely charged pieces of doped silicon being joined together to create a charge barrier in the resulting depletion layer between the n-type and p-type regions. A bipolar junction transistor (BJT) is effectively two diodes placed back to back, where a thin region of one charge type is sandwiched between two regions of the opposite charge (Figure 7.10).

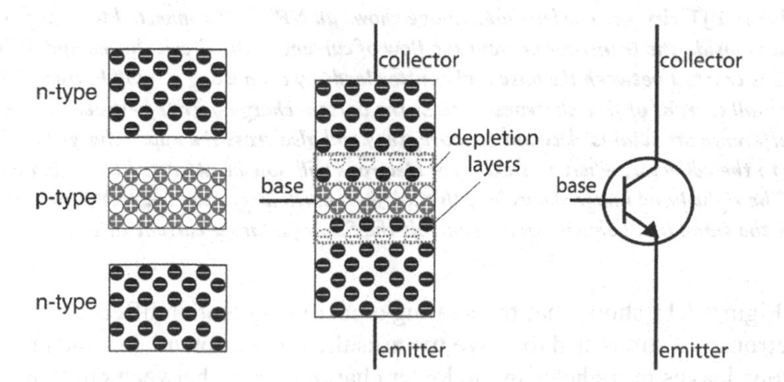

Figure 7.10 An NPN BJT transistor. ***The left-hand diagram shows how three pieces of doped silicon can be added together in an NPN (n-type/p-type/n-type) configuration. The middle diagram shows an NPN bipolar junction (BJT) transistor, where the base lead is connected to the p-type region, with collector and emitter leads being connected to the n-type regions. Note the presence of two depletion layers in the NPN configuration: these represent two charge barriers to the flow of current. The right-hand diagram shows the standard schematic symbol for an NPN BJT transistor where the arrow indicates the direction of current flow from collector to emitter – the mnemonic 'not pointing in' is often used for NPN.***

In the diagram, an NPN BJT is shown. **NPN indicates a n-type/p-type/n-type** configuration (a PNP BJT has the opposite arrangement), where the leads are named to indicate how electrons are emitted from the bottom n-type region to be collected at the top. It is important to remember the direction of **conventional current flow from collector to emitter** that is indicated by the **arrow pointing outwards towards the emitter** (the mnemonic '*not pointing in*' is often used for NPN) – a PNP has the opposite direction of current flow. Regardless of its configuration as either NPN or PNP, the stacking of three regions in a BJT creates two depletion layers and therefore two charge barriers to current flow (middle diagram). This means that, like a diode, in a normal state current cannot flow between any of the three leads (base, collector and emitter).

If a current source (e.g. a battery) is connected across the emitter and collector no current will flow, but if a second source (like an audio input signal) is then connected to the base it creates a positive potential difference between the base and the emitter. This positive potential difference attracts electrons from the lower n-type region which will also cross the bottom charge barrier (Figure 7.11).

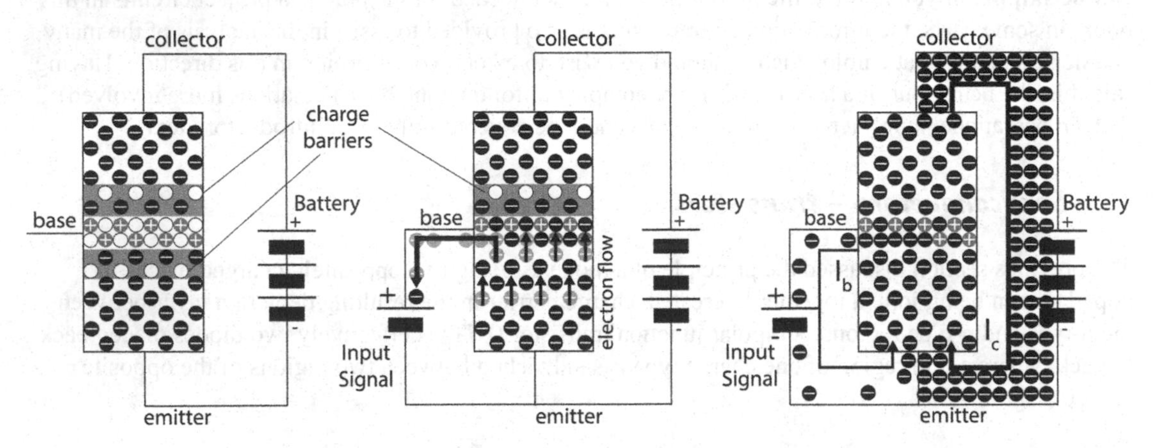

Figure 7.11 An NPN BJT circuit. ***The left-hand image shows an NPN BJT connected to a large battery, where the two charge barriers inside the transistor prevent the flow of current in the circuit. In the middle image, a positive potential difference is created between the base and emitter leads by connecting a small battery (i.e. an input signal), which attracts a small current of free electrons to cross the bottom charge barrier between emitter and base. Once this potential difference stimulates electrons to move, they will also cross the upper charge barrier to the n-type region connected to the collector, where this excess of electrons will now be attracted towards the positive collector battery terminal. The right-hand image shows how this allows a much larger current (Ic) to flow through the emitter, thus the small input signal current controls the flow of a larger current through the BJT.***

The diagram in Figure 7.11 shows that by creating a positive potential difference between the base and emitter, electrons are stimulated to move towards the higher potential of the base. This small current of electrons leaves more holes in the lower charge barrier (between emitter and base) which then allow other electrons to move and effectively overcome the barrier – the middle diagram shows only one charge barrier remaining. Although a small number of these electrons will leave the transistor through the base (as the base current I_b), the majority will now fill the holes in the upper charge barrier (between base and collector) and so this barrier also breaks down. Now there are an excess of electrons in the collector region, and because there is no charge barrier attracting them back towards the p-type region (to keep them in place) they can move towards the positive potential of the battery – this causes a current to flow between emitter and collector (I_c). This highlights the second crucial point about BJT transistors – the **collector potential must be more positive than the base** (and the emitter) to attract free electrons towards it. In functional terms, the small base current causes a larger current to flow from emitter to collector – the transistor becomes an **amplifier** (Figure 7.12):

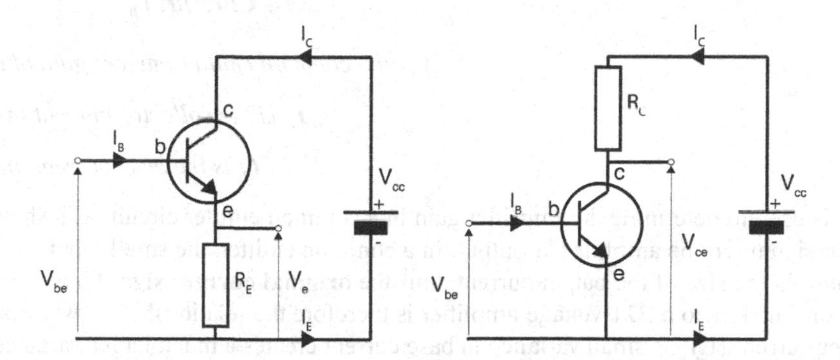

Figure 7.12 BJT common collector and common emitter amplifier circuits. ***The left-hand diagram shows an NPN transistor being used within a common collector circuit. In a common collector, the base is the input circuit, the emitter is the output circuit and the collector is common to both. Conversely, the right-hand diagram shows a common emitter circuit, where the base is the input circuit, the collector is the output circuit and the emitter is common to both. Note the presence of the load resistor in each case, which can initially be confusing as the collector resistor (R_C) controls the current flow (and hence the gain) in a common emitter circuit, whilst the emitter resistor (R_E) controls the common collector.***

The diagram shows two BJT circuits, where the output signal is measured across a load resistor (either R_E or R_C). The left-hand circuit is a **common collector** BJT amplifier (also known as an emitter follower because it does not provide voltage gain), which is often used for power amplification as it can provide high current gain to drive low-impedance loads like loudspeakers. The right-hand circuit is a **common emitter** amplifier, because the input signal is applied to the base and the output is taken from the collector – so the emitter is common to both loops. A common emitter circuit can amplify both current and also small voltages when operating in active mode (discussed below), which makes it a good choice for audio amplification where it can amplify signals at all stages of a system from input (buffering), through filtering and effects to other preamplification stages prior to output.

The diagram in Figure 7.12 shows base (I_B), collector (I_C) and emitter (I_E) currents, where the emitter current is the combined base and collector currents:

$$\textit{Emitter Current}, I_E = I_C + I_B \qquad (7.3)$$

I_E is the emitter current in amperes (A)

I_C is the collector current in amperes (A)

I_B is the base current in Amperes (A)

For a BJT transistor, the base current is small so the emitter current is effectively equivalent to the collector current once current begins to flow through the transistor. This is how common collector and common emitter circuits can amplify current signals for output – collector current (I_C) and emitter current (I_E) are both proportional to the input base current (I_B). For a common emitter circuit, the current gain (β) is thus defined as the **ratio of collector and base** currents:

$$\textit{Common Emitter Current Gain}, \beta = \frac{\textit{Collector Current}, I_C}{\textit{Base Current}, I_B} \tag{7.4}$$

β is the common emitter current gain of the amplifier

I_C is the collector current in amperes (A)

I_B is the base current in amperes (A)

This equation is used to determine the amplifier gain in a common emitter circuit, as it shows how much the input signal will be amplified at output. In a common emitter, the small input current applied to the base controls the size of the output current, thus the **original current signal has been amplified** by the transistor. The key to a BJT voltage amplifier is therefore the relationship between base current (I_b) and emitter current (I_e) – a small variance in base current creates a much larger variance in collector current output (Figure 7.13).

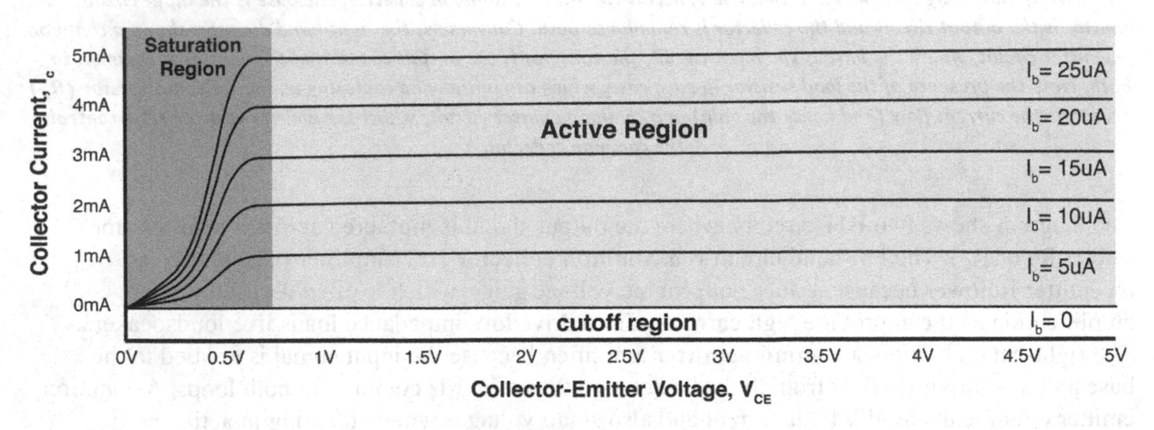

Figure 7.13 BJT current amplification curves. ***The diagram shows a graph of collector current (I_c) versus collector emitter voltage (V_{ce}) for a typical BJT transistor (R_C = 100Ω). Each line on the graph represents a specific base current (I_b) – values are listed on the right. The dark grey area highlights the saturation region, where the transistor is said to be fully on. The light grey area highlights the active region of the transistor, where the base current (I_b) controls the collector current (I_c) and the BJT acts as an amplifier. The bottom area highlights the cutoff region, where the transistor is said to be fully off.***

The diagram shows characteristic output curves for a BJT, where the base current (I_b) at the transistor input controls the size of the collector current (I_c) – the BJT acts as a current amplifier. When there is no base current (I_b= 0) the **transistor is off**, as indicated by the **cutoff region** (in white) at the bottom of the graph. Like a diode, the collector current initially increases in proportion to the collector voltage

(V_{CE}) until the potential difference needed to stimulate electron flow between base and emitter reaches 0.7V and the **transistor is on** – this is known as the **saturation region** (shown in dark grey). After this point, the transistor begins to operate in the **active region** (in grey), where changes in voltage between collector and emitter do not significantly impact on the current flowing through the transistor – **the input current dictates the output current** (Figure 7.14).

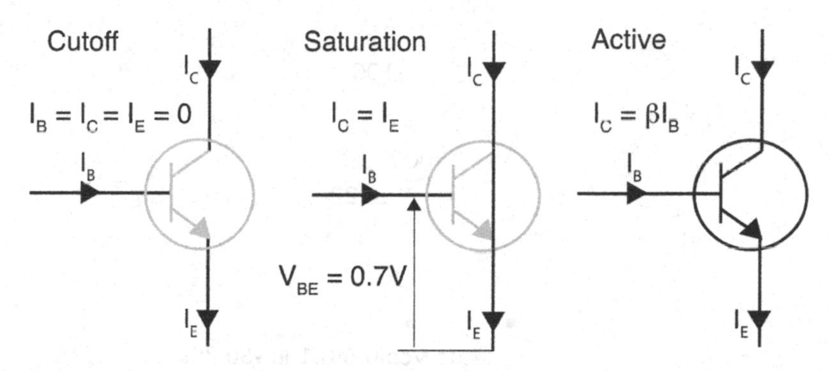

Figure 7.14 Transistor operational regions. ***The diagram shows the three defined regions of BJT operation. In the left-hand diagram, zero base current (I_B) means no current can flow between collector (I_C) and emitter (I_E) in the cutoff region. In the middle diagram, when a current begins to flow into the base and the base emitter potential (V_{BE}) reaches 0.7V then current flows from collector to emitter – the transistor becomes saturated with full current ($I_C = I_E$). In the right-hand diagram, once the base emitter potential is above 0.7V the transistor amplifies the base current ($I_C = \beta\ I_B$) – this is the active region.***

The diagrams show the three main operating regions of a BJT transistor (an additional reverse mode is not commonly used and so is omitted from this discussion). For the first two regions, the **transistor effectively acts as a digital switch**. When the input base current is zero, the transistor is in cutoff mode and no current flows. Chapters 4 and 5 showed how to use C code to program the Arduino for digital control. There are millions of transistors inside the ATMega328P chip that executes the programming code, where each transistor will equate cutoff mode to a logic 0 when working with digital signals. When current begins to flow into the base (I_B) to create a potential difference between base then emitter electrons start to cross the depletion layers between collector and emitter (in conventional current flow). Once this potential difference reaches 0.7V (like a diode) the transistor moves into active mode where current will flow freely from collector to emitter ($I_C = I_E$) – this equates to a logic 1 in a digital signal. These two operating regions are the fundamental building blocks of digital electronic circuits, where a transistor can be held at either logic 0 or 1 and then combined with other transistors to create more complex logical processes.

Although chapters 4 and 5 discussed the basics of digital signals when learning how to program the Arduino, the main focus of this book is the controlled processing of an analogue audio signal. In this instance, a **transistor can be used to amplify an input signal when in the active region** – where $I_C = \beta I_B$. To look at this active region in more detail, an LTspice circuit can be used to simulate a range of base current inputs to see how they change the collector current at output.

7.3.1 Worked example – simulating BJT characteristic curves using LTspice

This example will simulate a common emitter BJT amplifier circuit based on the LTspice schematic in Figure 7.15.

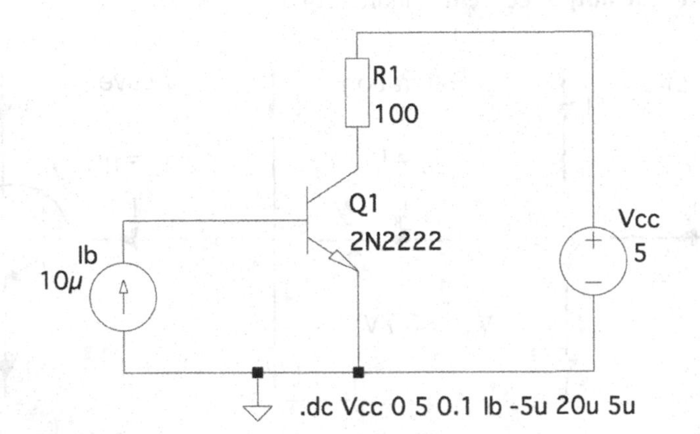

Figure 7.15 Common emitter amplifier circuit. ***In the diagram, a current source is used in a common emitter BJT amplifier circuit to step through a range of base current inputs. These inputs create a range of collector current outputs, which can be used to plot the characteristic curves shown in Figure 7.13.***

To build this circuit in LTspice, use component values of R1 = 100Ω, Transistor Q1 = 2N2222, Current Source I_B = 10μA and Voltage Source V_{CC} = 5V. Create a new circuit in LTspice and name it chp7_example1. The process for creating the schematic is as follows:

1. Lay out the components as shown in Figure 7.15: Current Source (on left), Voltage Source (on right), Transistor (middle), Resistor (in line with collector terminal) and the **GND node** (bottom). For the transistor, choose <npn – Bipolar NPN Transistor> from the component list.
2. Connect wires to form a series loop between Ib, Q1 base (NPN Transistor), Q1 emitter and GND.
3. Connect wires to form a series loop between V_{CC}, R1, Q1 collector (NPN Transistor), Q1 emitter and GND.
4. Configure the Current Source to be a DC value of 10μA. Configure the Voltage Source to be a DC value of 5V. Set the Resistor R1 = 100Ω.
5. **Name the Current Source (Ib) and the Voltage Source (Vcc)** so they can be accessed by the Spice Directive.
6. Add a Spice Directive to perform a **nested DC step analysis**, where VCC is incremented from 0 to 5V (in steps of 0.1V) every time Ib is incremented from −5μA to 20μA (in steps of 5μA). The analysis command is **.dc Vcc 0 5 0.1 Ib -5u 20u 5u** – *this will be discussed below.*

The LTspice circuit in Figure 7.15 shows a Spice Directive command that performs a nested DC step analysis. Nested analysis is similar to computer programming where a for loop can run inside another for loop, where each iteration of the outer loop will also iterate the entire inner loop from beginning to end (Figure 7.16):

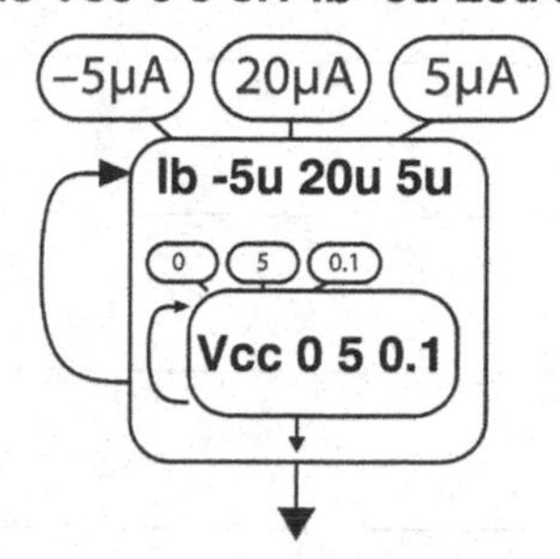

Figure 7.16 Nested step analysis in LTspice. ***The diagram shows how LTspice will process a Spice directive that contains a nested DC analysis loop. the outer analysis loop increases the base current (Ib) value from −5μA to 20μA in 5μA increments. Each time the base current increments, the inner loop increases the collector voltage (Vcc) from 0 to 5V in 0.1V steps for that base current. This allows a series of output simulation curves to be plotted that show different collector currents over a collector voltage range.***

The Spice directive listed above performs a DC analysis of the circuit, allowing the input current (Ib) and transistor supply (collector voltage Vcc) to be stepped through a range of values. In so doing, the resulting **current output at the collector** terminal of the BJT can be measured during LTspice simulation to show the characteristic output curves of a BJT (Figure 7.17).

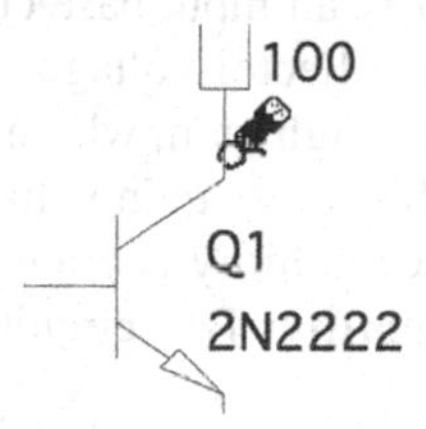

Figure 7.17 Measuring BJT characteristic output curves. ***In the diagram, running the simulation allows the current to be probed at the collector junction of the BJT in the circuit. To access the current probe, move the default voltage probe over the wire until it changes to the icon shown above; now current will be measured instead. The accompanying Spice directive allows both base (Ib) and collector (Ic) current for a range of supply voltages (Vcc) to be measured, which will generate the characteristic output curves for a BJT shown in Figure 7.13.***

The simulation output of the current probe should produce a graph similar to the one in Figure 7.18).

The graph in the diagram shows a series of collector currents, where $I_C = \beta I_B$. The current source sweep ranged from −5μA to 20μA, so the value of the highest output curve in the active region (Ic = 4mA) shows the collector current for a base current of Ib = 20μA. These values can be used with equation (7.5), to derive the common emitter gain for this circuit:

$$\textit{Base Current}, I_B = \mathbf{20\mu A}, \ \textit{Collector Current}, I_C = \mathbf{4mA}$$

$$\textit{Common Emitter Current Gain}, \beta = \frac{I_C}{I_B} = \frac{0.004}{0.00002} = \mathbf{200}$$

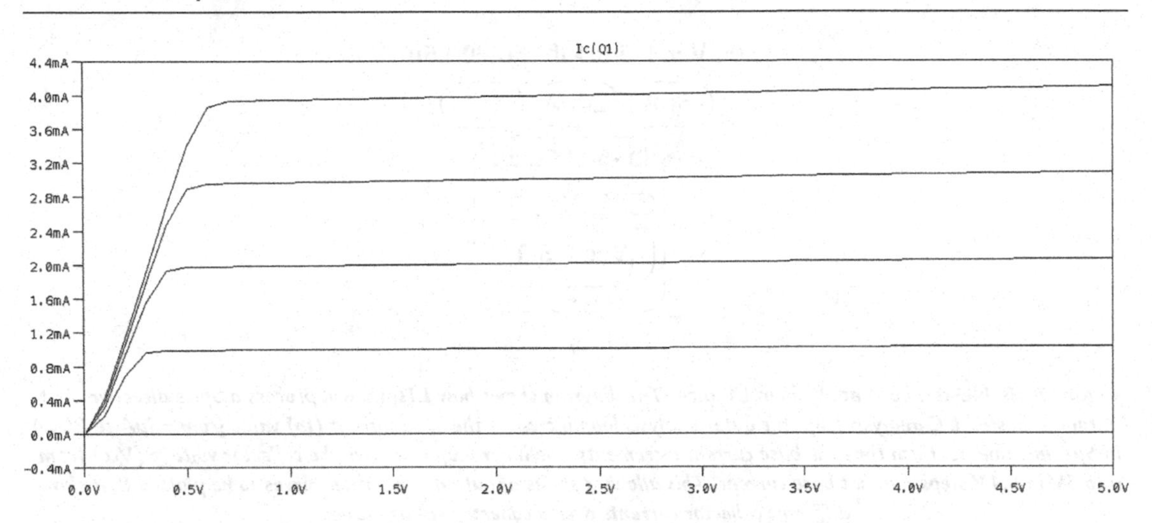

Figure 7.18 Simulation graph of characteristic output curves. ***The y axis shows a series of collector currents (Ic,) for a range of collector voltages (Vcc) along the x-axis. Each curve is produced by a specific base current ranging from −5μA to 20μA, where the −5μA is used to show the cutoff region of the BJT. The curves also show a gradient from 0V to ~0.7V for saturation, after which the active region of the BJT allows amplification to occur where $I_C = \beta I_B$.***

The BJT circuit in this example will amplify an input base current by a factor of 200 – the characteristic output curves show that the collector voltage makes no difference to this value within the active region. BJT devices can produce high gain, where practical common emitter gain values can range from 100 to 300. Having said this, these gain values can also vary widely even amongst the same type of transistor and so amplifier stability is often an issue. For this reason, additional steps are needed to configure a **common emitter** BJT circuit for use as a **voltage amplifier** for small input signals.

This use of a transistor may initially seem strange given that the BJT is primarily a current amplifier – why does Figure 7.12 define input and output signals as voltages? To understand this, Ohm's Law can be used to remember that any resistor will create a potential difference across it by resisting the flow of current, and because this resistance is fixed then any **increase in current will also increase the voltage** drop across the resistor. This means a common emitter circuit can be used within the active region to create a voltage amplifier, where changes to the base-emitter voltage (V_{be}) will vary the base current (I_b) and thus create a larger collector current (I_c). As the output current increases, so will the output voltage across the emitter (V_{ce}) – for small input signals the BJT will act as a voltage amplifier. The problem with a BJT voltage amplifier occurs when working with AC input signals like audio input from a microphone, where the **alternating current** of the input signal means the transistor will drop **out of the active region** for half of the input sine wave (Figure 7.19).

The diagram shows how the transistor only allows current to flow in one direction, so when the base current changes polarity in the bottom half of the sine wave cycle (where $I_b < 0$) the base becomes more negative than the emitter and thus no electrons will be attracted from the emitter towards it. If no electrons flow towards the base then none of them can cross the depletion layers to stimulate a collector current – **the transistor will not allow current to flow**. In a similar manner to diode

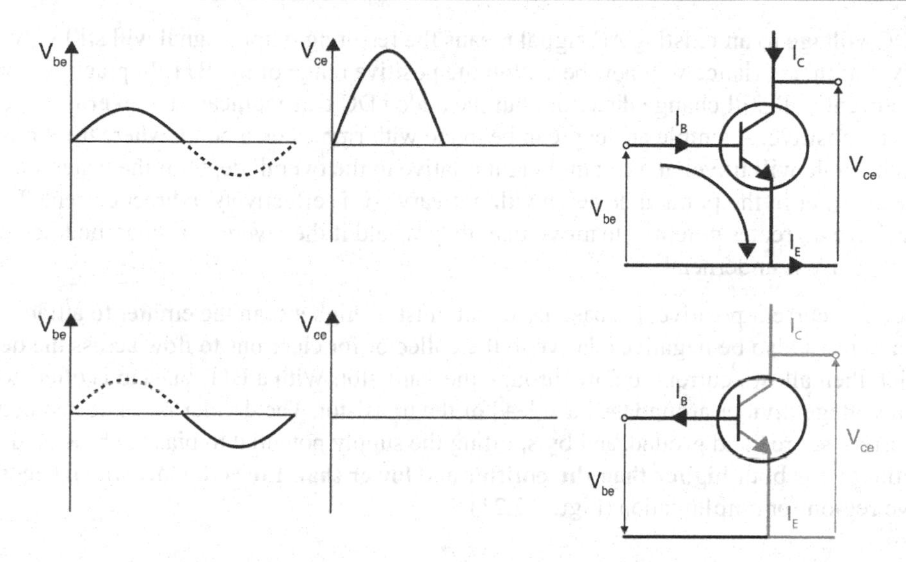

Figure 7.19 AC input signal BJT amplification. ***The top diagram shows the positive half of an input AC signal, where Ib flows into the transistor (conventional current) and the transistor is on – Vce is proportional to Vbe. In the negative half of the cycle, the input current reverses so Ib now flows out from the transistor – thus no current will flow from collector to emitter and Vce is zero.***

rectification in the previous section (Figure 7.8), only half of an input sine wave will be amplified by the transistor and so it will not work effectively with AC input signals.

To avoid this, the base-emitter voltage (V_{be}) of the transistor can be **biased** to always keep it within the active amplification region. This prevents the BJT from either switching off when the base current becomes negative during the second half of the sine wave cycle or dropping into the saturation region where the transistor is not linear. Biasing involves adding a DC component to the input signal to raise the overall level of the entire signal into the active amplification region (Figure 7.20).

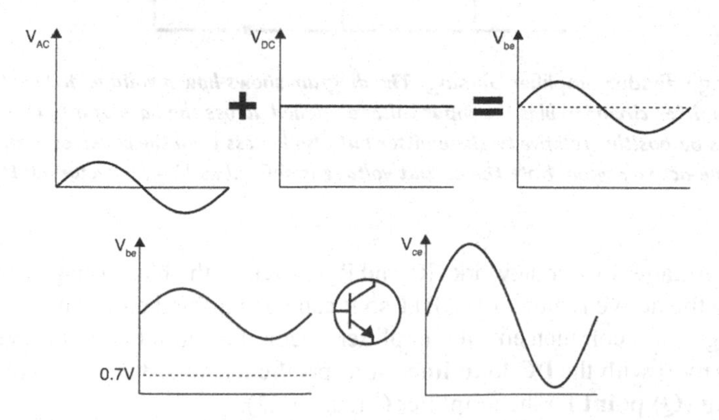

Figure 7.20 Biasing AC input signals for BJT amplification. ***The top diagram shows how an AC voltage signal can be added to a DC signal to raise the overall level of the alternating signal at output. The bottom diagram shows how this allows the voltage signal connected to the input of a transistor to be amplified fully by a common emitter BJT, as the input voltage (and hence current) never drops below 0.7V and so the BJT remains in the active region of amplification.***

Adding a DC voltage to an existing AC signal means the resulting output signal will still vary sinusoidally, but this variance will now be within the positive range of the BJT. In practical terms, the AC signal current will still change direction, but the added DC current means the **overall base current** will always be positive. A simple analogy can be made with ripples on a pond, where the sinusoidal wave of each ripple will travel at a certain height relative to the overall depth of the water, but the larger body of water in the pond underneath will not vary – it is effectively a direct current. Thus, the ripples now have a greater potential to move than they would if they were in a small puddle because of the larger DC current underneath.

To keep the base current positive, the base potential must be higher than the emitter to attract electrons towards it and must also be negative relative to the collector for electrons to flow across the depletion layers, which then allows current to flow through the transistor. With a BJT, biasing is often achieved by adding a voltage divider around the base lead of the transistor. The divider is connected between the collector current source and ground, and by splitting the supply potential to bias the base lead creates a base potential that is both **higher than the emitter** and **lower than the collector**, which keeps the BJT in the active region for amplification (Figure 7.21).

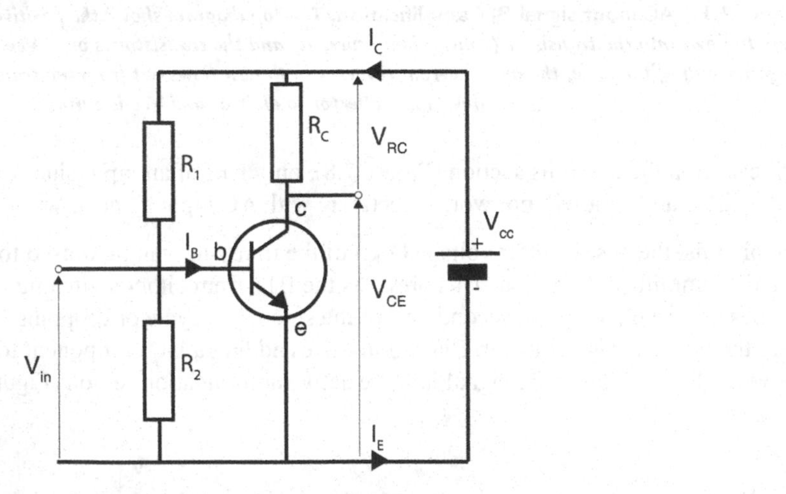

Figure 7.21 Voltage divider amplifier biasing. ***The diagram shows how a voltage divider (R_1 and R_2) can be added to a common emitter circuit to bias the input voltage applied across the base of a BJT. In so doing, the overall input signal will always be positive relative to the emitter but also be less than the collector – thus the BJT will always operate in the active region. Note the output voltage is defined as V_{CE} (collector-emitter voltage).***

In the diagram, the voltage divider network (R_1 and R_2) provides the bias voltage needed to lift the input signal into the active region of the BJT so it can be amplified correctly. To determine the optimum bias voltage for a common emitter amplifier circuit, the amplification curves for a BJT (see Figure 7.13) can be used with the **DC load line** for a specific load resistance (R_C) to determine the **quiescent operating (Q) point** for the amplifier (Figure 7.22).

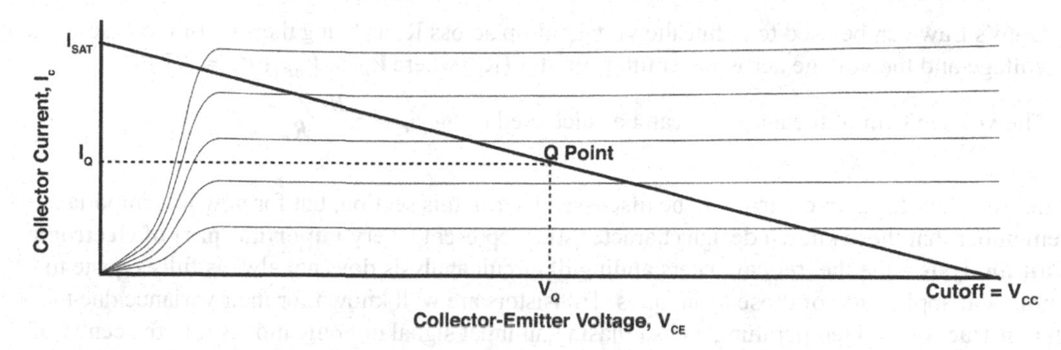

Figure 7.22 Determining the Q point to bias a common emitter amplifier. ***In the diagram, the Figure 7.13 graph of collector currents for specific input base currents is shown. The diagonal DC load line for a load resistance of 100Ω is overlaid on these curves to show where the Q point of the circuit lies (IQ). To determine the optimum bias point, a horizontal line is drawn at half of the saturation current (ISAT) to intersect with the DC load line, indicating the collector-emitter voltage (VQ) that will be needed.***

For a common emitter amplifier, the DC load line is defined for a specific load resistance of $R_C = 100\Omega$, and this is calculated as a line that **bisects the edge of the saturation and cutoff regions** of the BJT:

- **For cutoff**, when the transistor is fully off then $I_C = 0$ and thus the full potential of the supply voltage (V_{CC}) is presented across the emitter. This means that collector-emitter voltage $V_{CE} = V_{CC}$ and so the right-hand point on the DC load line must equal the supply voltage.
- **For saturation**, the region is defined from the point where the collector-emitter voltage $V_{CE} = 0$, although in practice a 0.7V potential difference is required to stimulate the flow of current across the depletion layers. If this collector-emitter voltage is 0, then the entire supply potential is presented across the collector resistor (R_C) and so $I_C = {}^{V_{CC}}\!/_{R_C}$.

With these two points calculated, the DC load line can then be drawn between them over the output curves to determine the optimum base and collector currents needed to maintain the transistor in the active region. To do this, the quiescent operating point (Q point) of the circuit must be found, which requires the use of several known BJT circuit design characteristics that have been derived by many years of collective practitioner experience:

- The Q point occurs when the collector current is half of the saturation current, $\boldsymbol{I_Q} = {}^{\boldsymbol{I_{SAT}}}\!/_{2}$ ($I_{SAT} = {}^{V_{CC}}\!/_{R_C}$).
- With I_C known, the base current (I_B) of the amplifier can be found using $\boldsymbol{\beta} = {}^{I_C}\!/_{\boldsymbol{I_B}}$ (equation 7.4) if a suitable value of β is assumed (the previous example gave β = 200).
- The value of the bias resistors in the voltage divider can be calculated assuming that the current flowing through R_2 should be at least 10 times the base current $(\boldsymbol{I_{R2} = 10I_B})$.
- Using KCL the current through R_1 should be 11 times the base current to produce two currents of $10I_B + I_B$ at the junction ($\boldsymbol{I_{R1} = 11I_B}$).
- The voltage drop across the emitter resistor should be around 10% of the supply voltage ($\boldsymbol{V_E} = {}^{\boldsymbol{V_{CC}}}\!/_{10}$).

- Ohm's Law can be used to define the voltage drop across R_2 as being the sum of the base-emitter voltage and the voltage across an emitter resistor (R_E) where $V_{R2} = V_{BE} + V_E = 0.7 + V_E$.
- The voltage gain of the amplifier can be calculated using $A_V = -{}^{R_C}\!/_{R_E}$.

Emitter resistors for gain control will be discussed later in this section, but for now it is important to remember that these known design characteristics represent a very **important part of electronic circuit analysis** – the theoretical understanding of circuit analysis does not always fully equate to the practical application of those techniques. Transistors are well known for their variance due to both construction and temperature, and so biasing an input signal not only moves it to the centre of the active amplification region but also helps to stabilize the gain characteristic of the amplifier. This is partly because there is now a fixed DC voltage being applied to the base of the BJT, and so any changes in the signal input will be proportionally much smaller than the overall biased signal – a 100mV AC input signal with a 1V bias signal represents a proportional variance of 5% (as the AC component will vary both upwards and downwards). As a result, there will be less change in the transistor's overall behaviour because it is amplifying a (primarily) fixed voltage – the AC variations represent a much smaller percentage of the signal. Although this is a fairly simplistic explanation of signal biasing for stability, it is important to note that biasing is also used to ensure the stable operation of the BJT in many amplification circuits.

Biasing an input signal is very common in amplification, where a signal must be moved into a valid input range to allow it to be amplified properly. Having said this, simply adding a DC voltage to an AC input signal is not sufficient and further steps must be taken to build a stable circuit. The first issue to be addressed is the additional current now supplied by the divider to the base of the BJT, which effectively 'swamps' the small current of the AC input signal. It is important to think through this process carefully, as previous usage of Kirchoff's Voltage Law (KVL) can create a mindset where current is perceived to flow in a single path that suits the analysis being performed. In practice, electrons will always flow down the path of least resistance, and so a larger current running through the voltage divider (which is taken from the larger collector supply) will flow towards the voltage input as the input current is much smaller at that point. To avoid this, the input to the BJT must be **AC coupled** to prevent the larger DC current from flowing into it (Figure 7.23).

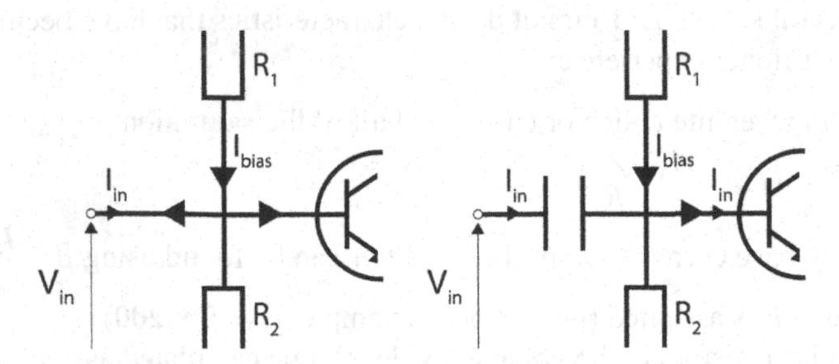

Figure 7.23 AC coupling of a BJT input signal. ***In the left-hand diagram, the larger current being provided by the bias resistors will flow in both directions out of the junction and flood the input signal unless it is prevented from doing so. For this reason, a series capacitor is added to the signal path in the right-hand diagram to prevent the flow of DC into the input (as capacitors block DC signals). In this way, the AC signal input is now coupled to the BJT input and can flow alongside the DC bias current.***

Having moved the input signal into a higher voltage range for amplification, this signal must also now be reduced at output to remove the DC component (which will also have been amplified). Once again, a coupling capacitor must be added at the output of the BJT to remove the DC bias signal that was added prior to amplification (Figure 7.24).

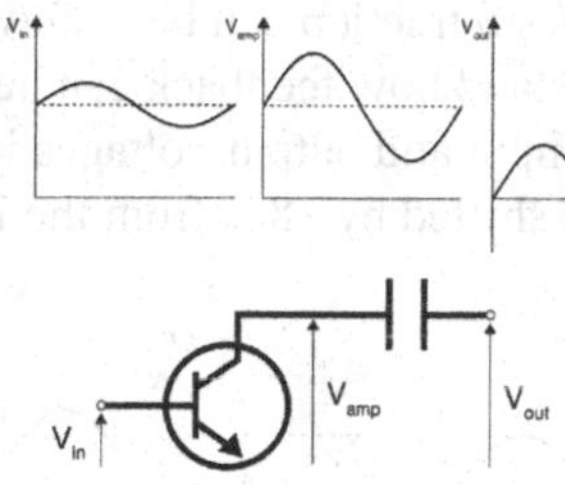

Figure 7.24 AC coupling of a BJT output signal. ***In the diagram, a coupling capacitor is added to the emitter output of a BJT amplifier. The biased input signal is amplified, but the DC component is then removed by the capacitor after amplification to retain only the amplified AC signal at the same relative potential as the input signal.***

Adding a coupling capacitor at the output of a BJT amplifier effectively 'de-biases' the signal to remove the DC component added prior to amplification. This allows the system to produce an amplified output signal which is referenced to the same level as the input (prior to it being moved by the DC bias). In this way, adding coupling capacitors at both common emitter input and output allows the input signal to be biased for amplification without the bias signal having any impact on the final output. Returning to the earlier analogy of ripples on a pond, the entire volume of water in the pond is not of interest – only the information in the ripple moving across the top of it.

Having dealt with the bias signal, the overall gain of the amplifier must also now be addressed. Controlling the gain of a BJT is important as the initial gain value (the previous example circuit gives $\beta = 200$) is not only very large, but also unstable and prone to variance due to construction variance and particularly temperature (which impacts on collector current). To avoid this, the gain of the transistor can be reduced by adding **negative feedback** to the circuit. Recall from chapter 2 (section 2.1) that feedback can be used to control a system, where the output of the system is used to change the input (Figure 7.25).

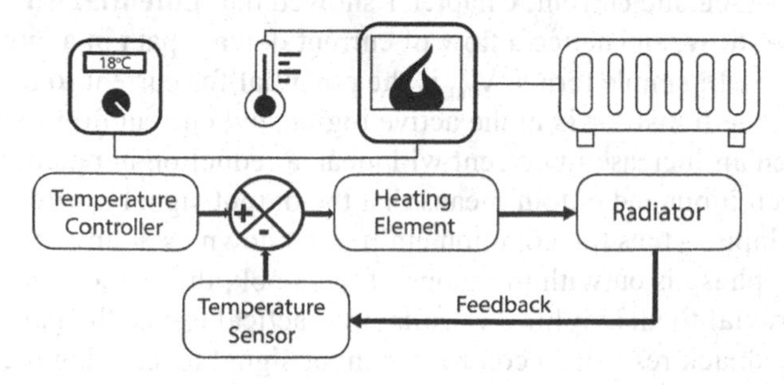

Figure 7.25 A closed-loop feedback system – taken from chapter 2, Figure 2.4. ***The diagram shows how a heating element is controlled by a combination of inputs. The temperature controller is used to set the heating element to a specific temperature, but the temperature sensor is used to feed back a signal to the heating element based on the output of the radiator. In this way, the temperature sensor is effectively used to reduce the input to the system based on the output the system produces.***

The diagram illustrates the use of feedback within a control system, where the temperature sensor will trigger when the radiator output reaches a certain temperature. The system diagram shows how a summing junction is used to combine the two inputs (controller and sensor) to the heating element, where the input from the temperature sensor is effectively subtracted from the input of the temperature controller. This **subtraction is called negative feedback** (as opposed to positive feedback, where the inputs are added), and in amplification this subtraction can be achieved by summing the input and output signals within the circuit. To understand how feedback can be used in a BJT amplifier circuit, the relationship between the phase of the input and output voltages in a common emitter circuit must be considered, where the output voltage is shifted by 180° from the input (Figure 7.26).

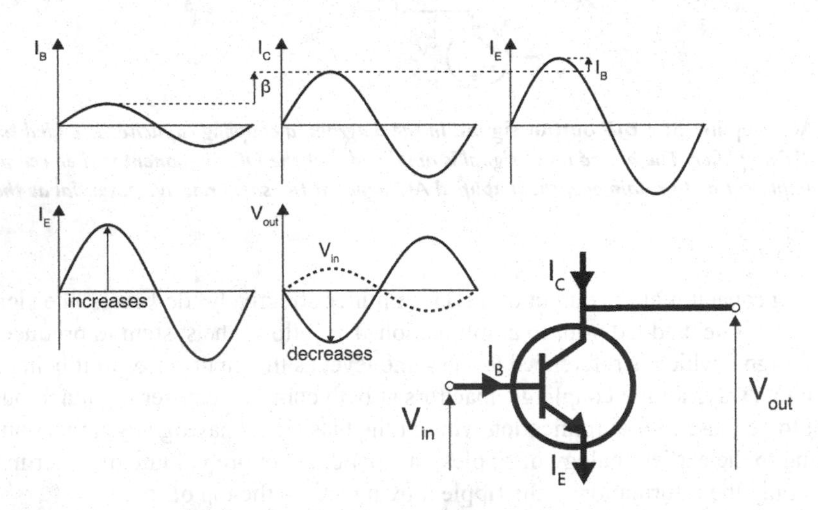

Figure 7.26 BJT output phase shift. ***The top half of the diagram shows how a BJT amplifies base current, which is increased (by a factor of β) at the collector, and hence the emitter current also increases ($I_E = I_C + I_B$). The bottom half the of the diagram shows that as the emitter current increases, so the potential across the emitter (V_{out}) must decrease, and hence a 180° lag is created between voltage input (dashed line) and output signals for a BJT. For this reason, a common emitter circuit is known as an inverting amplifier, where the output signal is effectively the inverse of the input.***

In the diagram, the key to understanding the phase shift between input and output voltage is the relationship between voltage and current. Chapter 1 showed that **potential difference is the potential for current to flow**, and hence a flow of current down a path in a circuit means that this potential no longer exists. In simple terms, V_{out} is the potential for current to flow from collector to emitter, and so when the transistor is in the active region and current flows through the emitter (where $I_E = I_C + I_B$) then an increase in current will mean a reduction in potential difference. The 180° phase shift between input and output means that the output signal is effectively inverted relative to the original input – thus the common emitter is known as an inverting amplifier. Although detailed analysis of AC phase is outwith the scope of this book, the concept of using a phase shift as negative feedback is crucial to all amplifier circuits. In practical terms, the gain of a transistor can be reduced by adding a feedback resistor to control the input signal level being presented at the base of the BJT (Figure 7.27).

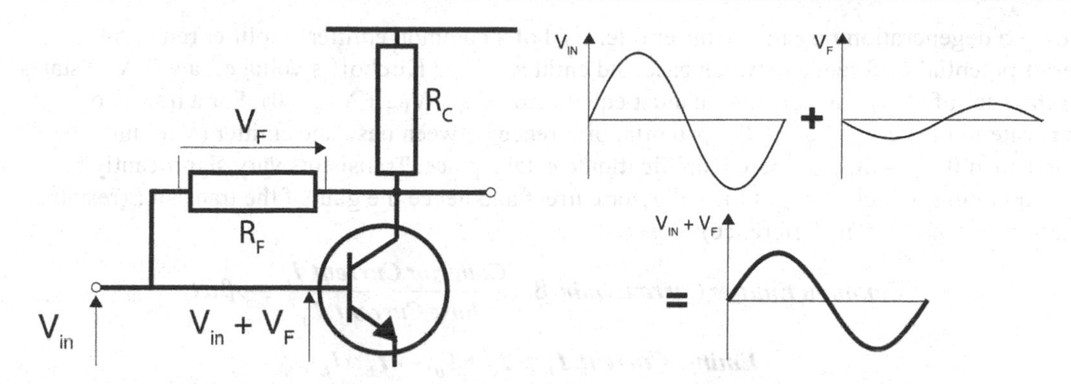

Figure 7.27 BJT amplifier feedback resistor. ***The diagram shows how the collector output of the BJT can be used to provide feedback to the input signal applied to the base junction. As this collector signal (V_F) is out of phase with the input (V_{in}), adding them together (V_{in} + V_F) effectively subtracts the collector signal from the input. The feedback resistor (R_F) forms a voltage divider with the collector resistor (R_C), which reduces the amount of output signal being summed with the input (as it would otherwise be larger than the input by a factor of β).***

The diagram in Figure 7.26 presents a visual overview of the use of feedback resistance, where the aim is to define the concept of adding a negative signal to reduce a positive one (rather than deriving the precise relationships involved). Negative feedback is also known as degeneration (as opposed to positive feedback, which is regeneration) and this idea of negative feedback is used throughout amplification to provide stability and more replicable circuit results, particularly when working with components that can vary as much as individual BJTs. Negative feedback will be discussed again in the following section on operational amplifiers, but for now it is important to note another much more common form of feedback resistor used with common emitter circuits. Although more complex to understand than simply adding an inverted waveform to reduce input, the addition of a degeneration resistor across the emitter of a common emitter BJT amplifier is often used to reduce the potential difference between the emitter and base of the BJT (Figure 7.28).

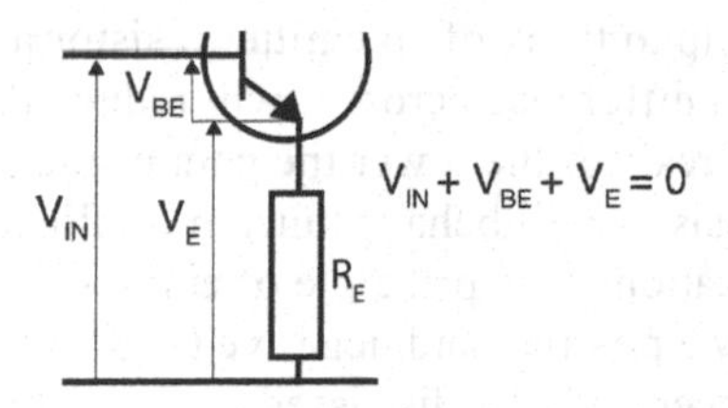

Figure 7.28 Adding a degeneration resistor to a BJT common emitter amplifier. ***The diagram shows how a resistor can be added to the emitter lead of a common emitter amplifier. In so doing, the potential difference between base and emitter is now reduced because the emitter resistor adds an extra voltage drop (V_{RE}). This reduction in potential between base and emitter reduces the gain of the BJT, but also stabilizes the overall amplifier circuit.***

Adding a degeneration resistor on the emitter lead of a common emitter amplifier reduces the overall potential difference between base and emitter, where Kirchoff's Voltage Law (KVL) states that the sum of all voltages in a loop must equal zero ($V_{IN} + V_{BE} + V_{RE} = 0$). For a transistor to operate in the active region, the potential difference between base and emitter (V_{BE}) must be greater than 0.7V – this is where amplification can take place. Transistors vary significantly based on temperature, which changes the collector current and hence the gain of the transistor (restating equations 7.4 and 7.5 for reference):

$$Common\ Emitter\ Current\ Gain, \beta = \frac{Collector\ Current\ I_C}{Base\ Current, I_B}, \rightarrow \beta \alpha I_C$$

$$Emitter\ Current, I_E = I_C + I_B, \rightarrow I_E \alpha I_C$$

$$KVL\ for\ BJT\ base\ loop, V_{IN} + V_{BE} + V_E = 0, \rightarrow V_E \alpha \frac{1}{V_{BE}}$$

The first equation shows how an increase in the collector current (due to temperature for example) effectively increases the gain of the circuit (the symbol ∝ means **proportional to**). The second equation shows that when the collector current increases, the emitter current will also increase, so by adding an emitter resistor (R_E), any increase in the emitter current will also increase the voltage across that emitter resistor. The third equation uses KVL to state that all voltages around the base loop of the BJT must sum to zero, and so as the emitter voltage increases the base-emitter voltage (V_{BE}) must decrease (because the input voltage does not change). The effect of increasing the emitter voltage means the base-emitter voltage reduces, and thus so must the current at the base of the BJT (I_B) – and hence the overall gain of the transistor is reduced. This is an important principle to remember:

Increasing the emitter potential reduces the gain of a BJT.

This type of degeneration feedback can be a little more difficult to understand, as unlike the collector resistor approach in Figure 7.27 **no actual signal is fed back** into the input – only the level of gain is reduced. It can help to think of an emitter resistor as a gain control for the BJT, where an increase in the potential difference across it will reduce the current flowing through the base – thus the larger the emitter resistor the lower the gain of the amplifier. Although the gain of the BJT is now lower, the transistor will behave much more linearly as a result of being held within a known range of amplification. This principle of emitter resistance is also used in a circuit known as a long tail pair to allow a positive and negative (180° out of phase) version of the signal to be summed together – this concept will be discussed again in the next section on differential amplification.

With all of the elements in a common emitter amplifier introduced, a final practical circuit can be designed (Figure 7.29).

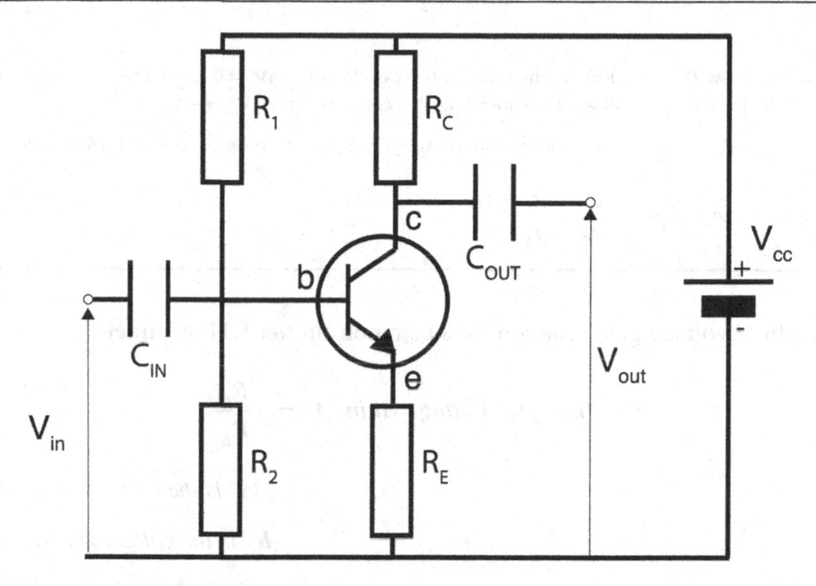

Figure 7.29 A common emitter BJT amplifier circuit. ***The diagram shows a common emitter amplifier with input biasing (R1 and R2), AC coupling (CIN and COUT) and a degeneration resistor (RE) to stabilize gain. This is a common transistor amplifier design pattern, and variations on this design are widely used in many audio electronics circuits.***

The diagram shows how an initial AC coupling capacitor is connected in series with the input (V_{in}) to prevent the current from the biasing network (R_1 and R_2) from flooding the input with a much larger current. The biasing network itself is used to add a DC bias signal to the initial AC input to ensure that the input signal is always within the active region of the BJT, where the quiescent operating point (Q point) can be determined from the characteristic BJT output curves that we simulated with LTspice in the previous worked example. A second coupling capacitor is connected at the collector output, to remove the DC bias signal that has been applied and return the output signal to the same voltage reference level as the input signal. The degeneration emitter resistor R_E is used to reduce the input signal level and thus reduce the gain of the amplifier, where the combination of the collector resistor R_C and the emitter resistor R_E effectively sets the voltage gain for a common emitter BJT amplifier.

Defining BJT voltage gain

Recalling from equation 7.1: ***Amplifier Voltage Gain,*** $A_V = \frac{V_{out}}{V_{in}}$.

The key to defining the voltage gain relationship for a BJT is the approximation that collector (I_C) and emitter (I_E) current are effectively equal, as the base current (I_B) is very small by comparison: $I_C \approx I_E$.

Then the common emitter voltage output (**taken at the collector resistor**) can be defined in terms of the emitter current (I_E): $V_{OUT} = V_C = I_C \times R_C = I_E \times R_C$.

Ignoring the biasing network, using KVL for the base loop means $V_{IN} + V_{BE} + V_E = 0$ which can be rearranged to $V_{IN} + V_{BE} = -V_E$. If the base emitter voltage is removed (as it is constant), then $V_{IN} \approx -V_E$.

Knowing that $V_C = I_E \times R_C$ and $V_E = I_E \times R_E$ allows the voltage gain equation to be restated in these terms:

$$\text{Amplifier Voltage Gain, } A_V = \frac{V_{out}}{V_{in}} = -\frac{V_C}{V_E} = -\frac{I_E \times R_C}{I_E \times R_E}$$

This leads to the final voltage gain equation for a common emitter BJT amplifier:

$$\text{Amplifier Voltage Gain, } A_V = -\frac{R_C}{R_E} \tag{7.5}$$

A_V is the voltage gain of the amplifier

R_C is the collector resistor in ohms (Ω)

R_E is the emitter resistor in ohms (Ω)

The equation above shows how the gain of a common emitter BJT amplifier can be controlled by the ratio of the collector (R_C) and emitter (R_E) resistors, providing a stable gain characteristic that is much more resistant to fluctuations in temperature (and other variances in BJT fabrication). The common emitter amplifier circuit shown in Figure 7.29 above adds a number of extra resistors and capacitors onto the initial design shown in Figure 7.12. This is an important point for your initial learning of electronics, as **the 'extra' components added to many circuit designs are often crucial to their effective practical operation**. It is often the case that it takes less time to learn how a BJT works, but significantly longer to understand how biasing, AC coupling and degeneration feedback function (and why they are necessary).

This book aims to provide a simple introduction to audio electronics, and so a more thorough discussion of transistor amplification theory is not provided. Having said this, a significant amount of theoretical explanation has been circumvented to provide a brief overview of the workings of a BJT amplifier – this does not mean that it is not important to your further learning. For now, a second transistor amplifier example is provided that includes biasing, AC coupling and feedback resistor concepts.

7.3.2 Worked example – simulating a common emitter amplifier with LTspice

In this worked example, a common emitter amplifier will be simulated using LTspice. The amplifier circuit will use a bias network to keep the BJT within the active region and combine a collector resistor (to control the BJT current) with an emitter resistor to provide degeneration feedback and increase circuit stability. AC coupling capacitors will be used to pass a sine wave input signal through the amplifier and the amplifier output will be connected to a load resistor. The BJT supply will use values from the Arduino (5V, 40mA), though this circuit will only be simulated (not built).

Before the schematic is designed, the previously stated BJT circuit design characteristics can be used to derive the component values for this circuit:

First find $I_Q = \frac{I_{SAT}}{2}$ (knowing $I_{SAT} = \frac{V_{CC}}{R_C}$) where the saturation current can be defined by the Arduino (maximum pin output 40mA for 5V supply) so $I_Q = \frac{40}{2} = 0.02 = \mathbf{20mA}$.

Assume β = 200, where $\beta = \frac{I_C}{I_B} \rightarrow I_B = \frac{I_C}{\beta} = \frac{0.02}{200} = 0.0001 = \mathbf{1\mu A}$.

The collector resistor must limit the current to this value, so rearranging Ohm's Law gives $\mathbf{R_C} = \frac{V_{CC}}{I_{SAT}} = \frac{5}{40} = 125\Omega$.

Now, emitter voltage should be around 10% of supply, so $V_E = \frac{V_{CC}}{10} = \frac{5}{10} = \mathbf{0.5V}$.

As base current is small assume $I_E \approx I_C, \mathbf{R_E} = \frac{V_E}{I_E} = \frac{0.5}{0.02} = 25\Omega$.

Now the voltage divider must bias the input signal, where $\mathbf{I_{R1}} = 11I_B = 11 \times 0.0001 = 0.0011 = \mathbf{1.1mA}$ and $\mathbf{I_{R2}} = 10I_B = 10 \times 0.0001 = \mathbf{1mA}$.

Knowing that $\mathbf{V_{BE} = 0.7V}$ to turn on the transistor (like a diode), $\mathbf{V_{R2}} = V_{BE} + V_E = 0.7 + V_E = 0.7 + 0.5 = \mathbf{1.2V}$.

From here, we can find $\mathbf{R_2} = \frac{V_{R2}}{I_{R2}} = \frac{1.2}{0.001} = 1200\Omega$.

As a voltage divider will split the supply ($V_{CC} = 5V$) between the resistors, $\mathbf{V_{R1}} = V_{CC} - V_{R2} = 5 - 1.2 = \mathbf{3.8V}$.

Now we can find $\mathbf{R_1} = \frac{V_{R1}}{I_{R1}} = \frac{3.8}{0.0011} = 3454\Omega$.

Restating all results for clarity when designing the schematic:

$$\mathbf{R_2 = 3454\Omega,\ R_2 = 1200\Omega,\ R_C = 125\Omega,\ R_E = 25\Omega}$$

The AC coupling capacitors are usually set to specific values based on the frequency of the signals involved, where the capacitor effectively forms a filter with the input (or output) resistance of the circuit. This example is already quite complex, and as filter circuits will be discussed in more detail in the following chapter capacitor values of 15µF (based around 100Hz) will be used to provide AC coupling on both the input and output of the circuit. A 100mV sine wave input will be used to measure the output signal from this circuit through a 10kΩ load resistance, using the schematic in Figure 7.30.

To build this circuit in LTspice, use component values of R1 = 3454Ω, R2 = 1200Ω, RC = 125Ω, RE = 25Ω, RL = 10kΩ, Cin = 15µF, Cout = 15µF, Transistor Q1 = 2N2222, Voltage Input Source Vin = 100mV and Voltage Supply Source V_{CC} = 5V. Create a new circuit in LTspice and name it chp7_example2. The process for creating the schematic is as follows:

1. Lay out the components as shown in Figure 7.30. Start with the Voltage Input Source (Vin on left), Voltage Supply Source (Vcc on right), Transistor (middle) and the **GND node** (bottom). For the

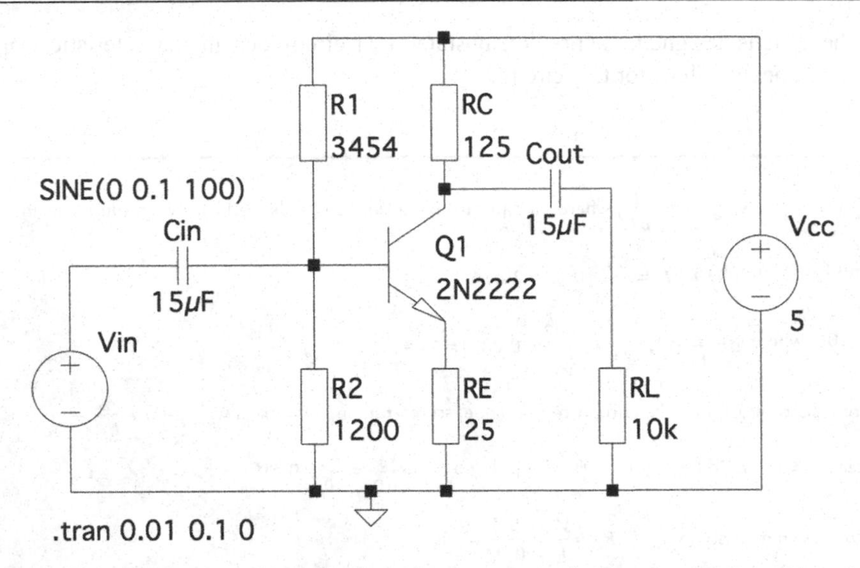

Figure 7.30 Common emitter amplifier circuit. ***In the diagram, a voltage source (Vin) provides a 100mV input sine wave of 100Hz to the amplifier circuit through an AC coupling capacitor (Cin). This signal is biased using the 5V BJT supply (Vcc) by a voltage divider network (R1, R2) to keep the input within the active region of amplification (based on the Q point). A collector resistor (RC) controls the current in the circuit, whilst an emitter resistor (RE) provides degeneration feedback to control the amplifier gain and improve the stability of the circuit. A second AC coupling capacitor (Cout) is connected to the common emitter output to debias the signal, which can then be measured across a load resistor (RL) of 10kΩ.***

transistor, choose <npn – Bipolar NPN Transistor> from the component list and then configure it for 2N2222.

2. Add the input AC coupling capacitor (Cin) in line with the base terminal (leave room for the bias network). Set its value to 15μF.
3. Add the bias resistors (R1 and R2) vertically on either side of the base terminal (on OSX, Ctrl R will rotate the component) after the AC coupling capacitor (Cin). Set their values to R1 = 3454Ω and R2 = 1200Ω.
4. Connect wires from the positive terminal of the Voltage Input Source (Vin) to Cin and the Q1 base (NPN Transistor).
5. Connect a wire from the negative terminal of the Voltage Input Source (Vin) to GND.
6. Connect wires from R1 to the Cin base wire, and from this wire to R2, then R2 to GND.
7. Add the collector resistor (RC) vertically above the collector terminal and connect wires between collector, RC and the positive terminal of the Voltage Supply Source (Vcc). Set its value to 125Ω.
8. Add the emitter resistor (RE) vertically below the emitter terminal and connect wires between emitter, RE and the negative terminal of the Voltage Supply Source (Vcc). Set its value to 25Ω.
9. Connect a wire from the negative terminal of the Voltage Supply Source (Vin) to GND (this should also connect to the emitter resistor).
10. Connect a wire from R1 to the positive terminal of the Voltage Supply Source (Vin) to provide power to the bias network.
11. Add the output AC coupling capacitor (Cout) between the collector terminal and RC. Set its value to 15μF.

12. Add the load resistor (RL) vertically between the right leg of Cout and GND. Set its value to 10kΩ.
13. Connect wires between collector terminal and Cout, from Cout to the load resistor (RL) and from RL to GND.
14. Connect wires to form a series loop between V_{CC}, R1, Q1 collector (NPN Transistor), Q1 emitter and GND
15. Configure the Voltage Input Source to be a **sine wave of amplitude 0.1 (100mV) and frequency 100Hz** (DC offset 0).
16. Configure the Voltage Supply Source to be a DC value of 5V.
17. Add a Spice Directive to perform a **transient analysis** between 0 and 0.1 seconds, with a step of 0.01 (**.tran 0.01 0.1 0**) – this will show 10 cycles of a 100Hz sine wave.

If the circuit has been built correctly in LTspice, the simulation should run and a voltage probe can be used to measure both the input signal (take a reading before capacitor Cin) and the output signal (take a reading after capacitor Cout, but before the load resistor RL). This should produce output traces similar to those in Figure 7.31.

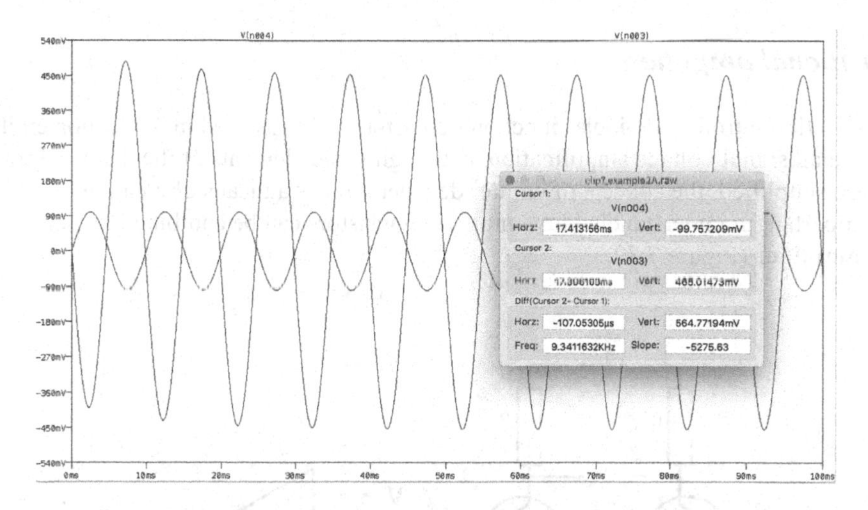

Figure 7.31 Simulation graph of a common emitter BJT amplifier circuit. ***The y axis shows both input and output voltage levels over a simulation time of 100 milliseconds (x-axis). The input is a 100mV 100Hz voltage signal, whilst the output is ~465mV 100Hz signal that is 180° out of phase with the input. This shows how the common emitter is an inverting amplifier circuit.***

The simulation graph shows both input and output voltage signals, where the output is now 180° out of phase with the input – this is an inverting amplifier. It is also interesting to note that the simulated voltage gain of the amplifier is lower than the value predicted when designing the circuit:

$$\textit{Predicted Amplifier Voltage Gain, } A_V^p = -\frac{R_C}{R_E} = -\frac{125}{25} = \mathbf{-5}$$

$$\textit{Simulated Amplifier Voltage Gain, } A_V^s = \frac{V_{out}}{V_{in}} = \frac{-465}{100} = \mathbf{-4.65}$$

There are several reasons for this discrepancy, such as the effect of the internal transistor resistance (often labelled R_e) on the circuit, the filtering performed by the input and output AC coupling capacitors and also the actual gain coefficient for the 2N2222 transistor (where $\beta \approx 259$) that gives a value of $I_B = 0.77\mu A$ (thus changing other values in the calculations performed in this example). These elements have been omitted to try and reduce the complexity of transistor theory, as the aim of this book is to introduce concepts and demonstrate how they combine within practical audio circuits rather than to provide a complete course on analogue electronics.

This book does not go into significant detail on transistor amplifier designs, as the operational amplifiers that are built from them (see next section) are arguably more stable and simpler to work with. For this reason, the common emitter circuit provided above aims to explain the typical components in transistor amplifier circuits as found in many audio effects pedals in an attempt to disambiguate the basic elements of an amplification circuit. There are many 'cookbook' circuits that can be copied and combined into more extensive audio circuits, and though it is recommended that a deeper understanding of transistor theory be pursued it is also noted that this is an area of analogue electronics that can take time to learn. In the next section, transistors will be combined to provide **differential amplification**, which is the key to understanding integrated circuits known as **operational amplifiers**.

7.4 Operational amplifiers

The previous section introduced bipolar junction transistors (BJT) as used in a common emitter amplifier for small signal voltage amplification. Although some elements of the theory of transistor operation were simplified, the concept of emitter degeneration as a means of creating negative feedback is important in understanding how multiple transistors can be combined to create a **differential amplifier** (Figure 7.32).

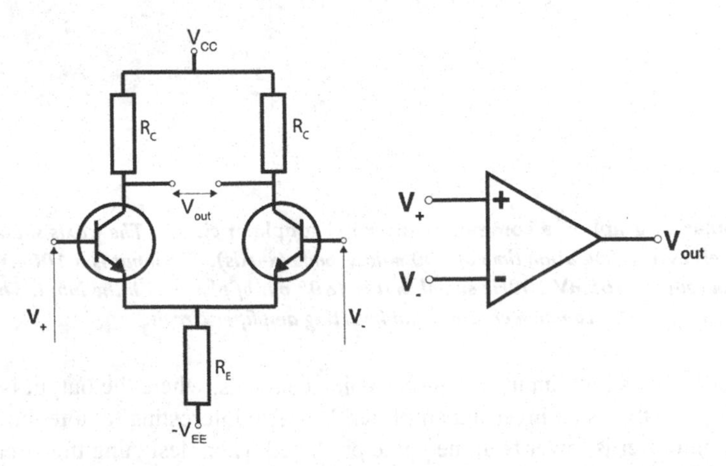

Figure 7.32 BJT differential amplifier and operational amplifier equivalent symbol. *In the left-hand diagram, two BJTs are connected together at the emitter junction (known as a long tail pair), where each half of the emitter acts as a voltage drop for the other. If both input signals are the same then degeneration caused by the emitter voltage from each BJT means that the gain of the other will be reduced and thus the amplifier will have zero output – known as common mode rejection. The right-hand diagram shows the equivalent operational amplifier symbol, where the circuit is simplified down to only the non-inverting (+) and inverting (−) inputs (and output) of the amplifier.*

The left-hand diagram in Figure 7.32 shows a long tail pair amplifier configuration, where two BJTs are connected together by a common emitter resistor and powered by a dual supply (V_{CC} and $-V_{EE}$). Setting aside the dual supply (which requires more advanced analysis), the concept of emitter degeneration feedback (see Figure 7.28) means that any increase in the voltage across one BJT emitter will decrease the gain of the other emitter – and vice versa. Both BJTs are effectively paired at the emitter (tail) where each one's output turns down the gain of the other, and in this way a long tail pair will **not amplify a signal that is common to both.** This principle is known as **common mode rejection**, where the long tail pair will reject any signal that is common to both inputs. The right-hand diagram shows the schematic symbol for an operational amplifier, which is used as a simplified representation of circuits such as the long tail pair. Analysis of more complex transistor circuits such as the long tail pair is outwith the scope of this book, and in some ways is not needed at an introductory level when such transistors are combined within a larger **integrated circuit** (IC). In practice, operational amplifiers may combine many more transistors to achieve specific input and output characteristics within a single IC (Figure 7.33).

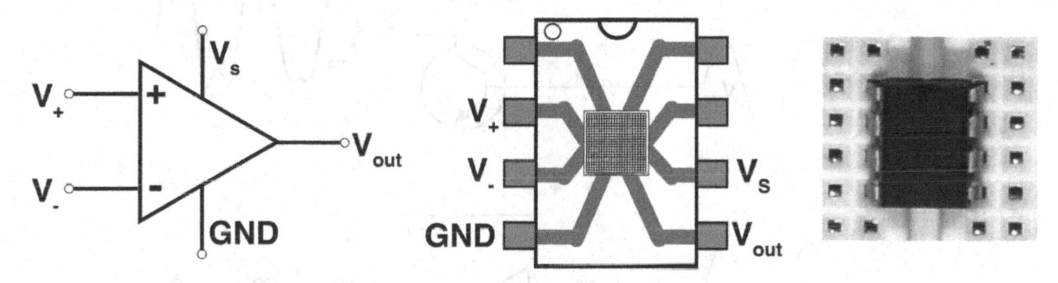

Figure 7.33 **An 8-pin integrated circuit.** ***The left-hand diagram shows the input and output connections on a schematic for a typical operational amplifier, with differential non-inverting (+) and inverting (−) inputs. The middle diagram shows how connections are made between pins outside the chip and a central die that combines many silicon components within a single integrated circuit. The right-hand image shows an LM386 operational amplifier mounted on a breadboard – note the small crescent indentation at the top of the chip that indicates its orientation (a small dot on the top left of the chip may sometimes be used instead).***

The connections to the schematic symbol shown in the left-hand diagram in Figure 7.33 are a generalized example of an operational amplifier, where a differential amplifier requires power connections for supply (V_S) and ground (in this case) alongside the input and output terminals. The plus and minus signs on the symbol indicate the connections for **non-inverting (+)** and **inverting (−)** amplifier inputs, which specify the resulting output phase of the input signal being amplified (inverting amplification shifts the signal by 180° at output). The GND connection indicates a single supply chip, where the input voltage is provided to one rail (positive) relative to the overall circuit ground. Many operational amplifiers are dual supply, where a positive and negative voltage (relative to a common ground) are provided. This is because an operational amplifier has **no internal ground reference**, so it is not directly connected to a ground terminal unless a single supply is used.

The 8-pin IC shown in the middle diagram shows how these connections would map to a typical chip, where the central die of the chip contains layers of silicon and copper that combine to create complex combinations of resistor and semiconductor components. The die is connected to the external pins of the IC, which allow it to be placed within larger electronic circuits. The right-hand image shows an LM386 audio amplifier as an example of an 8-pin IC – where the LM386 operational amplifier will be used in all projects for the remainder of this book. The middle diagram maps the main connections for the pins of an LM386, which are taken from the data sheet for that chip (this will be discussed in more detail in the chapter project).

The advantages of scale and flexibility provided by integrating large numbers of semiconductor components on a single IC chip has led to them becoming the fundamental building block of modern electronics, alongside their relative ease of use when compared within discrete component circuits. ICs come in many different form factors and configurations, depending on the task that they will be used for. In this book, a simple 8-pin amplifier like the LM386 allows audio amplification and filtering circuits to be built on breadboard without soldering (and with a minimal amount of electronic components). Although there are many uses for differential amplification, one of the most important for amplification (particularly in audio) is the use of **common mode rejection to reduce noise** within a circuit (Figure 7.34).

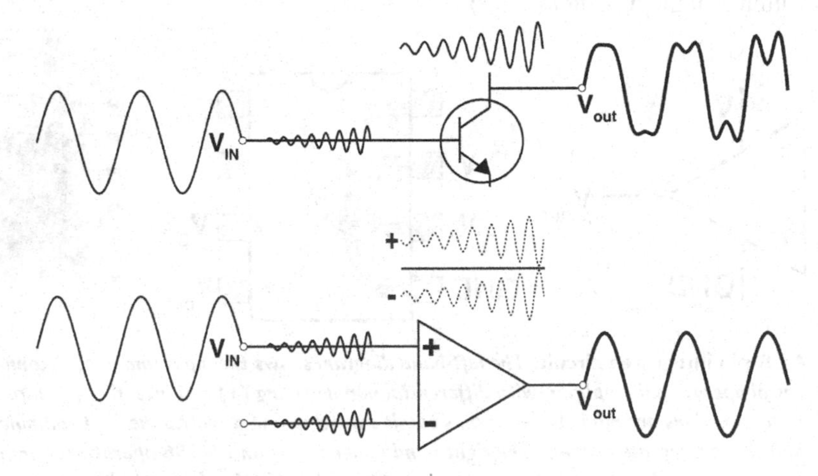

Figure 7.34 Common mode rejection noise reduction example. ***In the top diagram, an input signal is applied to a BJT as a common emitter amplifier, where noise in the circuit connections is also amplified at output. In the bottom diagram, a differential operational amplifier is used where circuit noise (theoretically common to both inputs) is now cancelled out due to common mode rejection – only the input signal is now amplified. As many sources of electrical noise may be present throughout a circuit, common mode rejection can be used to reduce (or even remove) some of them during amplification.***

The diagram shows how common mode rejection can be used to cancel out noise sources that are common to the entire amplification circuit. In the top example, the use of a BJT as a common emitter means that all sources (both signal and noise) will be amplified for output – any noise introduced into the input circuit by components, connections or the power supply will also be amplified and thus distort the output signal as a result. In the bottom example, a differential amplifier will only **amplify the difference** between the two inputs, and so any noise sources present in the input circuit will be common to both terminals. The non-inverting and inverting (out of phase by 180°) noise signals are effectively summed and **cancel each other out**, whilst the input signal (which is only connected to one of the input terminals) will be amplified as the difference between the two terminal inputs.

In practice, no common mode signal is completely cancelled out by this process and other factors (such as input signal frequency) will also impact on a differential amplifier's ability to cancel out unwanted noise signals. For this reason, the **common mode rejection ratio** (CMRR) of an amplifier is defined as a logarithmic ratio of the common mode gain (A_{cm}) relative to the differential gain (A_d):

$$\textit{Common Mode Rejection Ratio, CMRR} = 20log_{10}\left(\frac{A_d}{A_{cm}}\right) \tag{7.6}$$

A_d is the differential gain of the amplifier

A_{cm} is the common mode gain of the amplifier

The previous equation shows that a larger differential gain (compared to common mode gain) will result in a higher CMRR – thus operational amplifiers aim to have as a high a CMRR as possible (theoretically the CMMR should be infinite). The logarithmic scale used to represent CMRR reflects the difference in scale between the quantities involved, where a good amplifier should be able to carry signals of the order of volts with near zero common mode gain – i.e. in the nanovolt range (1×10^{-9}). As the scale of the relationship between these two quantities is so large, it is therefore best described using decibels for ease of comparison and calculation. A high CMRR can help to reduce the amount of noise that is subsequently amplified within the circuit, but it is not a single solution to circuit noise problems. The next section will discuss noise (and basic noise-reduction techniques) in more detail, but for now there are other design characteristics of operational amplifiers that must be considered.

In addition to common mode rejection, operational amplifiers are also designed to have very high input impedance and very low output impedance to ensure the highest possible efficiency during the amplification process. A detailed discussion of amplification efficiency is outwith the scope of this book, where audio amplifier classes such as A/B/AB/D relate to the efficiency of different methods used to amplify either all or part of the input signal. For introductory purposes, the efficiency of an amplifier broadly relates to the proportion of the input signal that is accurately preserved and scaled (i.e. amplified) for output. For this reason, the **input impedance of the amplifier must be as high as possible** (theoretically infinite) and the **output impedance of the amplifier must be as low as possible** (theoretically zero). To understand why this is the case, the operational amplifier circuit can be considered as an audio system that is designed to amplify a source to drive a load (Figure 7.35).

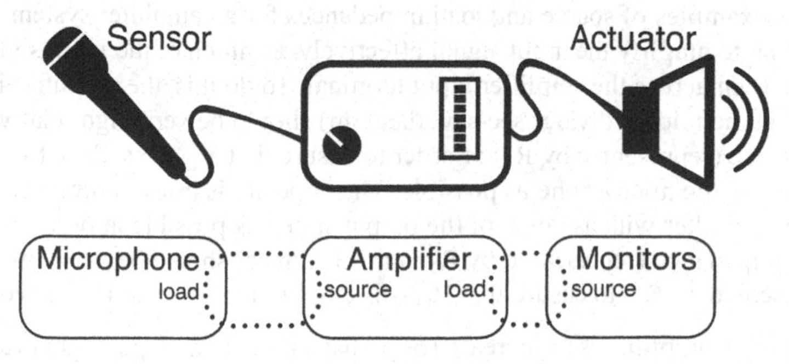

Figure 7.35 Source and load amplifier system example (adapted from chapter 2, Figures 2.6 and 2.7). ***The diagram shows a simple audio amplifier system that takes a microphone sensor input and amplifies it to drive a loudspeaker as an actuator output. The microphone impedance is the source the circuit must amplify, the loudspeaker impedance is the load the amplifier must drive.***

In a simple amplifier system like that in Figure 7.35, the amplifier circuit must amplify a source (sensor) to drive a load (actuator). To be an efficient amplifier system, the circuit should ideally not reduce the input signal whilst also maximizing the output signal that can be driven by the amplifier. This means that the amplifier should present very high impedance to the source and very low impedance to the load. The simplest way of understanding this is to consider both the input and output of the system as being part of separate voltage dividers (see chapter 3, section 3.3), where the relative size of the impedances involved will dictate the amount of the voltage signal across the amplifier input or output (Figure 7.36).

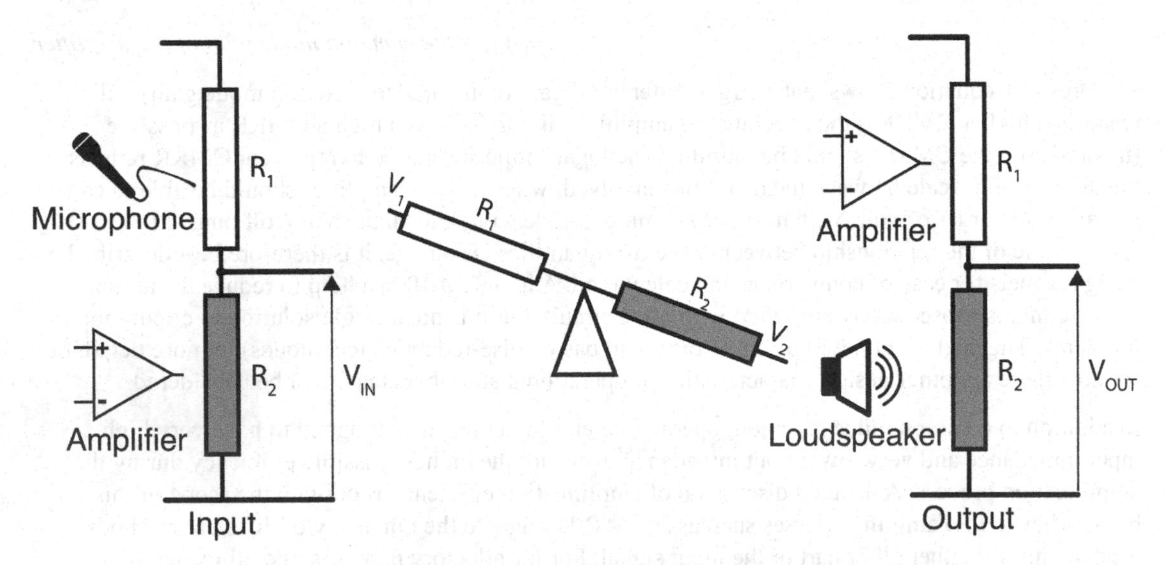

Figure 7.36 Source and load voltage divider examples. ***In the diagram, the left-hand divider shows how a microphone is connected to an amplifier. In this case, to maximize the input signal the amplifier impedance must be as high as possible when compared to the impedance of the microphone. Conversely, the right-hand divider shows how an amplifier is connected to a loudspeaker. In this case, the amplifier impedance must be as low as possible when compared to the impedance of the loudspeaker to maximize the output signal.***

The diagram shows examples of source and load impedances for an amplifier system. The left-hand divider illustrates that to amplify the input signal effectively, as much of the input signal voltage as possible must be seen across the amplifier input terminal. To do this, the amplifier impedance (represented by R_2 in the middle divider see-saw diagram) should be very high relative to the microphone impedance (represented by R_1) in order to ensure that V_{IN} is as close to the entire potential difference generated by the microphone as possible. The opposite is true at output, where the amplifier should drive the loudspeaker with as much of the output signal as possible in order to be efficient. Now the amplifier impedance (represented by R_1 in this instance) should be very low relative to the loudspeaker (represented by R_2) to ensure that V_{OUT} is close to the entire amplifier voltage signal.

The main function of an amplifier is to increase (or perhaps reduce) an input signal for output, and so the gain of the amplifier is its most important practical characteristic. The open-loop gain of an operational amplifier is defined as the uncontrolled (i.e. maximum) possible gain value, which in theory should be infinite – it should ideally be able to increase an input signal by any scaling factor required. In practice, many operational amplifiers will have an actual open-loop gain characteristic that is in the tens or even hundreds of thousands – still far too high for real amplification systems. As with the transistor amplifier in the previous section, some form of **negative feedback** is therefore needed to bring the gain of an operational amplifier down to a stable and manageable level (Figure 7.37).

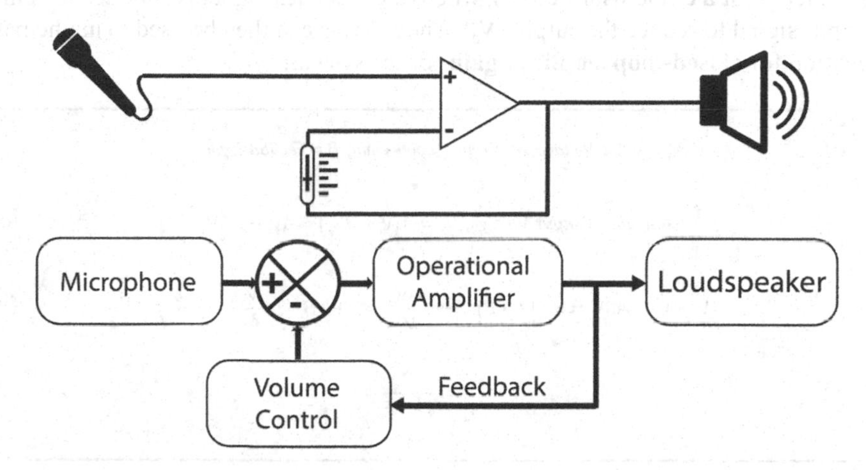

Figure 7.37 Negative feedback amplification example. ***In the diagram, the microphone signal is connected to the non-inverting input of an operational amplifier which will amplify this signal to drive a loudspeaker as a load (and hence emit sound). The open-loop gain of the operational amplifier would overload the speaker, so a volume control (i.e. potentiometer) is used to feed back some of the output signal to the inverting terminal of the amplifier, where it will be subtracted from the input.***

The system in this diagram uses negative feedback to reduce the input signal being applied to the non-inverting input of the operational amplifier. The feedback signal is taken from the output of the amplifier and connected to the inverting input, and because the inverting terminal is 180° out of phase it is subtracted from the non-inverting signal. In so doing, the signal is reduced by the amount of feedback provided, which is itself controlled by the potentiometer (volume control) that is connected in series with the inverting input. This combined gain characteristic is known as the closed-loop gain of the amplifier, where the open-loop gain (A) is reduced by the feedback gain (B) connected to the other differential input (Figure 7.38).

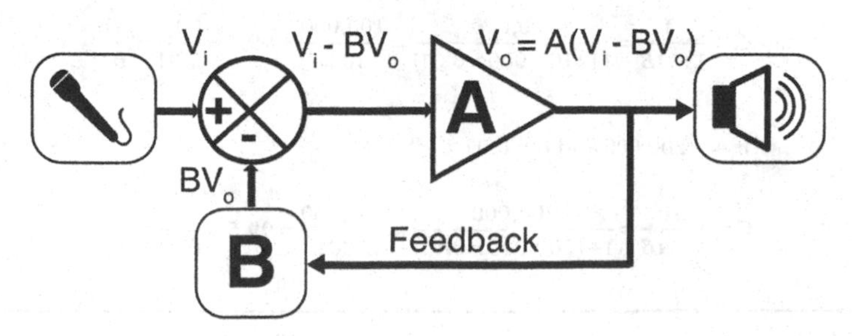

Figure 7.38 Operational amplifier feedback gain. ***The diagram shows how in system terms the feedback signal (BVo) is subtracted from the microphone input signal (Vi) at the summing junction (circle with cross), where the feedback loop gain (B) dictates the level of reduction performed. The open-loop gain (A) of the operational amplifier will amplify this summed signal to produce the output signal (Vo).***

The diagram in Figure 7.38 shows the system diagram for an operational amplifier with feedback that takes a microphone input signal (V_i) and amplifies it for output (V_o). The open-loop gain (A) and the feedback gain (B) of the amplifier are summed as differential inputs to the amplifier (represented as a summing junction of a circle with a cross), where the feedback signal is subtracted from the microphone input signal to reduce the output (V_o). These terms can then be used to mathematically derive the equation for **closed-loop amplifier gain** for the system:

$$V_i = \textit{Input Voltage},\ A = \textit{Open Loop Gain},\ B = \textit{Feedback gain}$$

$$\textit{Amplifier Output Voltage},\ V_o = A\left(V_i - BV_o\right) = AV_i - ABV_o$$

$$AV_i = V_o + ABV_o = V_o(1+AB) \rightarrow \frac{AV_i}{V_o} = 1+AB \updownarrow \frac{V_o}{AV_i} = \frac{1}{1+AB}$$

$$\therefore \frac{V_o}{V_i} = \frac{A}{1+AB}$$

Voltage gain is the ratio of output to input voltage, so the equation for closed-loop operational amplifier gain (G) is:

$$\textbf{\textit{Closed Loop Amplifier Gain, G}} = \frac{\boldsymbol{A}}{1+\boldsymbol{AB}} \tag{7.7}$$

***A** is the open-loop gain of the amplifier*

***B** is the feedback gain of the amplifier*

In this equation the open-loop gain (A) is actually unimportant, as can be shown with example values for (A) and (B):

If A = 100,000 and B = 0.01

$$G = \frac{A}{1+AB} = \frac{100{,}000}{1+(100{,}000\times 0.01)} = \frac{100{,}000}{1001} = 99.6 \approx \frac{1}{0.01} = \frac{1}{B}$$

If A = 200,000 and B = 0.01

$$G = \frac{A}{1+AB} = \frac{200{,}000}{1+(200{,}000\times 0.01)} = \frac{200{,}000}{2001} = 99.5 \approx \frac{1}{0.01} = \frac{1}{B}$$

This means that operational amplifier gain can be designed around the **reciprocal value of the feedback gain** ($\frac{1}{B}$). The system example in Figure 7.37 showed a potentiometer being used to reduce the amount of feedback being applied to the inverting terminal of the operational amplifier, but more commonly a **voltage divider** is used to determine the level of feedback (and hence the gain) in a **non-inverting operational amplifier** circuit (Figure 7.39).

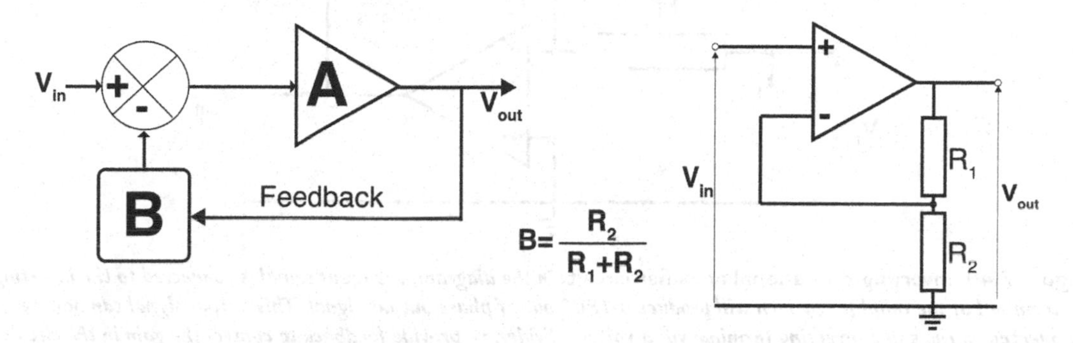

Figure 7.39 Non-inverting operational amplifier circuit. ***In the diagram, the input signal is connected to the non-inverting input of the operational amplifier. The feedback loop is connected from the amplifier output to the inverting terminal via a voltage divider, which determines the feedback gain (B) and hence the closed-loop gain ($\frac{1}{B}$) of the circuit.***

The schematic in Figure 7.39 shows how a non-inverting operational amplifier can be designed around the ratio of the feedback resistors (R_1 and R_2), where the **non-inverting amplifier gain** is effectively the **reciprocal of the feedback gain** (B) for the circuit. This feedback gain value is determined as the ratio of the voltage divider that supplies the inverting input:

$$B = \frac{R_2}{R_1 + R_2}$$

$$Non\text{-}Inverting\ Amplifier\ Gain = \frac{1}{B} = \frac{R_1 + R_2}{R_1} = 1 + \frac{R_2}{R_1}$$

By inverting the voltage divider equation then dividing throughout by R_1, the equation for non-inverting amplifier gain is:

$$\boldsymbol{Non - Inverting\ Amplifier\ Gain} = 1 + \frac{\boldsymbol{R_2}}{\boldsymbol{R_1}} \tag{7.8}$$

$\boldsymbol{R_1}$ is the input resistor connected to the non-inverting terminal

$\boldsymbol{R_2}$ is the feedback resistor connected to the non-inverting terminal

This is the standard gain equation for a non-inverting operational amplifier circuit, which is widely used in audio electronics circuits. The other commonly used configuration is the **inverting operational amplifier**, where connecting an input signal to the inverting terminal will already change the output phase by 180°, thus allowing both the input and feedback signals to be summed at the same inverting input terminal (Figure 7.40).

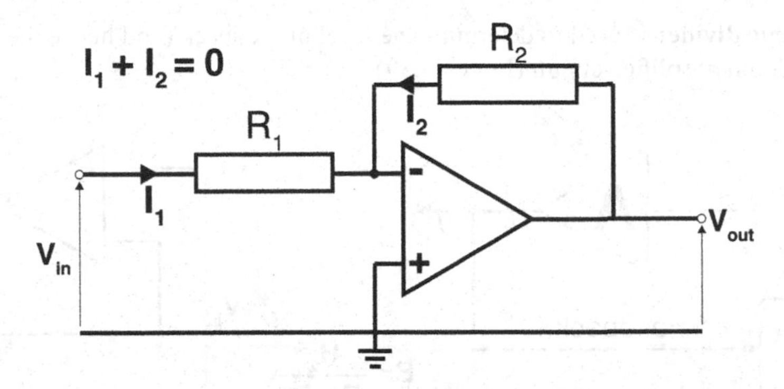

Figure 7.40 Inverting operational amplifier circuit. ***In the diagram, the input signal is connected to the inverting terminal of the amplifier, which will produce a 180° out-of-phase output signal. This output signal can now be connected to the same inverting terminal via a voltage divider, to provide feedback to control the gain in the circuit. Notice that the terminals of the amplifier have been reversed to indicate the inverting terminal is now at the top of the amplifier – the orientation of the terminals must always be checked when reading a schematic. The input (I_1) and feedback (I_2) currents are indicated to show how they sum to zero at the inverting input – this is because of the theoretically infinite input impedance of the operational amplifier.***

In this diagram, it is crucial to note that the **terminals of the operational amplifier have been reversed**. Whenever reading a schematic, one of the first tasks with operational amplifiers is to ensure that the orientation of the terminals is always in the same direction (this will be shown again below). The other significant difference in this circuit is the inclusion of the input (I_1) and feedback (I_2) currents to show how the inverting terminal has now become a virtual summing point within the circuit. The previous discussion on source and load showed why an operational amplifier should have as high an input impedance as possible – theoretically this value should be infinite. Working from this theoretical case, an infinite input impedance will allow zero current to flow into the circuit, and from Kirchoff's Current Law (KCL) it can be shown that the sum of all currents flowing into and out of the voltage divider node will equal zero. This principle of infinite resistance allowing zero current flow can then be used to mathematically derive the equation for **inverting amplifier gain**:

$$I_1 = \frac{V_{in}}{R_1},\ I_2 = \frac{V_{out}}{R_2},\ From\,KCL: I_1 + I_2 = 0$$

$$I_2 = -I_1 = -\frac{V_{in}}{R_1} = \frac{V_{out}}{R_2} \therefore V_{out} = -\frac{V_{in}R_2}{R_1}$$

Again dividing throughout by R_1, the equation for inverting amplifier gain can then be stated as:

$$\textbf{\textit{Inverting Amplifier Gain}} = -\frac{R_2}{R_1} \tag{7.9}$$

R_1 is the input resistor connected to the inverting terminal

R_2 is the feedback resistor connected to the inverting terminal

In many ways, operational amplifiers are theoretically much more straightforward than transistors, and the two basic configurations of non-inverting and inverting amplification are used widely in electronics. The swapping of terminals on the operational amplifier can often lead to confusion, and so it can be useful to rearrange the previous non-inverting circuit to follow the same terminal configuration as an inverting amplifier (Figure 7.41).

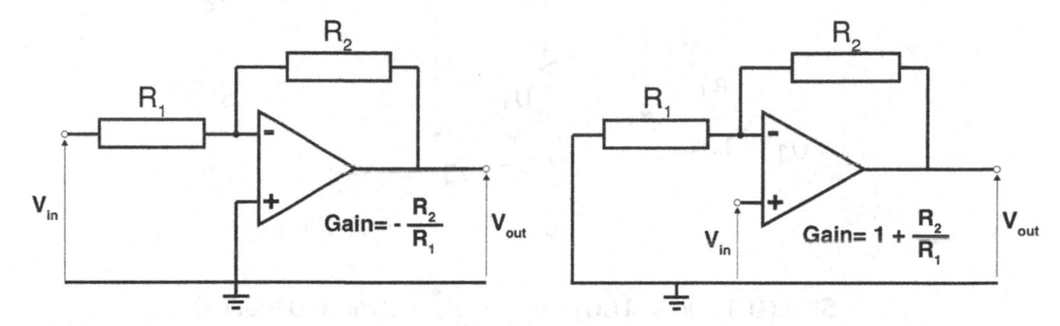

Figure 7.41 Reconfiguring operational amplifier layouts. ***In the diagram, both the inverting (left) and non-inverting (right) circuits are shown with the same terminal orientation (inverting top, non-inverting bottom). This layout can help to illustrate how conceptually similar both circuits are, though in practice the opposite terminal configuration (non-inverting top) is often used for non-inverting amplifier schematics. This diagram is provided both for understanding and also for reference, as some layouts may switch between terminal orientations within a single schematic.***

The diagram shows both inverting (left) and non-inverting (right) configurations for an operational amplifier, where the orientation of the input terminals is consistent (with inverting top, non-inverting bottom). Although many schematics will use the non-inverting layout shown in Figure 7.38, it is important to recognize the importance of terminal orientation when learning by analysing existing circuits. In the left-hand inverting amplifier schematic, both input and feedback signals are summed at the non-inverting input (which acts as a virtual summing point), whilst in the right-hand schematic a non-inverting input signal is fed from the output to the inverting terminal, where a 180° out-of-phase signal will now be summed as a differential input. The gain characteristics of both amplifiers are relatively similar, though a non-inverting amplifier can never have a gain of less than 1 – which prevents it from being used in specific circumstances where gain reduction is needed.

At this point, the two main operational amplifier configurations used in audio amplification have been introduced. Though many other operational amplifier designs exist (such as those using positive feedback to produce sound), the primary function of signal amplification will form the basis of the following worked example.

7.4.1 Worked example – simulating an inverting amplifier with LTspice

In this worked example, an inverting operational amplifier circuit will be simulated using LTspice. As with the previous BJT circuit, the supply will use values from the Arduino (5V, 40mA), though this circuit will only be simulated (not built). A 100mV sine wave input will be used to measure the output signal from this circuit through a 100kΩ load resistance, using the schematic shown in Figure 7.42.

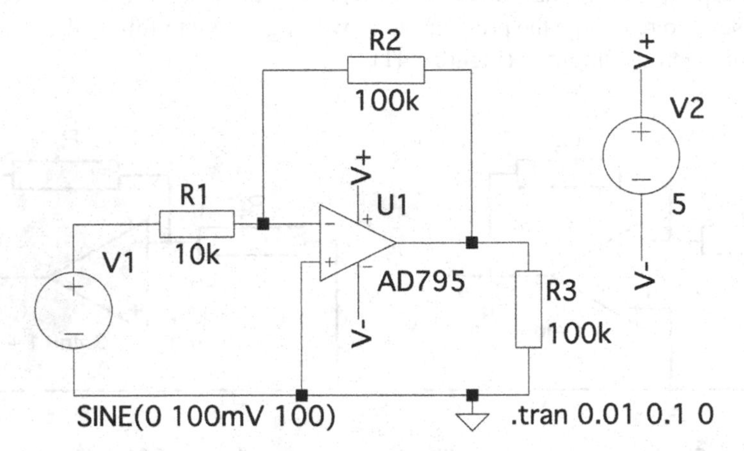

Figure 7.42 Inverting operational amplifier circuit. ***In the diagram, a voltage source (V1) provides a 100mV sine wave of 100Hz to the inverting input terminal of an AD795 operational amplifier. A voltage divider (R1 and R2) provides feedback to control the gain of the circuit, whilst R3 is a load resistance of 100kΩ. that allows the amplifier output signal to be measured. The second voltage source (V2) provides power for the operational amplifier via LTspice net names (V+/V−).***

The gain of the circuit can be set using R1 and R2, where

$Inverting\ Amplifier\ Gain = -{R_2}/{R_1} = -{100}/{10} = -10$. The output load resistance (used to allow LTspice to simulate a voltage level across it) is set to 100kΩ to match R2, though this is simply an approximation for circuit simulation – in practice, more advanced analysis would be needed to derive the equivalent Thevenin resistance for the operational amplifier to determine the optimum load needed to balance the circuit.

To build this circuit in LTspice, use component values of R1 = 10kΩ, R2 = 100kΩ and R3 = 100kΩ, Op-Amp U1 = AD795, Voltage Input Source Vin = 100mV and Voltage Supply Source V_{CC} = 5V. Create a new circuit in LTspice and name it chp7_example3. The process for creating the schematic is as follows:

1. Lay out the components as shown in Figure 7.42. Start with the Voltage Input Source (V1 on left), Voltage Supply Source (V2 on right), Operational Amplifier (middle) and the **GND node** (bottom). For the Op-Amp, select the [Opamps] folder and then scroll to find AD795.
2. Add resistor R1 between the top terminal of V1 and the inverting input of U1. Set the value to R1 = 10kΩ.
3. Add resistor R2 above the Op-Amp U1 and set the value to R1 = 100kΩ.
4. Add the load resistor (R3) vertically between the output of U1 and GND. Set its value to 100kΩ.
5. Connect wires between V1 and R1, R1 to the U1 inverting input, U1 inverting input (middle) to R2 and the non-inverting U1 input to GND.
6. Connect wires between U1 output and R3, U1 output (middle) and R2, R3 to GND.
7. Add a net name for V+ (same process as a GND connection), leave the Port Type as <none>. Place one above the positive terminal of V2, place the other above the positive power terminal of U1.

8. Add a net name for V−, again leave the Port Type as <none>. Place one below the negative terminal of V2, place the other below the negative power terminal of U1.
9. Connect wires between the V2 terminals and the net names, then between U1 terminals and net names.
10. Configure the Voltage Input Source (V1) to be a sine wave of amplitude 0.1 (100mV) and frequency 100Hz.
11. Configure the Voltage Supply Source (V2) to be a DC value of 5V.
12. Add a Spice Directive to perform a **transient analysis** between 0 and 0.1 seconds, with a step of 0.01 (**.tran 0.01 0.1 0**) – this will show 10 cycles of a 100Hz sine wave.

If the circuit has been built correctly in LTspice, the simulation should run and a voltage probe can be used to measure both the input signal (take a reading before R1) and the output signal (take a reading before R3). This should produce output traces similar to those in Figure 7.43.

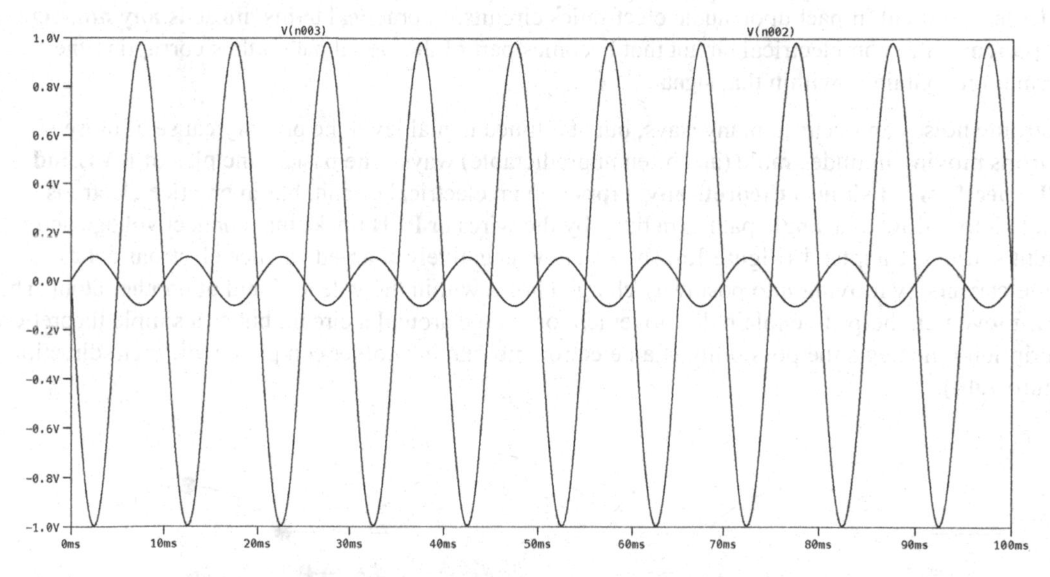

Figure 7.43 Simulation graph of an inverting operational amplifier circuit. ***The y axis shows both input and output voltage levels over a simulation time of 100 milliseconds (x-axis). The input is a 100mV 100Hz voltage signal, whilst the output is a 1V 100Hz signal that is 180° out of phase with the input – this corresponds to a gain value of −10.***

The simulation graph shows both input and output voltage signals for an inverting amplifier. The amplifier gain value was calculated as $-{R_2}/{R_1} = -10$ and the output trace is now a magnitude of 10 greater than the input but inverted so the output is also now 180° out of phase with the input. It is interesting to note that the design (and analysis) of this circuit was much simpler than that for the previous common emitter example (7.3.1), and also that the gain characteristic is much closer to the required value than was the case for a BJT amplifier circuit. In most modern circuits, operational amplifiers are preferred because of their greater stability and reliability – the calculated gain characteristic is much closer to the simulated value (Vin = 100mV, Vout = 1V, Vout(actual) = −982mV) than with the BJT amplifier. The addition of node names within LTspice allows the power rails of the operational amplifier to be connected up without making the schematic difficult to read. The next section will show how connections like power rails can introduce noise into a circuit that must be reduced, alongside compensating for the complex output impedance of a loudspeaker using a Zobel network and decoupling DC signals (in a similar manner to debiasing a BJT) to reduce the presence of unwanted noise signals in the amplified output.

7.5 DC blocking, power decoupling and Zobel networks

With an operational amplifier circuit simulated, it is now important to consider some of the additional practicalities involved in designing and constructing a functional circuit. This aspect of operational amplifier design can often be confusing for new learners, as in some respects it appears to be the introduction of a significant amount of additional components (mostly capacitors) that can make the relatively simple schematic of an operational amplifier seem much more complex. The primary reason for the addition of these components is the reduction of noise within the circuit, which has a particularly critical impact upon audio electronics circuits. In practical terms, **noise is any non-signal component** within an electrical circuit that becomes part of the signal path – thus corrupting the information contained within that signal.

Electronic noise can occur in many ways, but at a fundamental level the primary cause of noise is electrons moving in undesirable (and often unpredictable) ways. The basic principles of **KVL and KCL specify what should theoretically happen** in an electrical circuit, but in practice electrons do not simply flow in a single path as defined by the wires or PCB tracks that connect voltage and current sources. Chapter 1 (Figure 1.6) showed how negatively charged valence electrons act as charge carriers by moving into positively charged holes within the valence band of another atom. This linear movement helps to explain the movement of charge around a circuit, but as a simple theoretical description it neglects the possibility of an electron moving in another completely different direction (Figure 7.44).

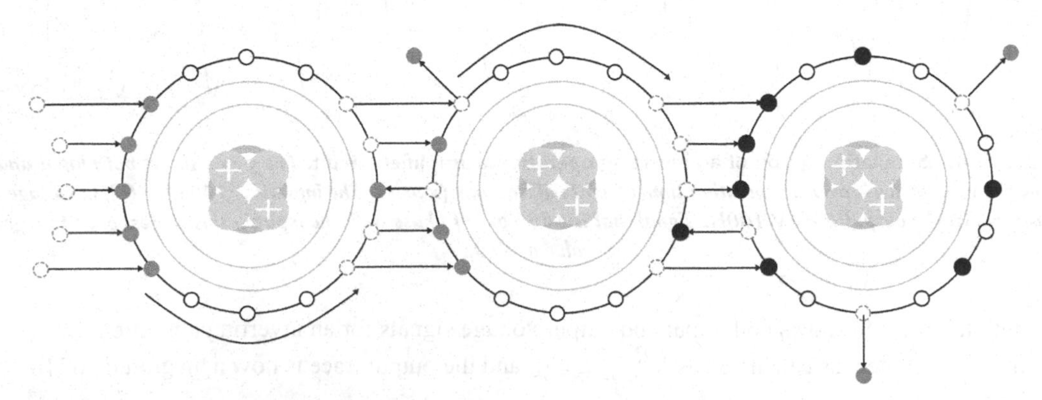

Figure 7.44 Unpredictable valence electron movement (adapted from chapter 1, Figure 1.6). ***In the diagram, the conceptually linear movement of valence electrons helps carry the change in net charge between atoms from left to right. In the top-left of the middle atom, a single electron leaves the nucleus for another point in space (potentially another atom) whilst in the middle-left of that atom a valence electron returns to one of the holes in the left-hand atom that it (theoretically) came from. The right-hand atom shows more examples of this non-linear movement of charge, illustrating that some electrons will not necessarily follow a linear path.***

In the diagram, the overall net movement of charge is still left to right, where the majority of charge carriers (electrons) have moved from one valence shell to the next. Having said this, some of these electrons (middle and right atoms) do not progress linearly and may move from right to left (against the overall net movement of charge) or in another direction entirely due to other factors (such as heat). Although circuit analysis techniques like KVL and KCL predict the linear movement of all electrons in a single direction around a path within a circuit, in practice some electrons will not follow this simple rule and will move in less predictable ways. This unplanned movement manifests itself as noise within an electronic circuit, where any factor that influences electrons to move outside the planned conductive path will create some level of noise signal. There are many different types of noise that can occur on an electronic signal path:

1. **Johnson-Nyquist (thermal) noise** – created by the thermal agitation of electrons within all resistive components
2. **Shot noise** – created by the movement of charge due to the discrete nature of electrons (more significant for small currents)
3. **$1/f$ (flicker) noise** – inversely proportional to frequency, created by imperfections in electronic components
4. **Burst (popcorn) noise** – low-frequency noise occurring in operational amplifiers and other semiconductors
5. **White (and other colours) noise** – noise colours are defined by the power and bandwidth of the signal involved, where white noise has equal power across the entire audio spectrum (pink noise is often used in room acoustics)

This list is provided to give a brief indication of both the complexity of noise and also the variety of sources that may generate it. Although a full discussion of this topic is outwith the scope of this book, the impact of noise can be significant in even the simplest circuits. As discussed in the previous section, operational amplifiers have a high CMRR and this can help to reduce the amount of noise within the circuit that is subsequently amplified, but this is not (in itself) a solution to circuit noise problems. Numerous audio manufacturers have devoted significant amounts of time (and research effort) to investigating methods of effectively reducing noise, but the simple fact is that noise cannot be completely removed from any signal path. Having said this, the level of noise in a circuit can be managed to keep its influence to a minimum and there are some component additions that can be made to an audio circuit to improve its noise performance.

Power sources in a circuit can introduce noise signals due to interference, where electrons on a power rail move across the insulating gaps on a printed circuit board into the signal path. In the previous section (Figure 7.23 and Figure 7.24), AC coupling capacitors were used in a BJT circuit to block the DC biasing component at both input and output. In operational amplifier circuits, this same concept is used to **prevent any DC noise signal** (due to power supply interference) from entering the amplifier in either direction (Figure 7.45).

By adding a blocking capacitor at both ends of the amplifier (often called a coupling cap, as they couple the AC component of the input), any DC signal that has been introduced into the signal path will be filtered out before it can reach the operational amplifier. This not only removes the DC noise, but also protects the amplifier from an increase in signal voltage that may prove too large for the internal gain of the circuit. The previous section showed how an input coupling capacitor can be used to prevent the larger bias current from flooding the much smaller input current to the BJT. In the same

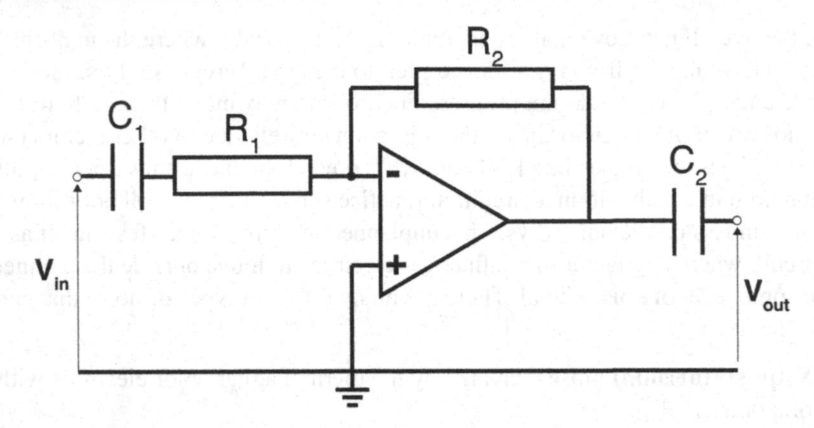

Figure 7.45 Operational amplifier DC blocking capacitors. ***In the diagram, a capacitor is added at both the inverting input (C_1) and amplifier output (C_2) of an inverting operational amplifier circuit. As with the biasing network in a BJT, DC current signals from other parts of an electronic system can flow back into the operational amplifier output and damage the component. Similarly, a DC signal component (as noise) that accompanies an AC input signal can easily damage an operational amplifier due to its high open-loop gain, and so must be blocked.***

manner, an output blocking capacitor isolates the signal path from other parts of the system that may work with larger currents which could potentially flow into the output of the operational amplifier and damage it.

When working with operational amplifier power rails, the **problem of AC noise** also occurs. Although internal power supplies are usually DC signals, noise from the AC mains power supply (which oscillates at 50Hz in Europe, 60Hz in North America) is also a common problem in recording studios. Section 7.2 (Figure 7.7 and Figure 7.8) showed how diodes can be used to rectify an AC mains power supply signal for DC output, but this simple circuit will not prevent all AC noise from entering the DC power supply to an audio circuit. An operational amplifier (which is composed of many transistors) requires a stable DC supply for amplification. In a BJT, the input signal to the base allows collector current to flow so an operational amplifier follows the same principle – any fluctuations in the collector supply will become part of the output signal. To reduce the impact of this noise, additional AC decoupling capacitors can be added to the power rails of the operational amplifier (Figure 7.46).

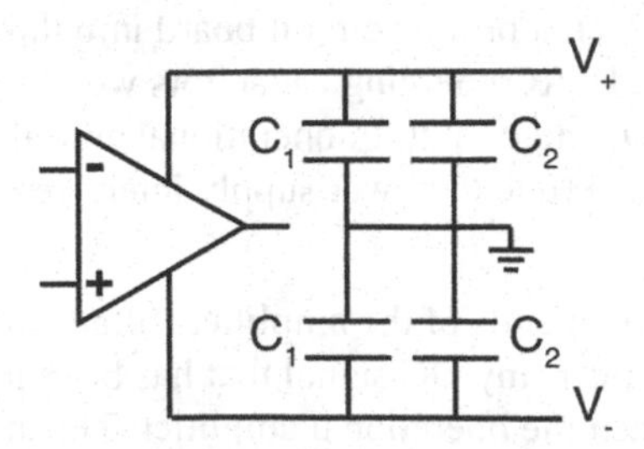

Figure 7.46 Operational amplifier power AC decoupling capacitors. ***In the diagram, additional capacitors have been added to decouple AC noise signals from both operational amplifier power supply rails. The capacitor values are different (e.g. C_1 = 1μF, C_2 = 0.1μF) to cover a wider range of AC frequencies that may be present (as noise) on the power rails. Input and output connections (and other components) have been omitted for clarity.***

In the diagram, AC decoupling capacitors have been added to both power supply rails to filter out noise components that may be present on the supply rails. This is particularly important if the circuit will be powered by an AC mains supply that must be stepped down (by a transformer) and then rectified (Figure 7.7) to produce a DC power signal. The diagram in Figure 7.46 represents one possible approach to AC decoupling, as in practice this is an area of noise reduction where more advanced techniques (and more extensive analysis) are required. For the purposes of this textbook (where all circuits will be built on breadboard) it is sufficient to add two decoupling capacitors across the power rails that are connected to the Arduino to reduce the noise introduced by the single rail 5V supply (Figure 7.47).

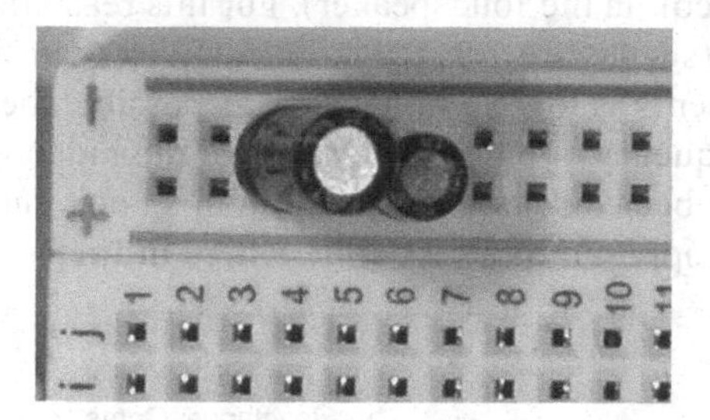

Figure 7.47 Adding AC decoupling capacitors to breadboard power rails. ***In the diagram, two AC decoupling capacitors (1μF and 0.1μF) are added between the positive and negative terminals of a breadboard. The circuits in this book are powered by a single-rail Arduino supply (5V to GND), so decoupling the breadboard rails helps to reduce the AC noise present in the circuit.***

This image shows how two capacitors can be added to a breadboard to provide a level of AC decoupling to the circuits being built. Although not a comprehensive strategy for dealing with power-supply noise, the prototyping nature of breadboard circuits means that some level of noise will be inherent across the entire circuit and so it is not practical to consider additional noise-reduction steps as a result. The final project in this chapter will build a simple audio amplifier, and it can be a useful test to add and remove these capacitors once the circuit is fully operational to see the effect they have (as their removal will not change the flow of power). As previously noted, significant efforts have been made to develop solutions for noise reduction in audio circuits, but the simple compromise shown in the image provides some level of reduction whilst acknowledging the shortcomings of breadboard prototyping (which in itself will never achieve high-fidelity results).

Another common addition to an audio operational amplifier is a **Zobel network**, which compensates for the internal inductance of a moving-coil loudspeaker (Figure 7.48).

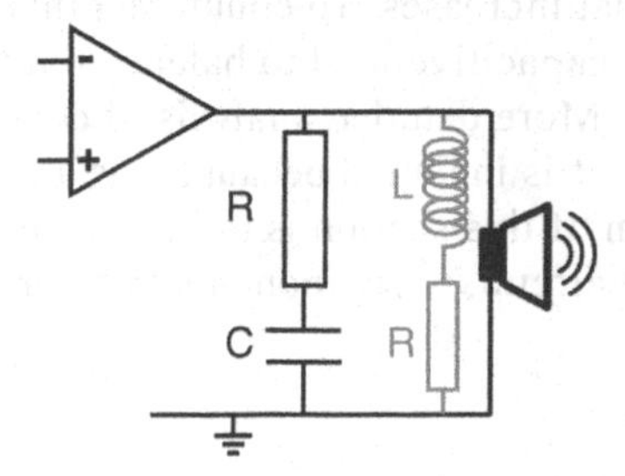

Figure 7.48 Connecting a Zobel network to an operational amplifier output. ***In the diagram, a resistor capacitor combination known as a Zobel network is added to the output of an operational amplifier to compensate for the inductance and resistance that are presented to the circuit by the copper coil within a loudspeaker (marked in grey).***

Otto Zobel published a 1923 paper for Bell Labs on impedance balancing that proposed the idea of image impedance, which in broad terms aims to match the impedance of one side of a network (known as a port into a network) with the other. In audio, this Zobel network takes the form of what is also known as a Boucherot cell, named after the French Railway engineer who first proposed using a series resistor and capacitor (or multiple capacitors) to cancel out the reactive component of an inductive load. In the diagram, the Zobel network of a resistor and capacitor is added to balance the inductor resistor combination (in grey) that represents a loudspeaker. The impedance of a loudspeaker is effectively a combination of both a **resistive and an inductive** component (due to the voice coil in the loudspeaker). For this reason, the concept of a loudspeaker as an output impedance load (see Figure 7.36) is more complex than simply ensuring that the operational amplifier presents as low an impedance as possible, because the loudspeaker impedance will vary with frequency due to the presence of the inductive component. For reasons of brevity, inductors have not been discussed in detail in this book but the basic concept of an inductor is in many ways the opposite of a capacitor, where **inductive reactance increases** with frequency (Figure 7.49).

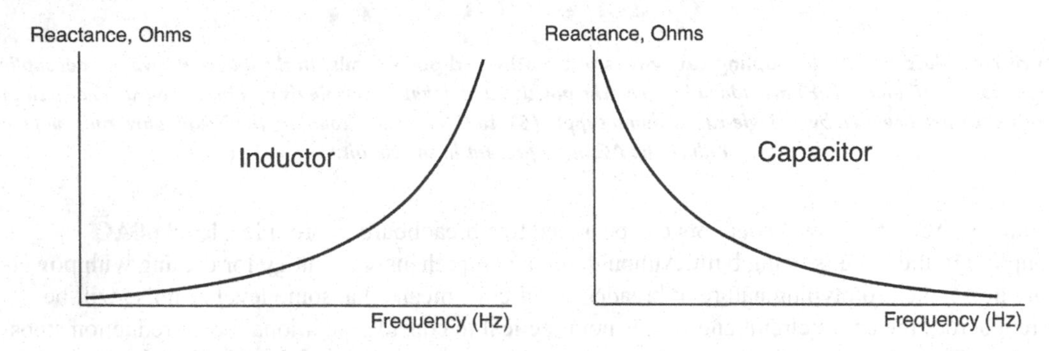

Figure 7.49 Graphs of capacitive and inductive reactance relative to frequency. ***The left-hand graph shows how inductive reactance increases with frequency, whilst the right-hand graph shows how capacitive reactance decreases as frequency increases. Although not a symmetrical relationship, a capacitor can often be used in a Zobel network to balance out the changing inductive reactance of a loudspeaker.***

The diagram shows how inductive and capacitive reactance are inversely related (though not directly proportional) to one another. In effect, the rising inductive reactance due to frequency will mean that any loudspeaker connected to an operational amplifier will vary its impedance as the frequency of the output signal increases. To counteract this non-linear response, a Zobel network introduces a resistive and capacitive load to balance out the inductive (and resistive) load presented by the loudspeaker. More detailed analysis of concepts such as Zobel networks is outwith the scope of this book, but it is included because it is a common element in most audio amplifier designs. The primary aim of this section is to introduce the 'extra' components that are often added to audio amplification circuits – components that can easily confuse the new learner (Figure 7.50).

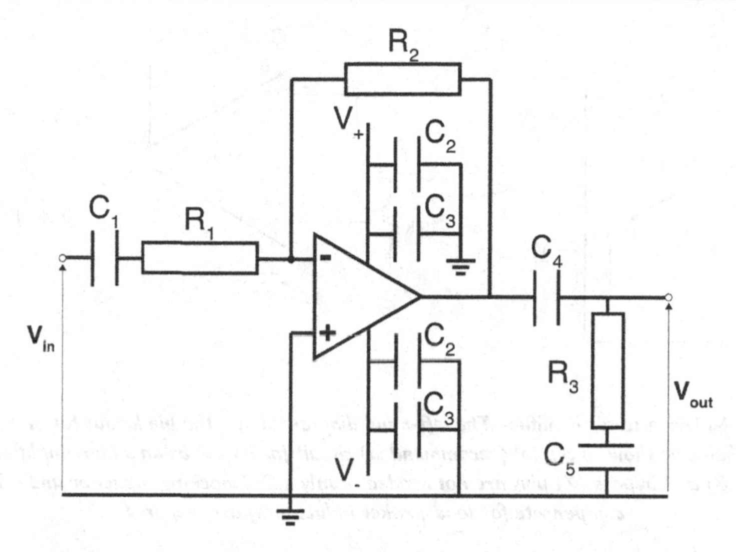

Figure 7.50 A typical audio inverting operational amplifier circuit. ***The diagram shows an inverting operational amplifier with negative feedback (R_1 and R_2) that includes DC blocking capacitors (C_1 and C_4), AC decoupling capacitors (C_2 and C_3) and a Zobel network (R_3 and C_5) that compensates for the inductive component of a loudspeaker connected to the amplifier output. This type of circuit design is very common in audio electronics equipment and can be varied (and augmented) for a variety of practical purposes.***

At this point, the inverting amplifier circuit from Figure 7.41 has now been augmented with a significant number of extra components (mostly capacitors). This is common in electronic circuits (particularly in audio) to remove DC signals, reduce AC noise and also balance the output of the amplifier when connected to a loudspeaker. It can often be very confusing for the new learner to see real schematics of audio equipment and try to map these back to the initial concepts of BJT or operational amplifiers that they have been built on. Whilst this section has provided a brief overview of noise-reduction techniques, in practice many audio electronics designers spend significant time and effort trying to derive new (and more efficient) ways of reducing noise in a circuit. For now, a more straightforward circuit example is provided in the chapter project, which focusses on building a simple audio amplifier.

7.6 Example project: building an audio amplifier

This project will use an LM386 operational amplifier to build a functional audio amplification circuit. The LM386 is a popular amplifier for prototyping and learning, which can achieve reasonable results for a relatively small component count. The LM386 is a low-power amplifier that works from a single-rail supply, which allows it to be powered by the Arduino 5V rail (though a 9V battery can also be used). In addition, the LM386 has an internally fixed gain level of 20 (which can be varied up to 200), which removes the need for an external resistor feedback network (Figure 7.51).

The LM386 is fairly straightforward to work with, as there are only three additional components (one resistor and two capacitors) required for the example circuit shown in the right-hand schematic in Figure 7.51. The schematic uses a Zobel network (0.05μF capacitor and 10Ω resistor) to balance the resistive and inductive load of the loudspeaker, alongside a DC blocking capacitor (250μF) to prevent any DC signals flowing into (or out of) the LM386 amplifier. A variable resistor (10kΩ) is shown

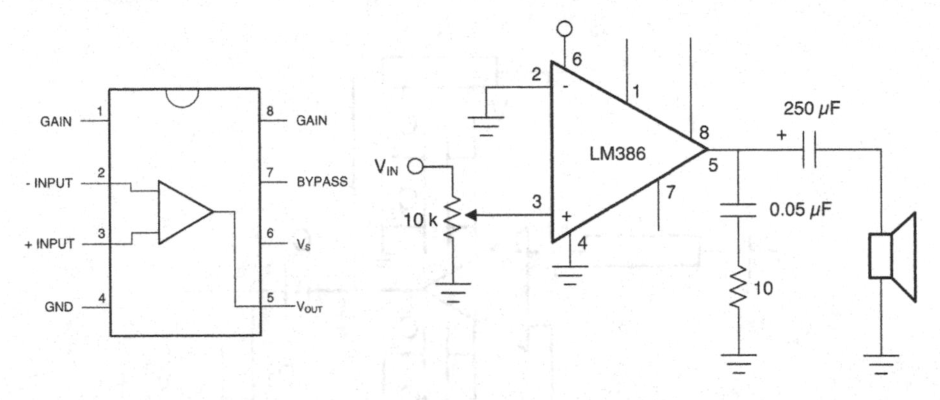

Figure 7.51 The LM386 audio amplifier. ***The left-hand diagram shows the pin layout for an LM386 chip, whilst the right-hand schematic shows a typical (recommended) circuit for its use as an audio amplifier. For a minimal circuit, the gain (1,8) and bypass (7) pins are not needed – only a DC blocking capacitor and a Zobel network (to compensate for loudspeaker inductance) are required.***

connected to the non-inverting input of the amplifier to act as a volume control by reducing the voltage seen across the input, but this can be omitted for prototyping purposes to keep the component count down (the project build will show both options). The power supply pin 6 (V_s) can be connected to a 5V source like the Arduino, which can also provide a ground signal (GND pin 4) for this chip. Pins 1 and 8 control the gain of the circuit, and so can be left unconnected – as can the bypass (pin 7), which is only used if the operational amplifier is not internally biased (Figure 7.52).

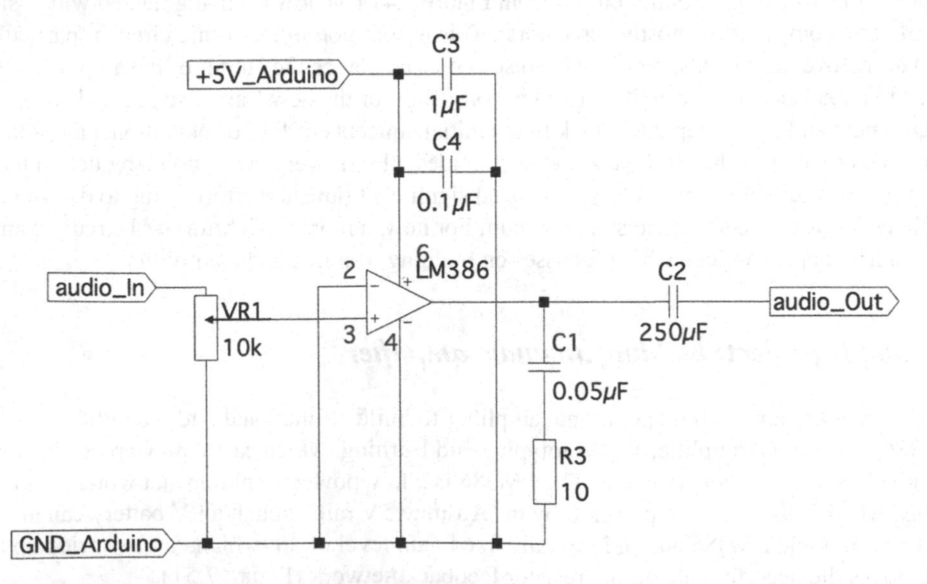

Figure 7.52 Final audio amplifier schematic. ***In the diagram, the audio input signal is connected via a 10kΩ potentiometer to the non-inverting input of an LM386 amplifier. A Zobel network (C_1, R_3) and DC blocking capacitor (C_2) are connected at the output of the LM386, alongside AC decoupling capacitors (C_3, C_4) that are connected across the power rails. The amplifier is powered by the +5V supply from the Arduino, which also provides the GND connection for the circuit.***

The project will use the same non-inverting configuration as shown in the LM386 datasheet in Figure 7.52, where the inverting input (pin 2) is tied to GND to provide a reference signal for differential amplification (where any circuit noise will ideally be reduced by CMRR). The only additional components to be added are AC decoupling capacitors ($C_1 = 1\mu F$, $C_2 = 0.1\mu F$) across the breadboard power rails (Figure 7.47), which will help to reduce any noise introduced by the Arduino supply and also by the construction of the breadboard power rails (breadboards do not provide high-quality electronic connections). Other capacitors (such as DC blocking on input) can be added for noise reduction, but as the aim of this project is to prototype a basic audio operational amplifier these have been omitted to keep the component count to a minimum. In this project (and the chapter 8 and 9 projects) the audio input and output are connected to the breadboard using PCB mount screw terminal blocks (Figure 7.53).

Figure 7.53 PCB mount screw terminal blocks. ***The images show a top (left) and side (right) view of a screw mount terminal block, which is useful in providing a stable breadboard connection for audio input and output cables. The right-hand image shows where the signal and ground wires are pushed into each terminal, which can then be screwed closed to hold the wire in place – thus preventing it being pulled out. The side view shows the connecting pins, which can then be mounted directly onto the breadboard.***

These blocks help to prevent the audio wires from moving and becoming detached from the breadboard (which can easily happen with a direct wire inserted into the breadboard). As shown in the chapter 5 MIDI drum trigger project, when connecting a wire to the block, it can help to pre-mount the terminal block to the breadboard to provide a solid surface to work with. In the case of two wires (like an audio connector), one issue can be too much exposed wire outside the terminal block that may potentially short out the signal and GND wires by allowing them to make contact with one another. To insert the wires, it can be easier to unscrew one terminal on the block, and with a wire in position (inside the terminal clamp) screw the terminal to clamp down on this wire and hold it in position. Repeat the process with the second wire, and this should provide a suitable audio connection for input or output to the breadboard.

Project steps

For this project, you will need: 1. 1 × LM386 operational amplifier 2. 4 × capacitors (250μF, 0.05μF, 1μF, 0.1μF), 1 × resistor (10Ω), 1 × potentiometer (optional 10kΩ) 3. 1 × audio input connector (3.5mm jack), 1 × audio output connector (loudspeaker) 4. 2 × connector cables and 6 × connector wires (images show short wires)	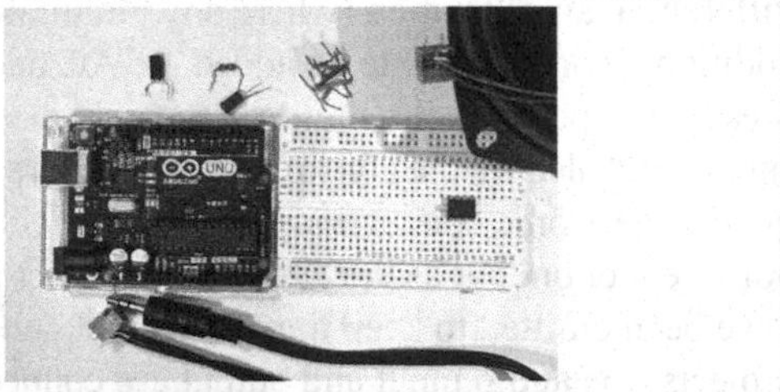
1. Add the LM386 across the column break with top left **pin 1 on e22** (the top of the chip has a notch) – bottom right pin 5 on f25. 2. Add an input screw connector block to **[B28 and B30]** and an output screw connector block to **[J28 and J30]**.	
Add a ground connector wire between input and output connectors from **[e30–f30]**.	
1. Connect the input and output GND to chip GND (pin 4) with a connector wire from **[c25–c30]**. 2. Connect the audio input to the LM386 non-inverting input (pin 3) with a connector wire from **[d24–d28]**.	
1. Connect the LM386 inverting input (pin 2) and GND (pin 4) with a connector wire from **[b23–b25]** and then connect both to the breadboard ground rail with a connector wire from **[a25–GND]**. 2. Connect the LM386 power supply Vs (pin 6) with a connector wire from **[j24–GND]**.	

Add a 250μF DC blocking capacitor from the LM386 output Vout (pin 5) to the audio output connector from pins **[h2--h28]**. The audio output connector may be slightly displaced by the capacitor, but don't force the capacitor as it can be angled if need be.	
1. Add a Zobel network to the LM386 output Vout (pin 5). This can be done in various ways, we will use a connector wire from pins **[g25–g19]** to provide space for the network. 2. Add a 0.05μF capacitor (or a 0.1μF if not available) between pins **[h19–i18]** then add a 10Ω resistor between pins **[j19–GND]**.	
Add AC decoupling capacitors (C_1 = 1μF, C_2 = 0.1μF) between the breadboard power and GND rails.	
1. Connect the Arduino 5V and GND to the breadboard power rails. 2. Bridge the breadboard GND rail to connect the LM386 GND connections (pins 2 and 4).	
Optional: replace the audio connector wire on **[d24–d28]** with a 10kΩ potentiometer – right wiper to audio **[d28]**, middle wiper to LM386 **[d24]** and left wiper to GND.	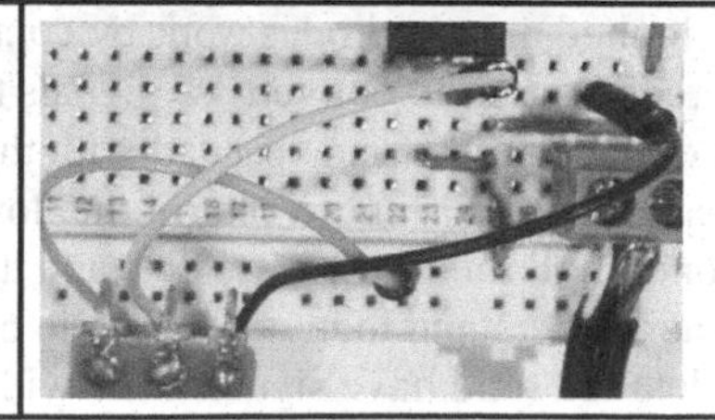

Connect the 3.5mm audio jack to an audio playback device – the headphone socket will provide a high enough voltage to drive a small (8Ω, <1W) loudspeaker when fed through the LM386 with a gain of 20. If everything has been connected properly, the LM386 should amplify the input signal from the audio playback device and output the electrical signal to be converted into sound by the loudspeaker.

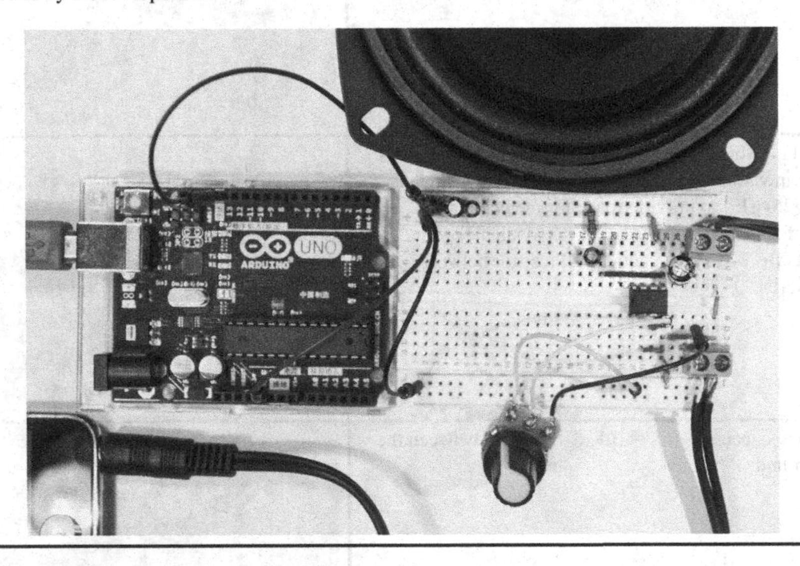

At this point, a functional audio amplifier circuit has been prototyped, with the option of including a potentiometer across the input connection to vary the voltage level at the non-inverting amplifier input. This circuit will be augmented in the following chapters to include audio filtering (chapter 8) and digital potentiometer control (chapter 9), but it is also possible to expand on the minimal amplifier circuit provided in the LM386 datasheet. The LM386 is a very versatile (and straightforward) operational amplifier chip that can be used for a variety of audio applications. The circuit shown above is effectively a building block for the projects in chapters 8 and 9, and can be used for your own projects outside this book.

7.7 Conclusions

This chapter looked at how diodes work by controlling the flow of charge, allowing them to be used in full wave rectification to convert AC signals to DC (and in LED circuits throughout this book). This control of current flow is expanded with the transistor, which can be used as an amplifier by varying the small base current to control the flow of a much larger collector current. Transistors are the cornerstone of all modern electronics, and this book provides an introduction to their workings to help the learner understand many of the audio circuits that use them. Having said this, transistors are much more detailed (and have significantly wider application) than discussed in this chapter, and so further study is recommended. Transistors can be combined within an integrated circuit (IC) such as an operational amplifier, which uses differential amplification to reduce circuit noise (through CMRR) and improve stability. The input impedance of an operational amplifier should be very high (theoretically infinite) and the output impedance very low (theoretically zero), to maximize the transfer of power between input and output signals (as sources and loads). In addition, an operational amplifier should have (theoretically) infinite gain, which must then be reduced using negative feedback.

The final project in this chapter combined a minimal audio amplifier chip (LM386) with additional components for noise reduction (DC blocking, AC power decoupling) alongside a Zobel network to balance the frequency-dependent output impedance of a connected loudspeaker (as an output transducer). The next chapter on audio filters will provide more detail in relation to frequency dependence, where a resistor capacitor network will be used to create both a low- and a high-pass filter. In so doing, the amplifier used in this chapter will be used as a building block for a more useful audio filter circuit – highlighting why a systems approach to audio circuits is important. This chapter has covered a lot of material (though not all of it in detail) and it should be noted that the main aim of this book is to learn the basics of introductory audio electronics circuits (and how to control them with the Arduino). It should not be considered as a replacement for a more advanced study of subjects like transistors and differential amplification, which require much more detailed investigation for current usage.

7.8 Self-study questions

The following questions are intended to reinforce your understanding of amplification and operational amplifiers (transistors are not included, as no practical transistor circuit is built in this book).

Amplification

Q1: *Calculate the power gain A_p and decibel power gain a_p (dB) of an audio amplifier with:*

(a) A DC input power of 2W and an output power of 4W.
(b) A DC input power of 1W and an output power of 5W.

Q2: *Calculate the voltage gain A_v and decibel voltage gain a_v (dB) of an audio amplifier with:*

(a) An input signal of 100mV and an output signal of 1V.
(b) An input signal of 100mV and an output signal of 0.5V.

Operational amplifiers

Use the following schematic for an inverting operational amplifier to answer the questions below:

Inline figure study1
Q3: *If V_{in} = 100mV and V_{out} = −0.5V, what are possible values of R_1 and R_2?*
Q4: *If V_{in} = 100mV and V_{out} = −2V, what are possible values of R_1 and R_2?*

All answers are provided for each chapter at the end of the book.

Chapter 7 summary

The following images may help you remember some of the key points in this chapter:

Relative decibel gains

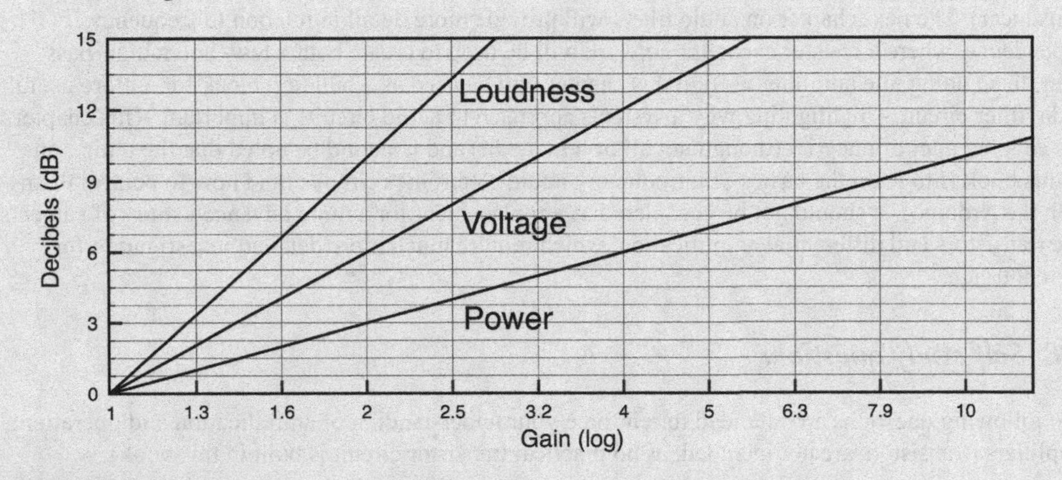

Full wave diode rectification

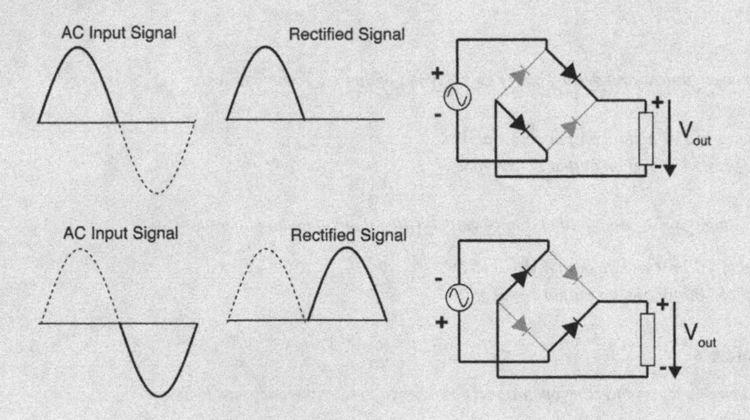

BJT common emitter amplifier

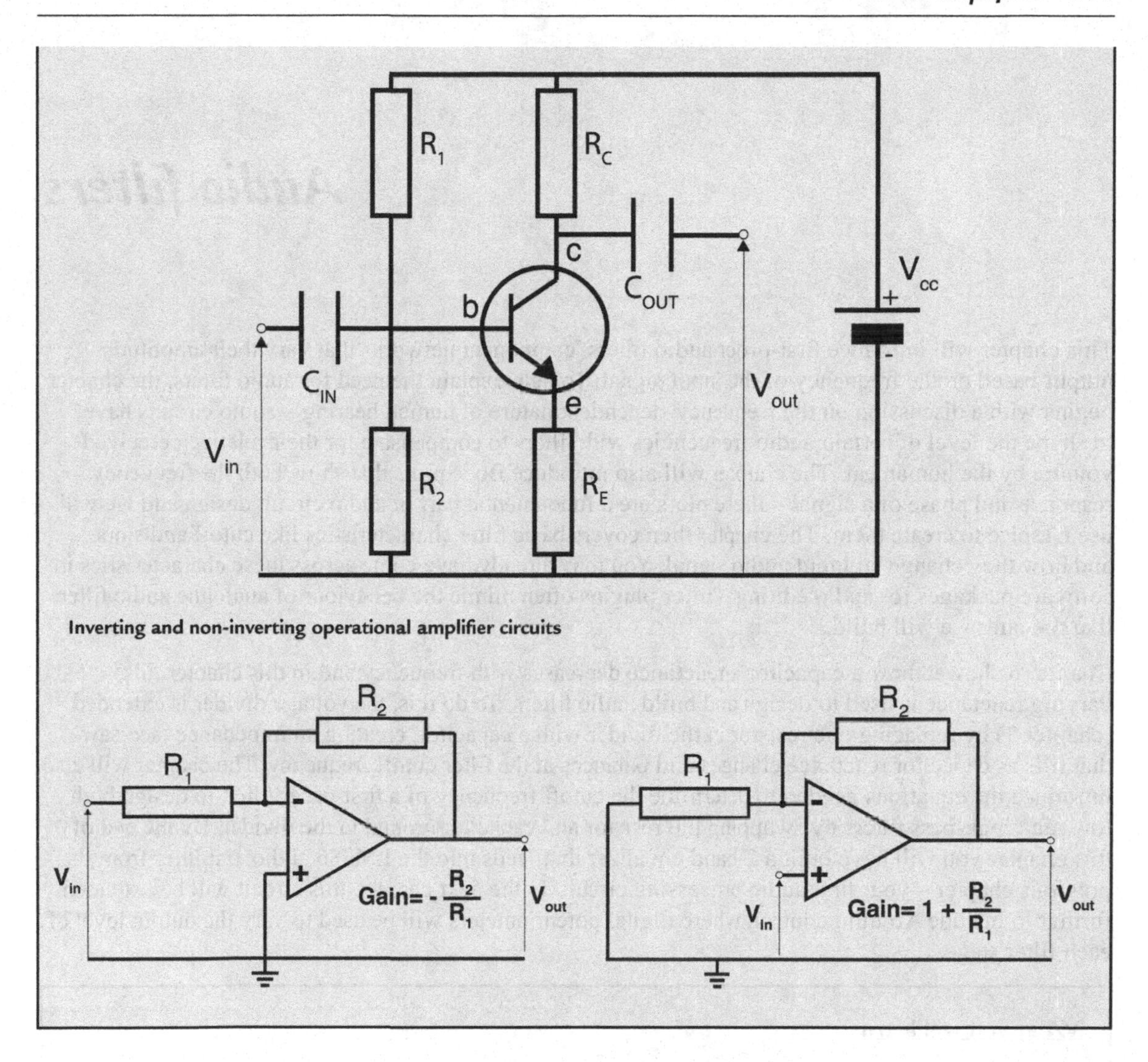

Inverting and non-inverting operational amplifier circuits

CHAPTER 8

Audio filters

This chapter will introduce first-order audio filters, component networks that vary their amplitude output based on the frequency of the input signal. To help explain the need for audio filters, the chapter begins with a discussion on the frequency-dependent nature of human hearing – audio circuits have to shape the level of certain audio frequencies with filters to compensate for their relative perceived volume by the human ear. The chapter will also introduce Bode plots that show both the frequency response and phase of a signal – these plots are a fundamental part of audio circuit design and we will use LTspice to create them. The chapter then covers basic filter characteristics like cutoff and slope and how they change an input audio signal. You may already have come across these characteristics in software packages for audio editing – filter plugins often mimic the behaviour of analogue audio filters like the ones we will build.

Chapter 6 showed how a capacitor's reactance decreases with frequency and in this chapter this varying reactance is used to design and build audio filters. To do this, the voltage divider is extended (chapter 3) by replacing one resistor in the divider with a capacitor, creating an impedance 'see-saw' that tilts as capacitor reactance changes and balances at the filter cutoff frequency. The chapter will also introduce the equations needed to determine the cutoff frequency of a first-order filter, to design both low- and high-pass filters by swapping the resistor and capacitor around in the divider. By the end of this chapter you will have built a 2-band equalizer that feeds into the LM386 audio amplifier from the previous chapter – your first audio processing circuit. In the next chapter, this circuit will be extended further to include Arduino control, where digital potentiometers will be used to vary the output level of each filter stage.

What you will learn

How the decibel scale is used in different ways in audio measurement
How to design first-order low- and high-pass filters
How to use LTspice to analyse possible circuit designs using Bode plots
How to build an audio amplifier with 2-band equalizer

8.1 Decibels and equal loudness

The human ear is an incredibly powerful *sensor* that detects the vibration of molecules (usually in air) and *transduces* these into nerve impulses that are then processed by the brain. The dynamic range of human hearing is very broad, capable of detecting very quiet sounds with physical intensities as low as 2×10^{-5} Pa (10^{-12} N/m^2) all the way up to loud sounds of higher than 20 Pa (1 N/m^2) where listening becomes painful. Similarly, most young people (without hearing loss) can detect audio frequencies of as low as 20Hz up to high frequencies of 20KHz – giving an overall range of 19,980Hz.

Defining pressure

A pascal is a unit of pressure that defines the force in newtons over a surface area of 1 square metre. Sound engineers will usually work with sound pressure level (defined below) whilst acousticians will mainly use sound intensity level, which defines the *energy* of a sound over a given area. Because the human ear detects changes in air pressure SPL (N/m^2) will be used, rather than SIL, which uses the quantity of watts per square metre (W/m^2).

Given the wide range of possible values that the ear can detect, it can prove difficult to represent audio measurements on a linear scale. As shown in chapter 7 when deriving amplifier gain, when working with a wide scale of values mathematicians often use *logarithms* to express exponential increases in a quantity, where the *exponent* is the power by which the quantity is raised (e.g. x^2, x^3, x^4, etc.). A decibel represents the ratio between two numbers on a logarithmic scale, and often one number is a reference value based on a known constant. Thus, to measure how quiet or loud a sound is the logarithmic decibel scale can be used, where sound pressure level (SPL) is defined as:

$$\boldsymbol{SPL} = \boldsymbol{20log_{10}}\left(\frac{p_1}{p_0}\right) dB_{SPL} \qquad (8.1)$$

p_0 *is a reference sound pressure (usually the lowest audible sound of* $10^{-12}\ N/m^2$ *or* $20\mu Pa$*)*

p_1 *is the pressure of the sound source being measured*

dB_{SPL} *is the unit of sound pressure level – measured in decibels*

By defining the lowest audible level as a $0dB_{SPL}$ reference, we can then compare example values for conversation ($40–60dB_{SPL}$) or recommended maximum safe levels for long-term (according to ISO 1999:2013, this is $85dB_{SPL}$ over 8 hours) and short-term ($120dB_{SPL}$ instantaneous) noise exposure. Though none of the circuits described in this book can drive high-output transducers to any significant levels, SPL gives an indication of how loud sound systems can safely operate.

Sound pressure level (SPL) can be used to objectively measure the relative pressure of a sound, but the human ear does not respond linearly to all sounds across the audio spectrum. The subjective term of loudness is often used to describe how a sound is perceived by a human listener, but it is important to note that this is not a directly measurable quantity – in particular hearing response can vary widely with age (and repeated exposure to high-SPL sounds). In general terms, the primary means of human communication is speech and so the ear is optimized for the main frequency range of speech intelligibility (around 300Hz to 4kHz). This creates a variance in how humans perceive loudness based on frequency, which was originally tested by Fletcher and Munson in the 1930s and subsequently refined in further experiments to the current ISO standard (ISO 226:2003) for equal loudness curves (Figure 8.1).

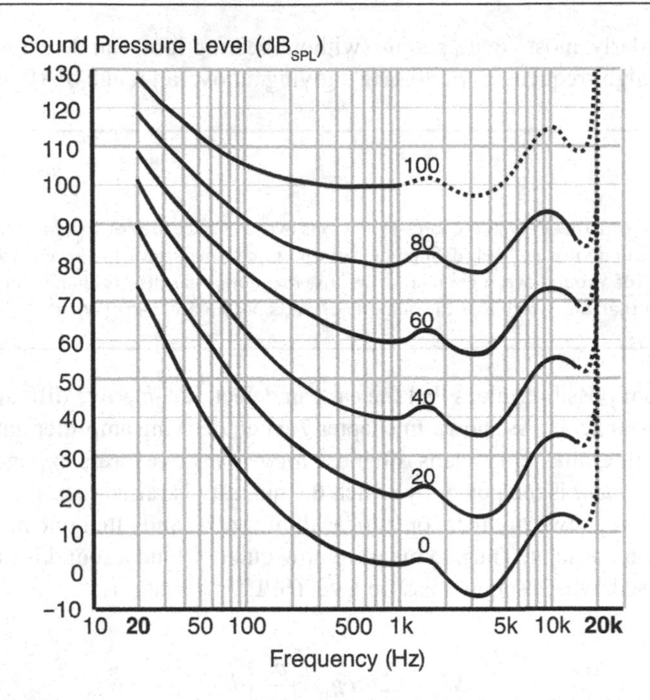

Figure 8.1 Equal loudness curves, adapted from ISO 226:2013. ***In the diagram above, the loudness of sounds at different frequencies (in Hertz) as perceived by human listeners (in phons) is shown for a range of sound pressure levels (in dB$_{SPL}$) relative to a reference sound at 1kHz (the dotted lines show estimated values for very high frequencies). The graph shows that sounds between 3 and 6kHz require less SPL to be perceived as equally loud, but above this level a higher SPL is needed to maintain the same perceived level of loudness. At lower frequencies (particularly below 100Hz), a significantly higher SPL level is required for a sound to be perceived at the same loudness level as the 1kHz reference tone. On the x-axis, decade values (10/100/1k/10kHz) for frequency are shown, alongside half decades (50/500/5kHz) and the limits of normal human hearing (20/20kHz).***

This graph shows how the perceived loudness of a sound (in **phons**) varies with frequency in relation to its actual sound pressure level (SPL). The x-axis shows a logarithmic frequency scale, which as discussed in chapter 7 (equation 7.2) is used to define the wide range of frequencies within the human hearing range (as a linear axis would be too large). On this axis, each vertical grey line represents a doubling in frequency, where the tick marks for each frequency decade of 10Hz, 100Hz, 1kHz, 10kHz are all separated by ten lines that subdivide that frequency range exponentially. In the graph, perceived loudness is defined for a range of frequencies within the audio spectrum relative to the perceived loudness of a 1kHz test tone. This means that the 1kHz reference tone is used as a subjective comparison with tones of other frequencies, to determine whether these other tones are perceived as being louder (or quieter) than the 1kHz test tone. For example – a 60 dB$_{SPL}$ tone at 1kHz is defined as 60 phons, but when a 100Hz tone is played as a comparison it must be *raised* to around 78 dB$_{SPL}$ before it is perceived by a human listener to be at the same loudness.

It can be seen from the curves on the graph that lower frequencies (particularly under 100Hz) are hardest for humans to hear – requiring a much higher SPL than sounds in the human speech range for the same perceived level of loudness. It is also interesting to note that a lower SPL is typically required for sounds between 2kHz and 6kHz, before hearing performance again reduces as the frequency increases to 10kHz and above (there are also noticeable 'bumps' at around 1.5kHz and 10kHz).
A full discussion of human hearing, perception and cognition is outwith the scope of this book, but the diagram in Figure 8.1 can help to illustrate why building a simple linear amplifier (like the one

in the previous chapter) will not produce a useful audio system. This is an important point for audio electronics design, as the non-linear response of the human ear requires circuits that can compensate in some way for different frequency ranges – a process known as **filtering**.

8.2 Filter characteristics and Bode plots

An audio filter can vary the gain level for different frequencies, allowing an input signal to be shaped to boost (increase) or cut (decrease) different frequency ranges as required. In simple terms, a filter will pass certain frequencies more easily than others, thus filtering out those that are not of interest. An audio filter has several important characteristics, which can be illustrated using a general example of a bandpass filter (Figure 8.2).

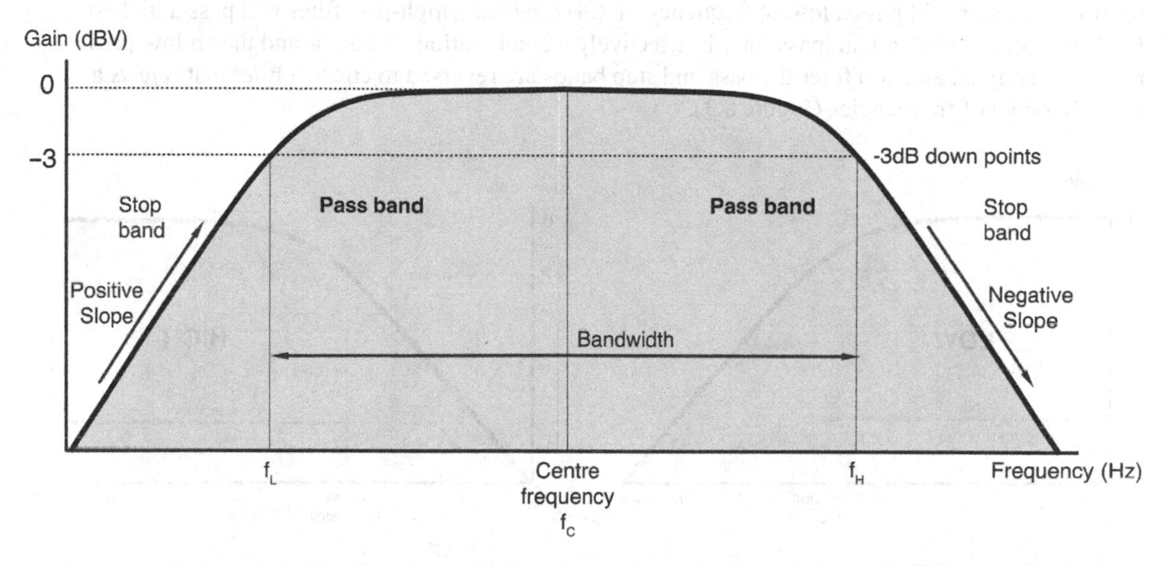

Figure 8.2 Bode magnitude plot of a general bandpass filter. ***The diagram shows the output gain (in dBV) relative to a logarithmic frequency scale (in Hertz) for a theoretical bandpass filter. The lower (fL) and upper (fH) cutoff frequencies are shown, alongside the centre frequency (fc). Filter bandwidth is defined as the range between the lower and upper cutoffs, while the slopes of the filter are also indicated (positive and negative), which determine the order of the filter. The stop bands indicate which frequency ranges will be significantly attenuated (reduced) by the filter, though in practice any real filter will reduce the level of all frequencies to some extent.***

A Bode magnitude plot (named after the scientist Hendrik Bode) shows the output gain of a system in dBV over a **logarithmic** frequency scale (for audio, this scale is centred on the human hearing range of 20–20kHz). A bandpass filter is named because it will pass a specific range of frequencies known as the **pass band**. It also has two stop bands below and above the filter's operating frequency range, where the output gain is attenuated to reduce those frequencies. The plot in Figure 8.2 shows the generalized frequency response for a 1V reference input signal, which has a value of 0dBV. then The filter cutoff frequency is then defined as the point where the output gain has dropped by 3dB (also known as the **−3dB down point**). For a bandpass filter there are both low (f_L) and high (f_H) cutoff frequencies, alongside a centre frequency (f_C) where the filter will have **unity gain** (i.e. the output is **0dBV**). Filter bandwidth is measured as the distance between the low and high cutoff frequencies, whilst the **slope** of the filter determines how much **attenuation** occurs in the stop band (the steeper the filter slope, the lower the gain outside of the pass band region).

Describing filters

Pass band – the frequencies that are **amplified** (to some level) by the filter
Stop band – the frequencies that are **not amplified** by the filter
Cutoff frequency – one (or more) frequencies at which the filter gain has **reduced by 3dB** (*−3dB down point*)
Centre frequency – the frequency at which the filter has **unity gain** (for a 1V reference, this is 0dB)
Bandwidth – the **frequency range** between cutoff frequencies
Slope – the **gradient** of attenuation of the filter (*the steeper the slope, the more the gain is reduced*)

There are three other common filter types: low pass, high pass and band stop. In the case of low-pass and high-pass filters, the bandwidth of the filter is defined relative to the limits of human hearing (a low-pass filter will pass a lowest frequency of 20Hz, whilst a high-pass filter will pass a highest frequency of 20kHz). A bandpass filter is effectively a combination of a high- and then a low-pass filter, whilst in a bandstop filter the pass and stop bands are reversed to create a filter that *rejects* a specific range of frequencies (Figure 8.3).

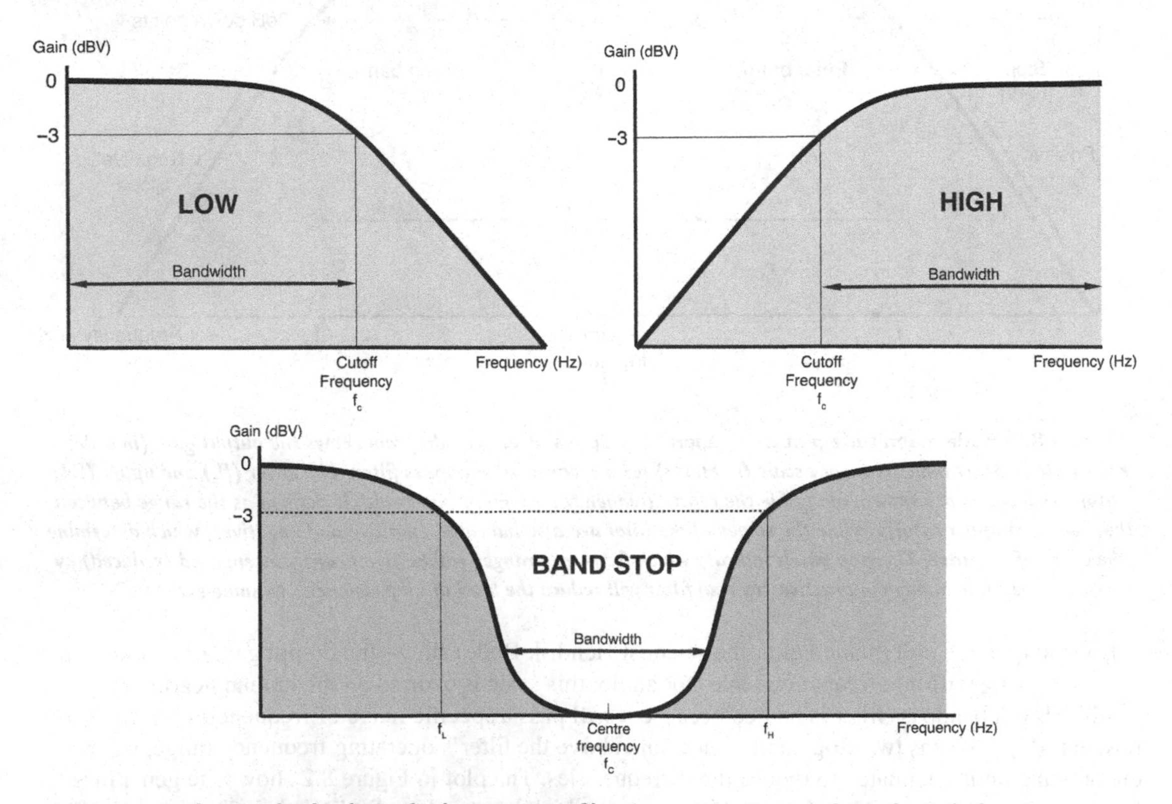

Figure 8.3 Bode magnitude plots of other common filter types. ***The diagram shows theoretical response curves for low, high and bandstop filter types. For a low-pass filter, the lower bandwidth limit is 20Hz whilst for a high-pass filter the upper bandwidth limit is 20kHz – relative to the limits of human hearing (20–20kHz). This does not mean a low-pass filter must have a low cutoff frequency, as the name only defines the direction of the filter slope (low has a negative slope, high has a positive). Like a bandpass, a bandstop filter is made by combining low and high-pass filters, where the bandwidth is now defined by the range of frequencies that are rejected by the filter (between f_L and f_H).***

The diagram in Figure 8.3 shows the frequency response characteristics for idealized filters, where the name of the filter describes the **direction of the slope, not cutoff frequency**. This is an important point, as a low-pass filter with a cutoff frequency of several kilohertz or a high-pass filter with a cutoff around 80–100Hz is often used in audio production. For example, when recording a drum kit it is typical to place additional microphones near individual drums to enhance particular elements of the signal taken from an overhead stereo pair. In this case, a snare drum microphone could be fed through a low-pass filter with a cutoff around 3kHz to reduce the level of the cymbals being recorded on that channel (see Figure 6.4 for a chart of instrument frequency ranges). Similarly, a guitar microphone on a live recording may pick up significant amounts of bass drum and so a high-pass filter with a cutoff around 100Hz will help to reduce this. In practice, any audio recording or live production will employ many different filtering strategies (there is no single method!) involving more complex filters that are effectively combinations of low- and high-pass filters – all of which have a cumulative impact on the overall sound. It is important to remember that audio **filters attenuate all frequencies in a signal to some extent** by virtue of the components involved (resistors and capacitors impede the flow of current), and so audio recording problems cannot be solved by simply adding more filtering processes.

The theoretical Bode plots shown in the previous diagrams define magnitude relative to frequency, but do not take the phase of the output signal (relative to the input) into account. Chapter 6 introduced the basics of signal phase (Figure 6.14), and the effect of phase cancellation is a known issue when working with more than one signal (e.g. stereo signals, effects send/returns). Although the circuits in this book work with one signal, it is still important to be aware of the phase shifts that occur due to the introduction of electronics components into the signal path (the previous chapter showed how an inverting amplifier will shift the phase of the output signal by 180°). Thus, the standard Bode plot combines both *magnitude* and *phase* plots to give a more complete expression of a system's frequency response characteristics (Figure 8.4).

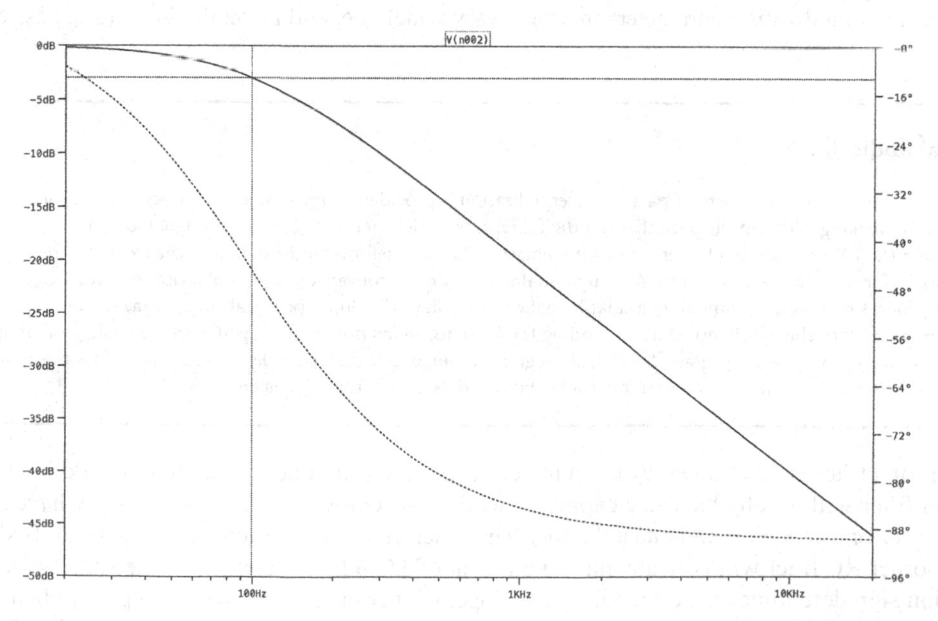

Figure 8.4 Standard Bode plot of a first-order low-pass filter. ***The diagram shows a standard Bode plot of frequency on the x-axis versus both the magnitude (solid line) on the primary (left) y-axis and phase (dashed line) on the secondary (right) y-axis. This plot shows the frequency sweep between 20 and 20kHz for a 1V reference input signal through a low-pass filter with cutoff frequency fc = 100Hz. The cursor marks the −3dB down point (primary y-axis), which corresponds to a phase angle of −45°(secondary y-axis).***

You will very likely have come across audio filters already, as they are used in most types of analogue and digital audio equipment in some form (Figure 8.5).

Figure 8.5 Audio filter examples. ***From left: tone control for Stratocaster-style electric guitar, guitar amplifier 3-band equalizer and preamplifier channel parametric equalizer. The tone control is effectively a low-pass filter, whilst the other examples combine more complex higher-order filtering stages.***

These images show examples of audio equalization stages that range from a simple guitar tone control, through the standard 3-band equalizer found in most audio processing stages to a more complex preamplifier channel equalizer that provides variable control over both filter shape and cutoff. Most electric guitars (and basses) have at least one tone control that is effectively a low-pass filter – active versions combine a small preamplifier to boost the signal and reduce the impact of filter attenuation. Most instrument amplifiers will usually provide at least two filter controls, which are often referred to as bass and treble – bass meaning low frequency and treble meaning high. A much more common setup for audio amplification incorporates an additional mid control, which is effectively a bandpass filter with a centre frequency in the middle of the audio spectrum. The more advanced waveshaping of a dedicated preamplifier channel is designed for specific applications like vocals which will involve highly specific equalization parameters that may vary widely depending on the input (e.g. bass or soprano).

Digital audio filters

In the digital domain, some form of parametric equalization is provided by most audio playback applications and plugins. By working with sample data that is a digital representation of the original audio signal, digital signal processing (DSP) methods can boost or cut frequency bands by transforming the data into the frequency domain using some form of Fourier transform. Although analogue to digital conversion is possible with the Arduino, the limited processing power and memory available make it difficult to develop a practical circuit. Having said this, chapters 4 and 5 of this book introduced C coding for Arduino, which does provide you with the basic grounding needed to begin learning to program Audio DSP. A great resource for developing digital audio plugins is the cross-platform Juice Framework – you can get free tools and tutorials from https://juce.com/.

The order of a filter is determined by the number of reactive components required to build it – a first-order filter will usually have one capacitor, whilst a second-order filter will typically have two capacitors (or one capacitor and one inductor). The order of a filter controls the gradient of its slope so a first-order RC filter will typically have a gradient of +/− **6dB/octave**, where the addition or subtraction sign determines the direction of the slope. Higher-order filters have steeper gradients (i.e. +/−12, +/−18, +/−24dB/octave) and may also have more complex response curves (e.g. a bandpass or bandstop filter). The following sections will show how to design first-order low-pass and high-pass filters and then simulate them using LTspice.

Decades vs octaves

Most electronics applications do not work with audio signals, and one area where this becomes apparent is in relation to frequency response. It is common to define the gradient of filters in terms of the decade - each frequency point represents an increase of ***ten*** from the previous frequency. This is fine for other forms of electronics, but in the previous chapter it was noted that ***music*** theory works with the ***octave*** as being the fundamental interval from which all others are derived - each octave ***doubles*** the frequency of the note below it. Other areas of audio (e.g. acoustics) use the octave as a more appropriate measure of frequency response for the human listener, and this is also the case for audio filters. It is important to remember that gradients will differ based on the range that you measure them over, so a **−6dB/octave** gradient corresponds to a **−20dB/decade** gradient in terms of their slope. As this is an audio electronics text, *the octave will be used* in all audio measurements.

8.3 First-order low-pass filter

A low-pass filter allows **frequencies that are lower than the cutoff** through the passband region (Figure 8.6).

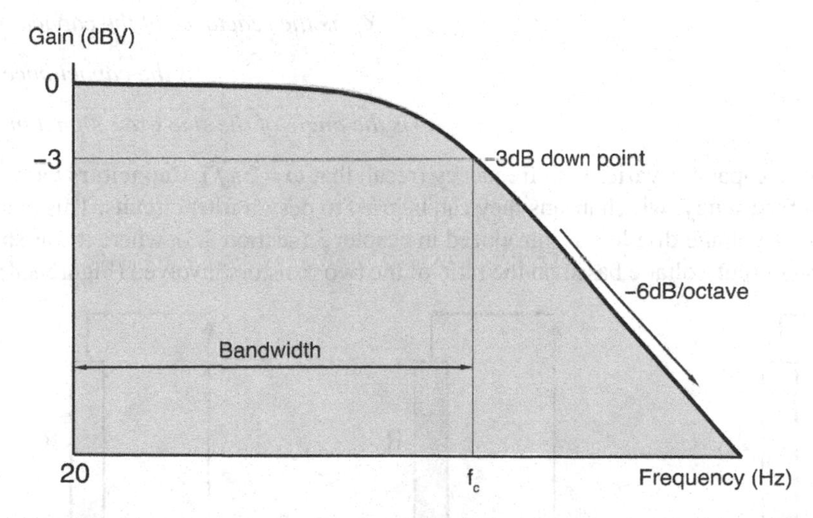

Figure 8.6 Bode magnitude plot of a first-order low-pass filter. ***The filter bandwidth begins at the lowest frequency of human hearing (20Hz) and ends at the cutoff frequency (fc) which is at the −3dB down point. After the cutoff frequency, the slope of the filter decreases at −6dB/octave – it has a negative slope.***

The diagram shows the bode magnitude plot for a first-order low-pass filter, which allows frequencies below the cutoff to pass unattenuated. Notice that the diagram defines the gradient of the slope of the filter as −6dB/octave – this is usually an indicator of an audio filter (rather than decade response being shown). The cutoff frequency occurs when the input signal frequency is attenuated by approximately 30% (a **voltage gain of ~0.701 gives the −3dB** down point), after which it will reduce by a further −6dB/octave. This reduction determines the slope of the filter, and although higher-order filter analysis is outwith the scope of this book, it is also important to note that a **−6dB/octave slope** means there is only one reactive component (e.g. a capacitor) – this gradient of slope is typical of a **first-order** low-pass filter. The direction of the slope distinguishes a low-pass (negative slope) from its high-pass (positive) counterpart (Figure 8.7).

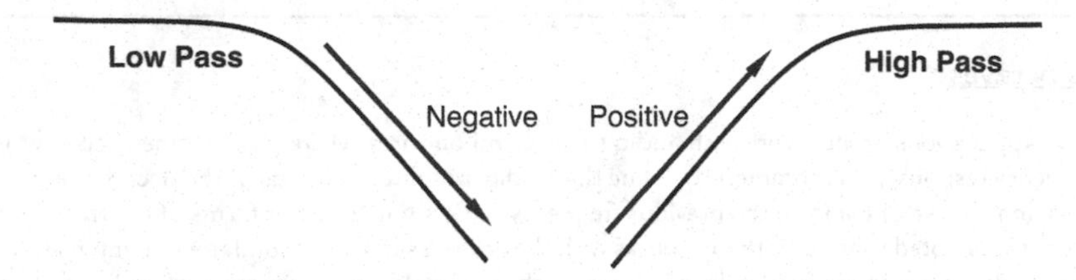

Figure 8.7 Low-pass and high-pass filter slopes. ***The diagram shows the slope direction for both low- and high-pass filters. The left-hand diagram shows a low-pass filter which has a negative slope, whilst the right-hand diagram shows a positive-slope high-pass filter.***

To create this filter response curve, the output voltage gain must be varied based on the frequency of the input signal. Chapter 6 introduced the capacitor as a frequency-dependent component, where section 6.3 (equation 6.10) defined the equation for capacitive reactance (X_C):

$$\textit{Capacitive Reactance}, X_C = \frac{1}{\omega C} \quad (8.2)$$

X_C is the reactance of the capacitor in ohms (Ω)

C is the capacitance in farads (F)

ω is the angle of the sine wave signal in radians (rad)

The reactance of a capacitor varies with frequency (recall that $\omega = 2\pi f$). **Capacitors vary their reactance with frequency**, which means they can be used to design filter circuits. This means returning to the definition of voltage dividers as introduced in chapter 3 (section 3.3), where it was shown that a divider splits the output voltage based on the ratio of the two resistors involved (Figure 8.8).

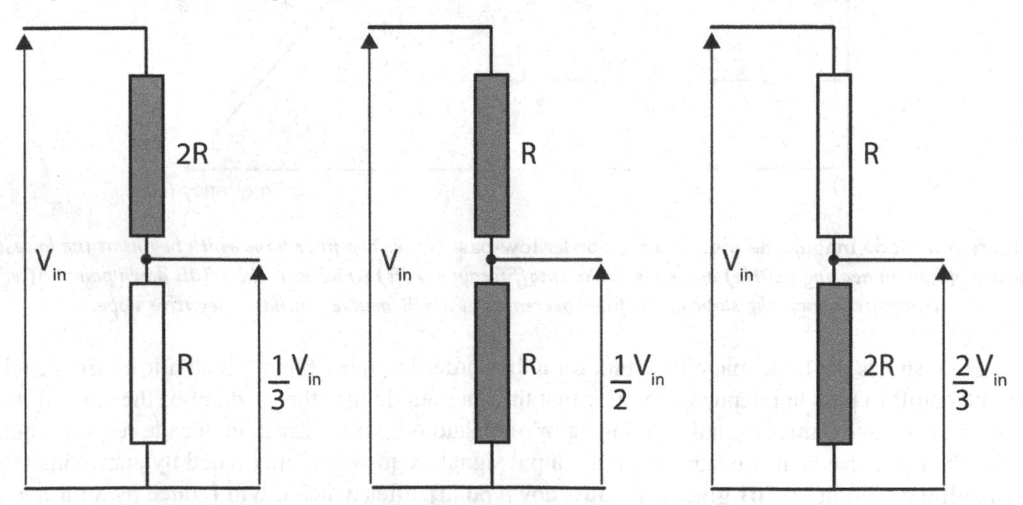

Figure 8.8 A voltage divider (adapted from chapter 3, Figure 3.6). ***The diagram shows how connecting two resistors in series allows the input voltage (V_{in}) to be divided between them. The left-hand divider has a larger resistor on the bottom, and so more of the input voltage is dropped across it. The middle diagram shows two equal resistances, where half the input voltage is dropped across each one. The right-hand diagram shows a larger top resistor, which means the voltage dropped across the second resistor is reduced.***

In this diagram, three voltage dividers are shown to illustrate the relationship between the ratio of the resistors and output voltage across the bottom resistor. The left-hand diagram shows how a larger bottom resistor ($2R$) will drop more of the input voltage (V_{in}) across it, whilst a smaller bottom resistor (when the top resistor is $2R$) will have less. The middle diagram shows what happens when the ratio of both resistors is equal, and in this case half of the input voltage will now be dropped across each resistor. Chapter 3 (equation 3.4) also derived the relationship between the output voltage (V_{OUT}) across the bottom resistor (R_2) relative to the ratio of the resistors and the input voltage (V_{in}):

$$\textit{Output Voltage, } V_{out} = V_{in}\left(\frac{R_2}{R_1 + R_2}\right) \quad (8.3)$$

R_1 is the resistance of the top divider resistor in ohms (Ω))

R_2 is the resistance of the bottom divider resistor in ohms (Ω)

V_{in} is the input voltage to the divider in volts (V)

V_{out} is the output voltage from the bottom of the divider in volts (V)

By solving the voltage divider equations above it can be seen that the ratio of the two resistors dictates the *proportion* of the input voltage dropped across them. As discussed in chapter 7, It can be useful to think of a divider as a seesaw that is balanced by the ratio of the resistor values involved (Figure 8.9).

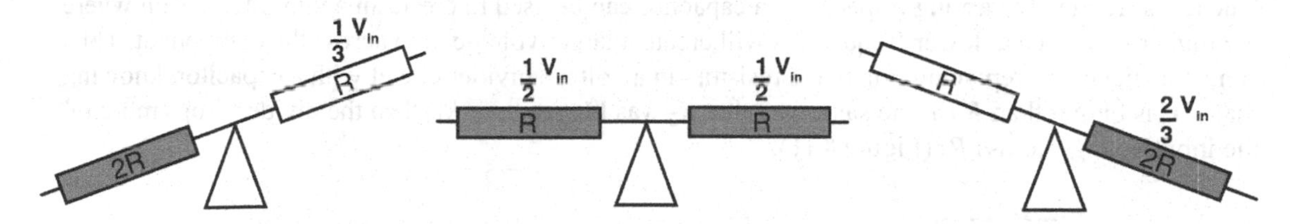

Figure 8.9 Seesaw analogy for voltage divider ratios (adapted from chapter 3). ***The diagram above illustrates the concept of balancing resistance values within a voltage divider, and how this ratio changes the output voltage seen across both components. The left-hand divider has more resistance at the top of the divider, so the output voltage is lower (the converse is true for the right-hand divider). The middle divider shows when both divider resistances are balanced, where half of the output voltage is dropped across each resistor.***

This diagram reiterates the seesaw divider analogy from chapter 3, primarily to highlight the importance of the case where both divider resistors are evenly balanced. In this instance, equal divider resistances lead to equal voltage being dropped across both ($\frac{1}{2} V_{in}$) and this point is crucial to understanding filters. For voltage dividers, as R_2 decreases the output voltage V_{out} decreases – so now this divider concept can be used to **vary the output voltage based on the frequency** of the input signal. This requires a component that changes its impedance based on frequency – the most common being the capacitor (though inductors were frequently used in older circuits). As discussed in the previous chapter **(see also the Appendix on AC equations)**, capacitors reduce their reactance (impedance) as the frequency across them increases (Figure 8.10).

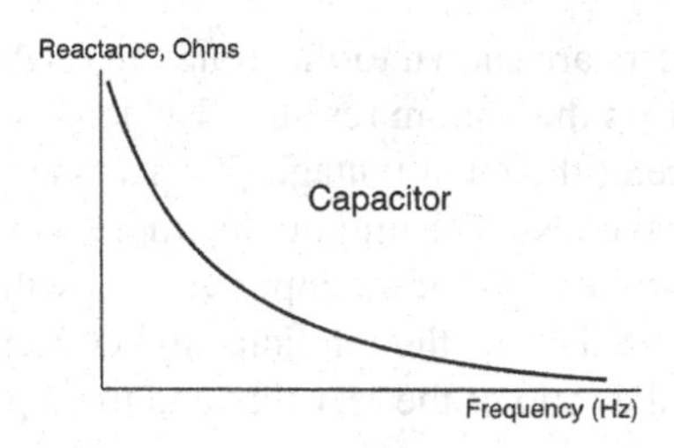

Figure 8.10 Graphs of capacitive reactance relative to frequency. ***The graph shows how capacitive reactance decreases with frequency, based on the inverse relationship between reactance and capacitance derived in the equation 6.4.***

This graph is derived using the equation for capacitive reactance (*Xc*), that is used in filter calculations:

$$\textit{Capacitive Reactance, } Xc = \frac{1}{2\pi fC} \tag{8.4}$$

X_C is the capacitive reactance of the capacitor (in ohms, Ω)

2π is the scaling factor of the frequency dependent capacitor

f is the frequency of the input signal (in hertz, Hz)

C is the value of the capacitor (in farads, F)

Due to its frequency-varying impedance, a capacitor can be used to create an audio filter circuit where the higher reactance at lower frequencies will create a larger voltage drop across the component. This can be achieved by replacing one of the resistors in a voltage divider circuit with a capacitor, knowing that if $\boldsymbol{C}$ is bigger than $\boldsymbol{R}$ (in the same way that $\boldsymbol{R_2}$ was **bigger** than R_1) then the divider drops more of the input voltage across R_2 (Figure 8.11).

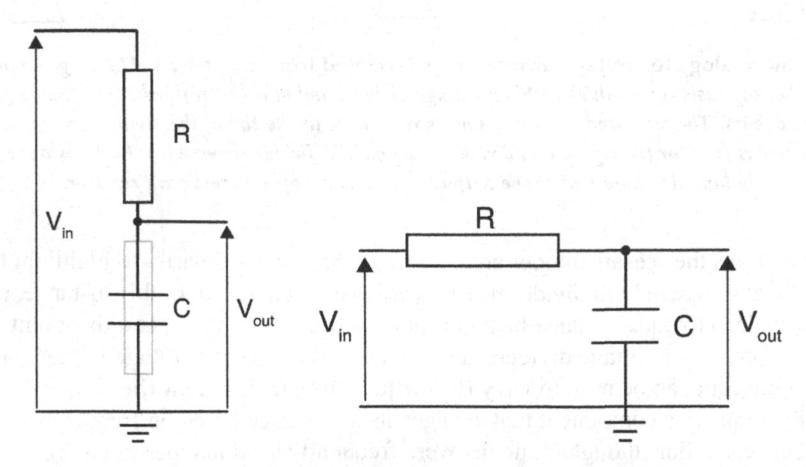

Figure 8.11 First-order low-pass filter circuit. ***The left-hand circuit shows how a voltage divider can become a filter when a capacitor replaces the bottom resistor, due to the frequency-dependent reactance of the capacitor. The right-hand circuit shows the standard component arrangement for this filter circuit (often referred to as an RC filter).***

This diagram shows how a voltage divider can be used as a filter if one of the resistors is replaced by a frequency-dependent component like a capacitor. For a low-pass filter, the bottom resistor in the divider is replaced with a capacitor – the specific resistor (top or bottom) being replaced dictates the direction of the filter slope (see Figure 8.7). The voltage divider equation (8.3) can now be restated for a low-pass filter in terms of the magnitude of the output voltage V_{out} relative to the magnitude of the divider impedance (resistor and capacitor) and the input voltage V_{in}:

$$\textbf{\textit{Low Pass Filter Output Voltage, }} V_{out} = V_{in}\left(\frac{Xc}{\sqrt{\left(R^2 + X_C^2\right)}}\right) \tag{8.5}$$

V_{out} is the output voltage from the low-pass filter (in volts, V)

V_{in} is the input voltage to the low-pass filter (in volts, V)

R is the impedance of the resistor (in ohms, Ω)

X_C is the capacitive reactance of the capacitor (in ohms, Ω)

This divider equation is based on the capacitor as the output load impedance. Chapter 6 (section 6.4) showed that impedance is a complex quantity that takes into account both the magnitude and phase of the component involved. Impedance is required for AC components like capacitors that change their reactance with frequency, rather than the DC value of resistance that was used in previous voltage divider equations. The difference now is that the **total impedance** for the circuit is calculated as the **square root of the squares** of the magnitudes of both component impedances ($\sqrt{\left(R^2 + X_C^2\right)}$), whereas the DC voltage divider calculated the total circuit resistance by simply adding all the resistances in the circuit ($R_1 + R_2$). Chapter 6 (section 6.3, equation 6.8) also defined the time constant as an exponential quantity that defines the time for a capacitor to reach ~63% of the supply voltage:

$$\textbf{\textit{Time Constant, }} \tau = RC \tag{8.6}$$

τ is the time constant in seconds (s)

R is the resistance in ohms (Ω)

C is the capacitance in farads (F)

The appendix on AC equations notes that the time constant is derived using a differential equation involving both the resistor and capacitor voltages in an RC circuit (although the full derivation is outwith the scope of an introductory text). Having said this, the equation effectively determines when the resistive and reactive components in the circuit are equal, which allows the cutoff frequency f_c (or −3dB down point) of the filter to be defined:

$$\textbf{\textit{Cutoff Frequency, }} fc = \frac{1}{2\pi RC} \tag{8.7}$$

f_c is the cutoff frequency of the circuit (in hertz, Hz)

2π is the scaling factor of the frequency dependent capacitor

R is the value of the resistor (in ohms, Ω)

C is the value of the capacitor (in farads, F)

The cutoff frequency is defined as the inverse of the time constant which is then scaled by a factor of 2π – this converts the chronological time value into a sinusoidal frequency (see chapter 6, section 6.2). When the resistive and capacitive components are balanced, the overall output voltage across the capacitor will be $\frac{1}{2}V_{in}$, which equates to a −3dB reduction in voltage terms. In addition, the phase shift introduced by the reactive component (capacitor) will also be balanced by the resistor, so a full phase shift of 90° (which would be expected for a capacitor) will be halved to produce a 45° shift at the cutoff frequency. Although phase is not discussed in detail in this book, the overall indication of a halfway point for −3dBV attenuation is characteristic of a filter cutoff.

This equation is used extensively in audio electronics, as it allows the input signal to be shaped relative to the specific non-linear audio components involved. The examples in Figure 8.5 show a range of audio filters that are commonly found in many types of audio equipment, where the simplest guitar tone control employs the same principles and components as more complex preamplifier channel filters. In all cases, the filter cutoff frequency equation shown above is a useful starting point in designing such audio processing systems (though higher-order filters require more detailed analysis due to the interaction of greater component numbers). In visual terms, the seesaw analogy from Figure 8.9 can once again be used to get a simple overview of how the output voltage V_{out} varies based on the frequency of the input signal V_{in} (Figure 8.12).

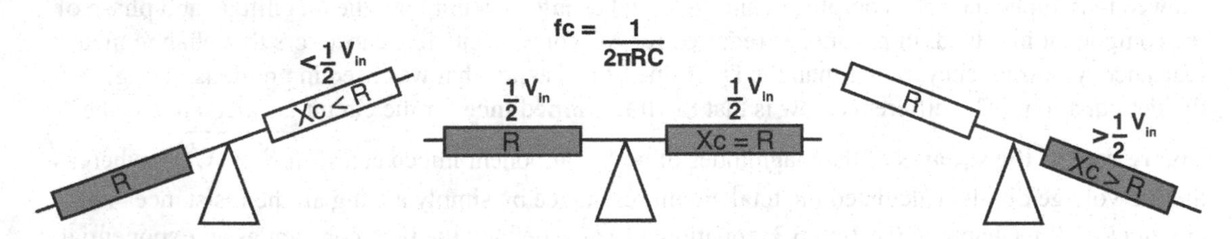

Figure 8.12 Seesaw analogy for a first-order low-pass filter. ***The diagram shows how changes in the reactance of the capacitor (X_c) will change the balance of the divider (i.e. the RC filter circuit). Thus changes in the frequency of the input voltage (V_{in}) from low to high will change the output voltage of the divider (across the capacitor), balanced around the cutoff frequency (fc).***

Now the behaviour of a low-pass filter has been discussed in conceptual terms, the previous equations can be applied to a first-order low-pass filter example.

8.3.1 Worked example – designing a first-order low-pass filter

An RC circuit is configured as a first-order low-pass filter, with a 160kΩ resistor and a 10nF capacitor.

Q1: What is the cutoff frequency of the filter?

$$\textit{Cutoff Frequency}\ fc = \frac{1}{2\pi RC} = \frac{1}{2\pi \times (160 \times 10^3) \times (10 \times 10^{-9})}$$

$$= \frac{1}{6.28 \times 160{,}000 \times 0.00000001} = \frac{1}{0.010048} = 99.52Hz \approx \mathbf{100Hz}$$

Answer: The cutoff frequency of a first-order low-pass filter with R = 160kΩ and C = 10nF is *approximately* 100Hz.

Notes:

The value for 2π is 6.28 (to 2 decimal places) - this is accurate enough for manual calculations.

The resistance value of 160kΩ has a scaling factor of 10^3, which gives 160,000Ω.

The capacitance value of 10nF has a scaling factor of 10^{-9} which gives 0.00000001F.

Follow these steps when calculating the cutoff frequency equation:

1. Write out all the quantities in the equation - *don't skip this step, it will only cost you time if an error is made.*
2. Convert all units to include scaling factors - *this avoids making errors when combining different 10^x values.*
3. Calculate the denominator of the equation - *this avoids getting confused when calculating the reciprocal.*
4. Calculate the reciprocal value (1 divided by the denominator) - *you can use the $\frac{1}{x}$ button on a calculator for this.*
5. State the answer to 2 decimal places - *with filter frequencies, you can then round up to the nearest whole number.*

***Q2**: What is the output voltage (V_{out}) at frequencies of 20Hz, 100Hz, 1kHz and 10kHz for a reference input signal of 1V?*

For **f = 20Hz**:

Firstly, calculate the capacitive reactance:

$$\text{Capacitive Reactance, } Xc = \frac{1}{2\pi fC} = \frac{1}{6.28 \times \mathbf{20} \times (10 \times 10^{-9})} = \frac{1}{6.28 \times \mathbf{20} \times 0.00000001} = \frac{1}{0.000001256}$$

$$= \mathbf{796178.34\Omega}$$

Now calculate the output voltage for an input voltage of 1V (0dBV reference level):

$$\text{Output Voltage, } V_{out} = V_{in}\left(\frac{Xc}{\sqrt{(R^2 + X_C^2)}}\right) = 1 \times \left(\frac{796178.34}{\sqrt{(160{,}000)^2 + (796178.34)^2}}\right)$$

$$= \frac{796178.34}{\sqrt{25600000000 + 633899949085.15}} = \frac{796178.34}{\sqrt{659499949085.15}} = \frac{796178.34}{812096.02} = \mathbf{0.98V}$$

For **f = 100Hz**:

$$\text{Capacitive Reactance, } Xc = \frac{1}{2\pi fc} = \frac{1}{6.28 \times \mathbf{100} \times (10 \times 10^{-9})} = \frac{1}{6.28 \times \mathbf{100} \times 0.00000001} = \frac{1}{0.00000628} = \mathbf{159235.67\Omega}$$

$$\text{Output Voltage, } V_{out} = V_{in}\left(\frac{Xc}{\sqrt{(R^2 + X_C^2)}}\right) = 1 \times \left(\frac{159235.67}{\sqrt{(160{,}000)^2 + (159235.67)^2}}\right)$$

$$= \frac{159235.67}{\sqrt{25600000000 + 25355998214.94}} = \frac{159235.67}{\sqrt{50955998214.94}} = \frac{159235.67}{225734.35} = \mathbf{0.7V}$$

For **f = 1kHz**:

$$Capacitive\ Reactance, \mathbf{Xc} = \frac{1}{2\pi fC} = \frac{1}{6.28 \times \mathbf{1000} \times (10 \times 10^{-9})} = \mathbf{15923.57\Omega}$$

$$= \frac{1}{6.28 \times \mathbf{1000} \times 0.00000001} = \frac{1}{0.0000628}$$

$$Output\ Voltage, \mathbf{V_{out}} = V_{in}\left(\frac{Xc}{\sqrt{(R^2 + X_C^2)}}\right) = 1 \times \left(\frac{15923.57}{\sqrt{(160{,}000)^2 + (15923.57)^2}}\right)$$

$$= \frac{15923.57}{\sqrt{25600000000 + 253560081.54}} = \frac{15923.57}{\sqrt{25853560081.54}} = \frac{15923.57}{160790.42} = \mathbf{0.1V}$$

For **f = 10kHz**:

$$Capacitive\ Reactance, Xc = \frac{1}{2\pi fC} = \frac{1}{6.28 \times \mathbf{10{,}000} \times (10 \times 10^{-9})}$$

$$= \frac{1}{6.28 \times \mathbf{10{,}000} \times 0.00000001} = \frac{1}{0.000628} = \mathbf{1592.36\Omega}$$

$$Output\ Voltage, \mathbf{V_{out}} = V_{in}\left(\frac{Xc}{\sqrt{(R^2 + X_C^2)}}\right) = 1 \times \left(\frac{1592.36}{\sqrt{(160{,}000)^2 + (1592.36)^2}}\right)$$

$$= \frac{1592.36}{\sqrt{25600000000 + 2535599.82}} = \frac{1592.36}{\sqrt{25602535599.82}} = \frac{1592.36}{160007.92} = \mathbf{0.01V}$$

Answer: For a reference input signal of 1V, the output voltages of a first-order low-pass filter with R = 160kΩ and C = 10nF are listed in the table below for frequencies of 20Hz, 100Hz, 1kHz and 10kHz.

Frequency (Hz)	Output voltage (V)
20	0.98
100	0.7
1000	0.1
10000	0.01

Notes:

The value for capacitance reactance, X_c is the only element that changes for different frequencies – the table includes the values for reactance and total impedance to illustrate how the capacitor has less effect as frequency increases:

Frequency (Hz)	Output voltage (V)	Capacitive reactance (Ω)	Total impedance (Ω)
20	0.98	796178.34	812096.02
100	0.7	159235.67	225734.35
1000	0.1	15923.57	160790.42
10000	0.01	1592.36	16007.92

Notice that the 10kHz value is effectively the 1kHz value reduced by a factor of 10 – after the cutoff frequency, a filter will attenuate the output voltage linearly (the slope of the filter). This is one of the reasons why most non-audio filter calculations will use a dB/decade range – the filter slope gradient can be calculated quickly using two decade values (a factor of 10 decrease is −10dB/decade). Although audio filters work with octaves, the above example has used decade values for frequencies above the cutoff (rather than octaves of 2kHz and 4kHz) to quickly show how reactance decreases linearly, demonstrating the slope of a first-order low-pass filter.

When working with large whole numbers, it is useful to use a comma to make the number of zeros easier to read. Thus, 160kΩ becomes 160,000Ω which is easier to count than 160000Ω.

It can also be helpful to make use of the memory settings on your calculator (if you are using one) when working with very small or large numbers to avoid making mistakes when transcribing a value from one equation to another (like capacitive reactance in the example). Copy and paste with software calculators performs the same function – as with coding (see the chapter 5 tutorial on reading/writing code), always aim to copy/paste a known value whenever possible to reduce the chance of introducing transcribing errors into your calculations.

With an understanding of the behaviour of a first-order low-pass filter, LTspice can be used to simulate the same circuit from the previous calculations. In previous chapters, the transient analysis command was used to view a specific period in time, but this type of analysis does not provide information on a range of frequencies like the Bode plot shown in Figure 8.4. To simulate a first-order low-pass filter circuit in LTspice, a different Spice Directive is needed to create a Bode plot that will show both the frequency and phase response over a given range (20–20kHz). The AC analysis command creates step commands, to step through a range of frequencies of interest where LTspice will calculate the signal parameters required. This is effectively doing the same thing as the worked example above, but over a wider range of values (and with far less chance of making arithmetic errors!).

The LTspice AC analysis command

LTspice defines the syntax of the AC analysis command as follows:

.ac <oct, dec, lin> <Nsteps> <StartFreq> <EndFreq>;

<oct, dec, lin> defines the type of frequency sweep as either octaves, decades or linear;

<Nsteps> is the number of analysis steps to plot within the frequency range specified (LTspice will determine these data points automatically);

<StartFreq> <EndFreq> the beginning and end of the frequency range to be simulated for analysis.

Thus, to perform **AC analysis** of a signal over the human hearing range (**20–20kHz**) use **octaves** with **200** steps (more steps means more calculation time):

.ac oct 200 20 20K

The AC analysis command is very powerful, and of direct use in filter design (and other audio electronics work) so it is worth becoming familiar with this structure so you can type it quickly as a Spice directive in a new circuit. In the next worked example, AC analysis is used to generate a Bode plot of an RC filter circuit to produce the same output curves shown in Figure 8.4.

8.3.2 Worked example – simulating a first-order low-pass filter using LTspice

This example will simulate a first-order low-pass filter circuit based on the LTspice schematic shown in Figure 8.13.

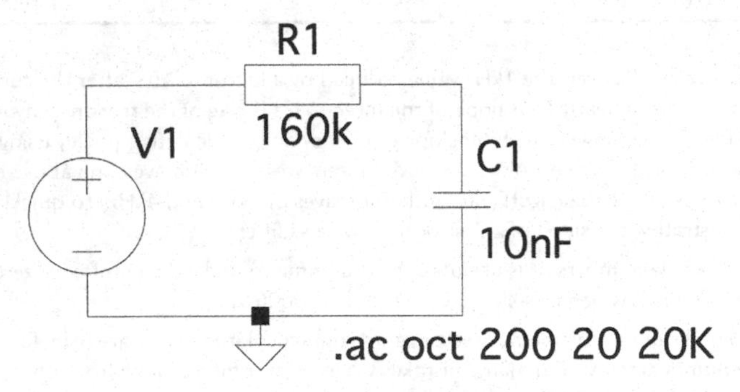

Figure 8.13 First-order low-pass filter circuit. *In the diagram, a voltage source is used as the input signal to a first-order low-pass RC filter. The values for the resistor (160kΩ) and capacitor (10nF) were chosen to provide a cutoff frequency of 100Hz, which should correspond to a reduction of −3dB in gain for that frequency (with a reduction of −6dB/octave thereafter).*

To build this circuit in LTspice, use component values of R1 = 160kΩ and C1 = 10nF with a voltage source that has an **AC amplitude of 1V**. Create a new circuit in LTspice and name it chp8_example1. The process for creating the schematic is as follows:

1. Lay out the components as shown in Figure 8.13: voltage source (on left), resistor, capacitor and the **GND node** (bottom).
2. Connect wires to form a series loop between V1, R1, C1 and GND.
3. Configure the voltage source to be an AC value of 1V. Set the resistor R1 = 160kΩ. Set the capacitor C1 = 10nF.
4. Add a Spice Directive to perform an **AC analysis**, where the frequency of the input signal (V1) is swept between 20 and 20kHz for a total of 200 analysis points. The analysis command is**.ac oct 200 20 20k.**

If everything in the schematic has been configured and connected correctly, applying a voltage probe to the junction between R1 and C1 should give the output trace shown in Figure 8.14.

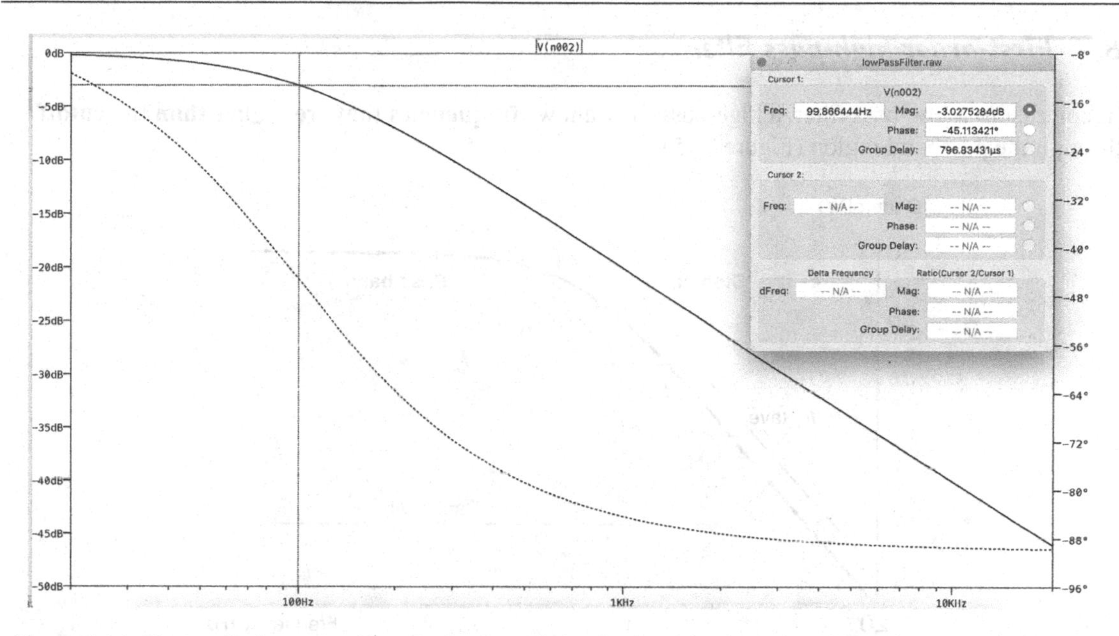

Figure 8.14 First-order low-pass filter Bode plot. ***The Bode plot shows both the magnitude and phase response for a first-order low-pass filter over a frequency range of 20–20kHz. The −3dBV down point is marked (based on a 1V reference input) showing the filter cutoff frequency of ˜100Hz. The dotted line (lower trace) shows the phase response, which equates to a −45° phase shift at the −3dBV down point (the overlaid inspector shows the actual simulation values for this trace).***

The Bode plot shown in this diagram combines both magnitude and phase, so it is important to remember that the **dashed line represents the phase response** of the circuit. The −3dBV down point of the circuit is the cutoff frequency of the filter, which the previous worked example calculated to be 100Hz. By clicking on the trace name (V(n002) in the diagram) a cursor can be attached to the plot, which can then be moved by dragging the mouse or using the arrow keys (this can take a long time but allows for accurate positioning!).

In the diagram, the exact cutoff value of 100Hz is between analysis steps in LTspice, but the overlaid cursor properties shows that a −3.02dBV reduction equates to an output phase shift of −45.11°. As noted in the cutoff frequency calculations (equation 8.7), the balance between resistive and reactive components occurs at the −3dBV down point, which is also when the phase shift due to the capacitor is at the halfway point. It is useful to know that a first-order filter with a single reactive component will introduce a phase shift into the output signal that will vary somewhere between 0 and 90° with a cutoff phase shift of −45° (in practice, 0° phase shift would only occur for a purely resistive circuit, thus the input frequency would be 0Hz!). In more complex filter circuits, the introduction of multiple phase shifts cannot be ignored, particularly when working with multiple channel inputs. An RC filter has now been designed and simulated, where the configuration of the components produces a low-pass filter characteristic. This is an important point about RC filters – by reconfiguring them you can create different response characteristics, as will now be shown with the first-order high-pass filter.

8.4 First-order high-pass filter

In contrast to a low-pass filter, a high-pass filter allows **frequencies that are higher than the cutoff** through the passband region (Figure 8.15).

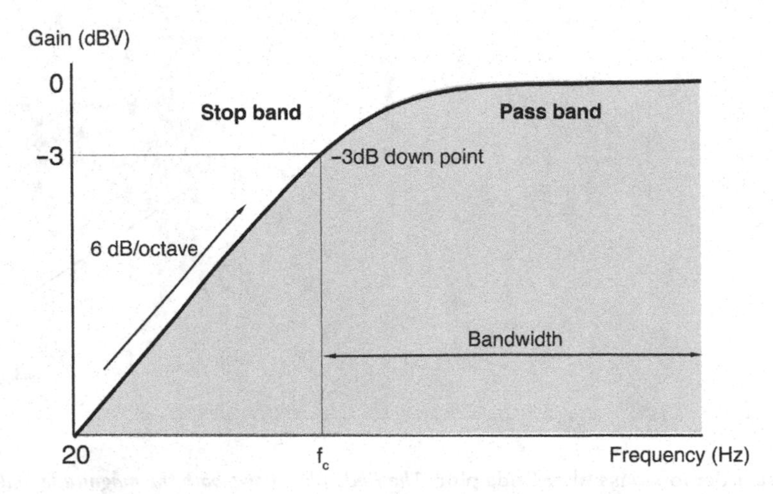

Figure 8.15 Bode magnitude plot of a first-order high-pass filter. ***The filter bandwidth begins at the cutoff frequency (fc) which is the −3dB down point and ends at the highest frequency of human hearing (20kHz). Before the cutoff frequency, the slope of the filter increases at 6dB/octave – it has a positive slope.***

A high-pass filter has the same slope gradient as a low-pass filter (6dB/octave), but this time it increases towards the cutoff frequency (f_c), after which the passband region begins. To design a first-order high-pass filter, the previous low-pass filter circuit can be adapted by simply swapping the components around (Figure 8.16).

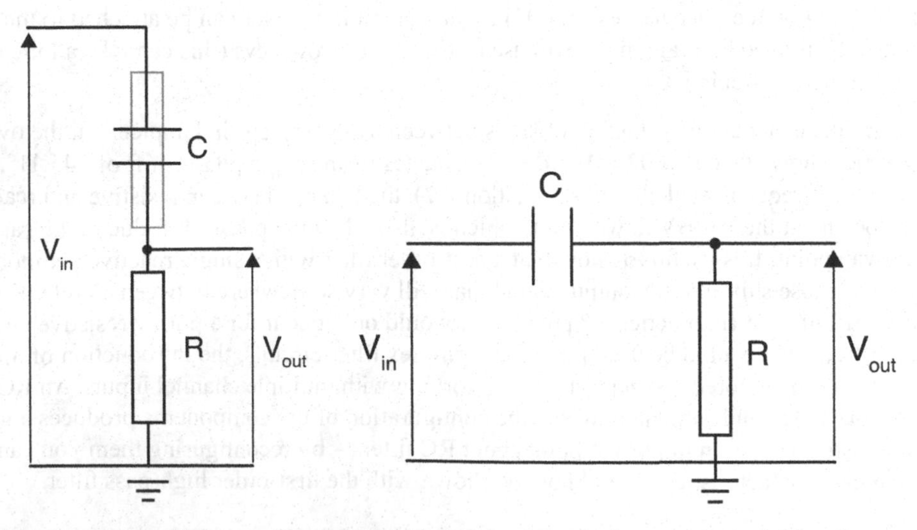

Figure 8.16 First-order high-pass filter circuit. ***The left-hand circuit shows a voltage divider where now the top resistor is replaced by a capacitor, whilst the right-hand circuit shows the standard CR filter component arrangement for a high-pass filter. As the reactive component of the capacitor decreases (as frequency increases) so the divider ratio changes until the resistor is the main impedance in the circuit (after the cutoff).***

The same filter cutoff equation can be used, but the voltage magnitude equation must be rearranged to take the reconfigured divider output impedance into account:

$$\textit{High Pass Filter Output Voltage, } V_{out} = V_{in}\left(\frac{R}{\sqrt{\left(R^2 + X_C^2\right)}}\right)$$

V_{out} is the output voltage from the circuit (in volts, V)

V_{in} is the input voltage to the circuit (in volts, V)

R is the impedance of the resistor (in ohms, Ω)

X_C is the capacitive reactance of the capacitor (in ohms, Ω)

It is important to note the difference in this equation, as it reflects the change in divider configuration from the previous low-pass filter. The total impedance for the circuit remains the same ($\sqrt{\left(R^2 + X_C^2\right)}$), but the **resistor** is now the **output load impedance**. The seesaw analogy can be used again to show how the varying capacitor will change the output voltage V_{out} for different frequencies of input voltage V_{in} (Figure 8.17).

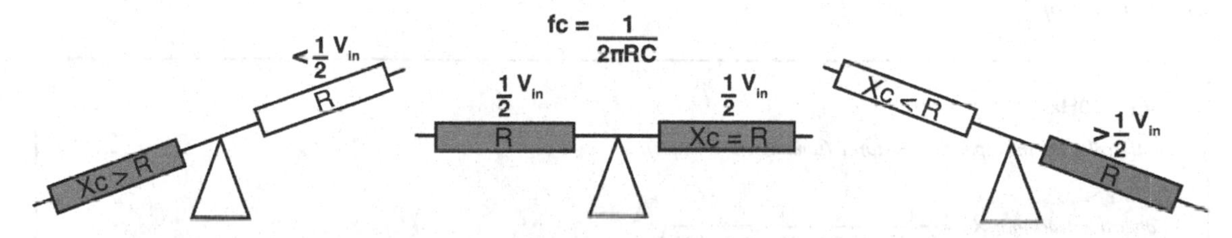

Figure 8.17 Seesaw analogy for a first-order high-pass filter. ***The diagram shows how changes in the frequency of the input voltage (Vin) from low to high change the reactance of the capacitor*** (X_C) ***and hence the balance of the divider. This variance changes the output voltage (Vout) across the load impedance of the divider, which for a high-pass filter is now the resistor R.***

As the previous section has already discussed RC filters, this time a first-order high-pass filter example will be designed based on octaves (rather than decades) to better understand how a filter circuit will behave when working with audio signals.

8.4.1 Worked example – designing a first-order high-pass filter

An RC circuit is configured as a first-order high-pass filter, with a 16kΩ resistor and a 10nF capacitor.

Q1: What is the cutoff frequency of the filter?

$$\textbf{\textit{Cutoff Frequency, fc}} = \frac{1}{2\pi RC} = \frac{1}{2p\times\left(16\times10^{3}\right)\times\left(10\times10^{-9}\right)}$$

$$= \frac{1}{6.28\times16{,}000\times0.00000001} = \frac{1}{0.0010048} = 995.23Hz \approx \mathbf{995Hz}$$

Answer: The cutoff frequency of a first-order high-pass filter with R = 16kΩ and C = 10nF is *approximately* 995Hz.

Notes:

The equation is the same as the previous example, but the resistance value is now 16kΩ. This value is used to show the effect of changing values – a factor of 10 reduction in resistance (or capacitance) will increase the cutoff frequency by a factor of 10 – due to the reciprocal in the equation $fc = \frac{1}{2\pi RC}$.

Chapter 7 looked at reciprocals in parallel resistor networks, using the simple maxim '*the higher the denominator the lower the result*'. This also means the lower the denominator the higher the result, so this example gives a further sense of how changing RC values will affect the cutoff frequency of the filter.

The resistor value of 16kΩ changes the cutoff frequency of the filter to around 995Hz, which is reasonably close to 1kHz (in the low-pass filter example, 160kΩ gave a 99.52Hz cutoff – rounding up to 100Hz). A resistor value of 15.9kΩ would be closer (giving a cutoff of 1001Hz), but this would require more resistor components which would increase noise (and costs).

***Q2**: What is the output voltage (V_{out}) at frequencies of 20Hz, 40Hz, 1280Hz and 2560Hz for a reference input signal of 1V?*

For **f = 20Hz:**

Firstly, calculate the capacitive reactance (same as LPF example):

$$\textit{Capacitive Reactance}, \boldsymbol{Xc} = \frac{1}{2\pi fC} = \frac{1}{6.28 \times \mathbf{20} \times (10 \times 10^{-9})}$$

$$= \frac{1}{6.28 \times \mathbf{20} \times 0.00000001} = \frac{1}{0.000001256}$$

=796178.34Ω

Now calculate the output voltage for an input voltage of 1V (0dBV reference level):

$$\textit{Output Voltage}, \boldsymbol{V_{out}} = V_{in}\left(\frac{\boldsymbol{R}}{\sqrt{(\boldsymbol{R}^2 + X_C^2)}}\right) = 1 \times \left(\frac{16{,}000}{\sqrt{(16{,}000)^2 + (796178.34)^2}}\right)$$

$$= \frac{16{,}000}{\sqrt{256000000 + 633899949085.15}} = \frac{16{,}000}{\sqrt{634155949085.15}} = \frac{16{,}000}{796339.09} = \mathbf{0.02V}$$

For **f = 40Hz:**

$$\textit{Capacitive Reactance}, \boldsymbol{Xc} = \frac{1}{2\pi fC} = \frac{1}{6.28 \times \mathbf{40} \times (10 \times 10^{-9})}$$

$$= \frac{1}{6.28 \times \mathbf{40} \times 0.00000001} = \frac{1}{0.000002512} = \mathbf{398089.17\Omega}$$

$$\textit{Output Voltage, } \boldsymbol{V_{out}} = V_{in}\left(\frac{\boldsymbol{R}}{\sqrt{\left(\boldsymbol{R}^2 + X_C^2\right)}}\right) = 1\times\left(\frac{16{,}000}{\sqrt{(16{,}000)^2 + (398089.17)^2}}\right)$$

$$= \frac{16{,}000}{\sqrt{256000000 + 158474988843.36}} = \frac{16{,}000}{\sqrt{158730988843.36}} = \frac{16{,}000}{398410.58} = \mathbf{0.04V}$$

For **f = 1280Hz**:

$$\textit{Capacitive Reactance, } \boldsymbol{Xc} = \frac{1}{2\pi fC} = \frac{1}{6.28\times\mathbf{1280}\times\left(10\times10^{-9}\right)}$$

$$= \frac{1}{6.28\times\mathbf{1280}\times0.00000001} = \frac{1}{0.000080384} = \mathbf{12440.29\Omega}$$

$$\textit{Output Voltage, } V_{out} = V_{in}\left(\frac{\boldsymbol{R}}{\sqrt{\left(\boldsymbol{R}^2 + X_C^2\right)}}\right) = 1\times\left(\frac{16{,}000}{\sqrt{(16{,}000)^2 + (12440.29)^2}}\right)$$

$$= \frac{16{,}000}{\sqrt{256000000 + 154760731.29}} = \frac{16{,}000}{\sqrt{410760731.29}} = \frac{16{,}000}{20267.23} = \mathbf{0.79V}$$

For **f = 2560Hz**:

$$\textit{Capacitive Reactance, } \boldsymbol{Xc} = \frac{1}{2\pi fC} = \frac{1}{6.28\times\mathbf{2560}\times\left(10\times10^{-9}\right)}$$

$$= \frac{1}{6.28\times\mathbf{2560}\times0.00000001} = \frac{1}{0.000160768} = \mathbf{6220.14\Omega}$$

$$\textit{Output Voltage, } \boldsymbol{V_{out}} = V_{in}\left(\frac{\boldsymbol{R}}{\sqrt{\left(\boldsymbol{R}^2 + X_C^2\right)}}\right) = 1\times\left(\frac{16{,}000}{\sqrt{(16{,}000)^2 + (6220.14)^2}}\right)$$

$$= \frac{16{,}000}{\sqrt{256000000 + 38690182.82}} = \frac{16{,}000}{\sqrt{294690182.82}} = \frac{16{,}000}{17166.54} = \mathbf{0.93V}$$

Answer: For a reference input signal of 1V, the output voltages of a first-order high-pass filter with R = 16kΩ and C = 10nF are listed in the table below for frequencies of 20Hz, 40Hz, 1280Hz and 2560Hz.

Frequency (hertz)	Output voltage (volts)
20	0.02
40	0.04
1280	0.79
2560	0.93

Notes:

This time the filter output voltage was calculated for some of the frequencies that are octaves of 20Hz, but these do not include the actual cutoff frequency of the RC filter (approximately 1kHz). This means the value at the nearest octave (1280Hz) is over 10% higher than 0.7V – the cutoff frequency value is expected to be the −3dB down point in the circuit. A 16kΩ resistor was chosen to show the effect of reducing resistance (or reactance) – a factor of 10 decrease in resistance means a factor of 10 increase in the cutoff frequency. The interesting point is that the filter frequencies used don't directly align with the octaves of the human hearing range – as will be seen in the chapter project, the practical values obtained with electronic components can sometimes be different.

Having worked through the derivation of some output values for a high-pass filter, it can now be simulated in LTspice – this will include using audio files to audition the output of the filter circuit.

8.4.2 Worked example – simulating a first-order high-pass filter with LTspice

In the first part of this example, the previous filter components are swapped around in a new LTspice circuit to create a first-order high-pass filter – the balance of the divider now moves in the opposite direction based on frequency. A simulation creates a Bode plot to learn how to change the magnitude axis scale to verify that the experimental calculations in the previous example are correct. A first-order high-pass filter circuit will be simulated based on the LTspice schematic in Figure 8.18.

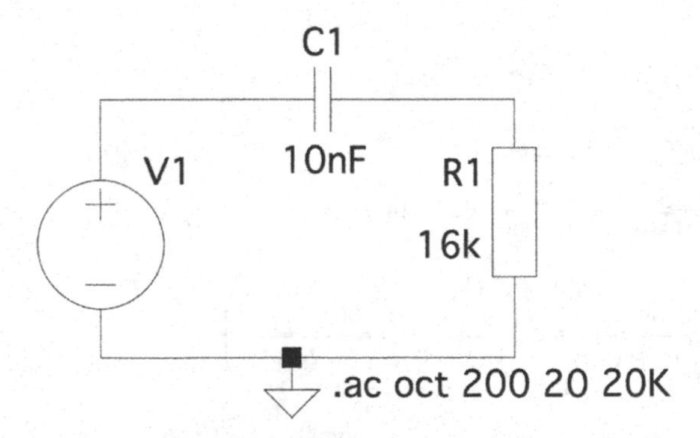

Figure 8.18 First-order high-pass filter circuit. ***In the diagram, a voltage source is used as the input signal to a first-order high-pass CR filter. The values for the resistor (16kΩ) and capacitor (10nF) were chosen to provide a cutoff frequency of 1kHz, which should correspond to a reduction of −3dB in gain for that frequency (with a reduction of −6dB/octave thereafter).***

To build this circuit in LTspice, use component values of R1 = 16kΩ and C1 = 10nF with a voltage source that has an **AC amplitude of 1V**. Create a new circuit in LTspice and name it chp8_example2. The process for creating the schematic is as follows:

5. Lay out the components as shown in Figure 8.13: voltage source (on left), capacitor, resistor and the **GND node** (bottom).
6. Connect wires to form a series loop between V1, C1, R1 and GND.

7. Configure the voltage source to be an AC value of 1V. Set the resistor R1 = 16kΩ. Set the capacitor C1 = 10nF.
8. Add a Spice Directive to perform an **AC analysis**, where the frequency of the input signal (V1) is swept between 20 and 20kHz for a total of 200 analysis points. The analysis command is **.ac oct 200 20 20k**

If everything in the schematic has been configured and connected correctly, applying a voltage probe to the junction between C1 and R1 should give the output trace shown in Figure 8.19.

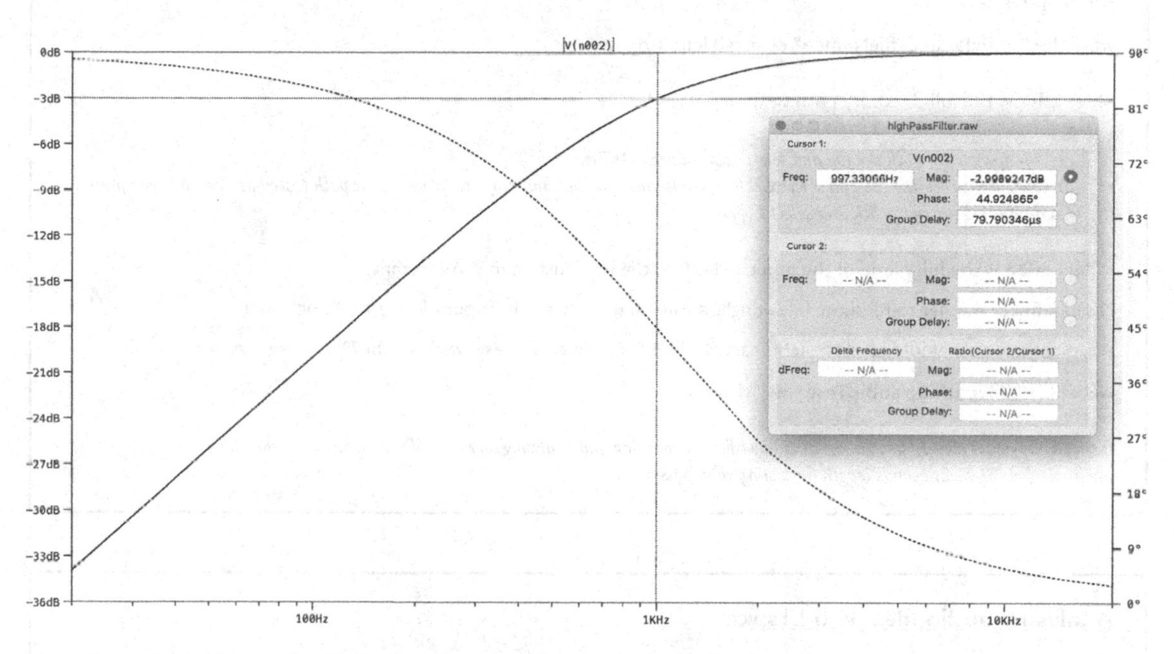

Figure 8.19 First-order high-pass filter Bode plot. ***The Bode plot shows both the magnitude and phase response for a first-order high-pass filter over a frequency range of 20–20kHz. The −3dBV down point is marked (based on a 1V reference input) showing the filter cutoff frequency of ~1kHz, which also gives a −45° phase shift at this point.***

As with the low-pass filter simulation, the diagram shows the exact cutoff value is calculated to be 997.33Hz for a −2.99dBV reduction in output (and a phase shift of −44.92°). The shape of the filter can be seen from the Bode plot, but this only visually illustrates what a first-order high-pass filter with a cutoff frequency of 1kHz will sound like. To address this, LTspice provides commands to read in an audio file as input to the circuit (through the voltage source). A transient analysis command can then be used to write out another audio file containing the output simulation data from our circuit. This can be a very useful way of auditioning possible circuit designs quickly to hear how they will change the input signal – though it cannot replace the Bode plot for accurate frequency analysis of an audio signal. The next example will show how to set up LTspice for **audio input and output**.

8.4.3 Worked example – reading and writing audio files with LTspice

Reading audio files with LTspice

LTspice allows you to pass digital audio data (in PCM WAV format) into a voltage source as a DC signal parameter. As there is only one output from a voltage source, monophonic audio data must be used - for either a mono or a stereo input file, LTspice requires the channel number to read from (left is 0, right is 1). The syntax for the file read command is as follows:

wavefile="/<filePath>/<fileName>" chan=<0 left, 1 right>

<filePath> is the full directory path to the input audio file:

- *on Windows, right-click the file and select Copy Address As Text*
- *on OSX, right-click the file and select Get Info, triple-click (or click and drag) to select the file path (listed as Where:) then press Command+C (or right-click and select Copy)*

<fileName> is the full name of the input audio file (the file must be in WAV format)

chan defines whether to read the left or right audio channel from the input file - left is 0, right is 1

A typical example would load the **left** channel of a file called **audioTest.wav** into the **DC** voltage source:

wavefile="/filepath..../audioTest.wav" chan=0

It should be noted that LTspice is not an audio editor, so reading and analysing an audio file can take time – for this reason, it is best to use audio files of 1–2 seconds length for testing when possible.

Analysing audio files with LTspice

If a wave file has been set as input to a DC voltage source, the audio file can be checked by analysing its data. This can be done with a transient Spice Directive that will analyse the circuit over a specified time period. LTspice defines the syntax of the transient command as follows:

.tran <Tstep> <Tstop> [Tstart [dTmax]] [modifiers]
<Tstep> is the time **increment** for each analysis step - LTspice performs its own compression on the wave data and calculates the step size at that time, so the step size can be **set to 0**
<Tstop> the **length** of time to perform the analysis for - this should be set to the length of the audio file
[Tstart [dTmax]] [modifiers] - these parameters are not needed for audio analysis, so can be **omitted**

A typical example would specify a **transient** analysis window of **2 seconds** (audio file length):

.tran 0 2

If the audio file is loaded correctly, running the LTspice simulation will open up an analysis window with a time axis (as opposed to the logarithmic frequency scale in the previous example). Clicking the probe on the output of the voltage source will generate a time domain plot of how the audio signal changes over time (Figure 8.20).

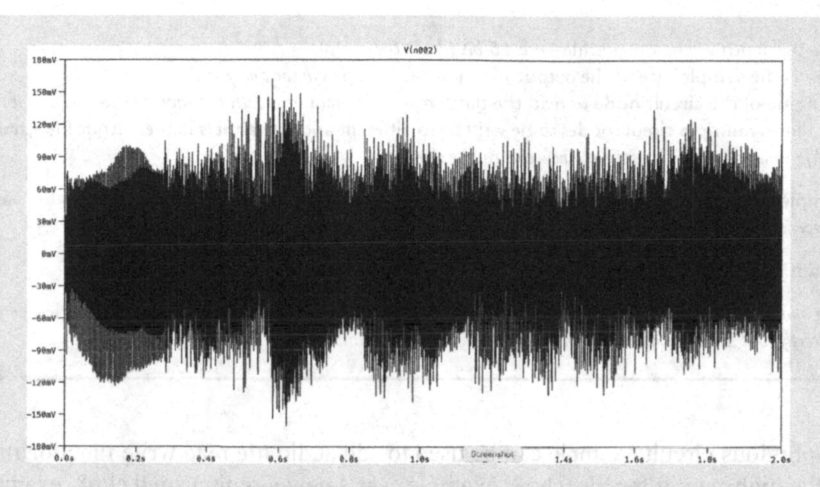

Figure 8.20 LTspice transient analysis window for audio input data (from LTspice by analog devices). ***The waveform is drawn over a 2-second analysis window to show changes in voltage over time. The input file is specified as the DC value for the voltage source, where the data from an audio wave file (.wav) can be loaded into the circuit simulation for analysis.***

Writing simulation output to an audio file with LTspice

By using a wave file as the DC voltage source input, the signal can then be read at other points in the circuit. To do this, a new Net Name must be added as an output Node from the circuit (Figure 8.21).

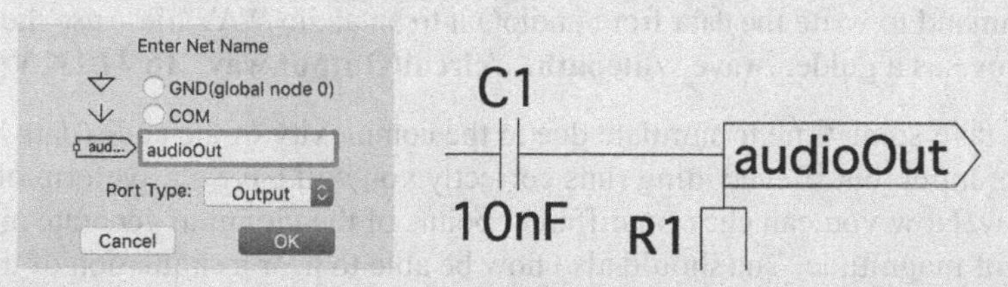

Figure 8.21 Adding an output node net name in LTspice (from LTspice by Analog Devices). ***The left-hand image shows a new net name being configured as an output port (with name audioOut). The right-hand image shows how this output node can be attached to a circuit as a point where simulation data can be obtained.***

The net name can be placed at any point in the circuit, allowing data to be read from it during analysis. Another Spice Directive can then be used to write the analysis data from this node to an audio file, to listen to a simulation of the audio circuits. The syntax for the file write command is as follows:

.wave /<filePath>/<filename.wav> <Nbits> <SampleRate> V(out) [V(out2) ...]

<filePath> is the full directory path to the input audio file – it is usually a good idea to put the output file in the same folder as the input file to avoid confusion when auditioning the output of a circuit (as a form of AB referencing)

<fileName> is the full name of the input audio file (the file must be in WAV format)

<Nbits> is the bit depth of the output file – *use **16-bit** files in the analysis*
<SampleRate> is the sample rate of the output file – *use **44.1kHz** files in the analysis*
V(out) is the name of the **circuit node** to read the data from – *this must be exactly the **same text** as the circuit node*
[V(out2) ...] allows **multiple** circuit nodes to be written to different audio **channels** in the output file – ***stereo** signals will not be used for analysis, but it is useful to know this can be simulated*

A typical example would write the data from a **node** labelled **audioOut** to a **16-bit**, **44.1kHz** audio file named **circuitOutput.wav**:

.wave "/filepath.../circuitOutput.wav" 16 44.1K V(audioOut)

As with reading audio files, it is important to remember that LTspice is an electronic circuit simulation package. It will perform reading and writing correctly, but it will not do so as quickly as a dedicated audio editor.

A copy of the previous circuit example can be used to add audio file read/write functionality. In LTspice save the high-pass filter file (chp8_example2) as a new circuit named chp8_example3 – it can now be edited without changing the Bode plot analysis performed in example 2:

1. Right-click on the voltage source and delete the 1V AC signal parameter. Then, type the following into the 'DC Value [V]:' textbox: **wavefile="/filepath..../audioEdit.wav" chan=0**. In this command, "/filepath.../" should be the specific location of the test audio file on your own machine, and 'audioEdit.wav' is the full name of the file to be used for auditioning the circuit (bottom).
2. Add an output node using a net name (as you would for a ground label) – name it audioOut. Place the node between the capacitor and resistor to measure Vout.

Add two Spice directives:

3. A transient analysis command of 2 seconds (or the length of your test audio file). .tran 0 2.
4. A .wave command to write the data from audioOut to an audio WAV file – use the command explained above as a guide. **.wave "/filepath.../circuitOutput.wav" 16 44.1K V(audioOut).**

The analysis will take some time to simulate due to the complexity of the audio data being used as the voltage source input, but if everything runs correctly you will have a waveform of your audio file in the plot window. Now you can click on different points of the circuit to generate multiple plots for visual indicators of magnitude. You should also now be able to hear a simulation of the circuit output by clicking on the audio file named circuitOutput.wav – if you load both your input and output file into an audio player, you can audition them both side by side to hear the effect of your first-order high-pass filter.

8.5 Controlling audio filters

In the chapter 7 project, the circuit had an optional potentiometer to control the output voltage from a simple audio amplifier. This was achieved by connecting the third potentiometer terminal to ground, which effectively creates a voltage divider between the other two terminals (Figure 8.22).

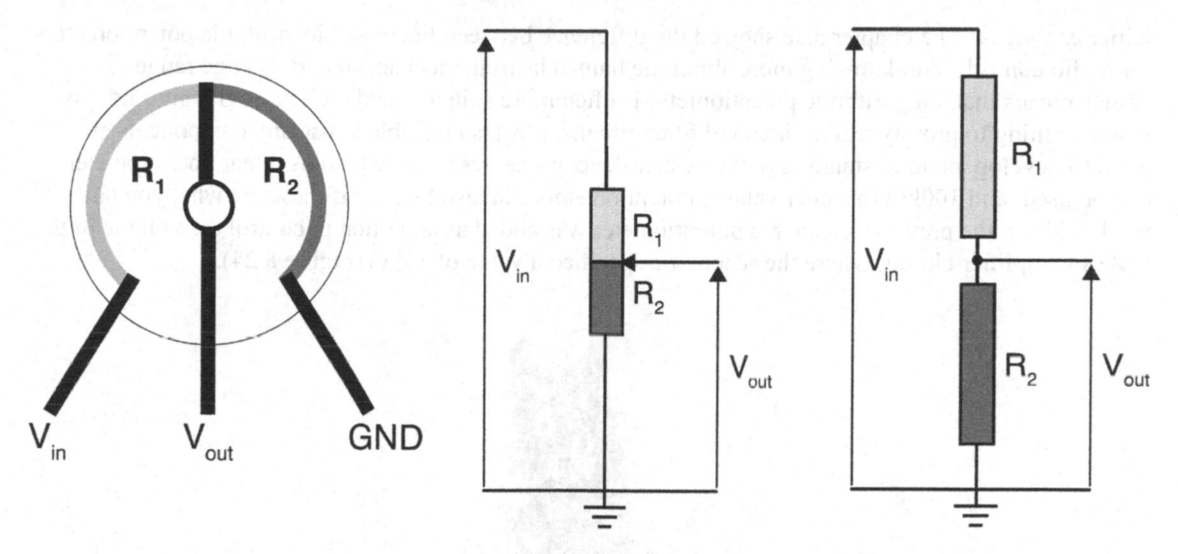

Figure 8.22 Voltage divider potentiometer. ***By connecting the third potentiometer terminal to ground, two resistances are created – one before the wiper (light grey) and one after the wiper (dark grey). A schematic with the potentiometer symbol (resistor with arrow) shows how a single resistor is effectively split by the position of the wiper, with the standard voltage divider circuit shown on the right for comparison.***

The diagram shows how varying the position of the potentiometer wiper changes the ratio of R_2 relative to R_1. As with the low- and high-pass filters, changing the impedances in a divider changes the ratio of the output voltage based on the equation in the figure above. Thus to manually control the output voltage level of a filter, a potentiometer can be added to the output of the RC filter stage. The updated circuit diagram for a first-order low-pass filter with manual potentiometer control would then be as shown in Figure 8.23.

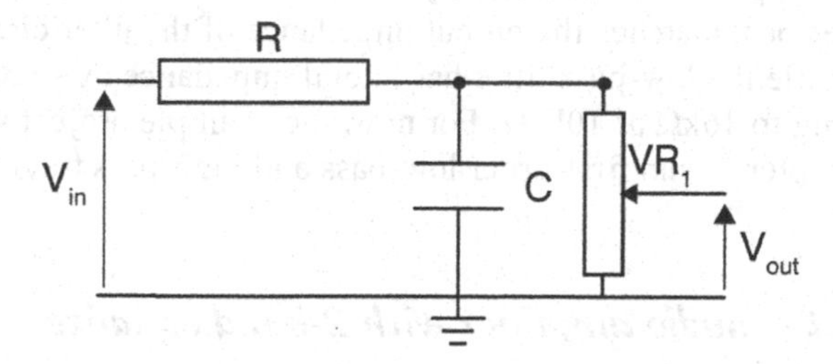

Figure 8.23 First-order low-pass filter with potentiometer control. ***The schematic shows a first-order low-pass filter with a potentiometer (variable resistor, VR_1) connected to control the output voltage. The filter output is presented across VR_1, which acts as a load resistance, where varying this resistance will either increase or reduce the magnitude of V_{out}.***

The value of the potentiometer (its maximum resistance) effectively dictates how quickly it changes the balance of the voltage divider. As discussed in chapter 2 (section 2.5, Figure 2.23), using a higher potentiometer value changes the range of output voltages it will provide relative to the distance the wiper can move. The chapter also showed the difference between linear and logarithmic potentiometers for audio controls, and knowing more about the human hearing mechanism and its huge range of values means that a logarithmic potentiometer is often more suited to audio circuits. Having said this, when learning to prototype amplifier and filter circuits, it is also feasible to use any components at hand to develop an understanding of the system-level processes involved – thus linear potentiometers can be used, and 100kΩ (or other values) potentiometers can also be used if these are what you have to work with. In the previous chapter, a potentiometer was added as an option to control the volume of the LM386 amplifier circuit, where the schematic specified a value of 10kΩ (Figure 8.24).

Figure 8.24 Example wiring for a 10kΩ potentiometer. ***The image shows a potentiometer wired for use as a voltage divider. The output wire is on the right (dark wire), with the input wire in the middle and ground wire on the right. This wiring configuration is used for all circuits in this book.***

To save on costs, 10kΩ pots will be used to control the low- and high-pass filter outputs in the example project below – allowing you to purchase one and then use it in multiple circuits throughout chapters 7–9. In practice, further analysis of the circuit is needed to determine what range of resistance values best matches the output impedance of the filter circuit involved, given that in the previous example the low-pass filter has a total impedance of ~226kΩ at the 100Hz cutoff frequency, dropping to 16kΩ at 10kHz. For now, the example project will combine the audio amplifier circuit from chapter 7 with first-order low-pass and high-pass filter stages to build a 2-band equalizer.

8.6 Example project – audio amplifier with 2-band equalizer

This project will combine two filter stages (low and high pass) with an LM386 operational amplifier to build an audio amplifier with 2-band equalization (often called a tone control). The circuit for this project extends the previous LM386 audio amplifier from the previous chapter (see Figure 7.52) to include first-order low-pass (fc = 100Hz) and high-pass (fc = 1kHz) filter stages that are recombined via potentiometers (VR1, VR2) at the non-inverting input (pin 3) of the LM386 (Figure 8.25).

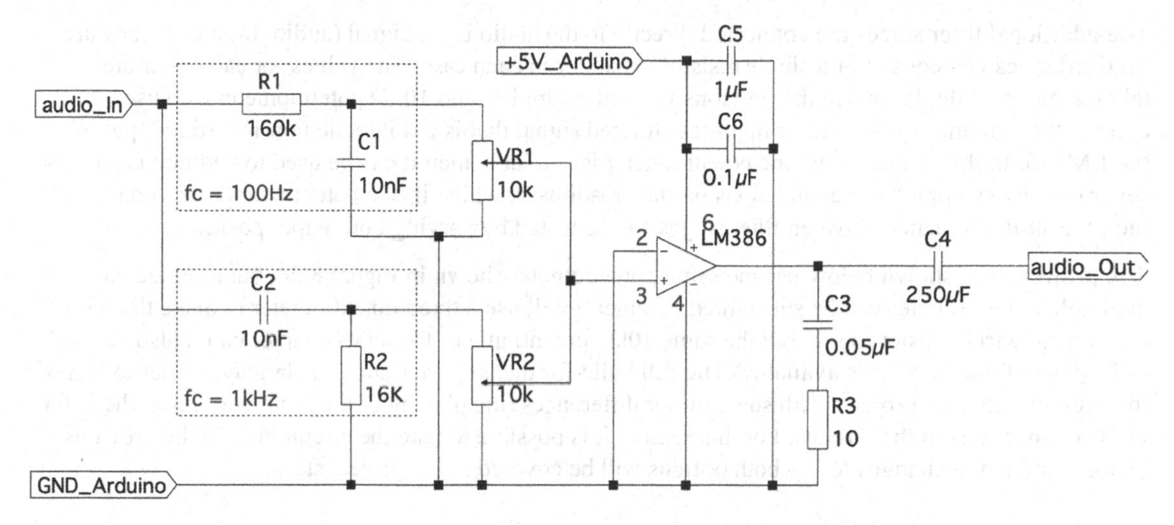

Figure 8.25 Two-band equalizer audio amplifier circuit. ***The schematic shows two filter stages, where the filter cutoffs are set at 100Hz (low pass) and 1kHz (high pass). The filter stages are connected via 10k potentiometers to the non-inverting input of an LM386 audio amplifier chip, which combines a DC blocking capacitor (C4) and a Zobel network (C4, R3) with power decoupling (C5, C6). The Arduino can provide +5V and GND connections for this circuit, as indicated by the nodes on the schematic (this circuit cannot actually be simulated using LTspice).***

The schematic in Figure 8.25 shows one of the simplest ways of building a 2-band equalizer, which minimizes the component count and keeps the signal path short. The disadvantage of this approach is the significant attenuation introduced due to the combination of two signal paths (low and high pass), which is a problem for all passive filter networks.

Practical circuits

Whilst an active filter design (where each filter is built around an operational amplifier) provides much better performance, the increase in both the complexity of the circuit and also the component count precludes its inclusion within this introductory text. In practical circuits, passive filters introduce significant degradation to the input signal that requires the use of other stages (e.g. buffering) or active designs to avoid poor results. For now, the schematic allows the learning material on amplification and filtering to be combined in a single breadboard circuit for analysis and prototyping, knowing that the output audio quality will be reduced as a result of working with a passive design (and breadboard connections).

The broader point in relation to components is that you can use whatever you have available to build some sort of filter – the 220Ω resistors used in the MIDI circuits in chapters 4 and 5 can be repurposed within a filter to keep purchase costs down. A 220Ω resistor could be combined with a 10μF capacitor to build a first-order low-pass RC filter with a cutoff frequency of ~72.4Hz, or a 1μF capacitor to build a first-order high-pass CR filter with a cutoff frequency of ~720Hz. Although these filter frequencies would not be very useful in a real 2-band equalizer, the aim of this book is to introduce concepts using practical circuit construction and so it is more important to build something and find out how it works rather than trying to design the perfect filter using a first-order design. The same principle applies for components like potentiometers – if one is available, swap the connections to audition each filter stage in turn. If you only have access to a 50kΩ or 100kΩ potentiometer (logarithmic or linear) then use it! It is more important to know the limitations of the components you are prototyping with so you can progress your learning quickly: purchasing significant amounts of extra capacitors and resistors is not necessary at this introductory stage.

The additional filter stages are connected directly to the audio input signal (audio_In) and as they are first-order designs consist of a single resistor/capacitor in each case. The values for each filter are taken from the calculations in the previous worked examples, and 10kΩ potentiometers are used to control the amount of low- and/or high-pass filtered signal that is fed into the non-inverting input of the LM386. In this way, if only one potentiometer is available then it can be used to audition each filter separately by swapping the connections on the breadboard, whilst if two potentiometers are connected then the shift in balance between filter stages can be tested by varying both wiper positions.

The project steps shown below use the same potentiometer shown in Figure 8.24, but as noted any available potentiometer can be substituted (chapter 9 will use a fixed potentiometer to make the rest of the control wiring easier to see, but the same 10kΩ potentiometer from this chapter could also be used in its place if that is what is available).The full build for the amplifier circuit is largely similar to that of the previous chapter project, with some minor differences (the placing of the input connector, the extra GND connections in the circuit). For this reason, it is possible to take the circuit built in the previous chapter and make changes to it – both options will be covered in the project steps.

Project steps

For this project, you will need: 1. 1 × LM386 operational amplifier 2. 6 × capacitors (2 × 10nF, 250μF, 0.05μF, 1μF, 0.1μF), 3 × resistors (160kΩ, 16kΩ, 10Ω), 2x potentiometer (optional single potentiometer both 10kΩ) 3. 1 × audio input connector (3.5mm jack), 1 × audio output connector (loudspeaker) 4. 4 × connector cables (for Arduino) and 9 × connector wires (connector cables can be used if not available)	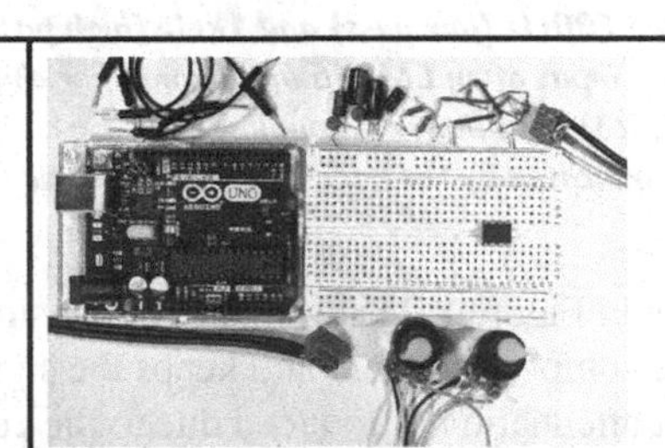
1. Add the LM386 across the column break with top left **pin 1 on e22** (the top of the chip has a notch) – bottom right pin 5 on f25. If working from the previous circuit, remove the input and output connector blocks and ground wire on **[e30–f30]**. 2. Connect the LM386 inverting input (pin 2) and GND (pin 4) with a connector wire from **[b23–b25]** and then connect both to the breadboard ground rail with a connector wire from **[a23–GND]**. 3. Connect the LM386 power supply Vs (pin 6) with a connector wire from **[j24–GND]**. 4. Add a 250μF DC blocking capacitor from the LM386 output Vout (pin 5) to the audio output connector from pins **[h25–h28]**. 5. Add a Zobel network to the LM386 output Vout (pin 5). Use a connector wire from pins **[g25–g19]** add a 0.05μF capacitor between pins **[h19–i18]** then add a 10Ω resistor between pins **[j19–GND]**.	
1. Add a ground connector wire to the bottom ground rail **[GND-f30]**. This will ground the output connector block. 2. Add signal wires from the LM386 non-inverting input (pin 3) **[c17–c24]**, **[c13–c7]** and **[d3–d7]**. These will connect the filter stages to the amplifier input.	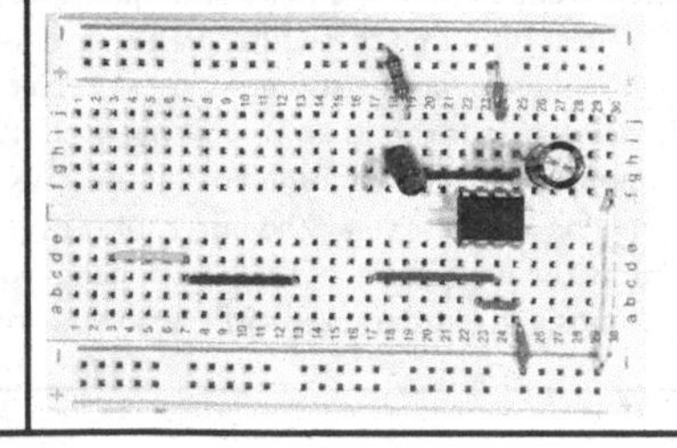

1. Add the input connector block on **[b1–b3]**. 2. Add the output connector block on **[j28–j30]**.	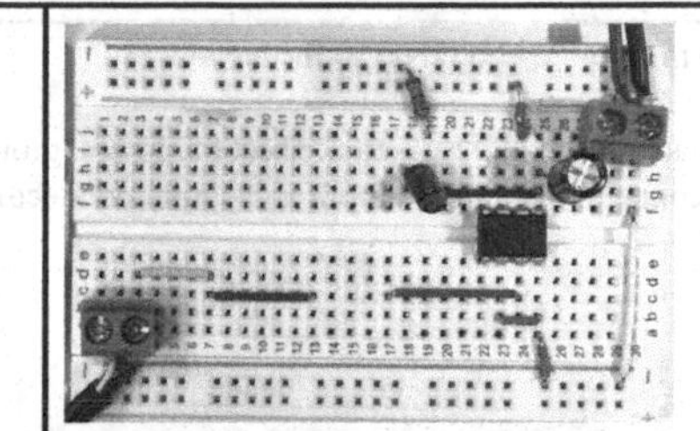
1. Connect the input ground to the bottom ground rail **[c1–GND]**. 2. Connect the low-pass filter 160kΩ resistor between **[e7–e11]**. 3. Connect the low-pass filter 10nF capacitor between **[a10–GND]**.	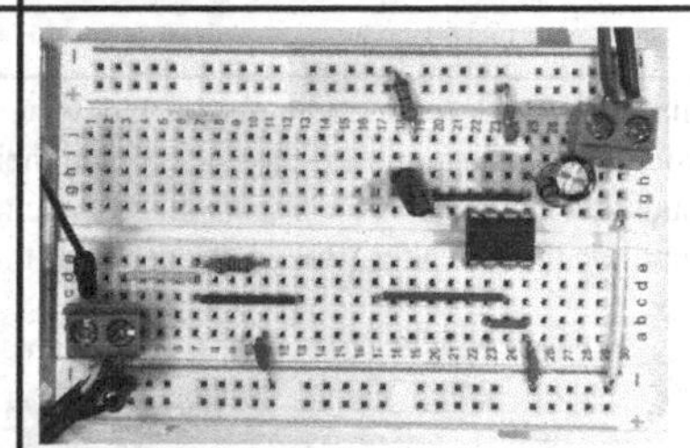
1. Connect a 10kΩ potentiometer (or replacement) with the input (middle wire) to the low-pass filter **[VR1in–d11]**. 2. Connect the output of the potentiometer to the non-inverting amplifier input **[VR1out–b17].** 3. Connect the potentiometer ground wire to the bottom ground rail **[VR1gnd–GND]**.	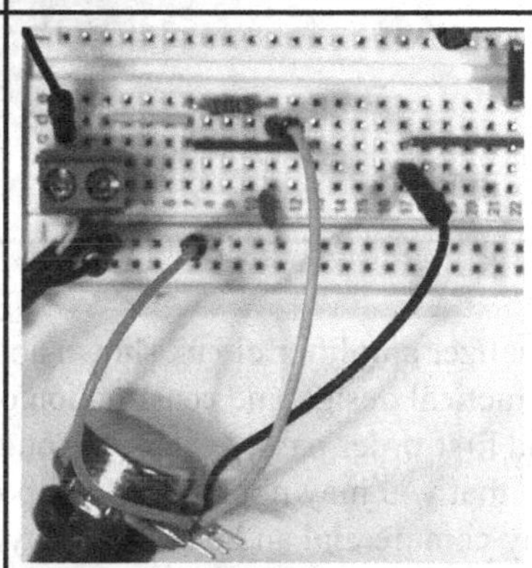
1. Connect a signal wire from **[c7–c13]**. This will connect the audio input to the high-pass filter. **2.** Connect the high-pass filter 10nF capacitor between **[e13–e15]**. 3. Connect the high-pass filter 16kΩ resistor between **[b15–GND]**.	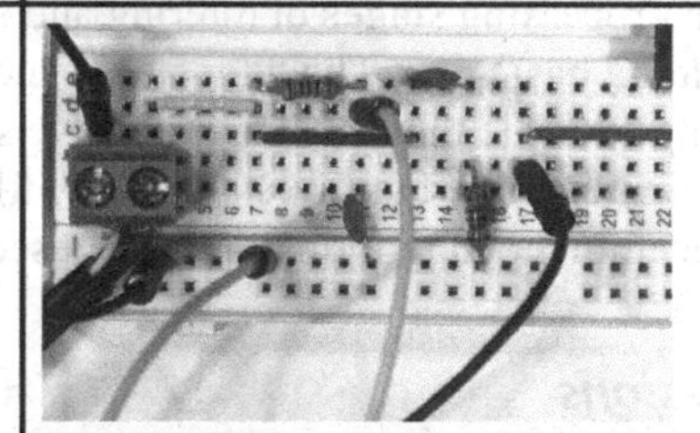
1. Connect a second 10kΩ potentiometer with the input (middle wire) to the high-pass filter **[VR2in–c15]**. 2. Connect the output of the potentiometer to the non-inverting amplifier input **[VR2out–c17].** 3. Connect the potentiometer ground wire to the bottom ground rail **[VR2gnd–GND]**.	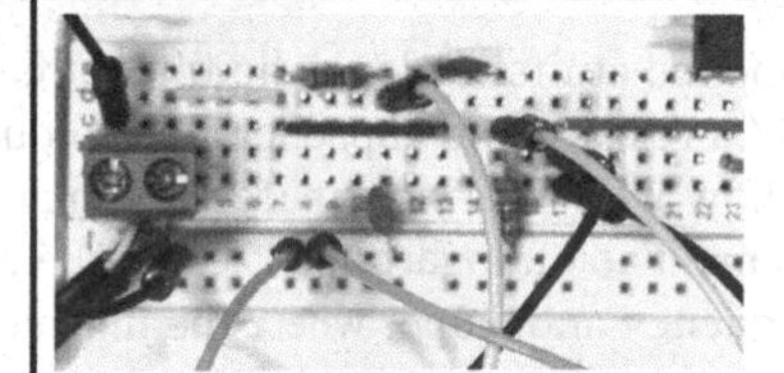
Add AC decoupling capacitors ($C_1 = 1\mu F$, $C_2 = 0.1\mu F$) between the breadboard power and GND rails.	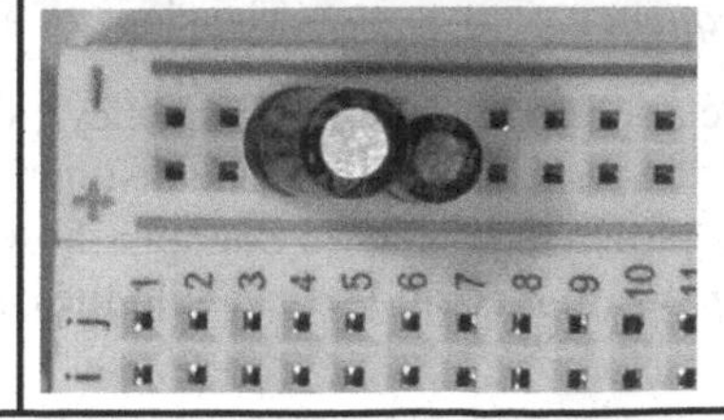

<table>
<tr>
<td>1. Connect the Arduino 5V and GND to the breadboard power rails.
2. Bridge the breadboard GND rails to connect the filters' grounds, LM386 ground connections and the input and output connector grounds.</td>
<td>
</td>
</tr>
<tr>
<td colspan="2">Connect the 3.5mm audio jack to an audio playback device. If everything has been connected properly, the audio input signal should be fed through the low- and high-pass filters, and their combined output should be amplified by the LM386 for output to a connected loudspeaker. Varying the wipers on the potentiometers should vary the amount of low and high frequencies being heard through the loudspeaker (though the effect of the high-pass filter will be much easier to hear).
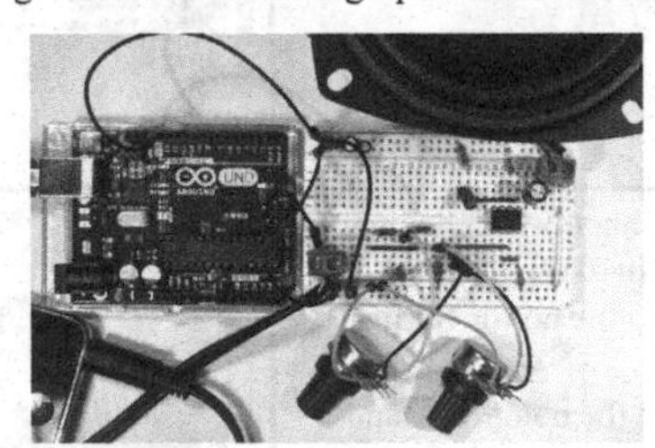</td>
</tr>
</table>

The 2-band equalizer amplifier circuit shown above combines a lot of theoretical elements on filtering, alongside the practical design and construction of an audio amplifier circuit. The limitations of breadboard (and first-order passive filters), alongside the limitations of the minimal LM386 audio amplifier, mean that you may not be overly impressed by the result of this project (particularly when compared to any commercial audio equipment you may own!). The point of this project is to show how different audio processing stages of filtering and amplification can be combined to build a functioning system – albeit it a basic one. The filter stages shown here are an introduction to audio filtering, which is a complex topic that requires extensive study. Having said this, the circuit above can easily be augmented and extended to investigate such filter concepts and like the amplifier in the previous chapter it is recommended that this circuit be used as a building block for more advanced filter designs.

8.7 Conclusions

This chapter looked at first-order audio filter circuits and how they can be constructed using a combination of a resistor and capacitor. The frequency-dependent response of human hearing was discussed to show why filtering is necessary in audio, and also how to work with decibels that define audio measurements on a logarithmic scale. The chapter continued by introducing filter characteristics like cutoff frequency and slope, where the time constant of an RC circuit is used to derive the equation for *Cutoff Frequency* $fc = \frac{1}{2\pi RC}$. The slope of a first-order filter has a gradient of −6dB/octave (where each octave is a doubling of frequency), which means there is only one reactive component (a capacitor) – the direction of the slope distinguishes a **low-pass (negative slope) from a high-pass (positive slope)** filter. The worked example showed how to construct Bode plots in LTspice, and also how to use LTspice to read and write audio files as simulation data from a circuit – which allows filter designs to be auditioned prior to constructing them. The example project combined first-order low-pass and high-pass filter stages to build a 2-band equalizer, where the output signal level of each filter is controlled by a potentiometer. This equalizer is similar to other audio and musical equipment you may already have used, but as it is prototyped on breadboard it will not perform to the same standard as commercial equipment.

The next chapter combines everything that has been taught in this book to build an **Arduino controlled 2-band equalizer amplifier**, where analogRead() is used to measure the value on a potentiometer and scale it to control the resistance value on wo digital potentiometers that replace those used in the project in this chapter. In so doing, the combination of digital control systems discussed in chapters 4 and 5 with the audio amplification and filtering systems of chapters 7 and 8 shows the power of the Arduino as a control chip, alongside how it may be purposed for use with audio circuits.

8.8 Self-study questions

The following questions aim to give an indication of how specific first-order filters can be designed – two questions with known values, followed by two where one of the component values must be found:

Q1: An RC circuit is configured as a first-order low-pass filter, with a 1kΩ resistor and a 1μF capacitor. What is the cutoff frequency of the filter?

Q2: A CR circuit is configured as a first-order high-pass filter, with a 10kΩ resistor and a 10nF capacitor. What is the cutoff frequency of the filter?

Q3: Design a low-pass filter with a cutoff of 200Hz – how close can you get to this value with a resistor of 470Ω and a single capacitor?

Q4: Design a high-pass filter with a cutoff of 3kHz – how close can you get to this value with a single resistor and capacitor of 0.1μF?

All answers are provided for each chapter at the end of the book.

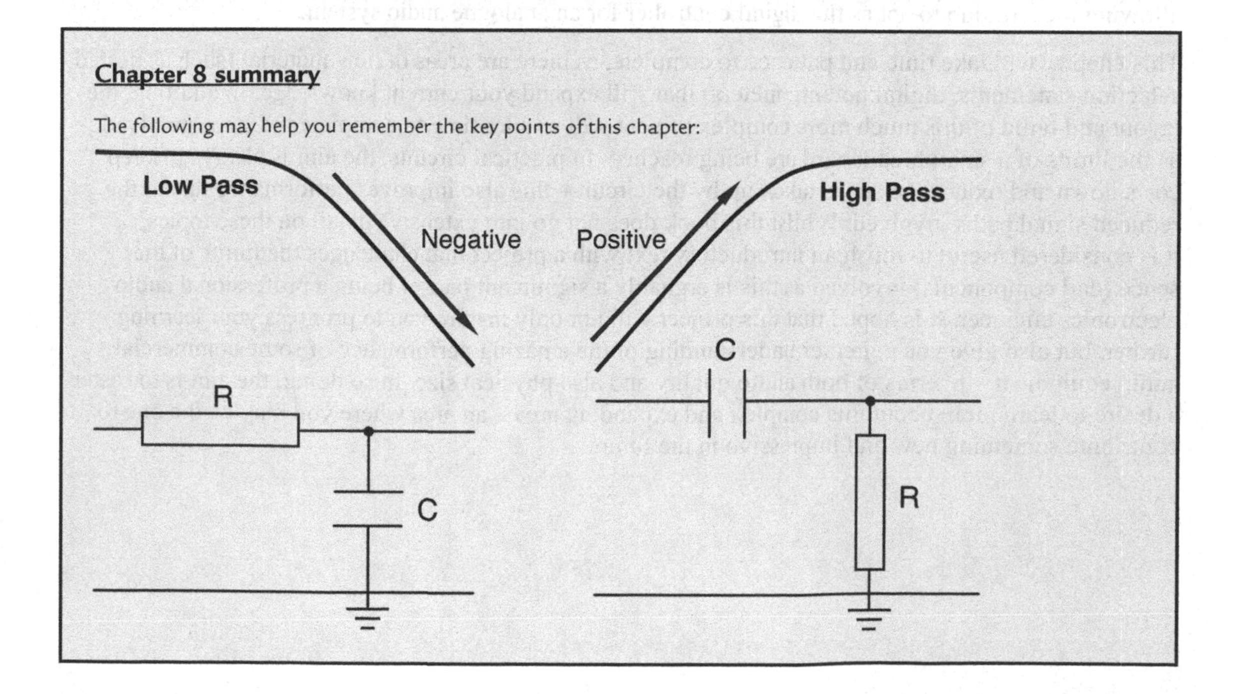

CHAPTER 9

Arduino audio control

This chapter focusses on the design and construction of a more significant project – an Arduino-controlled audio amplifier with 2-band equalizer. This project combines elements of your learning from previous chapter projects in this book:

1. **Chapter 5** – making decisions based on control inputs (i.e. switches and piezoelectric sensors)
2. **Chapter 7** – amplifying an audio signal (using the LM386 to drive a loudspeaker as an output actuator)
3. **Chapter 8** – filtering an audio signal for amplification (the 2-band equalizer amplifier circuit)

The final system uses digital (push button) and analogue (potentiometer) control inputs to change the state and data of the system; this requires adding nested selection statements to the Arduino code to make more complex decisions. The code will also be extended beyond the serial communication of MIDI to work with the serial peripheral interface (SPI), a standard digital communications protocol for controlling multiple integrated circuits. In this project, the Arduino will act as the controller for two MCP413-103 digital potentiometer chips that respond to control information sent over the SPI bus that connects all three together. In so doing, the digital potentiometers now act as the interface to an audio rendering system that combines the audio amplifier from chapter 7 with the 2-band equalizer from chapter 8. The digital potentiometers will control the level of filter output sent to the amplifier, allowing the Arduino to act as the digital controller for an analogue audio system.

This chapter will take time and patience to complete, as there are areas of new material (such as nested selection statements, digital potentiometers) that will expand your current knowledge. In addition, the layout and build of this much more complex project will require time to complete the steps involved, as the limits of a small breadboard are being reached. In practical circuits, the aim is always to keep costs down and reduce the space taken up by the circuit – this also improves performance due to the reduced signal paths involved. Whilst this book does not go into extensive detail on these topics, it is considered useful to finish an introductory text with a project that challenges the limits of the space (and components) involved as this is arguably a significant part of being a professional audio electronics engineer. It is hoped that this project will not only inspire you to progress your learning further, but also give you a greater understanding of the amazing performance of some commercial audio equipment –in terms of both audio quality and also physical size. In so doing, the aim is to foster a desire to learn more about this complex and expanding area – an area where you may be the one to contribute something new and impressive in the future.

What you will learn

How to use nested selection statements to work with digital and analogue control inputs
How to build state and data storage into a digital control system
How to use the Arduino to implement the SPI communications protocol
How to use digital potentiometer chips to control audio signals
How to combine a digital control system with an audio render system
How to build a project with physical space constraints
How to keep signal and power connections separate

9.1 Final project overview

The final project in this book focusses on a practical combination of the work performed in previous chapters, with the outcome being an Arduino-controlled audio amplifier with 2-band equalizer. To do this, the system in Figure 9.1 will be implemented.

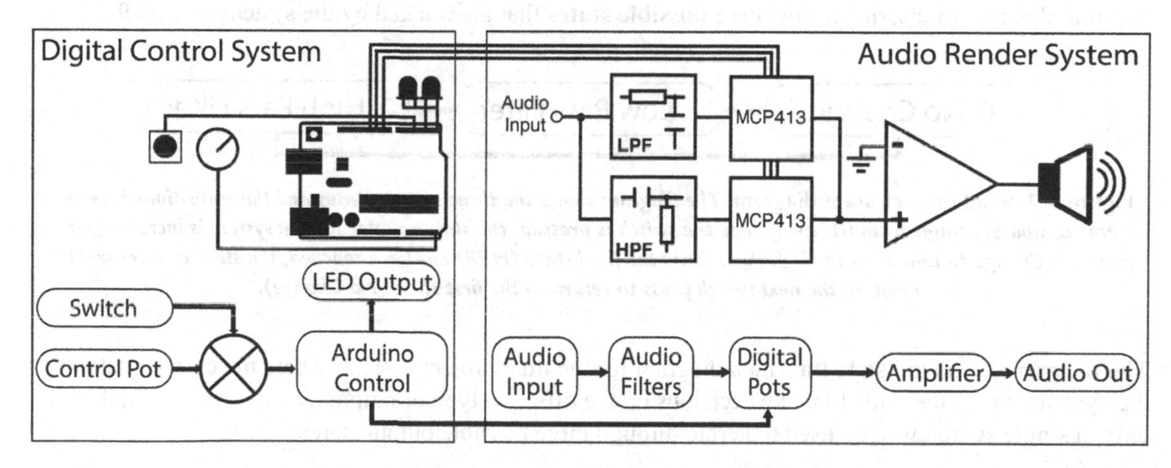

Figure 9.1 Final project system diagram. ***The diagram shows how a digital Arduino system is used to control an audio render system that extends the 2-band equalizer amplifier circuit of chapter 8. The Arduino processes sensor inputs from a switch (to select filter) and potentiometer (to set filter level) and then uses the serial bus to control the digital potentiometers (MCP413) as output. The LEDs indicate which filter is currently selected – based on the state of the system. The digital potentiometers replace the analogue rotary potentiometers used in chapter 8, allowing the filter levels to be set by the Arduino. The filter outputs are combined as inputs to the non-inverting terminal of an LM386 audio amplifier, which drives a loudspeaker as an actuator output from the audio render system.***

This system will be built in three stages, to allow the processes and functionality of the digital control system to be introduced prior to combining it with the audio render system derived from the 2-band equalizer amplifier circuit built in the previous chapter. The build stages are as follows:

- **Stage 1: Arduino state control** – this stage implements the push-button switch that controls the state of the system, which is also output to the state LED indicators. The analogue control potentiometer is also connected in stage 1A, to allow the data values for the digital potentiometers to be controlled in the second and third stages of the project.

- **Stage 2: Arduino digital filter control** – the second stage of the project connects the MCP413-103 digital potentiometers that are controlled by the serial peripheral interface (SPI) bus that can be implemented using the Arduino. The stage 2 circuit connects a further two LEDs to test varying the potentiometer output values using the analogue control potentiometer.
- **Stage 3: Arduino-controlled 2-band equalizer amplifier** – the final project stage combines the digital control system implemented in stages 1 and 2 with a modified version of the 2-band equalizer amplifier circuit built in chapter 8. One of the main challenges at this point in the build is fitting all components on a single breadboard, and so the chapter 8 circuit is rearranged slightly to accommodate this (hence the chapter 8 circuit cannot be directly extended in this project build).

The system above mimics the structure of a MIDI system (see chapter 2.4, Figure 2.17), where the Arduino now digitally controls the sound produced by the audio render system – which in this case is a 2-band equalizer amplifier circuit. The switch and potentiometer inputs define the **state and data** of the system, with LEDs being used as control outputs to indicate the current filter selected. State-based programming is a more advanced way of controlling code, which allows the previous processes and events of the system to be retained for future evaluation. In this project, the current filter being controlled is part of a sequence of **three possible states** that are tracked by the system (Figure 9.2).

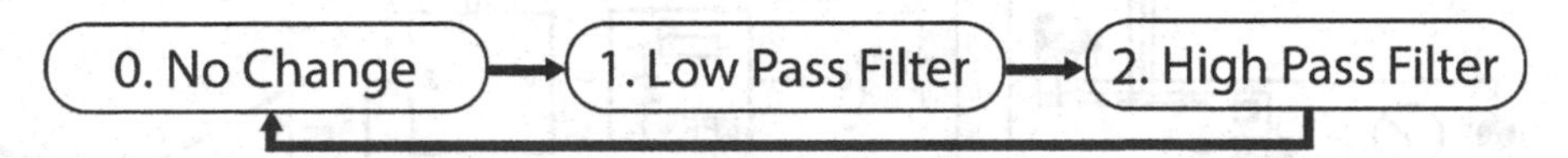

Figure 9.2 Final project state diagram. ***The diagram shows the three system states and the transitions between them (computers count from 0). Every time the switch is pressed, the state counter for the system is incremented – from No Change to Low Pass to High Pass. Once the final state (HPF) has been reached, the state counter will be reset on the next switch press to return to the first state (No Change).***

The diagram shows the state transition diagram for the final project system, where the current state of the system allows the code to make decisions on the different type of outputs it will produce. In this case, a single switch will be used to iterate through three possible output states:

0. **No-change output** – this default state prevents unintentional changes being made to a filter value
1. **LPF output** – when selected, this state will allow the low-pass filter (LPF) value to be controlled
2. **HPF output** – when selected, this state will allow the high-pass filter (HPF) value to be controlled

The current state of the system will be indicated by the LEDs connected to the control system. Each LED will light up for a specific filter state (LPF/HPF) and neither will light when the system is in the no-change state. The system state dictates the decisions it will make on outputs, but the data in the system provides the information needed to execute the processes resulting from these decisions. In simple terms, when a filter is selected the **value of that filter** must be defined – this is the **data** of the system. This means that for each state the Arduino will read the value of the potentiometer (as a control input) and use this to update the digital potentiometers connected to the audio output system. The state and data from the switch and potentiometer are thus used by the Arduino to control the audio render system, which is an amended version of the 2-band equalizer amplifier circuit built in chapter 8. By replacing the analogue potentiometers used to control the filter levels in the chapter 8 project circuit, digitally controlled potentiometers allow the Arduino to directly set the resistance levels of the filters (Figure 9.3).

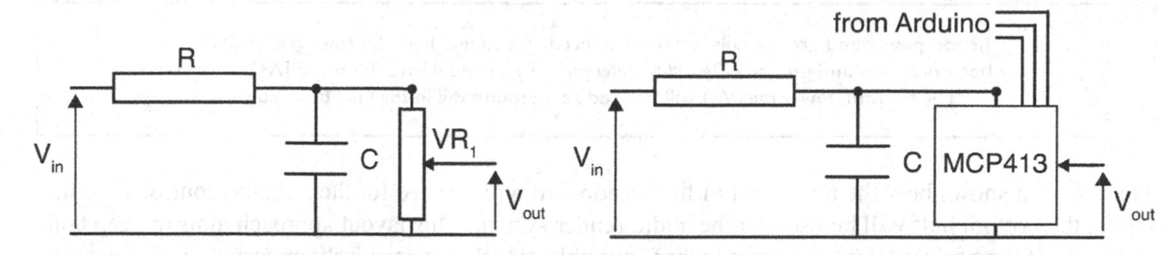

Figure 9.3 Controlling audio filters with digital potentiometers. ***The left-hand schematic shows a low-pass filter connected to an analogue potentiometer that controls the output level of the filter (Vout). In the right-hand schematic, a digital potentiometer (MCP413) is now used to set the resistance across the output (Vout). The MCP413 can be directly controlled by the Arduino to dynamically update the potentiometer value.***

The schematic shows how a digital potentiometer can replace an analogue rotary potentiometer as a control point for an audio filter circuit. The Microchip Technologies MCP413 digital potentiometer can be updated using a serial peripheral interface (SPI), allowing the Arduino to send information to one (or more) of these chips on a single data bus. The following sections of this chapter will cover the stages needed to build an Arduino-controlled audio amplifier with 2-band equalizer. The final circuit layout for this project will split the breadboard into two halves – the top half for the control system, the bottom half for audio render system (Figure 9.4).

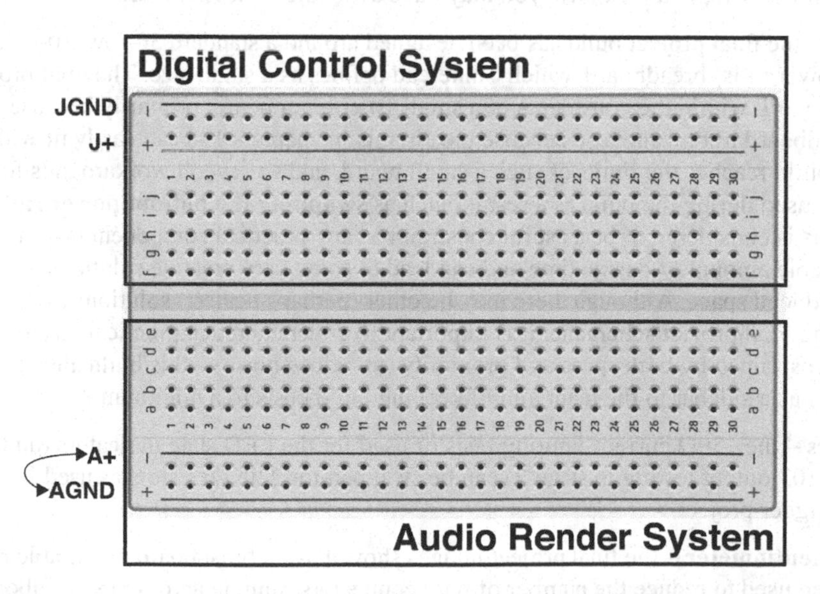

Figure 9.4 Final project breadboard layout areas. ***The diagram shows how the final project breadboard will split the components between the control system (top) and the audio render system (bottom). Note the labelling of the power and ground rails (J top, A bottom) – this will be used in all project build stages. In the final circuit, the centre column break will be used to mount the digital potentiometers (MCP413) and audio amplifier (LM386) ICs. Also note that the bottom power rails have been swapped around – this will allow components to be placed more easily in the final project stage.***

The **top power and ground rails** will be referenced by a capital letter for row j **[J+] [J GND]**
The **bottom power and ground rails** will be referenced by a capital letter for row a **[A+] [A GND]**
The bottom power rail (V+) will be used as a ground rail in the final build stage

The diagram shows how the upper half of the breadboard will be used for the Arduino control system, whilst the bottom half will be used for the audio render system. This layout approach aims to keep both elements of the project clearly separated where possible, which is particularly important in a project where layout space will be limited. A single breadboard will be used in the final project, which requires several workaround techniques to be used to fit all components onto a single board – the **bottom power rails are swapped** in the final build stage for this reason. These will be noted at each stage of the final project build, highlighting the need to carefully follow all instructions step by step throughout this chapter.

9.1.1 Project components

For the completed final project, you will need various components for each stage of the build. The project steps in each of the three build stages assume that the components from each previous stage are retained, as the circuit will be extended in each of these stages to add additional functionality. Some components can be swapped for others you may have available – these include:

Breadboard – the final project build has been designed around a standard 30-row, 10-column (excluding power rails) breadboard, which Tinkercad defines as a small size. The final project layout is designed to work within the constraints of a small 10-column board, primarily because this is the cheapest breadboard to buy and also because the circuits in chapters 1–8 can easily fit within it. The final project build reaches the limits of such a small board, and so several workarounds for component layout will be used during the build as a result (such as **swapping the bottom power rails**). In actual fact, this is considered to be a useful constraint as any practical component layout will take up a considerable amount of design time and can lead to some very creative solutions to the eternal problem of reduced space. Although there may be other (perhaps neater!) solutions to the layout that implements the final project schematic, it is important to understand that practical electronics circuits are heavily constrained by both space and proximity (to reduce noise) – this build aims to introduce these elements as a sidebar to the main aim of keeping build costs to a minimum.

LED resistors – the 150Ω current-limiting resistor used for the LED state indicators (and two others in the MCP413-103 output testing in stage 2) can be swapped for 220Ω resistors as used in the chapter 5 MIDI drum trigger project

Analogue potentiometer – the final project images show a 3-pin breadboard mountable potentiometer, which has been used to reduce the number of wire connectors running across the breadboard (as they make other components harder to see). Any other three-connector potentiometer (i.e. Vin, Vout, GND) can be used instead – such as those used in chapters 7 and 8.

Filter components – the audio filters in stage 3 of the build use the same components as the 2-band equalizer amplifier in chapter 8 (16kΩ, 160kΩ, 10nF). If these specific values are not available, other combinations can be used – as long as they are added in an RC (low-pass) and CR (high-pass)

configuration for the different filter inputs. If other values are used, equation 8.7 from chapter 8 can be used to determine the *Cutoff Frequency* $fc = \frac{1}{2\pi RC}$

Zobel network components – the Zobel network included in both the chapter 7 and 8 project builds is a standard configuration taken directly from the LM386 datasheet. Although analysis of Zobel networks is outwith the scope of this book, any combination of resistor/capacitor components that are around values of 10Ω and 0.05μF can be used.

Power-decoupling capacitors – capacitor values of 1μF and 0.1μF have been used in chapter 7 and 8 to decouple the breadboard power rails (to reduce noise), and in the final project one reason for splitting the board between the control and audio render systems is to reduce the amount of power-supply noise introduced between the input and output audio ground connectors. Additional capacitors added across the power rails will noticeably reduce noise in the circuit, but if values of 1μF and 0.1μF are not available then others in the same range can be substituted.

Final project component list (A full schematic and component list are also included in the Appendices)

1. 1 × Arduino Uno R3 (other boards may not conform to the SPI bus implementation used in this project)
2. 1 × small breadboard (30 row, 10 column)
3. 1 × push-button switch (SW1)
4. 2 × LEDs (plus an additional two for the MCP413-103 output test circuit in stage 2)
5. 1 × 10kΩ analogue potentiometer (VR1) – a 3-pin breadboard-mountable pot from RS is used
6. 1 × 10kΩ pull-down resistor – for the push-button switch (R1)
7. 1 × 150Ω current-limiting resistor (R2) – for LED state indicators (plus two additional for stage 2 test circuit)
8. 2 × MCP413-103 10kΩ digital potentiometers
9. Audio filter components – low-pass 160kΩ resistor (R3), high-pass resistor 16kΩ (R4), C1 = C2 = 10nF
10. 1 × LM386 operational amplifier
11. Zobel network – 10Ω resistor (R5), 0.05μF capacitor (C3)
12. DC blocking capacitor (LM386 output) – 250μF capacitor (C4)
13. Power-decoupling capacitors – 1μF capacitor (C5) and 0.1μF capacitor (C6)
14. 1 × audio input connector (3.5mm jack), 1 × audio output connector (loudspeaker)
15. 12 × connector cables (for Arduino) and 25 × connector wires – connector cables can be used if wires are not available, but this will significantly increase the visual complexity of the final layout
16. 2 × small cable ties – these are usually shipped with electronic cables such as power supplies and headphones, and whilst often consigned straight to the rubbish bin by the consumer are very useful to electronics engineers seeking to keep cables grouped in a larger circuit! The control system uses a tie to group the push-button switch input and LED state indicator output cables together (project stage 1), and a second tie to group the SPI bus connectors for $\overline{CS}$, SCK and SDI/SDO output pins. This may seem like a trivial addition, but if the circuit does not function correctly and troubleshooting begins then a jumble of connector wires can quickly become a problem in itself.

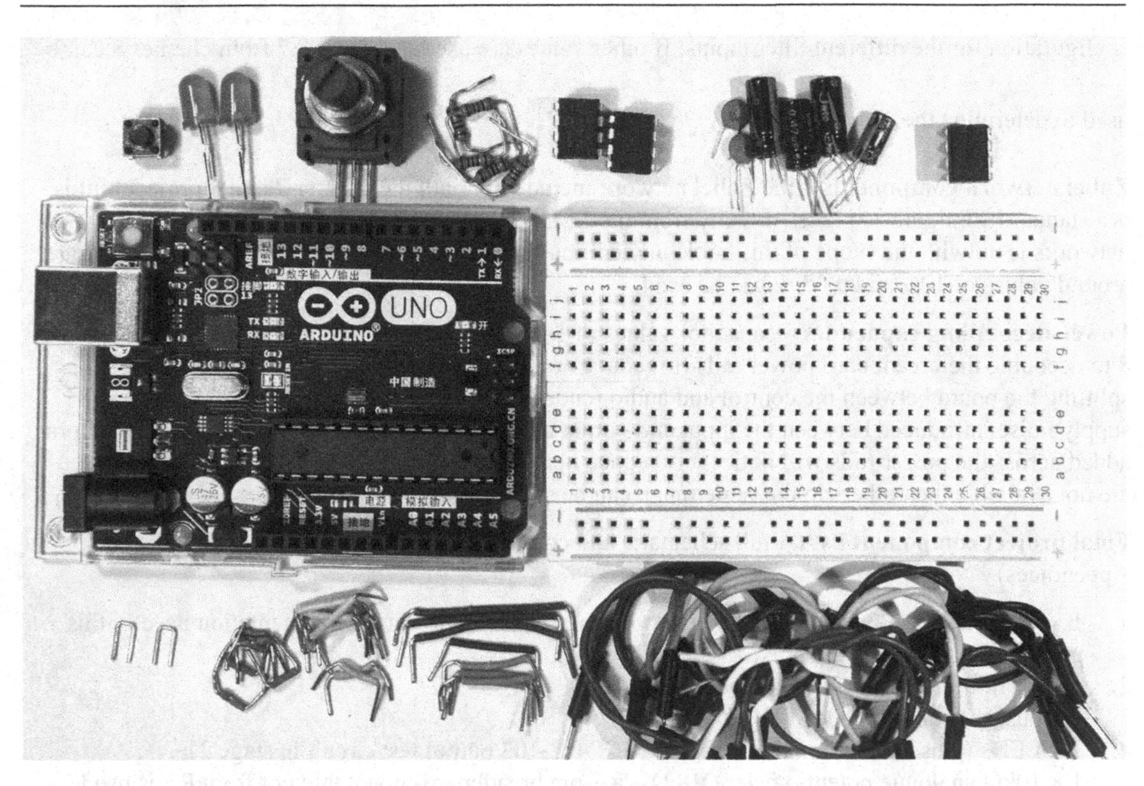

Figure 9.5 Final project components. ***The image shows (clockwise from top left) the connectors, wires, push-button switch, LEDs, analogue potentiometer, resistors, digital potentiometer chips (MCP413-103), capacitors and the LM386 audio amplifier chip used in the final project build. If sufficient space is available, it can be useful to set out these components near to the breadboard to avoid the confusion of searching for a specific resistor/capacitor value when trying to place components correctly. Having said this, the arrangement above is used to group the project components in a single image – this level of proximity would easily lead to components being mislaid or scattered when building each stage of the project.***

The layout in Figure 9.5 shows the full component list for the final project, where the significant amount of connectors and wires required is evident from the image. Most of these connectors will be used to implement the Arduino control system that includes the MCP413-103 digital potentiometers – these chips are the interface to the audio render system, which is broadly the same as the 2-band equalizer amplifier circuit built in chapter 8. It is recommended to source all of the components (and connectors) required prior to commencing the build, to avoid the confusion introduced by searching for them when also focussing on placing those components correctly on the breadboard.

Electronic breadboard layouts can quickly become complex and tiring to work with (particularly when a mistake has occurred), so it is important to spend a little time preparing your components and working area in advance to avoid displacing resistors and capacitors when placing other components on the board – long sleeves and long hair are particularly prone to snagging small electrical wires. This project will require **time and patience** to work through this much more complex breadboard circuit (where space becomes very limited). It is recommended that you **do not rush ahead** in this chapter, as many of the elements you have encountered in previous chapters will require small but specific changes in order to work within a larger digital control system. This familiarity can lead to mistakes, so following all processes and steps in the correct order is crucial to building the final circuit.

9.2 Arduino state control

The first section of this chapter will focus on the Arduino control system, combining control inputs for state and data with LED outputs for user interaction (Figure 9.6).

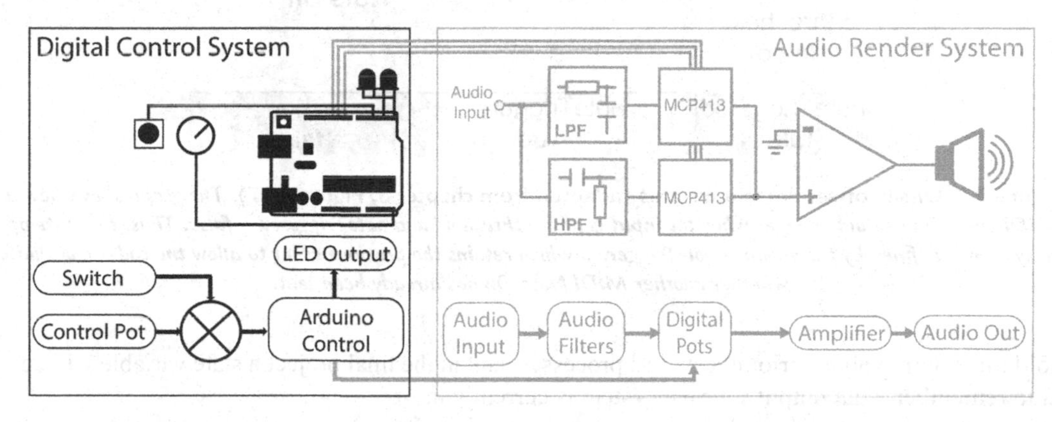

Figure 9.6 Final project stage 1 – Arduino control system diagram. ***The diagram shows the digital control system used in the final project, which is centred on the Arduino microcontroller. Control inputs from a push-button switch and analogue potentiometer will be combined to define the state and data of the system. In this first stage, LED outputs will be used to indicate the state of the system – alongside serial data output of the potentiometer wiper values in the simulator.***

The diagram in Figure 9.6 shows the digital control system of the final project, which will be used in later stages to control the audio render system. In this section, input from a push-button switch will be used to iterate through the states of the system (0–2), whilst an analogue potentiometer will provide the data values that the system will use in the next section to set the digital potentiometers. Two LEDs are used to indicate the current state of the system, which will be accompanied by serial output of the data values taken from the analogue potentiometer through the Tinkercad simulator. This is the only stage of the project build that can be simulated in Tinkercad, though it will be used for component layout diagrams for clarity (and consistency with previous projects).

Chapters 4 and 5 introduced the Arduino as a digital microcontroller, where chapter 5 showed how inputs can be processed by code instructions (written in C) to produce outputs. In the MIDI drum trigger project in chapter 5, an onset/offset detector was used to threshold the input from a piezoelectric sensor to determine when to send a MIDI Note On message representing a drum stroke. In addition to thresholding, the code also used a variable called noteTriggered to retain information about the last note event – the **previous state of the system** (Figure 9.7).

The noteTriggered variable is crucial to the correct operation of the drum trigger, as it prevents multiple MIDI Note On messages being sent without a corresponding MIDI Note Off. In programming terms, noteTriggered is called a **state variable because it retains information about the previous state** of the system (what MIDI Note was last sent). State variables are used throughout programming

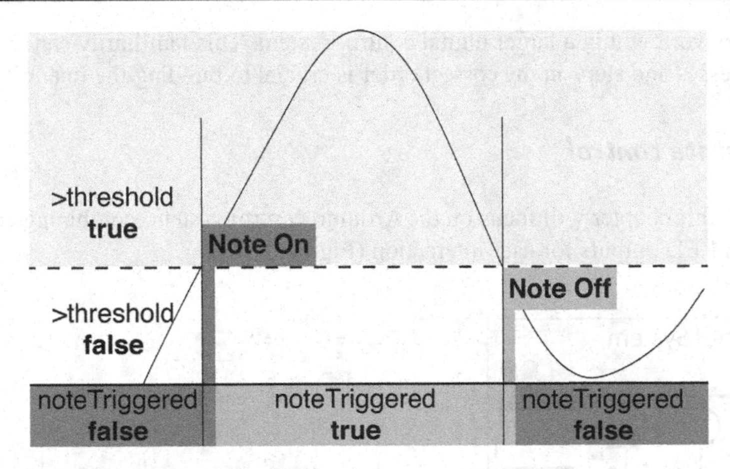

Figure 9.7 Onset/offset detection states (adapted from chapter 5, Figure 5.37). ***The graph shows how a MIDI Note On can only trigger when the input is above threshold and noteTriggered = false. Thus, the state of the system is defined by the variable noteTriggered, which retains the previous event to allow the code to evaluate whether another MIDI Note On has already been sent.***

to hold information about prior events and processes, and in the final project a state variable will be used to remember what output state the system is currently in.

In the chapter 5 MIDI switch controller project (section 5.8) two input switches were used, where each switch defined a single output (with else being the third Note Off output). This made the logic of the selection statement easier, but also required more components and breadboard space than will be available in this larger final project. For this reason, in the final project a single switch will be used to **iterate through multiple states**. In this way, one digital switch can be used to execute different Arduino processes by incrementing through each state whenever the switch is pressed. The Arduino will be used to control two digital potentiometers linked to a low-pass and high-pass filter. This means that there will be a total of three states in the system (Figure 9.8).

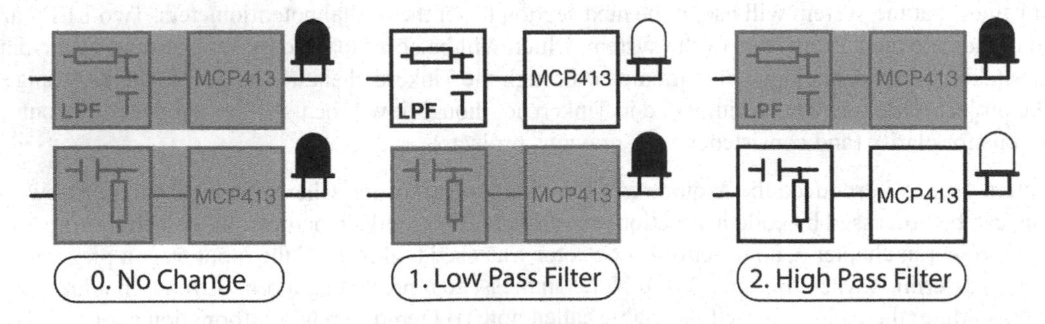

Figure 9.8 Final project system output states. ***The diagram shows how the system requires a total of three states, relating to both the two audio filters (low and high pass) and also the default state of no change. For each state, the system should be able to control the relevant filter (LPF/HPF) or do nothing – this corresponds to the else condition in a selection statement. LEDs are also used to indicate the current state of the system as a control output.***

The diagram shows the overall workflow of the control system in the final project, where a single switch will iterate through three possible states. The No Change (default) state effectively does nothing, but this also prevents unintentional changes and so is crucial to the operation of the system. The LPF and HPF states will read the input potentiometer control value and use this to update a specific MCP413 digital potentiometer. In both cases, an output LED will also light as an indicator of the current filter being controlled to help translate the system states into practical use.

To build multiple states around a single switch input, a selection statement must be amended to include a **counter variable** that keeps track of the current state. Although a thorough discussion of state variables (and their associated logic) is outwith the scope of this book, a brief explanation of how Arduino code can use a selection statement to retain state information will be given in the practical context of the control system for the final project. Taking a selection statement as an initial template, a counter variable can be added to keep track of how many times the selection statement has executed (Figure 9.9).

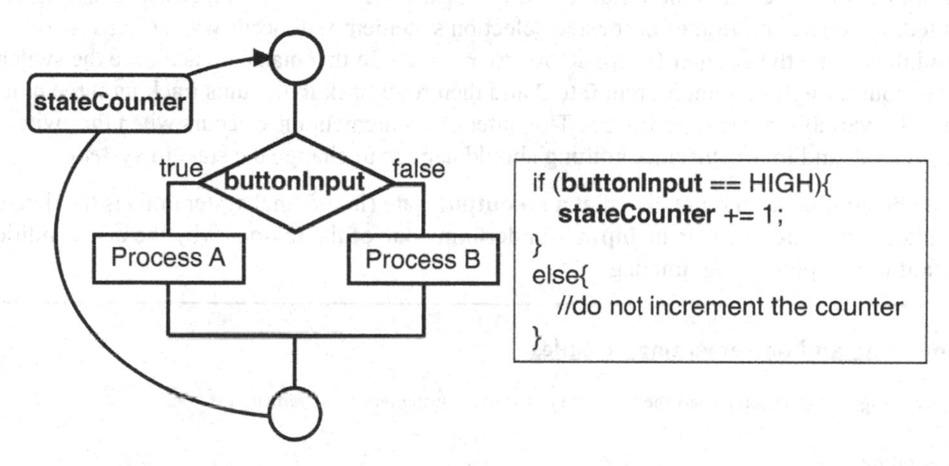

Figure 9.9 Adding a counter to a selection statement. ***In the flow diagram, the selection will execute to evaluate the condition of the variable buttonInput. When buttonInput is true (i.e. the switch has been pressed) the stateCounter variable is incremented (stateCounter += 1) to keep track of how many times the button has been pressed – if buttonInput is false, stateCounter is not incremented.***

The flow diagram shows how an additional `stateCounter` variable is used to keep track of how many times the switch has been pressed (see below for more information on incrementing with +=). This is a widely used programming technique, where the previous behaviour of the system is tracked by data to retain information about processes that have already been executed. For the final project, there are three states that must be counted by the code – after which the counter must then **reset to return to the first state**:

```
if (buttonInput == HIGH){
  //nested selection statement
  if(stateCounter < 2)
  {
       //within bounds - increment the variable using +=
       stateCounter += 1;
   }
```

```
        else{
            // stateCounter == 2 so reset the counter to return to state 0
            stateCounter = 0;
            }//end of nested if
    }
    else{
      //do nothing
    }//end of if
```

The code shows the outer selection statement for an input variable `buttonInput`, where the true condition (when the switch is pressed) now contains a **nested if statement**. This nested selection statement evaluates whether the variable `stateCounter` has reached 2 (`stateCounter < 2`) and if not then it increments the variable (`stateCounter += 1`). As `stateCounter` is incremented, the else condition of the nested selection statement will occur when (`stateCounter == 2`) and this resets the counter (`stateCounter = 0`). In this manner, each time the switch is pressed the counter will increment from 0 to 2 and then reset back to 0 – thus tracking three unique states with the variable `stateCounter`. The outer else statement only occurs when the switch has not been pressed, and in this instance nothing should happen to change the state of system.

It can often be confusing to remember that a **no output** state (in the final system this is the default No Change state) is **not the same as no input** – underlining one of the reasons why the else condition is so important in computer programming.

Incrementing and decrementing variables

Previous coding examples have used the standard assignment statement to increment a value:

```
int stateCounter = 0;
stateCounter = stateCounter + 1; // stateCounter == 1
```

the above code statement does the following (see also chapter 4, Figure 4.12):

1. Access the value of the variable `stateCounter` by copying
2. Add an integer value of 1 to the variable value
3. Reassign this new value to the variable `stateCounter`

When working with integer counter variables, a compound assignment statement is often used to reduce this syntax:

```
int stateCounter = 0;
//increment the variable using +=
stateCounter += 1; // stateCounter == 1
//decrement the variable using -=
stateCounter -= 1; // stateCounter == 0
```

Compound assignment statements perform the same operation as the standard assignment statement - but they are much quicker to type. You will often encounter these statements in practical code examples, and so the final project uses compound assignment to introduce this syntax - if you find it confusing then you can simply swap the statement with the standard version shown above.

A nested selection statement can increment a state variable using a single switch, allowing the code to use this variable to make more informed decisions about which processes to carry out. With a control input that changes state, outputs can be added to indicate the current state of the system. As noted with the use of a single input switch, to save on breadboard space **two LEDs can be connected in parallel** to indicate all three states of the system (Figure 9.10).

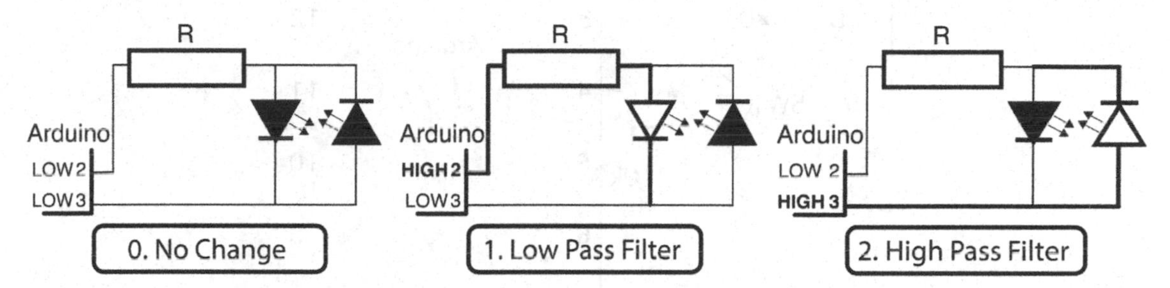

Figure 9.10 Swapping output polarity to change LEDs. ***The diagram shows how two LEDs can be connected in opposite directions in parallel, allowing each to be lit separately by the same connecting wires. The first case (No Change state) is indicated by no LED, whilst the second (LPF) is indicated by LED 1 and the third (HPF) by LED 2. To achieve this, the polarity of the Arduino output pins 2 and 3 is reversed to light each LED separately.***

The schematic examples show how reversing the polarity of the Arduino output pins (LOW/HIGH) can be used to light each LED separately, where the LOW pin acts as a GND connection for current to flow into (from +5V HIGH). An LED will only allow current to flow in a single direction, so if two LEDs are connected in parallel in opposite directions then only one of them will light when current flows. The Arduino digital output pins can be used to light the LEDs and indicate all three states:

0. **No output** – pin 2 (LOW), pin 3 (LOW)
1. **LPF output** – pin 2 (HIGH), pin 3 (LOW)
2. **HPF output** – pin 2 (LOW), pin 3 (HIGH)

At this point, a Tinkercad project can be built to test this input/output control system.

9.2.1 Worked example – Arduino state control

The system will require a push-button switch for input, and two LEDs connected in opposite directions in parallel for output (Figure 9.11).

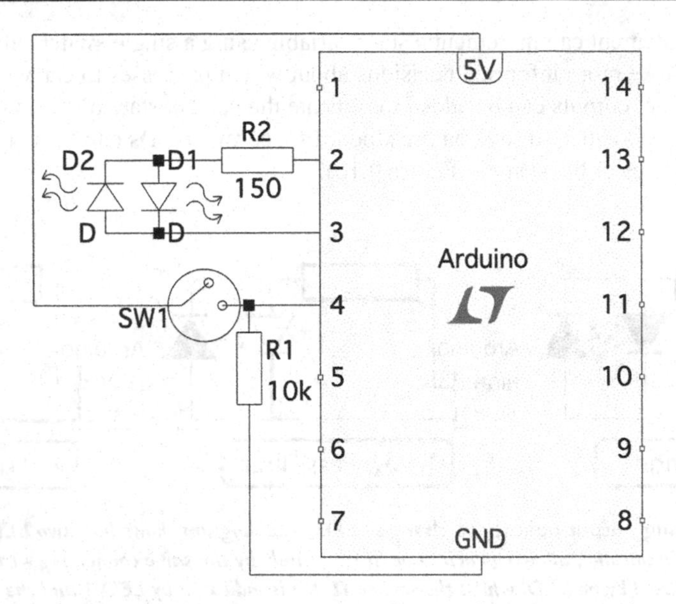

Figure 9.11 Arduino state control test circuit. ***The schematic shows how a single push-button switch (SW1) can be connected as an Arduino digital input, with two LEDs (D1, D2) as output indicators to show the state of the system. The polarity of the output pins (2 and 3) will be reversed in the Arduino code to light each LED separately.***

The schematic in Figure 9.11 shows a single switch input and two LEDs connected in parallel as outputs, which allows the setup code for the project to be defined:

```
const byte lowPassLED = 2;
const byte highPassLED = 3;
const byte buttonInputPin = 4;

int buttonInput = 0;          // button input will be LOW/HIGH
byte stateCounter = 0;         // current state of the system
bool pushedButton = false; // Boolean to stop multiple button triggering
void setup()
{
  pinMode(lowPassLED, OUTPUT);
  pinMode(highPassLED, OUTPUT);
  pinMode(buttonInputPin, INPUT);
}//end of setup
```

This code declares the input (`buttonInputPin`) and output (`lowPassLED, highPassLED`) pin numbers, alongside an integer to hold the buttonInput (the Arduino API specifies an integer for digitalRead()). The state counter variable (`stateCounter`) will only increment between 0 and 2, so a byte provides more than enough memory for this value range. The final variable is a Boolean (`pushedButton`), which is included to prevent multiple triggering on the push=button input (Figure 9.12).

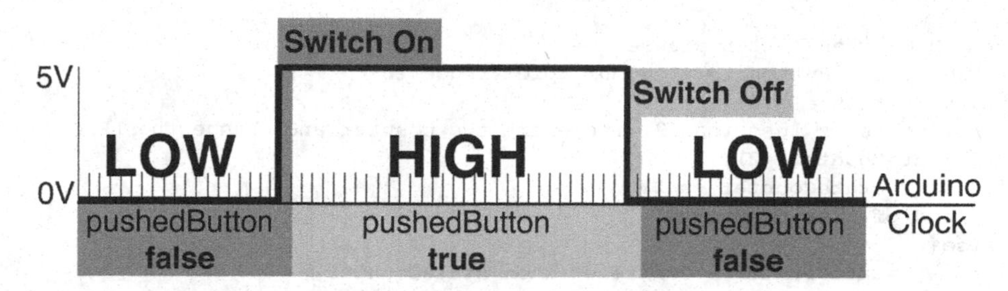

Figure 9.12 Multiple switch input triggering. ***The diagram shows how an input switch going from LOW to HIGH can trigger for multiple cycles of the Arduino loop() function (short vertical lines). The Boolean pushedButton can be used to separate out the first HIGH event (Switch On) and first LOW event (Switch Off) to prevent multiple triggering – note how pushedButton overlaps LOW and HIGH button events.***

In this instance, the clock speed of the Arduino means that the loop() function (where the value of `pushedButton` will be read) will execute at far too high a rate to leave time for any switch to transition from on back to off, leading to multiple cycles of loop() detecting the same push-button input event as a new input. To avoid this, the Boolean `pushedButton` variable acts in a similar manner to the noteTriggered variable to prevent multiple switch inputs from occurring:

```
  // read the push-button input pin:
buttonInput = digitalRead(buttonInputPin);
  if (buttonInput == HIGH && !pushedButton) {
  // HIGH when button pressed - set pushedButton to stop multiple triggering
  pushedButton = true;
}//end of buttonState == HIGH && !pushedButton
  else if (buttonInput == LOW && pushedButton){
  // LOW when button not pressed, clear pushedButton
  pushedButton = false;
}//end of buttonState == LOW && pushedButton
```

This selection statement tests a combination of two Boolean variables, where the first condition of interest is (`buttonInput == HIGH && !pushedButton`). The use of `!pushedButton` occurs when the button has initially been pushed for the first time, and the first instruction inside this condition must then set (`pushedButton = true;`) to prevent further triggering if the button is still HIGH when the next loop() cycle executes. Thus, the only time a valid button push occurs is when the input is HIGH and `pushedButton` is LOW.

The second condition of interest is when (`buttonInput == LOW && pushedButton`). In this case, the value of buttonInput has returned to LOW as the switch is now released, but again only the first one of these events is of interest – when (`pushedButton == true`). Now, the instruction inside this condition sets (`pushedButton = false;`) to prevent further triggering of the selection statement. This template condition can now be expanded to include the previous nested condition that increments the state counter:

```
void loop(){
  // read the push-button input pin:
  buttonInput = digitalRead(buttonInputPin);
    if (buttonInput == HIGH && !pushedButton) {
```

```
      // HIGH when button pressed
     //set pushedButton to stop multiple triggering
     pushedButton = true;
     //if state is less than 2, increment the counter and change the LED
     if(stateCounter <2){
             stateCounter += 1;
     }//end of stateCounter <2
     else{
            stateCounter = 0;//reset when stateCounter == 2
     }//end of stateCounter else
    }//end of buttonState == HIGH && !pushedButton
   else if (buttonInput == LOW && pushedButton){
    // LOW when button not pressed, clear pushedButton
       pushedButton = false;
    }//end of buttonState == LOW && pushedButton
 }//end of loop()
```

Now that the input switch control is incrementing the current state through `stateCounter`, this state variable can be used inside a function call to update the LED output pins accordingly:

```
void changeLEDState(){
  if(stateCounter == 1){//LPF
    digitalWrite(lowPassLED, HIGH);
    digitalWrite(highPassLED, LOW);
  }//end of stateCounter == 1
  else if(stateCounter == 2){//HPF
    digitalWrite(lowPassLED, LOW);
    digitalWrite(highPassLED, HIGH);
  }//end of stateCounter == 1
  else{
   //assume stateCounter == 0 (also cover for any other condition)
   digitalWrite(lowPassLED, LOW);
   digitalWrite(highPassLED, LOW);
  }//end of else stateCounter
}//end of changeLEDState
```

The function uses a selection statement to evaluate the current state and update the digital output pins that are connected to the LEDs accordingly. This function can then be called inside loop() after the nested selection statement that increments the state variable, to use the updated value to change the configuration of the LED output pins (2 and 3). This code listing can now be tested in Tinkercad, to ensure that it is functioning correctly. To do this, a prototype circuit for the schematic in Figure 9.11 can be built (Figure 9.13).

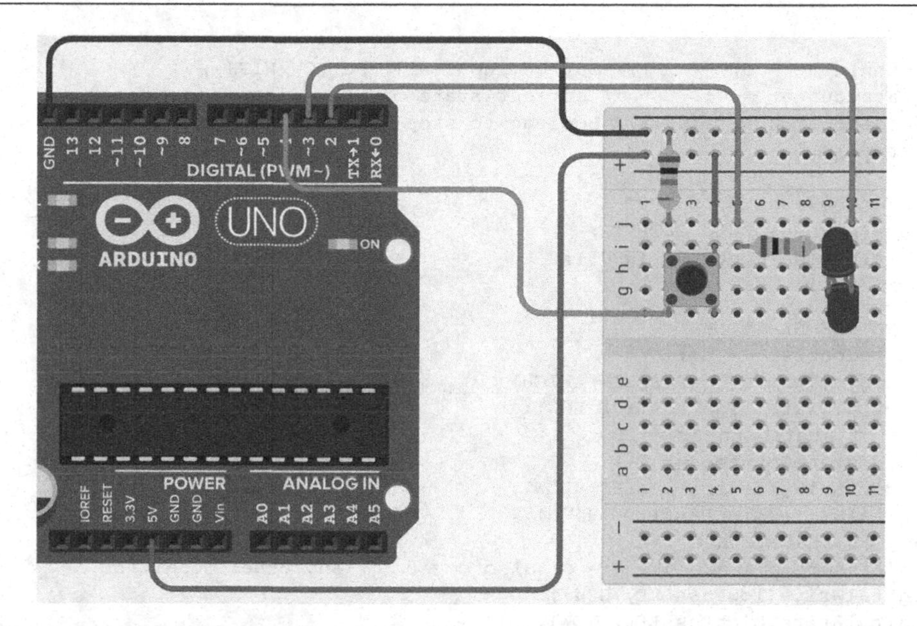

Figure 9.13 Arduino state control test circuit layout. ***The diagram shows a push-button switch input connected in series with a 10kΩ resistor to digital pin 4 on the Arduino. Digital pins 2 and 3 are configured for output, with each pin connected to either side of a parallel diode network (in series with a 150Ω resistor that limits current). As the output pins change polarity from LOW (0V) to HIGH (5V) so will each oppositely wired LED light in turn.***

The breadboard layout shown in Figure 9.13 is fairly straightforward, other than the placement of both diodes (in opposing directions) in parallel on pins [g9–g10] and [h9–h10] respectively. Although this looks a little strange in the Tinkercad simulator, in practice both LEDs will easily sit on the same columns with adequate space between them. This configuration is intended to save space on the larger project circuit, which also accounts for the connection of the 10kΩ pull-down resistor directly to the GND rail of the breadboard. Create a new circuit named finalProj_stateControl and place the components as follows:

1. Add a push-button switch between pins [i2–f4].
2. Add a 10kΩ resistor between the switch and top ground (JGND) between pins [j2–JGND].
3. Add a connector to the top positive rail [j4–JV+].
4. Add an input connector from the switch to Arduino digital pin 4 [f2–AD4].
5. Add a 150Ω resistor between pins [i5–i9].
6. Add an LED cathode-anode [h9–h10].
7. Add an LED anode-cathode [g9–g10].
8. Add an output connector from Arduino digital pin 2 to the 150Ω resistor [AD2–j5].
9. Add an output connector from Arduino digital pin 3 to the LEDs [AD3–j10].
10. Add a connector between the negative (ground) rail and the Arduino top left GND pin.
11. Add a connector between the positive rail and the bottom Arduino +5V pin.

At this point, the full code listing for the Arduino state control project can now be entered into Tinkercad:

```
const byte lowPassLED = 2;//LED output pin
const byte highPassLED = 3; //LED output pin
const byte buttonInputPin = 4; //Switch input pin
```

```
int buttonInput = 0;        // button input will be LOW/HIGH
byte stateCounter = 0;       // current state of the system
bool pushedButton = false; // Boolean to stop multiple triggering
void setup()
{
  pinMode(lowPassLED, OUTPUT);
  pinMode(highPassLED, OUTPUT);
  pinMode(buttonInputPin, INPUT);
}//end of setup

void changeLEDState(){
  if(stateCounter == 1){
       digitalWrite(lowPassLED, HIGH);
     digitalWrite(highPassLED, LOW);
  }//end of stateCounter == 1
  else if(stateCounter == 2){
       digitalWrite(lowPassLED, LOW);
    digitalWrite(highPassLED, HIGH);
  }//end of stateCounter == 1
  else{//assume stateCounter == 0 (also cover for any other condition)
    digitalWrite(lowPassLED, LOW);
    digitalWrite(highPassLED, LOW);
  }//end of else stateCounter
}//end of changeLEDState
void loop()
{
  // read the push-button input pin:
  buttonInput = digitalRead(buttonInputPin);
    if (buttonInput == HIGH && !pushedButton) {// HIGH when button pressed
      //set pushedButton to stop multiple triggering
      pushedButton = true;
      if(stateCounter <2){ //if state less than 2, increment counter & change LED
              stateCounter += 1;
      }//end of stateCounter <2
      else{
            stateCounter = 0;//reset when stateCounter == 2
      }//end of stateCounter else
      changeLEDState();//call function to update LED
    }//end of buttonState == HIGH && !pushedButton
  else if (buttonInput == LOW && pushedButton){
        pushedButton = false; // LOW when button not pressed, clear pushedButton
    }//end of buttonState == LOW && pushedButton
}//end of loop()
```

If the code compiles correctly the circuit should run in the simulator, allowing the push button to iterate through the states from 0 to 2 and then back to 0. The LEDs should light in sequence, with the No Change state being indicated when both are off. This stage of the project can now be built on breadboard, with the circuit being retained for later stages to be added to produce the final project build. The reference for top positive (J+) and top ground (JGND) should be noted when placing components.

Stage 1 – project steps

For Stage 1 of the project, you will need: 1. 1 × push-button switch (SW1) 2. 2 × LEDs (D1, D2) for state indicator output 3. 1 × 10kΩ pull-down resistor (R1) 4. 1 × 150Ω current-limiting resistor (R2) 5. 3 × connector cables, 1 × connector wire and 1 small cable tie	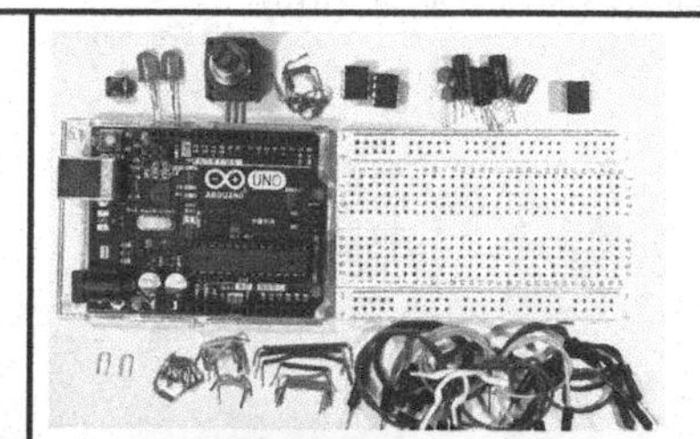
1. Add the push-button switch from pins [g2–i4] – ensuring the connectors are aligned A(j4) and B(i4) (see chapter 2, Figure 2.20). 2. Add the 10kΩ pull-down resistor (R1) directly to the top ground rail (to save space) between pins [j2–JGND]. 3. Add a connector wire between switch pin A (j4) and the top positive rail [j4–JV+]	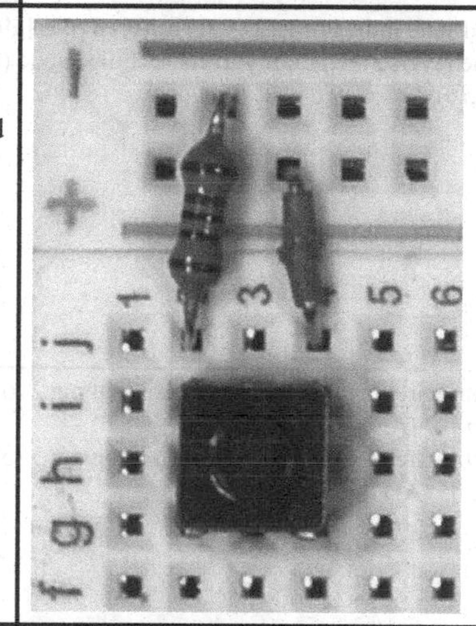
1. Add a signal wire from the switch output D(f2) to the Arduino digital pin 4 [f2–AD4].	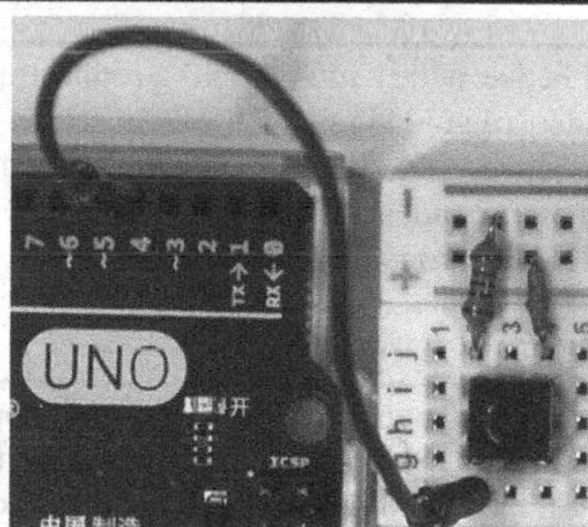
1. Add a 150Ω current-limiting resistor (R2) between pins [i5–i9].	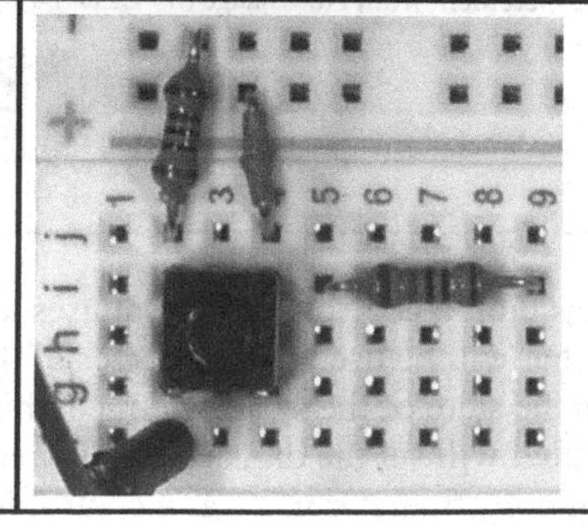

<table>
<tr><td>1. Add the first state indicator LED (D1) between pins [h9–h10] – connect the cathode (h9) and anode (h10).</td><td>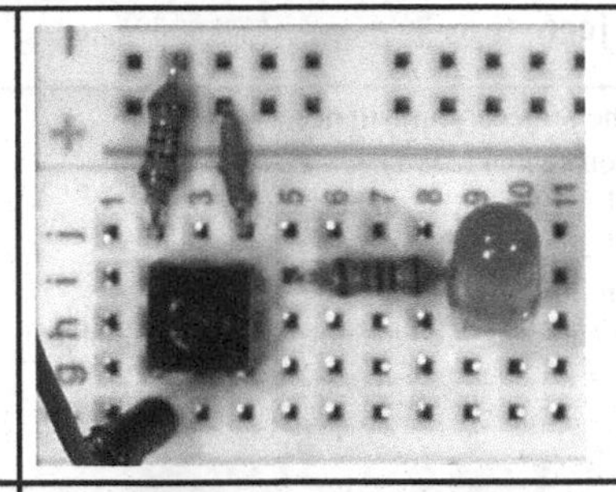</td></tr>
<tr><td>1. Add the second state indicator LED (D2) between pins [g9–g10] – this time reverse the polarity to connect the anode (g9) and cathode (g10).</td><td></td></tr>
<tr><td>1. Add a connector wire from Arduino digital output pin 2 to breadboard pin j10 [AD2–j10].
2. Add a connector wire from Arduino digital output pin 3 to breadboard pin j5 [AD3–j5].</td><td>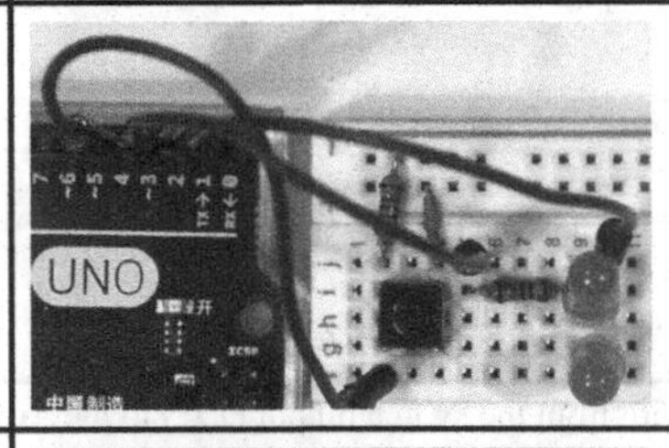
</td></tr>
<tr><td>1. Add a connector between the Arduino GND terminal (pin 14) and the breadboard ground rail.
2. Add a connector between the Arduino +5V terminal and the breadboard positive rail (not shown).
3. Use a small cable tie to group these cables together</td><td>
</td></tr>
<tr><td colspan="2">Copy and paste the code from the finalProj_stateControl project in Tinkercad into the Arduino IDE and save it as a new Arduino project named finalProj_Stage1-Control.ino. Compile the code to check for copy errors and then upload the sketch to the Arduino Uno and test the circuit. If everything compiles correctly, each time the push-button switch is pressed the LED state indicators should increment from No Change (no LEDs), through LPF (LED 1) to HPF (LED 2):
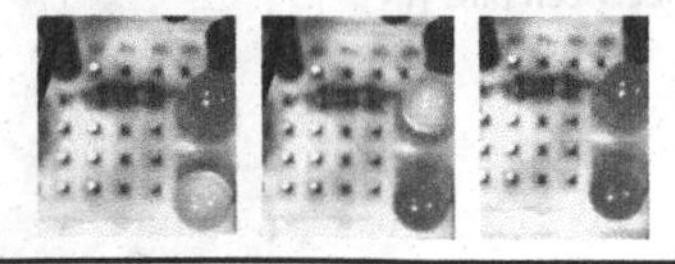</td></tr>
</table>

This project has built a control system that uses a push-button switch input to increment a system state variable that will be used in later stages of the project to determine which audio filter will be changed. With the state of the system defined, this circuit can now be extended to include an analogue potentiometer that will set the digital potentiometer data values in the final circuit. This allows a single analogue input potentiometer to act as a control input to the Arduino, which will read the value of the analogue potentiometer and use this to set the digital potentiometers connected to the low-pass and high-pass filters. The input potentiometer will be placed on the top right of the board in our existing Tinkercad circuit to allow other components to be added in the final layout (Figure 9.14).

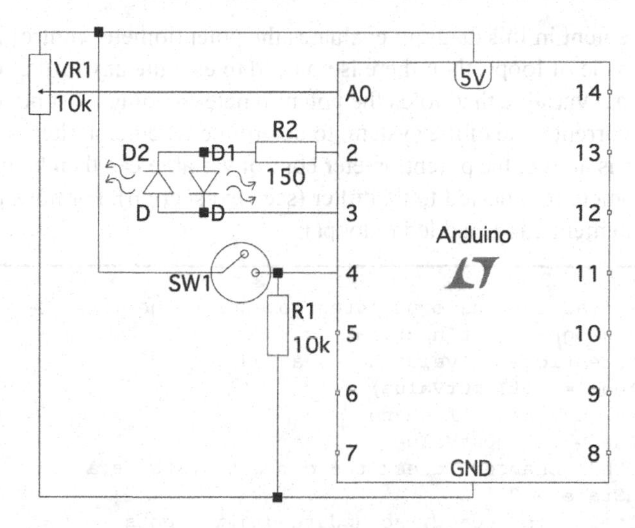

Figure 9.14 Arduino state control test circuit schematic with input potentiometer control. ***In the diagram, a potentiometer (VR1) has been added as a control input on analogue input pin A0. By combining the potentiometer reading with the state variable (set by the push-button switch) each digital potentiometer can be individually configured by the Arduino.***

The diagram shows the addition of a potentiometer as an analogue control input (on pin A0), which can then be used by the Arduino to set the value of each digital potentiometer. The Arduino will now require additional code to be added to the loop() function to respond to changes to the potentiometer input. The potentiometer control input has two criteria that can be framed as questions (to form selection statements) to determine when other processes should execute:

1. **Has the potentiometer value changed**? – there is no point in executing any other code if the wiper has not been moved by the user, as the data has not changed
2. **Does the current state require a potentiometer value**? – only the LPF and HPF states require potentiometer input values, so the state of the system must be greater than 0

These two criteria can be combined within a second nested selection statement that tests for both potentiometer input and also system state (Figure 9.15).

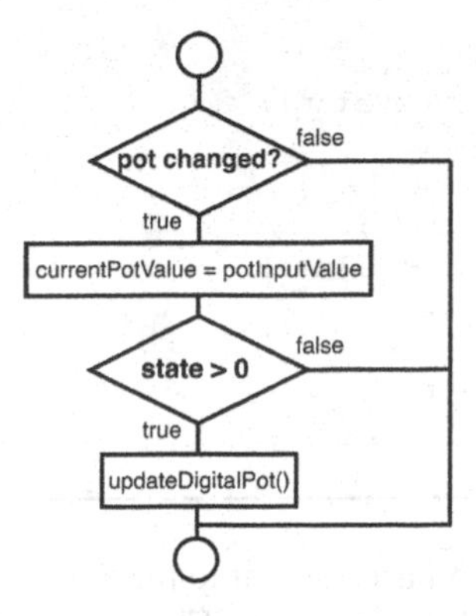

Figure 9.15 Nested selection statement flow diagram. ***The diagram shows a nested selection statement where the outer selection evaluates whether the potentiometer control input value has changed – if so then this data value is updated. The inner selection evaluates whether the current state is greater than zero (i.e. a filter has been selected) – if so then the relevant digital potentiometer can be updated.***

The outer selection statement in this diagram evaluates the potentiometer control input – if this has not changed since the last cycle of loop() then there is no need to execute any further code. If this value has changed, then the data variable that holds the potentiometer reading must be updated. The inner selection evaluates the current state of the system, to determine whether a filter is currently selected (i.e. state > 0). If a filter is active, the potentiometer control variable can then be used to update the relevant digital potentiometer connected to that filter (see next section). For now, the code to execute this nested selection statement can be added to loop():

```
//inside loop() - read the analogue potentiometer value
potInputValue = analogRead(potInputPin);
// check if the potentiometer value has changed
if((currentPotValue != potInputValue)){
    // only update the value if changed
    currentPotValue = potInputValue;
    // if the pot has changed, check the current system state
    if(lastButtonState > 0){
            //a filter is selected, so update digital pots
            updateDigitalPot();
    }//end of lastButtonState > 0
    else{
            //assume lastButtonState == 0 (also cover for any other condition)
    }//end of else lastButtonState > 0
}//end of currentPotValue != potInputValue
else{
} //end of else (currentPotValue != potInputValue)
```

The outer selection in this code evaluates whether the potentiometer input value has changed (`currentPotValue != potInputValue`) – if this is not the case then the selection ends. If the potentiometer has changed, the new value is updated (`currentPotValue = potInputValue;`) and the current system state is then evaluated in the inner selection statement (`lastButtonState > 0`). If a filter is selected (i.e. not state 0) then the digital potentiometers can be updated using the function call `updateDigitalPot()`, which uses the system state variable to determine what output is required:

```
void updateDigitalPot(){
  if(lastButtonState == 1){
     Serial.println("Pot 1 Value");
        Serial.println(currentPotValue);
  }//end of lastButtonState == 1
  else if(lastButtonState == 2){
     Serial.println("Pot 2 Value");
        Serial.println(currentPotValue);
  }//end of lastButtonState == 2
  else{//state 0 - no change
  }//end of else lastButtonState
}//end of updateDigitalPot
```

The `updateDigitalPot()` function uses output to the Serial Monitor (`Serial.println`) to check that the logic of the code is working correctly. This code can now be tested in Tinkercad, using the updated layout shown in Figure 9.16.

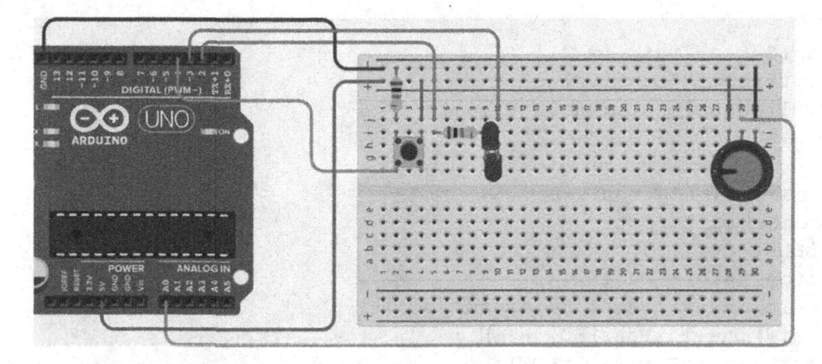

Figure 9.16 Arduino state control with input potentiometer control layout. *The diagram shows a +5V signal connected to a potentiometer, where this voltage is split by the potentiometer wiper as an input to Arduino analogue input pin A0. The value of the potentiometer can then be read as the data of the Arduino control system, which will be used to control the digital potentiometers in the final project. Note the positioning of the potentiometer at the top right of the board – this is to allow other control system components to be added in the following sections.*

To extend the previous state control circuit, save the file as finalProj_ stateControlPotInput and place the components as follows:

1. Add a 10kΩ potentiometer between pins [i28–i30] (rotate component 180° to keep the pin connections visible).
2. Add a connector to the top ground rail [j30–JGND].
3. Add a connector to the top positive rail [j28–JV+].
4. Add a control input connector from the potentiometer to the Arduino [j29–ADA0].

The full code listing for this system build can then be copied and pasted into the Tinkercad code window:

```
const byte lowPassLED = 2;
const byte highPassLED = 3;
const byte selectButtonPin = 4;
const byte potInputPin = A0;
int buttonState = 0;          // current state of the button
int lastButtonState = 0;      // previous state of the button
bool pushedButton = false;
int currentPotValue = 0;
int potInputValue = 0;
void setup()
{
  Serial.begin(9600);
  pinMode(lowPassLED, OUTPUT);
  pinMode(highPassLED, OUTPUT);
  pinMode(selectButtonPin, INPUT);
  pinMode(potInputPin, INPUT);
}//end of setup

void updateDigitalPot(){
  if(lastButtonState == 1){
     Serial.println("Pot 1 Value");
        Serial.println(currentPotValue);
  }//end of lastButtonState == 1
```

```
    else if(lastButtonState == 2){
       Serial.println("Pot 2 Value");
          Serial.println(currentPotValue);
    }//end of lastButtonState == 2
    else{
    }//end of else lastButtonState
  }//end of updateDigitalPot
  void changeLEDState(){
    if(lastButtonState == 1){
      Serial.println("Filter 1 Selected");
         digitalWrite(lowPassLED, HIGH);
      digitalWrite(highPassLED, LOW);
    }//end of lastButtonState == 1
    else if(lastButtonState == 2){
      Serial.println("Filter 2 Selected");
         digitalWrite(lowPassLED, LOW);
      digitalWrite(highPassLED, HIGH);
    }//end of lastButtonState == 1
    else{//assume lastButtonState == 0 (also cover for any other condition)
      Serial.println("No Filter Selected");
      digitalWrite(lowPassLED, LOW);
      digitalWrite(highPassLED, LOW);
    }//end of else lastButtonState
  }//end of changeLEDState
  void loop()
  {
    // read the push-button input pin:
    buttonState = digitalRead(selectButtonPin);
    //read the analogue potentiometer value
    potInputValue = analogRead(potInputPin);
    //now check if the value has changed
    if((currentPotValue != potInputValue)){
      currentPotValue = potInputValue;
      if(lastButtonState > 0){
        //update digital Pot if a filter is selected
               updateDigitalPot();
      }//end of lastButtonState > 0
      else{//assume lastButtonState == 0 (also cover for any other condition)
      }//end of else lastButtonState > 0
    }//end of currentPotValue != potInputValue
    else{
    } //end of else (currentPotValue != potInputValue)
     // if the state has changed, increment the counter
     if (buttonState == HIGH && !pushedButton) {
        // HIGH then button pressed
        pushedButton = true;
        //change the LED indicator
        if(lastButtonState <2){
               lastButtonState += 1;
              changeLEDState();//call function to update LED
        }//end of lastButtonState <2
        else{
              lastButtonState = 0;
              changeLEDState();//call function to update LED
        }//end of lastButtonState else
     }//end of buttonState == HIGH && !pushedButton
```

```
else if (buttonState == LOW && pushedButton){
     // LOW then button not pressed
        pushedButton = false;
   }//end of buttonState == LOW && pushedButton
}//end of loop()
```

Stage 1A – Project Steps

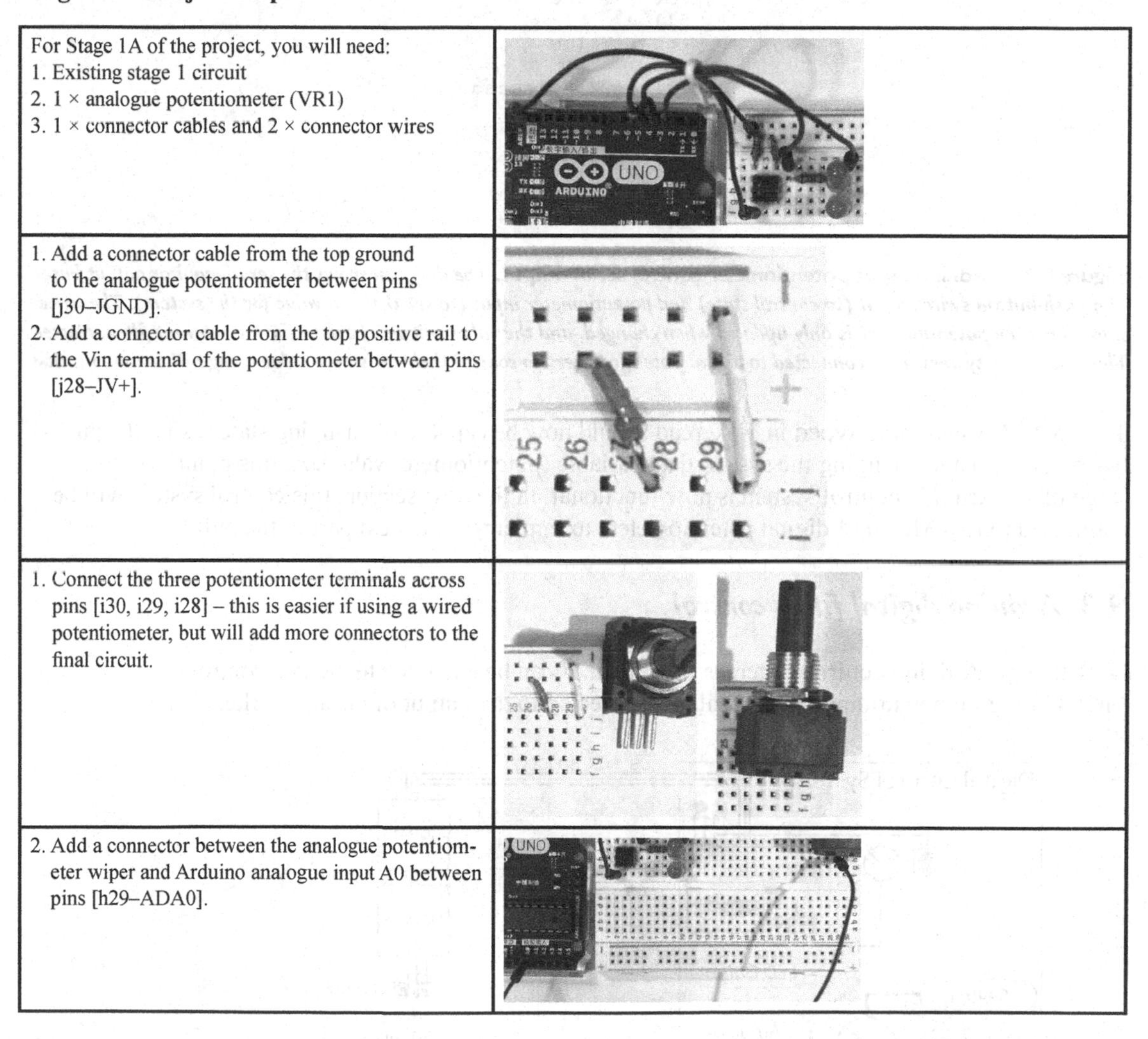

For Stage 1A of the project, you will need: 1. Existing stage 1 circuit 2. 1 × analogue potentiometer (VR1) 3. 1 × connector cables and 2 × connector wires	
1. Add a connector cable from the top ground to the analogue potentiometer between pins [j30–JGND]. 2. Add a connector cable from the top positive rail to the Vin terminal of the potentiometer between pins [j28–JV+].	
1. Connect the three potentiometer terminals across pins [i30, i29, i28] – this is easier if using a wired potentiometer, but will add more connectors to the final circuit.	
2. Add a connector between the analogue potentiometer wiper and Arduino analogue input A0 between pins [h29–ADA0].	

Copy and paste the code from the finalProj_ stateControlPotInput project in Tinkercad into the Arduino IDE and save it as a new Arduino project named finalProj_Stage1-ControlPot.ino. Compile the code to check for copy errors and then upload the sketch to the Arduino Uno and test the circuit. If the code compiles correctly, the circuit should simulate output when a filter is selected (i.e. the system is not in state 0). Each time the LPF (LED 1) or HPF (LED 2) states are selected with the push button, the analogue potentiometer value should be read by the Arduino and output to the serial monitor. The serial monitor statements should show the selected filter and the current value (Figure 9.17).

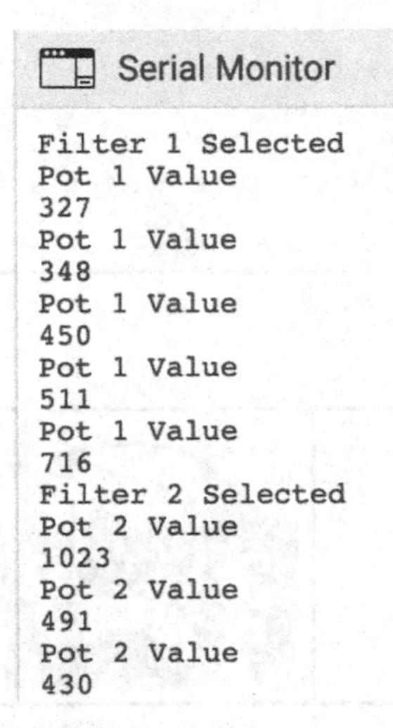

Figure 9.17 Arduino input potentiometer control serial output. ***The diagram shows the serial monitor output based on push-button switch input (to control state) and potentiometer input (to set the data value for the system). The trace shows how the potentiometer is only updated when changed, and the value will only be used for the relevant filter selected. This allows the system to be connected to digital potentiometers to control their resistance value using SPI communication.***

The control system prototyped in Tinkercad should now be capable of changing state (using the push-button switch) and changing the system data variable (potentiometer value). At this point, the first stage of the Arduino control system is now functional. In the next section, this control system will be connected to two MCP413 digital potentiometers to implement the next part of the build.

9.3 Arduino digital filter control

Now that the Arduino control system is functional, it can be extended to include control of the MCP413 digital potentiometers that will be connected to the output of the audio filters (Figure 9.18).

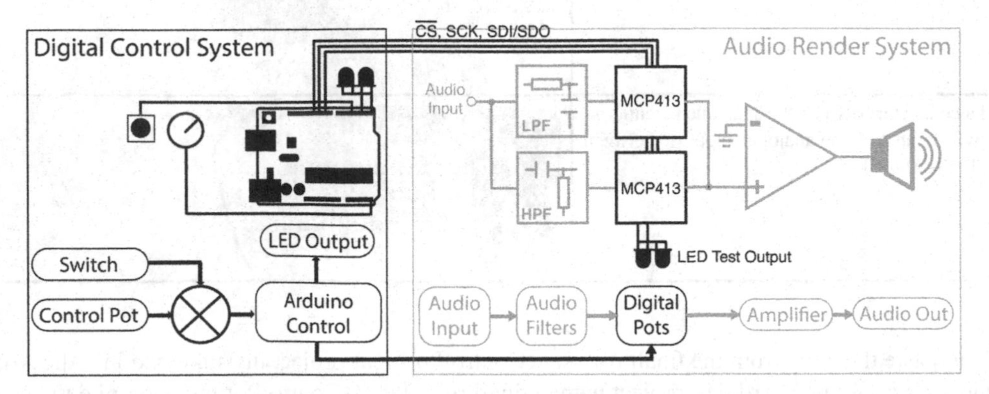

Figure 9.18 Final project stage 2 – digital potentiometer control system diagram. ***The diagram shows how the Arduino digital control system can now be extended to control the MCP413 digital potentiometers that will be used in the audio render system. The MCP413 chips can be controlled using the serial peripheral interface (SPI), indicated by the three connecting wires from the Arduino that branch across the digital control and audio render systems. An LED test output circuit is also shown, which will be used in this stage to verify the correct control and operation of the MCP413 chips.***

The diagram shows how the second project stage extends the Arduino control system to include connections to two MCP413 digital potentiometers. Although there are other ways to digitally control audio signals such as voltage-controlled amplifiers (VCAs), digital potentiometers have been chosen for this book primarily because they are relatively inexpensive to buy. In addition, the MCP413 is controlled via the serial peripheral interface (SPI), which is a useful protocol that works well with the Arduino, allowing prior work on serial data output (in chapter 5) to be extended for this project. Microchip Technologies provide a range of digital potentiometers that vary in complexity (and number of potentiometers), but for this project the MCP413-103 chip will be used (Figure 9.19).

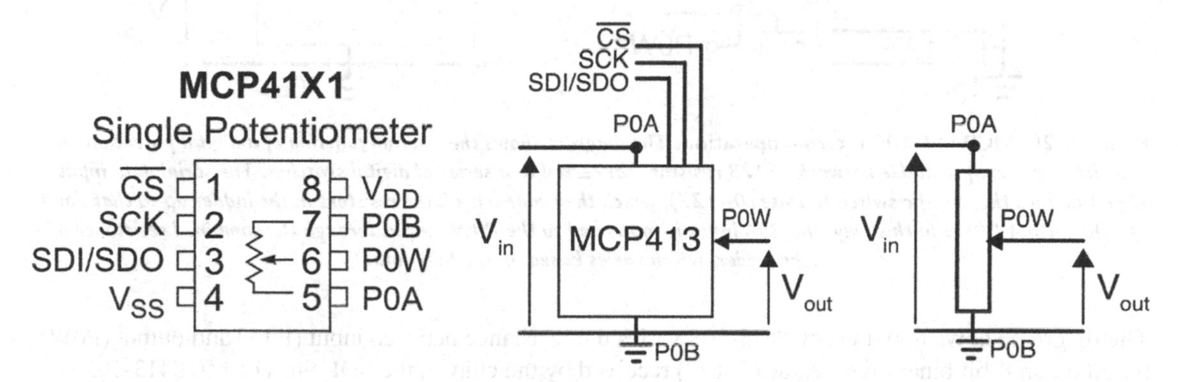

Figure 9.19 The MCP413-103 10kΩ digital potentiometer chip. ***The left-hand diagram shows the pin layout for the 8-pin MCP413-103 IC, where pin 1 denotes the chip select (CS) pin that is used to enable communication with the chip. The clock (SCK) and serial data input/output (SDI/SDO) pins are used to receive SPI data, with ground (Vss) and power (Vdd) pins being able to run from +5V Arduino supply. The middle diagram shows how the right-hand pins of the chip operate in the same manner as an analogue potentiometer (right-hand diagram for reference), where POA typically connects to the input signal, POW provides the wiper output and POB is connected to the circuit ground (GND).***

The Microchip Technologies MCP413-103 digital potentiometer is a very versatile chip and is relatively easy to work with once the distinction between digital control inputs and potentiometer connections has been made. The left-hand diagram shows the pin layout for the 8-pin IC package, where a broad distinction can be made between control inputs (left pins 1–3) and resistor outputs (right pins 5–7) alongside the power (V_{DD} pin 8) and ground (V_{SS} pin 4) pins. The middle diagram shows the schematic symbol used in this book, which aims to mimic an analogue potentiometer (shown in the right-hand diagram for reference) with additional control inputs connected to the top of the chip.

The chip is digitally controlled to select a specific resistance level from an internal ladder network of resistors, where the data value received on the SDI/SDO pin (0–128) closes one of a series of internal switches that corresponds to the resistance required (Figure 9.20).

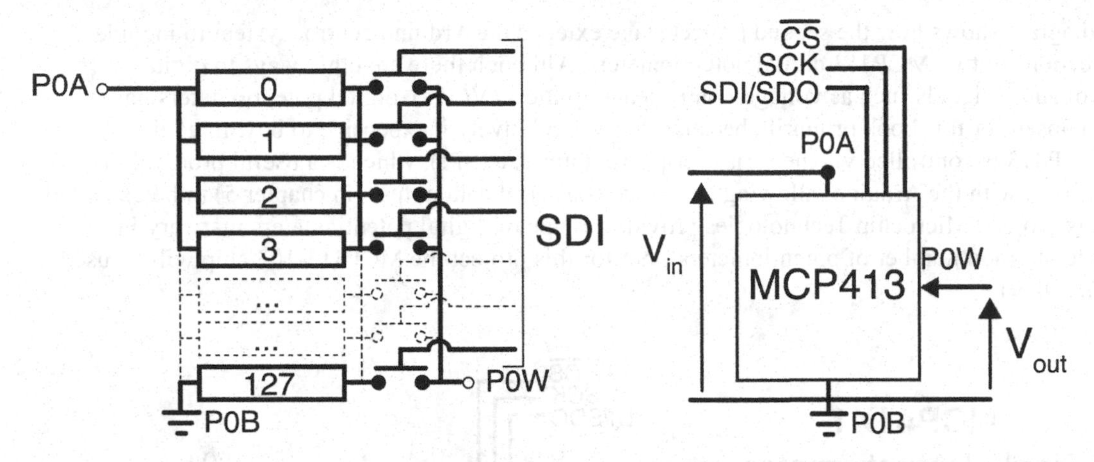

Figure 9.20 MCP413-103 internal operation. ***The diagram shows the internal function of a digital potentiometer, which consists of a ladder network of 128 resistors connected to a series of digital switches. The serial data input (SDI) defines the specific switch to close (0–127), which then connects all the resistors in the ladder up to that point to the output POW. In this way, the POA input is connected to the POW output through the combined resistance of the ladder, which varies based on the SDI data.***

The diagram shows how the MCP413-103 varies the resistance between input (P0A) and output (P0W) based on an 8-bit binary data value (0–127) received by the chip on the SDI pin. The MCP413-103 is updated using a serial peripheral interface (SPI), which can be provided by an Arduino when running a specific software library (<SPI.h>). For the MCP413-103 chip, the SPI interface operates using three pins:

1. **Chip select (CS)** – this pin tells the MCP413-103 to begin receiving data on the SDI pin
2. **Serial clock (SCK)** – this pin provides the synchronization clock for the data received (see below)
3. **Serial data input/serial data output (SDI/SDO)** – either separate data pins or combined (as with MCP413-103)

The overbar on the chip select pin (CS) indicates that it is an active low logic chip, which means that the pin must be set to digital LOW (0V) to enable the chip. Though a full discussion of these aspects of digital electronics is outwith the scope of this book, reasons for using active low logic include noise reduction and increased fanout (the number of devices that can be connected together) due to the low activation level. In this project, each chip select pin (CS) must be set to digital HIGH (+5V) when not in use, and only set LOW (0V) when serial data is being sent to that specific chip on the SDI/SDO pin.

The serial clock is needed to ensure that digital data sent by the Arduino is read correctly by the MCP413-103, effectively acting like a metronome to inform the chip **when** to read the incoming serial data (Figure 9.21).

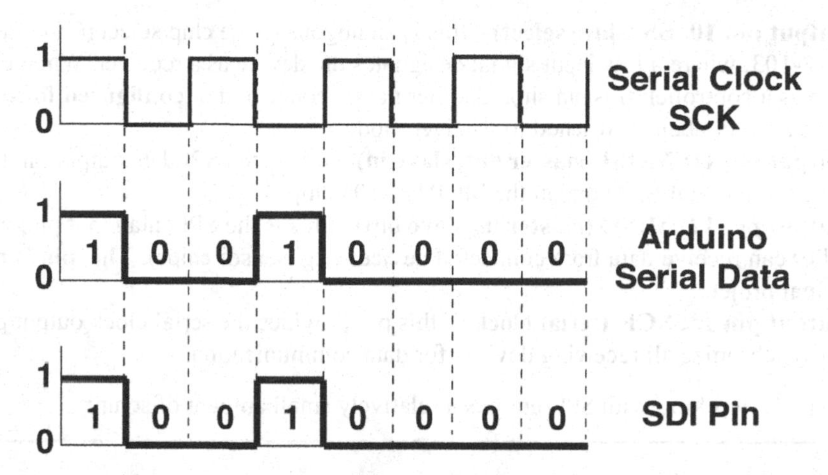

Figure 9.21 Digital clock synchronization example. ***In the example, the serial clock pin provides a synchronization pulse that tells all connected devices when to begin receiving data. If the Arduino provides the clock as the controller, any serial data it sends will be synchronized to this pulse. As a receiver, the MCP413-103 uses its SCK pin to synchronize with the clock and read the Arduino serial data correctly.***

The diagram shows a serial clock example, where the controller (Arduino) provides both the synchronization clock pulse (SCK) and also the serial data that will be used to configure the digital potentiometer chips. The MCP413-103 acts as a receiver for this data on its SDI pin, where the Arduino clock signal is also connected to its SCK pin to allow it to synchronize to the incoming serial data. This simple form of clocking is used widely in digital electronics, where connected devices share a common clock pulse to ensure that they all send and receive data in synchronicity. The MIDI output examples in chapter 5 used an older form of clocking known as a universal asynchronous receiver transmitter (UART), where timing is managed by a dedicated circuit connected to both the MIDI In and OUT. Although still very robust, MIDI synchronization is now largely performed by other methods (e.g. USB) that carry the MIDI data without the need for the 5-pin DIN connector defined in the original MIDI specification.

Communications terminology

In this book, the terms controller and receiver are used when talking about device communications – these are based on the standard terms of sender/receiver that are used throughout analogue and digital communications (and also widely used in human communication models). Controller is used (rather than sender) to indicate the device that manages the communication process (in this case the Arduino), as this may not necessarily be the sender – e.g. in the case of sensor devices connected to an Arduino controller. Having said this, most electronics data sheets will use the term master when referring to the controlling device and slave when referring to the devices that synchronize with it. Although these are also standard terms (and widely used), it is noted that this language is both problematic and outdated and so controller/receiver are preferred.

The specific pins needed to control the MCP413-103 digital potentiometer chip are now known, so the next task is to program the Arduino to control a serial peripheral interface (SPI). To do this, the SPI library (<SPI.h>) is used to configure the Arduino for serial communications based around a standard SPI 4-pin configuration for the Arduino Uno:

Digital output pin 10: SS (slave select) – this is analogous to the chip select ($\overline{CS}$) pin on the MCP413-103, where a low input signal designates the device as a receiver. When using the Arduino as a controller, this pin should either not be connected or **configured for output** to avoid the Arduino being switched to receiver mode.

Digital output pin 11: MOSI (master out, slave in) – this is the SPI data output pin that is connected to the SDI/SDO pin on the MCP413-103 chip.

Digital output pin 12: MISO (master in, slave out) – this is the SPI data input pin, where the controller can receive data from connected devices (e.g. sensor chips). **This pin is not needed** in the final project.

Digital output pin 13: SCK (serial clock) – this pin provides the serial clock output pulse that is used to synchronize all receiving devices for data communication.

In the Arduino code, working with SPI requires a relatively small amount of setup:

```
#include <SPI.h>//add the Arduino library that configure SPI communication
const byte lowPassPot = 9;//define the CS output pin for the LPF MCP413 chip
const byte highPassPot = 10;//define the CS output pin (override SS)
int potInputValue = 0;//input control pot variable that is sent as SPI data
const byte potWriteCommand = 0x00;//define the SPI write command
void setup()
{
  pinMode (lowPassPot, OUTPUT);//set digital pin for LPF CS output
  pinMode (highPassPot, OUTPUT);//set digital pin for HPF CS output
  digitalWrite(lowPassPot, HIGH);//set LPF CS signal to HIGH (do not receive)
  digitalWrite(highPassPot, HIGH);//set HPF CS signal to HIGH (do not receive)
  SPI.begin();//start the serial clock (SCK) on digital output pin 13
}//end of setup
```

The code listing above shows how Arduino digital output pins 9 and 10 are specifically configured to act as chip select ($\overline{CS}$) for more than one MCP413-103 chip (see below). The variable `potInputValue` was used in the previous control example to hold the data read from the analogue control potentiometer (on analogue input pin A0). The constant variable `potAddress` is used to define the write command the MCP413-103 chip expects when receiving a new potentiometer value, where `0x00` is an abbreviation of the full binary value `00000000` that corresponds to a potentiometer write command. The MCP413-103 chip can also respond to other commands like read potentiometer value and increment/decrement potentiometer value, but these are not needed in this project.

The setup() function configures pins 9 and 10 for output (as chip select), and then writes a digital HIGH (+5V) signal to both pins to ensure that the connected devices are not initially set to receive data – to prevent any unintentional changes to the potentiometer chips. The next line contains a function call to `SPI.begin()` which tells the Arduino to start the serial clock output on pin 13 – allowing all connected receivers to synchronize to this signal. By using a common clock, multiple chips can be connected to a single SPI bus that shares both the clock (SCK) and data (SDI) connections – only the chip select pin ($\overline{CS}$) must be unique for each connected chip. This is a very powerful element of protocols like SPI, where a single controller (Arduino) can control multiple receivers using a minimum amount of connections (Figure 9.22).

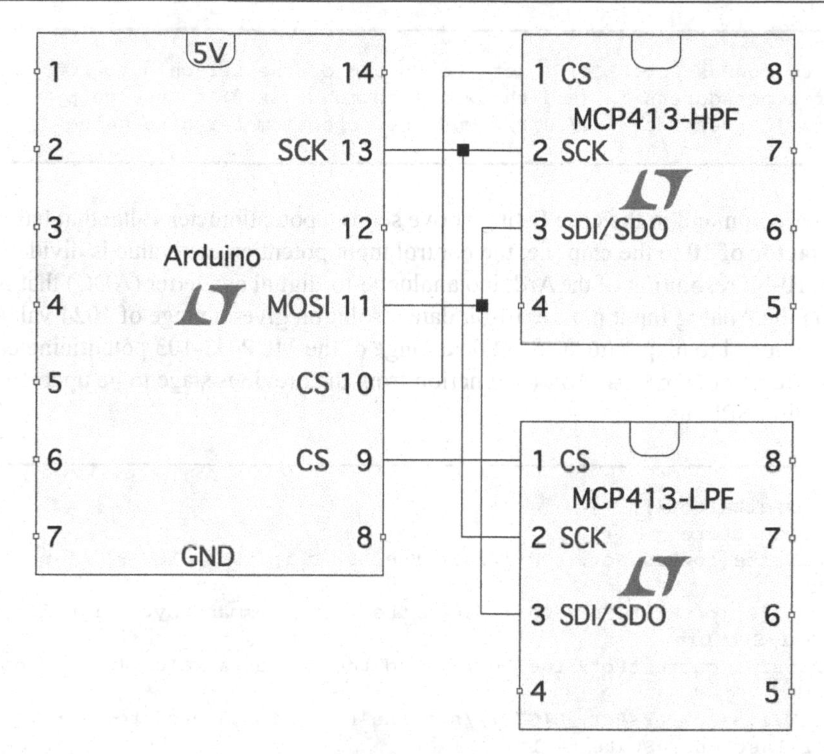

Figure 9.22 Arduino shared SPI bus example. ***The diagram shows the shared clock (SCK pin 13) and serial data (SDI/SDO pin 11) SPI bus connections that are used by the final project circuit, which allow multiple MCP413-103 chips to be connected to a single Arduino as the controller. When the Arduino needs to send to a specific chip, it can set its chip select ($\overline{CS}$) pin LOW whilst leaving the other chip select HIGH. Thus, to access the low-pass filter (LPF) the Arduino sets pin 9 LOW, whilst for the high-pass filter (HPF) it sets pin 10 LOW.***

The Arduino can control multiple chips from a single SDI/SDO data connection – alongside the shared serial clock running on the SCK pin. Pin 10 can be used for chip select ($\overline{CS}$) even though the Arduino UNO SPI configuration uses digital pin 10 as a receiver connection – the setup() command allocating it for digital output overrides this pin allocation. Each chip can now receive the same information, which allows the Arduino to run a single SPI instance, and also reduces the amount of connections needed – this is important when working with limited space on a single breadboard. To change the value on a specific digital potentiometer, three coding processes must be executed:

1. Take the $\overline{CS}$ pin for the relevant chip LOW to enable receiving data
2. Send the write command over the SPI bus to the chip (only a $\overline{CS}$ = LOW chip will receive the command)
3. Send the potentiometer data value over the SPI bus to the chip

These steps can be executed using a combination of `digitalWrite()` (to take the relevant $\overline{CS}$ pin LOW) and the transfer function (`SPI.transfer()`) that is included in the SPI library:

```
digitalWrite(lowPassPot, LOW);//set the CS pin of the LPF chip to LOW
SPI.transfer(potAddress);//send the write command (0x00) to the chip
SPI.transfer(currentPotValue/10);//send the potentiometer data value
```

The data transfer command in the code listing above sends a potentiometer value that has been **reduced by a factor of 10** to the chip (i.e. the control input potentiometer value is divided by 10). This is based on the 10-bit resolution of the Arduino analogue-to-digital converter (ADC) that is used to sample input on the Analog Input pins. A 10-bit data resolution gives a range of 1024 values (0–1023), which must be reduced to map onto the 8-bit data range of the MCP413-103 potentiometer (0–127). This allows the `updateDigitalPot()` function from the previous stage to be updated to output data on the Arduino SPI bus:

```
void updateDigitalPot(){
  if(lastButtonState == 1){
    digitalWrite(lowPassPot, LOW);//set the LPF chip CS to LOW so it will receive
   commands
    SPI.transfer(potWriteCommand);//send the write command byte from Arduino MOSI
   to the SDI/SDO pin
    SPI.transfer(currentPotValue/10);//send the pot data value (scaled down
   to 0-127)
    digitalWrite(lowPassPot, HIGH);//set the LPF chip CS back to HIGH
  }//end of lastButtonState == 1
  else if(lastButtonState == 2){
    digitalWrite(highPassPot, LOW);//set the HPF chip CS to LOW so it will
   receive commands
    SPI.transfer(potWriteCommand);//send the write command byte from Arduino MOSI
   to the SDI/SDO pin
    SPI.transfer(currentPotValue/10);//send the pot data value (scaled down
   to 0-127)
    digitalWrite(highPassPot, HIGH);//set the HPF chip CS back to HIGH
  }//end of lastButtonState == 2
  else{
  }//end of else lastButtonState
}//end of updateDigitalPot
```

Now the configuration and control of the MCP413-103 chips has been planned, a simple LED output circuit can be connected to each chip to test Arduino control of the digital potentiometer resistance value. To do this, the Arduino power and ground connections can be used as the input (P0A) and ground (P0B) for the MCP413-103 chip. In addition, the chips must also be powered (and grounded) in order to operate, which leads to a much more complex schematic (Figure 9.23).

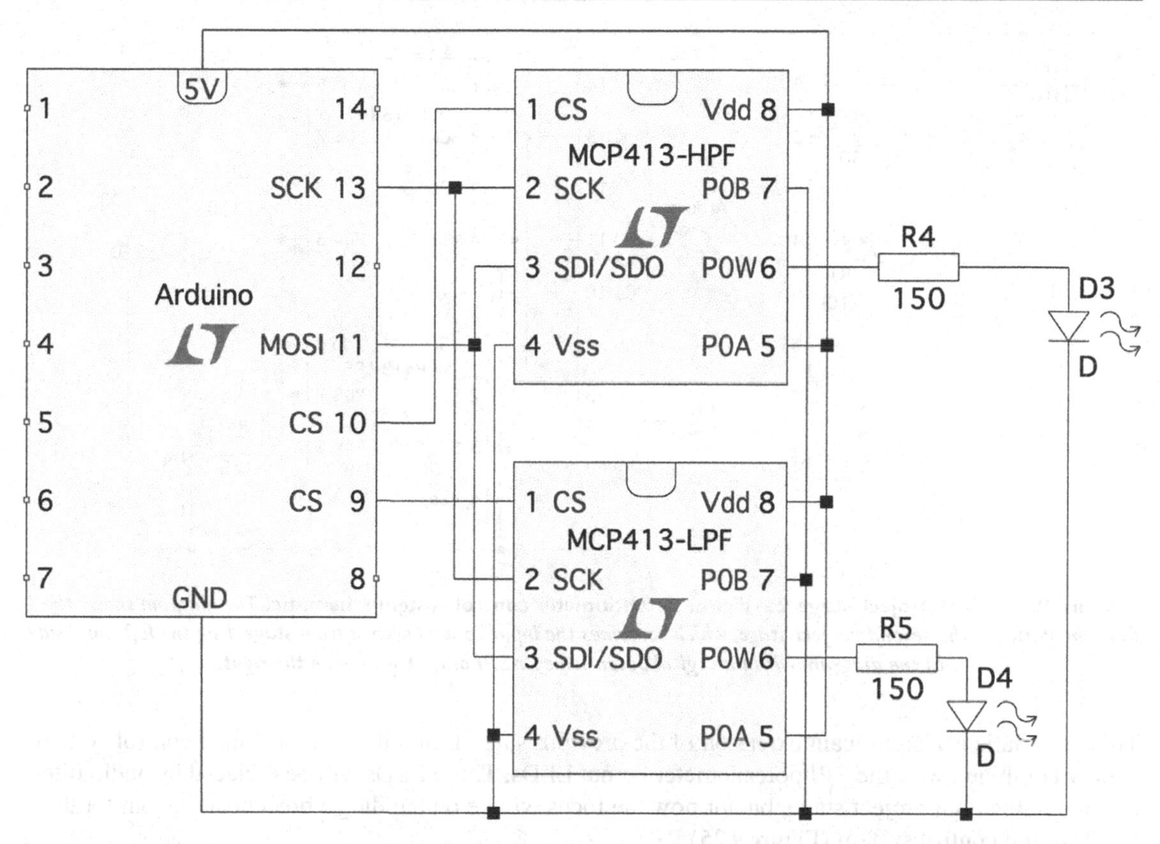

Figure 9.23 Arduino MCP413-103 output example. ***The diagram shows the full connections (pins 1–8) required to operate each MCP413-103 chip. The power (pin 8) and ground (pin 4) connections from the Arduino can also be used to provide power to an LED output circuit (pins 5 and 7) for each chip, which can be used to check that the resistance varies when updated by the SPI bus. The P0W wiper output (pin 6) is connected to a 150Ω current-limiting resistor and LED, which are also grounded by the Arduino.***

The schematic shows all the connections needed to control two digital potentiometer chips. The first thing to note is the number of bridged wires on the schematic, which can initially make it appear more complex than when arranged on breadboard (where connecting wires will be physically separated from each other). Comparing the left-hand side of the chips (pins 1–4) between Figure 9.22 and Figure 9.23 shows how the ground wire has been added to pin 4 (V_{SS}) of each chip. On the right-hand side, the Arduino power (+5V) supply is connected to pin 8 (V_{DD}) of each chip and also to pin 5 (P0A) as the input to a voltage divider. The Arduino ground (GND) is connected to pin 7 to provide the bottom connection for the voltage divider (P0B), with pin 6 (P0W) being the wiper position that is output to the resistor/LED circuit for each chip. The resistor/LED circuits can be used to indicate the output from the MCP413-103 chips, but because these chips cannot be simulated using Tinkercad (as no simulated component for this chip is provided) a breadboard prototype must be built. This prototype can be combined with the existing input control circuit that uses a push-button and potentiometer to set the state and data variables for the system, where the state is then output to indicator LEDs (Figure 9.24).

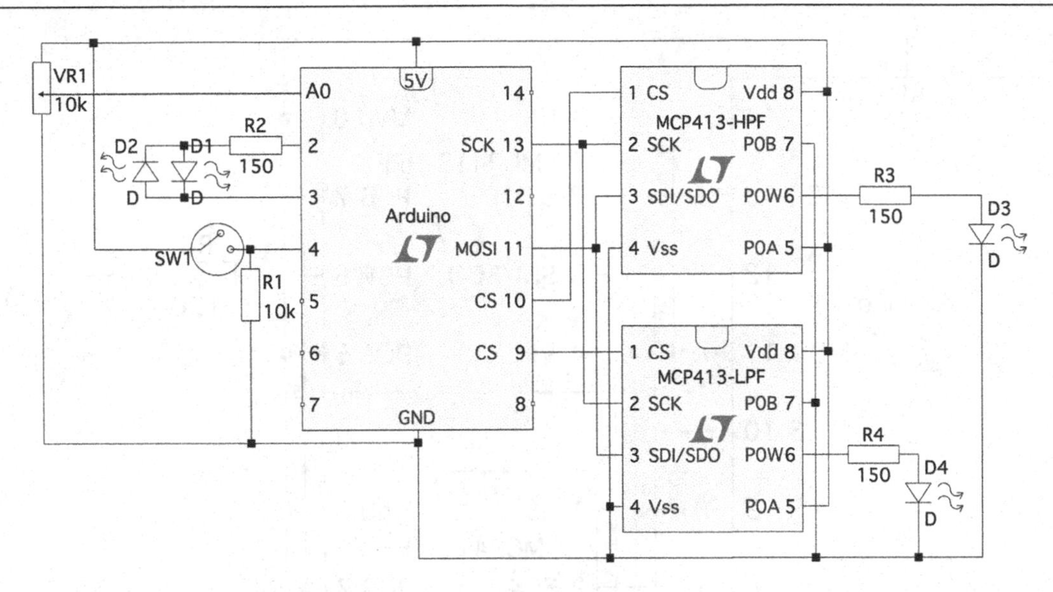

Figure 9.24 Final project stage 2 – digital potentiometer control system schematic. ***The diagram shows the full schematic for the second project stage, which combines the input control system from stage 1 on the left-hand side of the diagram with the digital potentiometer SPI output system on the right.***

This schematic is a significant extension of the previous stage 1 circuit, where the input control system is now combined with the SPI potentiometer output LEDs. These LEDs will be replaced by audio filter circuits in the final project stage, but for now the focus will be on building a breadboard layout for the full Arduino control system (Figure 9.25).

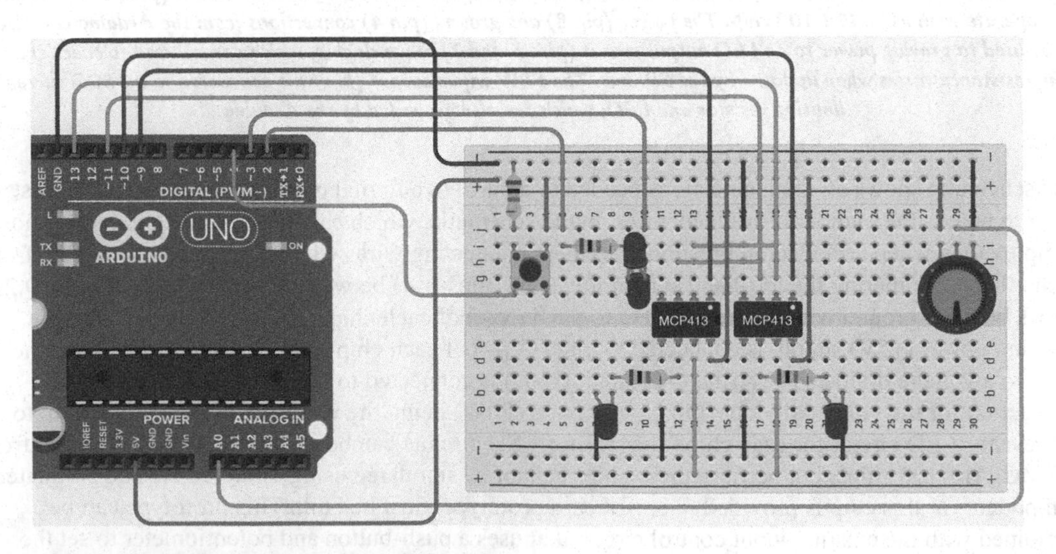

Figure 9.25 Final project stage 2 – digital potentiometer control system breadboard layout. ***In the diagram, the full layout and connections for the Arduino control system are shown in a Tinkercad mock-up. The actual layout for this stage will differ slightly for practical reasons, such as the wires on [b8 and b21], and the final project will replace the LEDs in the lower half of the board with the audio render system.***

This Tinkercad circuit is used for reference layout purposes only, as the simulator does not list an MCP413 chip that could be used to analyse potentiometer output. The main benefit of prototyping the layout is to show the control connections between the MCP413 chips and the Arduino, as these introduce a significant amount of wiring to the breadboard circuit. Open the Arduino IDE, and create a new circuit named finalProj_FullControlSystem.ino: copy the following code and paste it into the editor:

```
#include <SPI.h>//add the Arduino library that configure SPI communication
const byte lowPassLED = 2;//state indicator LED polarity pin
const byte highPassLED = 3;//state indicator LED polarity pin
const byte selectButtonPin = 4;//digital input pin to read push button switch
const byte potInputPin = A0;//analog input pin for the control potentiometer
const byte lowPassPot = 9;//define the CS output pin for the LPF MCP413 chip
const byte highPassPot = 10;//define the CS output pin for the HPF MCP413 chip
int buttonState = 0;         // current state of the button
int lastButtonState = 0;     // previous state of the button
bool pushedButton = false;//boolean used to prevent multiple switch input
   triggering
int currentPotValue = 0;// the stored value of the last potentiometer input (used
   to detect change)
int potInputValue = 0;//input control pot variable that is sent as SPI data
const byte potWriteCommand = 0x00;//define the SPI write command
void setup()
{
  pinMode(selectButtonPin, INPUT);
  pinMode(potInputPin, INPUT);
  pinMode(lowPassLED, OUTPUT);
  pinMode(highPassLED, OUTPUT);
  pinMode (lowPassPot, OUTPUT);//set digital pin for LPF CS output
  pinMode (highPassPot, OUTPUT);//set digital pin for HPF CS output
  digitalWrite(lowPassPot, HIGH);//set LPF CS signal to HIGH (do not receive)
  digitalWrite(highPassPot, HIGH);//set HPF CS signal to HIGH (do not receive)
  SPI.begin();//start the serial clock (SCK) on digital output pin 13
}//end of setup

void updateDigitalPot(){
  if(lastButtonState == 1){
    digitalWrite(lowPassPot, LOW);//set the LPF chip CS to LOW so it will receive
   commands
    SPI.transfer(potWriteCommand);//send the write command byte from Arduino MOSI
   to the SDI/SDO pin
    SPI.transfer(currentPotValue/10);//send the pot data value (scaled down
   to 0-127)
digitalWrite(lowPassPot, HIGH);//set the LPF chip CS back to HIGH
}//end of lastButtonState == 1
else if(lastButtonState == 2){
digitalWrite(highPassPot, LOW);//set the HPF chip CS to LOW so it will receive
   commands
SPI.transfer(potWriteCommand);//send the write command byte from Arduino MOSI to
   the SDI/SDO pin
SPI.transfer(currentPotValue/10);//send the pot data value (scaled down to 0-127)
digitalWrite(highPassPot, HIGH);//set the HPF chip CS back to HIGH
}//end of lastButtonState == 2
```

```
else{
}//end of else lastButtonState
}//end of updateDigitalPot
void changeLEDState(){
  if(lastButtonState == 1){
    digitalWrite(lowPassLED, HIGH);//light the first LED
    digitalWrite(highPassLED, LOW);//complete the circuit with GND (LOW)
  }//end of lastButtonState == 1
  else if(lastButtonState == 2){
    digitalWrite(lowPassLED, LOW);//light the second LED
    digitalWrite(highPassLED, HIGH);//complete the circuit with GND (LOW)
  }//end of lastButtonState == 1
  else{//assume lastButtonState == 0 (also cover for any other condition)
    digitalWrite(lowPassLED, LOW);//both LEDs off - set pins 2 & 3 to LOW (GND)
    digitalWrite(highPassLED, LOW);
  }//end of else lastButtonState
}//end of changeLEDState

void loop()
{
  // read the push-button input pin:
  buttonState = digitalRead(selectButtonPin);
  //read the analogue potentiometer value
  potInputValue = analogRead(potInputPin);
  //now check if the value has changed
  if((currentPotValue != potInputValue)){
     currentPotValue = potInputValue;
     if(lastButtonState > 0){
        //update digital Pot if a filter is selected
     updateDigitalPot();
     }//end of lastButtonState > 0
     else{//assume lastButtonState == 0 (also cover for any other condition)
     }//end of else lastButtonState > 0
   }//end of currentPotValue != potInputValue
   else{
   } //end of else (currentPotValue != potInputValue)
    // if the state has changed, increment the counter
    if (buttonState == HIGH && !pushedButton) {
     // HIGH then button pressed
     pushedButton = true;
     //change the LED indicator
     if(lastButtonState <2){
         lastButtonState += 1;
     }//end of lastButtonState <2
     else{
           lastButtonState = 0;
     }//end of lastButtonState else
     changeLEDState();//call function to update LED
    }//end of buttonState == HIGH && !pushedButton
  else if (buttonState == LOW && pushedButton){
       // LOW then button not pressed
          pushedButton = false;
    }//end of buttonState == LOW && pushedButton
}//end of loop()
```

Stage 2 – Project Steps

For stage 2 of the project, you will need: 1. Existing stage 1A circuit 2. 2 × MCP413-103 10kΩ digital potentiometers 3. 2 × LEDs (D3, D4) 4. 2 × 150Ω resistors (R3, R4) 5. 6 × connector cables and 10 × connector wires	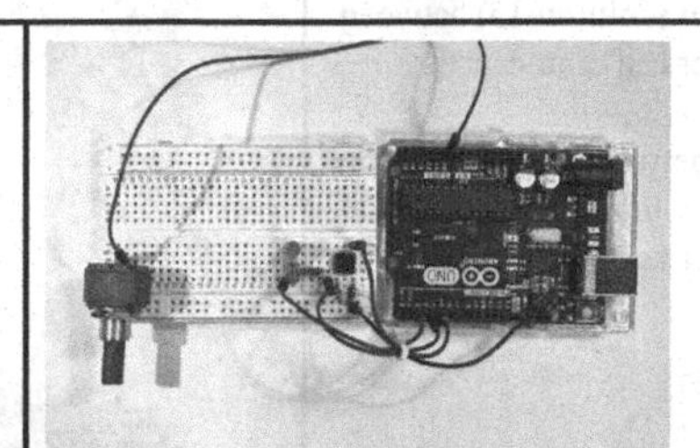
1. Add an MCP413-103 chip across the column break with top-left pin 1 (CS) in breadboard pin f14 [MCP1–f14] – this will be the LPF chip. 2. Add an MCP413-103 chip across the column break with top-left pin 1 (CS) in breadboard pin f19 [MCP1–f19] – this will be the HPF chip.	
1. Connect a connector wire between the Vss pin on the LPF chip (top-left f14) to the top ground rail [j11–JGND]. 2. Connect a connector wire between the Vss pin on the HPF chip (top-left f19) to the top ground rail [j16–JGND].	
1. Connect a connector wire between the SDI/SDO pins on both chips on pins [g12–g17]. 2. Connect a connector wire between the SCK pins on both chips on pins [h13–h18].	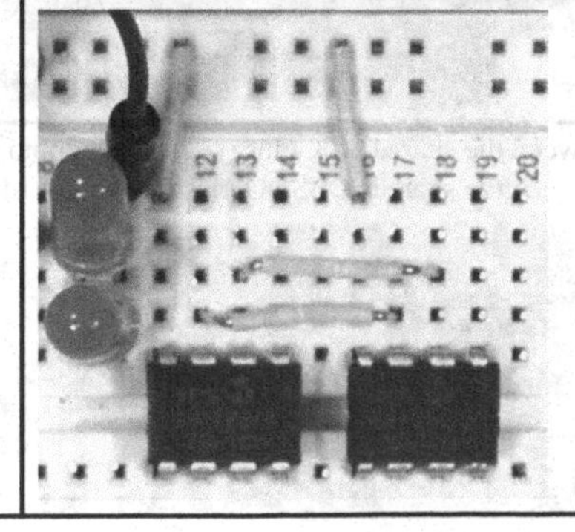

<table>
<tr><td>1. Connect a connector wire between the Vdd pin on the LPF chip and the gap (column 15) between the chips to allow the power rail connections [d14–d15].
2. Connect a connector wire between the Vdd pin on the HPF chip and the gap (column 20) between the chips to allow the power rail connections [d19–d20].</td><td>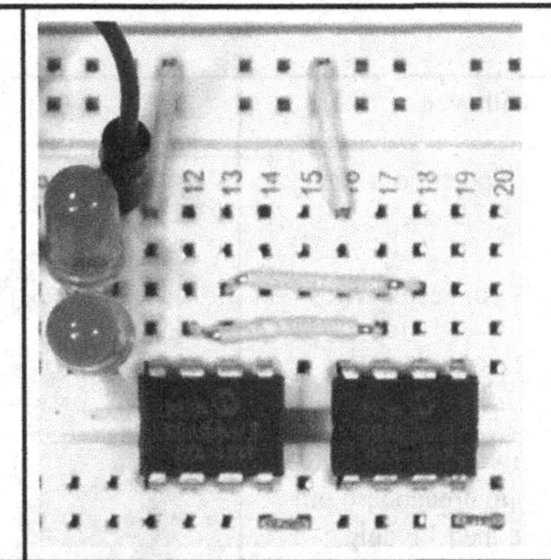</td></tr>
<tr><td>1. Connect a connector wire between the LPF chip Vdd pin and the top power rail [e15–JV+].
2. Connect a connector wire between the HPF chip Vdd pin and the top power rail [e20–JV+].</td><td>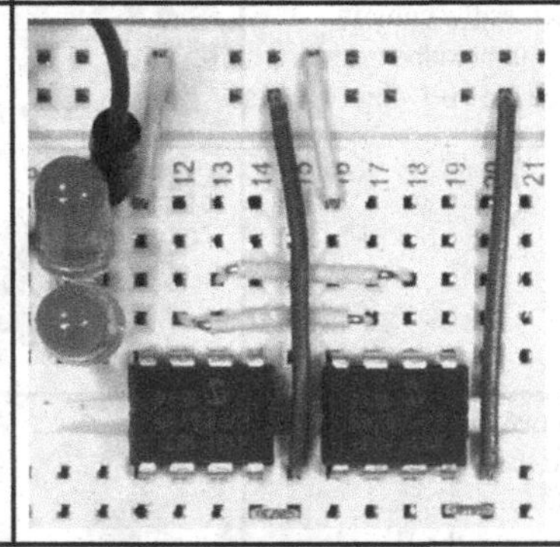</td></tr>
<tr><td colspan="2">1. Add a connecting wire between the LPF CS pin (j14) and Arduino digital output pin 9 [j14–AD9].
2. Add a connecting wire between the HPF CS pin (j19) and Arduino digital output pin 10 [j14–AD10].
Note: power-decoupling capacitors are shown in several of these images, but are not needed until stage 3.
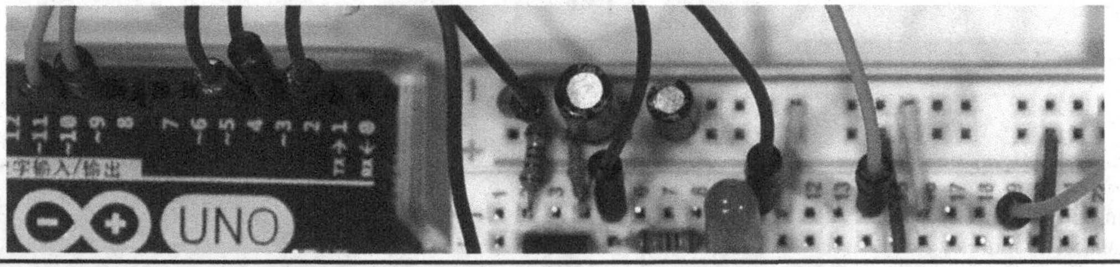</td></tr>
<tr><td colspan="2">1. Add a connecting wire between the shared SCK pins (j18) and Arduino SCK output digital pin 13 [j18–AD13].
2. Add a connecting wire between the shared SDI/SDO pins (j17) and Arduino MOSI output pin 11 [j17–AD11].
</td></tr>
</table>

<table>
<tr><td colspan="2">

1. Use a small cable tie to group the SPI communication cables together – the image shows a spare loop that appears conjoined but is actually separate from the switch/state LED wires.

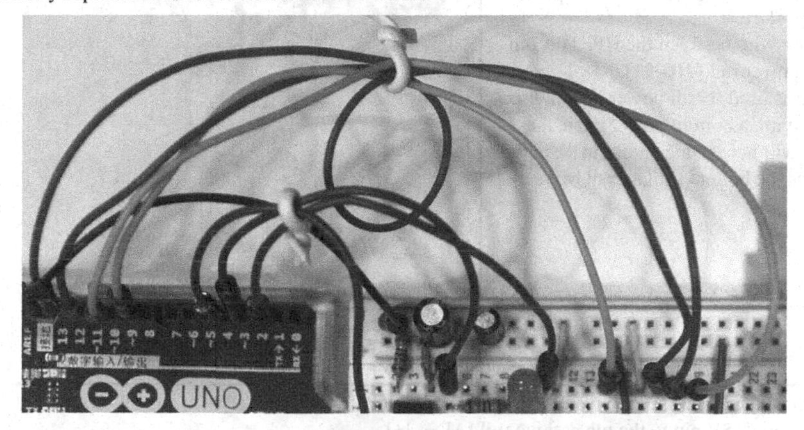

</td></tr>
<tr><td>

1. Connect a connector wire between the LPF chip P0B pin (pin 7) and the **bottom ground rail** [a13–AGND] – *the reference image is slightly blurred.*
2. Connect a connector wire between the HPF chip P0B pin (pin 7) and the **bottom ground rail** [a18–AGND].

</td><td>

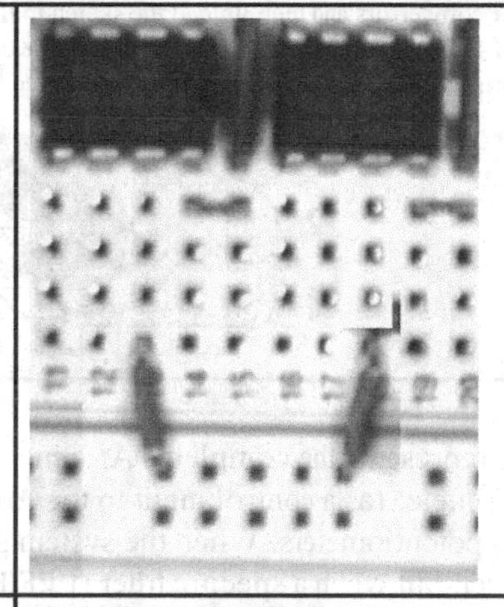

</td></tr>
<tr><td>

1. Connect a 150Ω current-limiting resistors across rows [b8–b12] – *the reference image is slightly blurred.*
2. Connect a second 150Ω current limiting resistor across rows [b17–b21].
3. Connect a potentiometer test LED to the **bottom ground rail** across rows [b8–AGND].
4. Connect a potentiometer test LED to the **bottom ground rail** across rows [b21–AGND].

</td><td>

</td></tr>
</table>

<table>
<tr><td>
1. Add a connecting wire between the LPF P0A pin and the bottom power rail [d11–J+] – the reference image is slightly blurred.
2. Add a connecting wire between the HPF P0A pin and the bottom power rail [d16–J+].
The top power rail is used for all power lines in this project to avoid having a bottom power rail in the final circuit – this will help reduce noise in the audio system, which uses the bottom ground rail between audio connectors.
</td><td>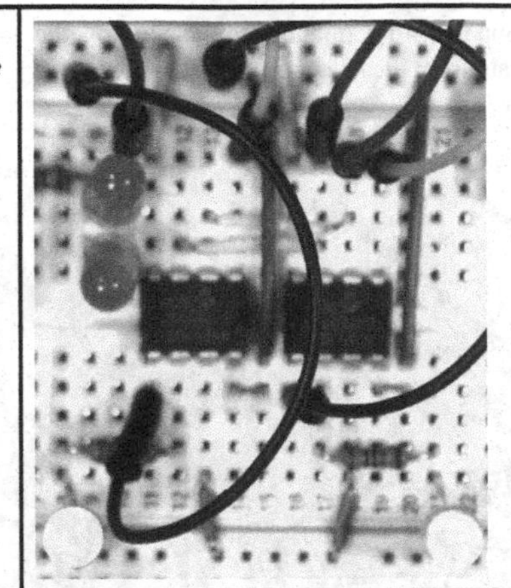</td></tr>
<tr><td colspan="2">
1. Add a bridging connector across the breadboard ground rails [JGND–AGND].
2. Connect the Arduino +5V pin to the top positive rail [AD+–J+].
Compile the code to check for copy errors and then upload the sketch to the Arduino Uno to test the circuit. If everything compiles correctly, each time the LPF (LED 1) or HPF (LED 2) states are selected with the push-button the analogue potentiometer value should change the brightness of the test output LEDs connected to the MCP413-103 digital potentiometers.
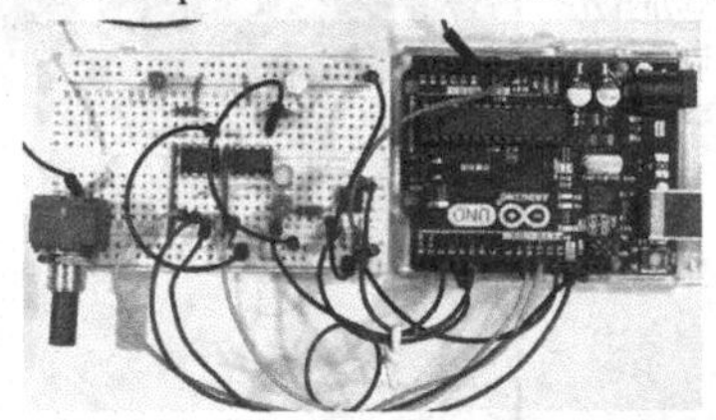
</td></tr>
</table>

The circuit shown above represents the completed Arduino control system for the final project, where a single analogue potentiometer (as a control input to the Arduino) is used to set the value of the two MCP413-103 digital potentiometers. When the system state is set to no change then the control potentiometer has no effect, but when a specific filter (LPF/HPF) is selected then the specific test output LEDs should increase and reduce in brightness based on the control potentiometer value. Note that when another filter is selected the current digital potentiometer value is retained – when the chip select pin ($\overline{CS}$) is taken HIGH the MCP413-103 will no longer receive write messages or data.

This circuit can now be extended to include the 2-band equalizer amplifier circuit that was first built in chapter 8. The layout of this circuit is broadly similar to that chapter, but some small changes are required to align the filter components with the digital potentiometer input pins (P0A, P0W, P0B). The next section will build the last stage of the final project in this book, so if you are completing this stage late at night (or do not have significant time to spare) then it is recommended to take a short break at this point – to avoid potential errors that may be introduced to an already functional circuit.

9.4 Final project – Arduino-controlled audio amplifier with 2-band equalizer

The final stage of this project combines the Arduino digital control system built in stages 1 and 2 with a 2-band equalizer amplifier derived from the circuit built in chapter 8. To avoid confusion when working with this more complex system, the Audio render system can be broken down into sections – the first of which shows how the MCP413-103 digital potentiometers are used to control the audio filter inputs in the circuit (Figure 9.26).

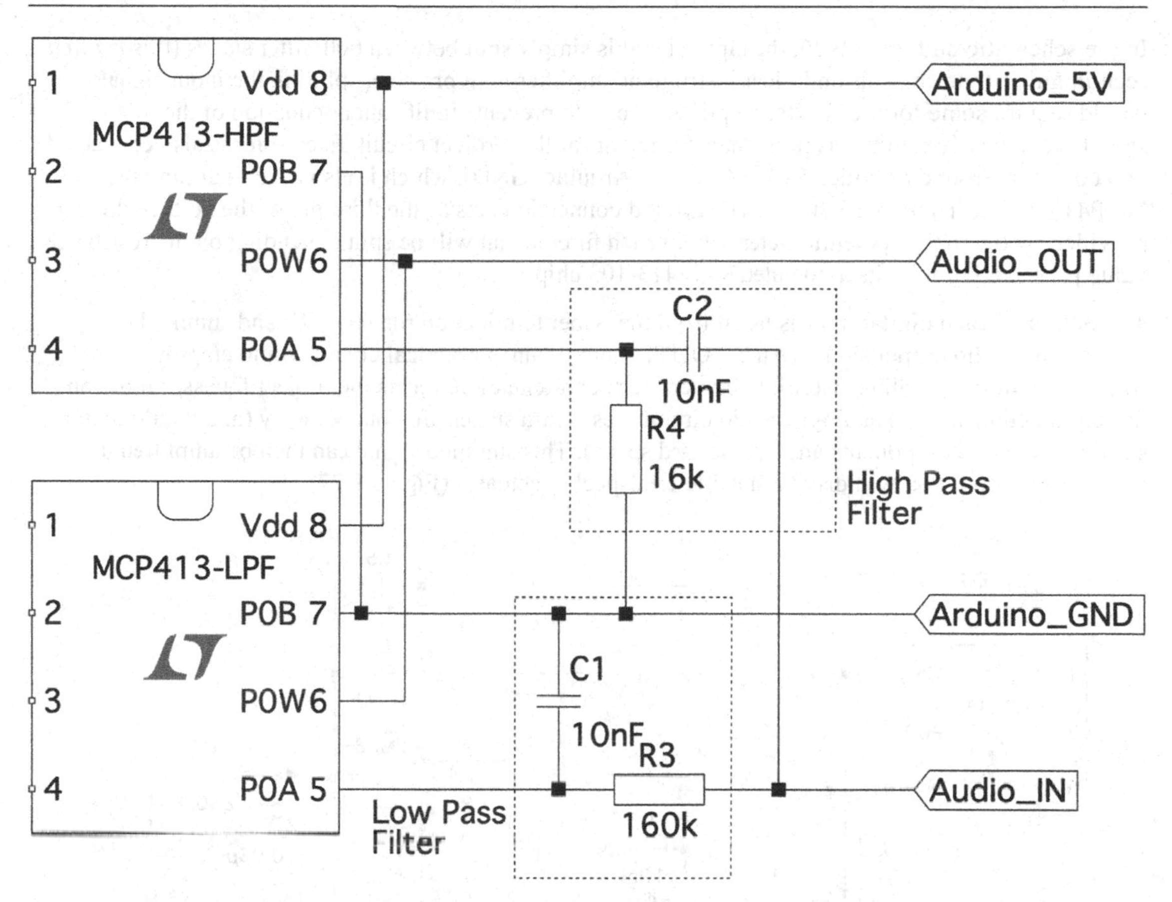

Figure 9.26 Digital potentiometer audio filter control. ***The right-hand side of the schematic shows how the audio input connection (Audio_IN) is split between two filter stages – the bottom low-pass filter (RC network) and the top high-pass filter (CR network) which are both connected to the common Arduino GND. The left-hand of the schematic shows the two MCP413-103 digital potentiometer chips that take these filter signals as inputs to a voltage divider (between P0A and P0B) where the combined pin 6 (P0W) outputs provide the Audio_OUT signal that will be connected to the LM386 amplifier.***

In the schematic, the digital potentiometers from the circuit shown in Figure 9.22 and Figure 9.23 are now connected to low- and high-pass audio filters. The audio input (Audio_IN) connection is split between both filters, where the output of each filter is connected to pin 5 (P0A) of the MCP413-103 chips. It may initially be difficult to follow this circuit path, as the previous chapter on audio filters used the conventional linear arrangement of components that is most commonly seen in audio schematics. The rearranged layout shown in the diagram in Figure 9.26, where both filters are now connected from right to left, still represents exactly the same circuit paths as those shown in chapter 8. The additional change for the low-pass filter, where the filter is effectively flipped over so that the ground connection is at the top, also makes no difference to the actual current flow through the filter. In each case, it can be useful to think of the signal path of each filter in the following order:

Low-pass filter – signal passes through resistor, output connection before capacitor, capacitor goes to ground
High-pass filter – signal passes through capacitor, output connection before resistor, resistor goes to ground

In the schematic on Figure 9.26, the input signal is simply split between both filter stages (this is often seen in guitar tone controls and older instrument amplifiers). In practice, splitting the input signal would require some form of buffer amplifier circuit to prevent significant attenuation of the input signal, but a passive splitter is more than sufficient for this project circuit. Each filter is also connected to a common ground provided by the Arduino (Arduino_GND), which is also used to ground the MCP413-103 chips on pin 7 (P0B). The ground connection acts as the third pin of the voltage divider provided by the digital potentiometer, where each filter signal will be split depending on the resistance value presented to it by its associated MCP413-103 chip.

The output of each digital filter is taken from the wiper terminal on pin 6 (P0W) and summed to provide the audio output signal (Audio_OUT). Once again, a practical circuit would employ some form of summing amplifier circuit to prevent further attenuation, but as the project focusses more on the digital control of an analogue audio circuit these extra stages are not necessary (and would require significant extra components and breadboard space). This summed signal can then be amplified to provide an output that will drive a suitable loudspeaker actuator (Figure 9.27).

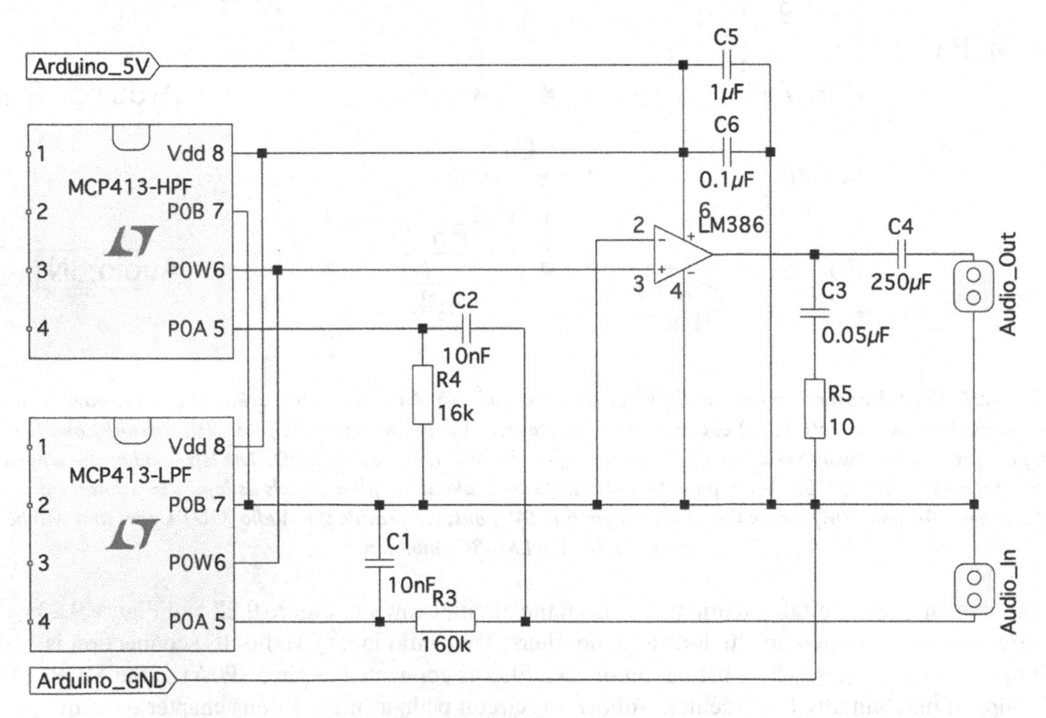

Figure 9.27 Digital potentiometer 2-band equalizer amplifier circuit. ***The schematic shows an extension of the previous Figure 9.26 schematic, where the combined audio output from the POW wiper pins (pin 6) of both MCP413-103 chips is fed into the non-inverting input of an LM386 audio amplifier. AC decoupling capacitors (C5, C6) are connected across the power rails to reduce noise, whilst a DC blocking capacitor (C4) is connected at the LM386 output to protect the chip. A Zobel network (C3, R5) is also connected in parallel to the amplifier output to balance the inductive impedance of any loudspeaker connected to Audio_OUT. Notice the ground connection between the Audio_IN and Audio_OUT connectors that is then connected to the Arduino as the common ground for the circuit – this shorter ground path helps to reduce circuit noise.***

The right-hand side of the circuit shown in Figure 9.27 represents the complete audio render system for the final project, which combines two audio filter networks with an LM386 audio amplifier. This circuit takes the P0W wiper outputs (pin 6) from each digital potentiometer and connects them to the non-inverting input of the LM386 audio amplifier – the inverting input on pin 2 is connected to the common system ground from the Arduino. The output of the amplifier is connected in series with a DC blocking capacitor (C4) to the Audio_OUT connector, which protects the amplifier from any current surges coming back into the circuit. In addition, a Zobel network (C3, R5) is connected in parallel with the audio output to help balance the inductive impedance load of any loudspeaker that may be connected to the circuit (where the inductive reactance will increase with frequency).

AC decoupling capacitors (C5, C6) are also connected across the power rails to reduce noise in the circuit, which is particularly important when combining analogue audio and digital control signals (from the Arduino) on a breadboard with common power rails. Digital signals operate at a far higher level (+5V) than a typical analogue audio input such as a consumer line level at around $0.3V_{RMS}$ (see chapter 7, Table 7.1). This discrepancy in signal level is compounded when working with communications protocols like the SPI bus, where rapid fluctuations in voltage levels during to the transmission of binary data (+5/0V) can create significant noise throughout the circuit. It should be noted that whilst the Arduino is a very flexible microcontroller, it is primarily a digital device that either samples analogue input (to convert it to digital data) or employs techniques like pulse wave modulation (PWM) to mimic analogue voltage output levels (to convert digital to analogue) – it is not an analogue device. In addition, digital noise reduction has a very different set of requirements and thresholds as it will never be conveyed for human listening, and so microcontrollers like the Arduino are not optimized to work directly with analogue audio signals.

This shared access to power rails is also the reason for the additional ground connection between the Audio_IN and Audio_OUT connectors shown in the schematic above. Although all ground terminals in the circuit are connected to the Arduino, a short ground path between the audio connectors in the circuit will help to reduce noise introduced by other elements – as the audio signals have a ground path that does not pass directly through the digital control circuit. It is good practice to keep audio ground paths as short as possible when working with combined systems and, as a further element of noise reduction, the bottom +5V power rail [J+] is not connected in the final breadboard layout to help isolate all audio render system ground connections from the Arduino supply (this requires the LM386 to be oriented in the other direction on the breadboard). Although these types of design consideration are arguably more advanced than required for an introductory text that works with breadboard prototypes, and are also not in themselves a robust or comprehensive noise-reduction strategy, they are included to help illustrate the need for power-supply isolation in more advanced schematics that you may encounter as your learning progresses beyond this book.

With the audio amplifier circuit added to the digitally controlled filter stages, the final project circuit for an Arduino controlled 2-band equalizer amplifier can be produced (Figure 9.28).

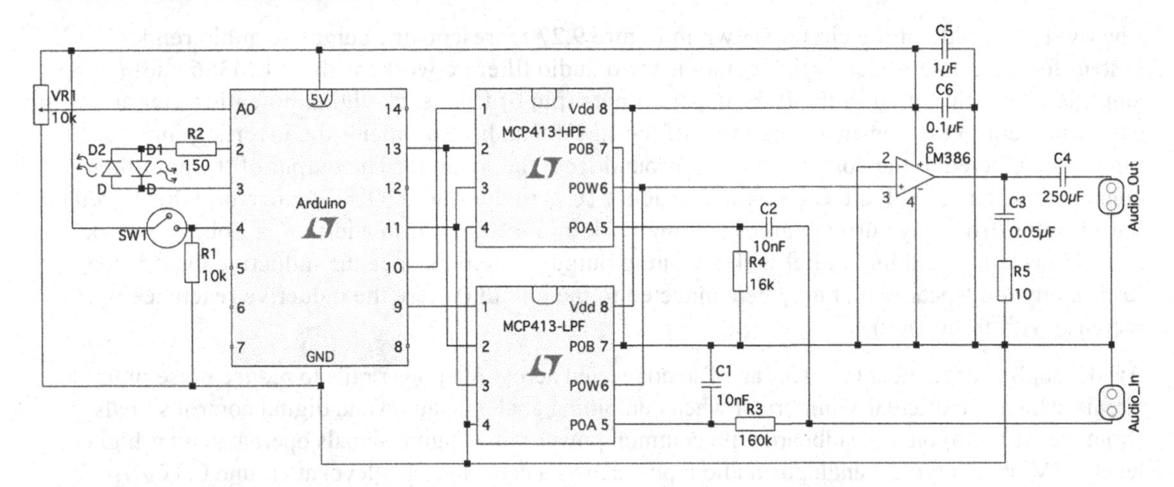

Figure 9.28 Final Arduino-controlled 2-band equalizer amplifier circuit. ***The schematic shows how the previous Arduino control system (developed in stages 1 and 2 of the project) can now be connected to two audio filter stages, where the filter cutoffs are set at 100Hz (low pass) and 1kHz (high pass). These filters are connected to the MCP413-103 digital potentiometers, whose wiper output on POW (pin 6) is then combined and connected to the non-inverting input of an LM386 audio amplifier. The minimal design of the LM386 is derived from its datasheet to include a Zobel network and DC output blocking (C4), and additional AC decoupling capacitors (C5, C6) are also added to reduce noise on the power rails for the system.***

This schematic combines the Arduino digital control system (left-hand side) and the audio render system (right-hand side) within a single circuit. The control inputs on the left (push-button switch and analogue potentiometer) are used to set the state and data of the system, where each filter output is defined as a separate state (with no filter output being the default). The Arduino controls two MCP413-103 digital potentiometers using an SPI bus (Arduino pins 9, 10 11, 13) that can set the value of each potentiometer using the $\overline{\text{CS}}$ control connections on Arduino pins 9 and 10 (MCP413 pin 1) to select each chip. On the right-hand side of the schematic, the audio input (Audio_IN) is split between a low- and a high-pass filter that are then connected to the potentiometer P0A input pins (pin 5) of the MCP413-103 chips. The output P0W wiper pins (pin 6) are then recombined at the non-inverting input to an LM386 audio amplifier. This amplifier also has AC decoupling (C5, C6), a Zobel network (C3, R5) and DC blocking (C4) to stabilize and protect the audio output (Audio_OUT) from the circuit, which has an additional ground connection to the audio input for noise reduction.

The Arduino provides both power and ground connections, and, as noted at the beginning of this chapter, the bottom power rails will be swapped over in this final build stage. This can initially be confusing, but it is important to remember that breadboard rails are simply marked for reference – the rail itself does not change the connections to it! Swapping the power rails at the bottom of the breadboard allows resistor components to be directly connected between columns and ground without the need for additional bridging wires, which saves much-needed space in the final circuit. For this reason it is important to note that:

the bottom power rail (V+) will be used as a ground rail (AGND) in the final build stage

Using the bottom rail for the audio render system ground connections also allows the audio ground rail to be kept further apart from the Arduino control rails at the top of the breadboard, helping to minimize noise by giving both audio connectors a ground path that does not directly include the Arduino power supply. Whilst this is not a comprehensive noise-reduction strategy (given that this is a breadboard prototype circuit), it is again noted that keeping audio signal paths (and ground connections) as isolated as possible can help to reduce the amount of noise introduced into the signal by control system processes like the SPI bus.

Although this circuit cannot be fully simulated using Tinkercad (which is primarily designed for control system simulation), the layout for the project can be presented as a mock-up for reference purposes when building the final circuit (Figure 9.29).

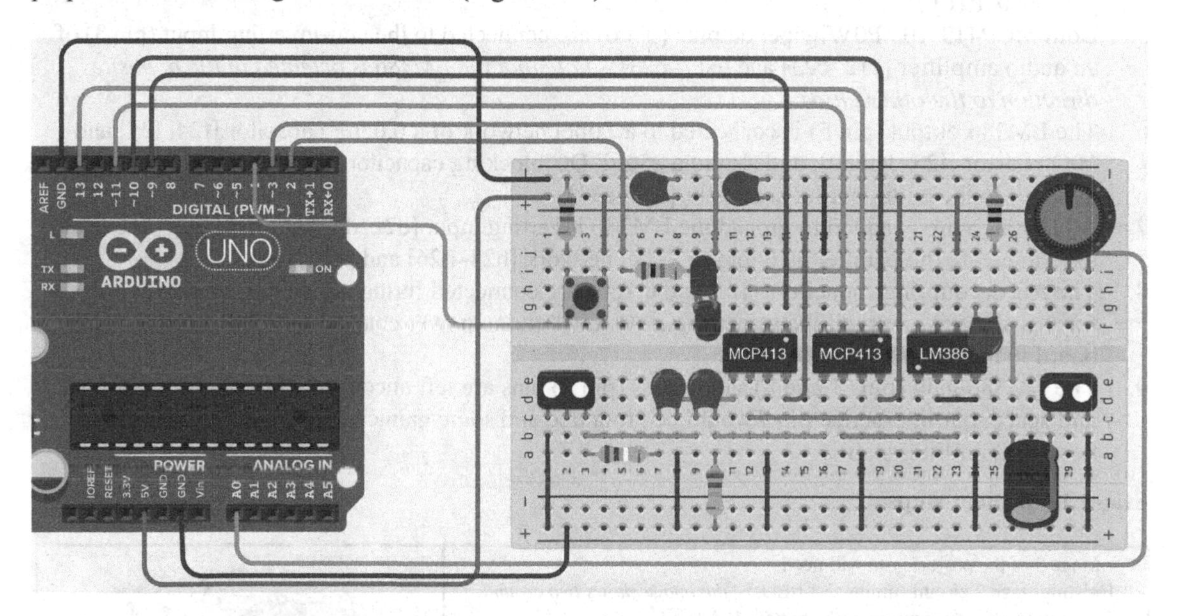

Figure 9.29 Final project stage 3 – digital potentiometer control system breadboard layout. ***In the diagram, the top half of the breadboard shows the existing Arduino control system from stages 1 and 2 where additional AC decoupling capacitors on the top power rails are spaced further apart and the analogue potentiometer has been rotated for visual clarity. The bottom half of the board is the audio render system, where the audio input [c1–c3] is connected to two filter stages whose outputs are fed into the P0A input (pin 5) of the MCP413-103 digital potentiometers. The P0W outputs of these potentiometers (pin 6) are connected to the non-inverting input of an LM386 audio amplifier (note pin 1 of the amplifier is oriented in the opposite direction to the potentiometer chips). The amplifier output circuit is broadly similar to those of chapters 7 and 8, where a Zobel network and DC blocking capacitor [b26–b28] are used to balance the connection to an output loudspeaker.***

The breadboard layout shown in Figure 9.29 extends the previous Arduino digital control system layout of stage 2 (where the LED test output circuits have now been removed). There are several elements to note before beginning to build the final stage of the project:

1. The audio input connector on pins [c1–c3] is connected to the bottom ground rail (which is labelled V+) that is in turn connected to the lower ground connectors on the Arduino – this also keeps the layout less cluttered by not adding a bridging wire across the rails (as in previous projects).
2. The audio input is split at pin c3 to go through a low-pass RC filter 160kΩ resistor [a3–a7] and 10nF capacitor [c7–c8], and a high-pass CR filter 10nF capacitor [c9–c10] and 16kΩ resistor [a10–AGND].
3. The low-pass filter output is connected to the P0A (pin 5) of the first MCP413-103 chip on pins [d7–d11] – *this is not clearly visible in the layout above.*
4. The high-pass filter output is connected to the P0A (pin 5) of the second MCP413-103 chip on pins [b10–b16].
5. Both MCP413-103 P0W wiper outputs (pin 6) are connected to the non-inverting input (pin 3) of an audio amplifier [c12–c23] and [b17–b23] – *note that the LM386 is oriented in the opposite direction to the other chips.*
6. The LM386 output (pin 5) is connected to a Zobel network of a 0.05μF capacitor [f24–f25] and 10Ω resistor [j25–JGND], and also to a 250μF DC blocking capacitor [b26–b28] that is connected in series to the audio output connector [c28–c30].
7. A bridging wire is added to ground the LM386 inverting input [d22–d24] and bridging wire is also used to create the parallel path for the Zobel network [h24–h26] and [e26–f26].
8. The AC decoupling capacitors (1μF and 0.1μF) are connected further down the top power rails for visual clarity, but will be placed closer to the Arduino power connections on the top left of the board in the actual circuit build.
9. The LM386 gain (pins 1 and 8) and bypass (pin 7) pins are left unconnected to save components and space, but in practice pin 7 would be grounded and some gain control would be applied to keep the amplifier stable.

Stage 3 – project steps

For Stage 3 of the project, you will need: 1. Existing stage 2 circuit (minus test LEDs) – *the image shows two ground connectors already running to the bottom AGND* 2. Two filter resistors – low-pass 160kΩ resistor (R3), high-pass resistor 16kΩ (R4) 3. Two filter capacitors C1 = C2 = 10nF 4. 1 × LM386 operational amplifier 5. Zobel network – 10Ω resistor (R5), 0.05μF capacitor (C3) 6. DC blocking capacitor (LM386 output) – 250μF capacitor (C4) 7. Power-decoupling capacitors – 1μF capacitor (C5) and 0.1μF capacitor (C6) 8. 1 × audio input connector (3.5mm jack), 1 × audio output connector (loud-speaker) 9. 2 × connector cables and 12 × connector wires	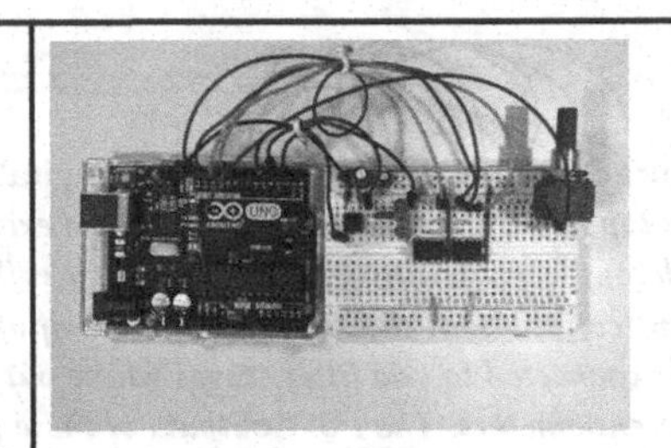
1. Add an audio input connector across pins [c1–c3]. 2. Ground the input connector terminal by connecting its ground (pin 1) to the bottom circuit ground which is marked as the V+ rail between pins [b1–AGND].	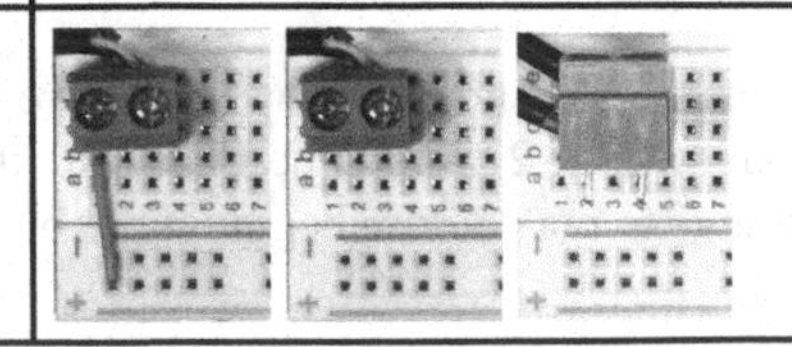

1. Connect the 160kΩ low-pass filter resistor (R3) across pins [a3–a7]. 2. Connect the 10nF low-pass filter capacitor (C1) across pins [c7–c8]. 3. Connect the network to ground [a8–AGND].	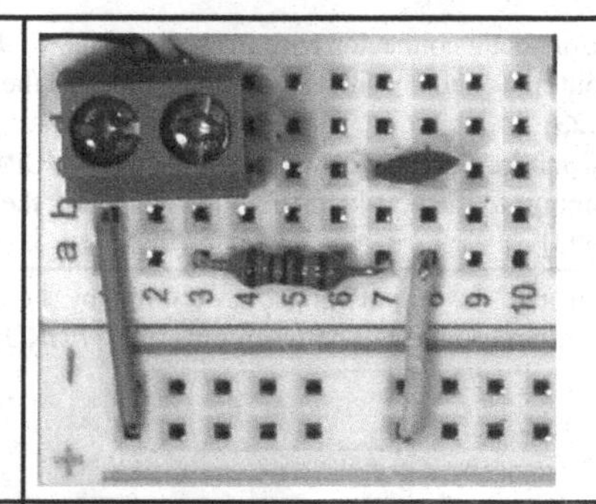
1. Connect a connector wire from the audio input to the high-pass filter network [b3–b9] – *pin partially obscured by capacitor.* 2. Connect the 10nF high-pass filter capacitor (C2) across pins [c9–c10]. 3. Connect the 16kΩ high-pass filter resistor (R4) to ground across pins [a10–AGND].	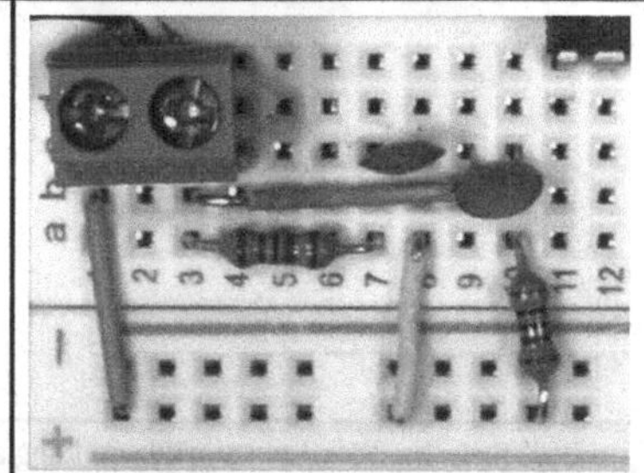
1. Connect a connector wire from the low-pass filter output to the P0A input (pin 5) of the 1st MCP413-103 chip [b3-b9] – *wire partially obscured by capacitors.* 2. Connect a connector wire from the high-pass filter output to the P0A input (pin 5) of the 2nd MCP413-103 chip [b10–b16]. 3. Connect pin 3 of the LPF MCP413-103 chip to ground with a connector wire between pins [a13–AGND].	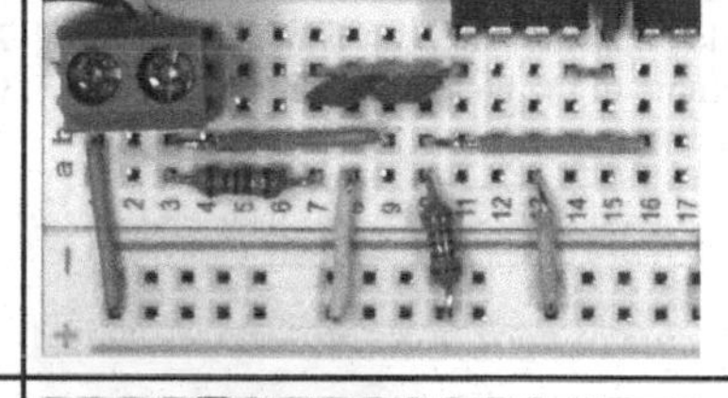
1. Connect pin 3 of the HPF MCP413-103 chip to ground with a connector wire between pins [a18–AGND]. 2. Add the LM386 audio amplifier with pin 1 on breadboard pin e21 – *note this is in the opposite orientation to the digital potentiometer chips.* 3. Connect a connector wire from the inverting input (pin 2) to the LM386 ground pin (pin 4) [d22–d24]. 4. Connect the LM386 ground pin (pin 4) to the breadboard ground across pins [a24–AGND].	
1. Connect the LM386 Vs (pin 6) to the top power rail [j23-J+]. *The LM386 Vs is oriented with pin 6 facing the top of the breadboard to keep all power connections on the top half of the board. This helps to reduce the overall noise in the circuit, as no power connection is made to the bottom of the board where the audio connectors are grounded.*	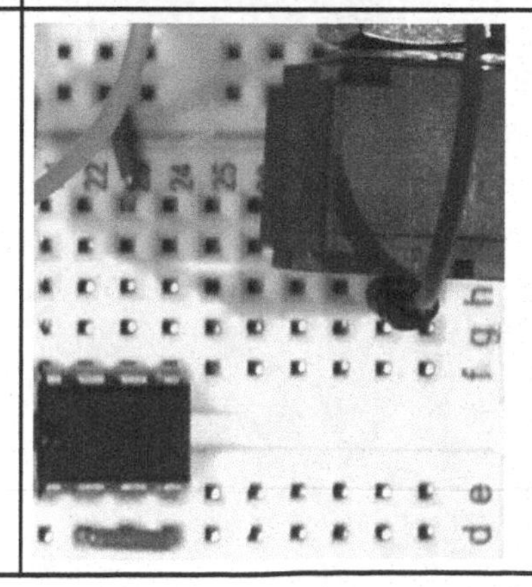

1. Connect a connector wire from the LPF MCP413-103 chip P0W wiper output to the non-inverting input (pin 3) of the LM386 between pins [c12–c23]. 2. Connect a connector wire from the HPF MCP413-103 chip P0W wiper output to the non-inverting input (pin 3) of the LM386 between pins [b17–b23].	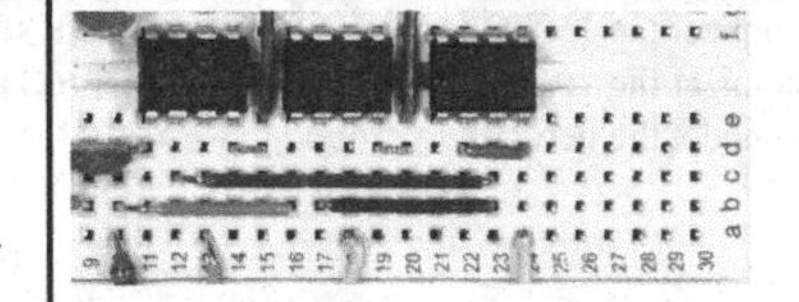
1. Ground the output connector terminal by connecting its ground pin to the bottom ground rail between pins [b30–AGND].	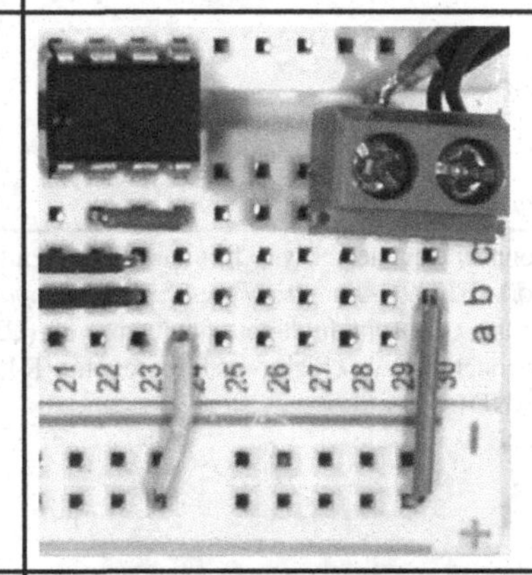
1. Begin building the Zobel network by adding a connector wire between pins [h24–h26] to link to the LM386 output (pin 5). 2. Add a second connector wire between pins [e26–f26] to bridge the centre column of the breadboard and link to the LM386 output.	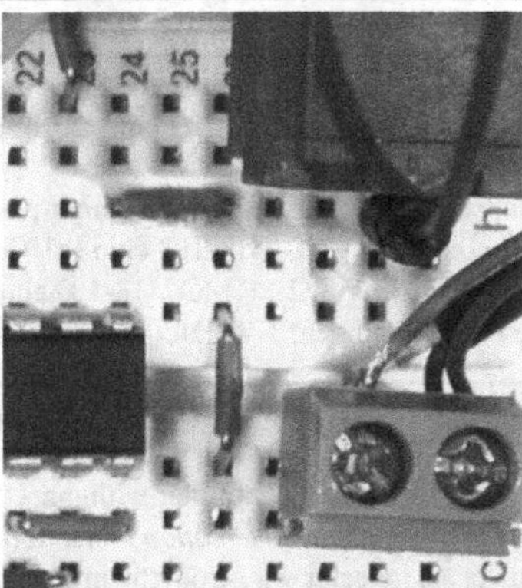
1. Add the Zobel network 0.05μF capacitor [f24–f25] 2. Complete the Zobel network with a 10Ω resistor [j25–JGND]	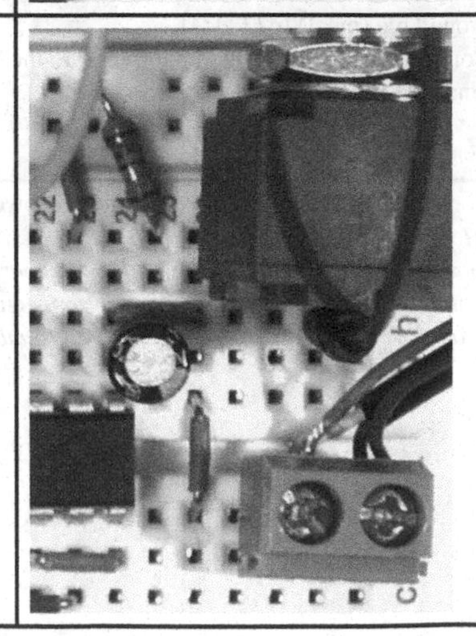

1. Add a 250µF DC blocking capacitor to connect the LM386 output to the audio output connector on pins [b26–b28].	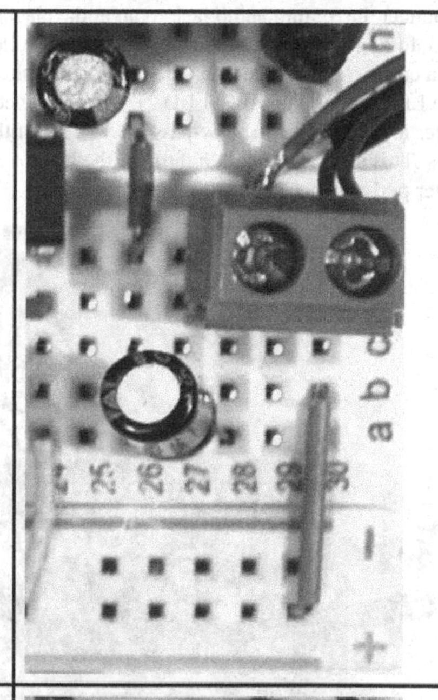
1. Add a 1µF AC decoupling capacitor across the top power rails, as close as possible to the Arduino connector wires – the image shows the fourth column of the power rails 2. Add a 0.1µF AC decoupling capacitor across the top power rails, as close as possible to the Arduino connector wires – the image shows the seventh column of the power rails.	
1. Add a ground connector cable from the bottom Arduino ground pin [ADGND] to the bottom breadboard ground rail, which is **V+ in this circuit** [ADGND–AGND]. 2. Add a power connector cable from the Arduino +5V pin to the top breadboard power rail [AD+–AV+]. **Note: the bottom power rail (V+) will be used as a ground rail (AGND) in the final build stage.**	

Connect the 3.5mm audio jack to an audio playback device. If everything has been connected properly, the audio input signal should be fed through the low- and high-pass filters, and their combined output should be amplified by the LM386 for output to a connected loudspeaker. The previously uploaded code from stage 2 should still be running on the Arduino, so each time the LPF (LED 1) or HPF (LED 2) states are selected with the push-button the analogue potentiometer value should change the filter level of either the low-pass filter (LPF) or the high-pass filter (HPF) connected to the MCP413-103 digital potentiometers. This should be audible through the connected loudspeaker, though given the prototype nature of the project high-quality filter performance should not be expected.

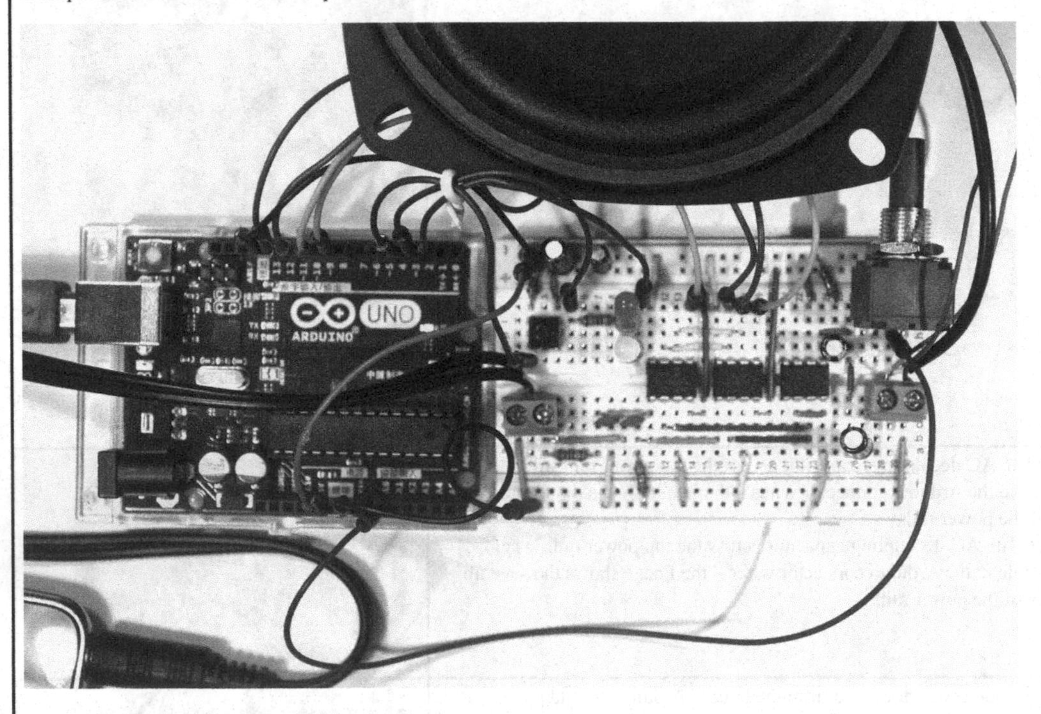

9.5 Conclusions

This chapter has combined concepts from previous chapters to build a more expansive electronics circuit that combines both analogue audio rendering and Arduino digital control:

1. **Chapter 5** – making decisions based on control inputs (i.e. switches and piezoelectric sensors)
2. **Chapter 7** – amplifying an audio signal (using the LM386 to drive a loudspeaker as an output actuator)
3. **Chapter 8** – filtering an audio signal for amplification (the 2-band equalizer amplifier circuit)

In the final project, the Arduino control system uses a push-button switch to control the state of the system, whilst an analogue input potentiometer controls the data used to vary the digital potentiometers. These digital potentiometers are connected to the Arduino using an SPI bus, which acts as the interface between the Arduino control system and the audio render system. The audio render system combines low-pass (LPF) and high-pass (HPF) filter stages that are controlled by the digital potentiometers and connects their output to an LM386 audio amplifier to drive a loudspeaker. In so doing, an Arduino-controlled audio amplifier with 2-band equalizer has been built.

This project represents a significant increase in time and complexity from previous chapter projects, where the combination of the Arduino control system with an audio render system introduced a lot of additional practical constraints. The breadboard layout for this project is much more involved, and many workaround techniques (such as swapping the power and ground rails at the bottom of the board) were employed to fit everything together within a small working area. This was done for two reasons: to keep costs down, and to demonstrate some of the practical constraints of circuit design and layout (where space is a cost in itself). More advanced texts will discuss the importance of short signal paths and the related problem of component heat management (due to proximity), but for this book the summative point is the combination of analogue audio rendering with digital Arduino control.

The final chapter provides a short summary of all that you have learned during this book, to link these elements together and also make some suggestions for how you may now progress your learning beyond the scope of this introductory text. It is important to read this chapter to remind yourself of the **distance you have travelled** in completing this book. It is hoped that the practical nature of the circuits involved will help to inspire future interest and investigation within this subject area.

CHAPTER 10

Conclusions

Congratulations – you have now completed the material in this book! The last nine chapters have looked at three core areas of electronics concepts, Arduino control and audio circuits (Figure 10.1).

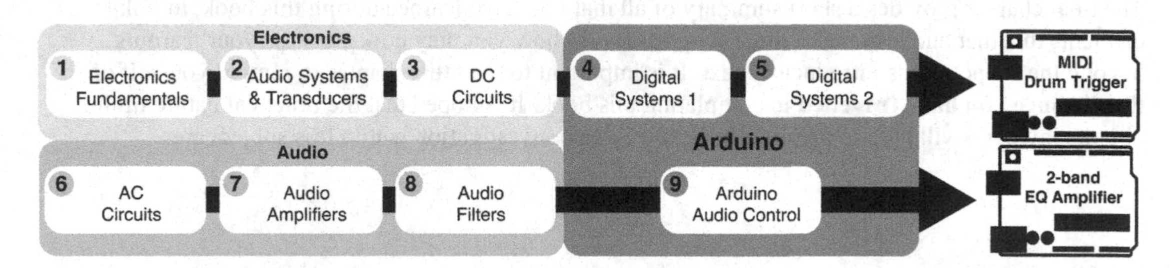

Figure 10.1 Textbook organizational structure. ***This diagram is taken from the introduction chapter and shows the three main areas of this textbook: electronics concepts, Arduino control and audio circuits. The first three chapters introduced systems and DC circuit theory, leading to chapters 4 and 5, which covered programming the Arduino to build the milestone project of a MIDI drum trigger. Chapters 6–8 then introduced audio-related topics of AC circuit theory, amplification and filtering. The final project in chapter 9 combined all of these areas to build an Arduino-controlled 2-band equalizer amplifier.***

This book has aimed to provide you with an introduction to audio electronics and Arduino control, and has covered a lot of material within a single introductory text. In each chapter, practical projects were used to underpin the concepts and methods involved to help ground your learning. The first three chapters focussed on electronics concepts, where chapter 1 introduced the fundamental quantities of current, resistance and voltage that are used throughout electronic circuits. This chapter also discussed how some of the scales and the number of quantities involved can often confuse the new learner – even though the actual equations used in electronics (e.g. Ohm's Law, Kirchoff's Laws) are often quite simple arithmetic operations. The next chapter looked at electronic systems to show how sensors (inputs) are processed by microcontrollers like the Arduino to produce actuator (outputs). Chapter 3 then introduced Ohm's Law and Kirchoff's Laws as the fundamental equations that govern direct current (DC) circuits. At this point, the important concept of the voltage divider was also introduced – which is used in operational amplifier biasing, feedback, audio filtering and also digital potentiometers.

Chapters 4 and 5 moved on to look at Arduino control, where the ATmega328p microcontroller can be programmed using the C (or C++) language – though it was noted that these chapters provide an overview of programming instructions and methods, as a full programming course is outwith the scope of this book. Chapter 4 looked at variables, functions and Arduino output (using PWM) while chapter 5 moved on to arrays, digital/analogue input, and working with external libraries to set up serial communication through a standard 5-pin DIN MIDI interface. The chapter 5 project used this interface with a piezo sensor to build a MIDI drum trigger system – representing the first milestone of the book.

Chapter 6 introduced alternating current (AC) signals, which vary over time. Sinusoids are used extensively in audio analysis, and so understanding the time-varying response of capacitors is key to working with audio circuits. Chapter 7 discussed amplification as a means of increasing the small signals from sensors like microphones to a level that can drive a load like a loudspeaker. The chapter looked at how semiconductors are used in amplification, providing an overview of BJT amplification as a reference for many audio schematics. The combination of BJTs within an integrated circuit creates an operational amplifier, which is much more stable to work with. The chapter project built a minimal LM386 audio amplifier, showing how capacitors can be used for DC blocking, power decoupling and adding a Zobel network to balance the inductive load of a loudspeaker output. Chapter 8 used this amplifier project as the basis of a 2-band equalizer project that combines first-order low- and high-pass filters in a passive design – noting the limitations of such circuits in practical terms.

The final chapter 9 project adapts this 2-band equalizer amplifier circuit and combines it with digital potentiometers to allow the Arduino to control the filter levels within the amplifier circuit. In so doing, a much more complex build process was required that uses the entire space available on a 30-row breadboard. This project covered the use of the serial peripheral interface (SPI) protocol to allow the Arduino to control MCP413-103 digital potentiometer chips that were connected to first-order passive low- and high-pass filter stages. These control points represent the second milestone of the book – where Arduino digital control has been integrated with an analogue audio circuit for filtering and amplification.

10.1 Future work

As an introductory text, the aim of this book was to provide you with enough information to begin practical audio electronics work without becoming too detailed in areas that will require further time and study. This is most notable in relation to both computer programming and AC circuit theory, where the introductory nature of these chapters is apparent (and necessary for brevity). Thus, chapters 4, 5 and 6 should not be considered as comprehensive and further investigation will be required by the reader to develop a deeper understanding of these topics. For now, the three core areas of the book can be considered as initial directions for further study – there is always more to learn!

10.1.1 Electronics fundamentals

The fundamental concepts of current, resistance and voltage can take time to become fully familiar with. Even though you may now understand what voltage and current are and how they behave at a basic level, synthesizing these quantities into a working understanding of circuit behaviour requires both further learning and practical experience – the more circuits you build, the more you will become familiar with the practicalities of the movement of electrons. Although Ohm's Law and Kirchoff's Laws are both simple and powerful, they describe ideal conditions that do not always manifest themselves in practice. The effects of small variances (such as resistor tolerances or battery current) can all combine to produce less than expected results, and so practising with circuits that combine series and parallel components is recommended to study how they deviate from the expected values. LTspice can quickly simulate a circuit that can then be transferred to breadboard for analysis, where a multimeter can then be used to measure the actual levels in the circuit.

In addition, as circuits become more complex, an equivalent circuit is often used to reduce the number of the components involved in the analysis of the current stage of the system. A Thevenin circuit will represent all the components before a specific node in the circuit as a single voltage source and resistor that is their equivalent. Similarly, a Norton equivalent would represent these components as a single current source with a resistor – either can be used depending on the type of signal analysis being performed on the following stage of the circuit. In practical terms, the audio filter stages in the chapter 8 project could have been modelled as a Thevenin equivalent circuit to determine the input source impedance (and signal voltage level) seen by the LM386 audio amplifier. Though not overly complex as a technique, Thevenin and Norton equivalence was omitted to keep the focus of chapter 8 on audio filtering only – nevertheless they are very useful techniques that are worth learning.

Audio electronics can become quite complex due to the use of AC signals, and it is important to understand that this book has not provided a detailed discussion of several important areas. Although the theoretical complexity of AC signals has been reduced where possible to allow time-varying signals to be used, there are significant elements (notably phase) that require further study if you wish to extend your knowledge of audio electronics. For phase calculations, complex numbers can be daunting for many learners (the clue is in the title!) but they are a very elegant solution to the problem of representing a dimension of variance (in this case magnitude over time) within a set of data. When AC circuit theory was first proposed, the use of imaginary numbers was a very clever way of linking the magnitude and phase of a signal together. In this system, the symbol j represents the value of $\sqrt{-1}$ (in mathematics this is the symbol i, but as i is used for current in electronics the next letter j is used), which helps to define the portions of a sine wave where the amplitude is negative. Although initially more difficult mathematically, complex numbers are nevertheless essential to progressing your knowledge of audio electronics beyond the basic usage discussed in this book.

In the introduction to this book, it was noted that no soldered components (other than the analogue potentiometers in chapter 8) are used. This decision was made mainly to keep costs down (a soldering station can be expensive) and reduce the working area needed to complete the projects – but it was also made for safety reasons. Soldering is an essential technique that every electronics engineer should know, and it should not be feared (no more than learning how to safely use a hob to boil a pot of pasta or rice), but it can often be forgotten that solder melts at ~180°C for lead alloy and ~240°C for lead-free (the iron will typically be ~350°C), which are much hotter temperatures than those typically found in most home kitchens. For these reasons, soldering was avoided in this introductory text but learning how to solder a jack connector onto an audio cable is a useful skill to learn and it is recommended that you investigate this further.

Recommended further study

Sensors that detect different modalities (e.g. movement, sound) and how to combine them
Soldering, to be able to make and repair audio connectors (and eventually build your own PCBs)
Using complex numbers to analyse AC phase in polar, rectangular or exponential form
Thevenin and Norton equivalent circuits for circuit analysis, to help break more complex circuits down

10.1.2 Arduino control

This book has covered the basic use of the Arduino as a microcontroller, but there are a plethora of other resources available that look at many of the other possibilities of this board. The Arduino community is large and very innovative, comprising everyone from electronics educators and students through to hobbyists and those from completely different areas who wish to add electronic control to their activities. There are also numerous resources available for learning computer programming that will help to extend your knowledge of what the Arduino is capable of. For modern programming (which is primarily focussed on mobile or 'always on' devices) a more thorough investigation of event-driven programming (EDP) is recommended to better understand how a microcontroller like the ATmega328p will respond to inputs within the code – the loop() function is effectively the 'heartbeat' of the Arduino.

It is also recommended to progress your learning of programming through a strictly typed language like C/C#/C++, as these languages enforce memory management (unlike languages such as JAVA that provide automatic garbage collection). When working with microcontrollers, maximizing the available resources is crucial and so learning to think of data in terms of its size is a very useful approach. In addition, if you wish to progress into audio digital signal processing (DSP) then memory allocation (and pointers) becomes fundamental to processing and rendering – the C language is still widely used in audio callback structures due to its simplicity and speed. There are many Arduino audio DSP examples available, but this book has avoided working directly with digital audio as the Arduino is arguably not powerful enough for any useful DSP processing of a digital input signal. For more advanced audio programming, the JUCE platform (https://juce.com/) provides open-source, cross-platform resources for audio DSP – alongside a growing developer community.

The chapter 5 project implemented a simple MIDI drum trigger system based on a single piezo sensor input. In practice, this illustrates the beginning of significant possibilities for the Arduino as part of a MIDI control system. The MIDI protocol is capable of much more than Note On and Note Off messages (though this was its primary design), and considering the many different ways in which MIDI control is used in practice opens up the Arduino to numerous possibilities. MIDI is robust, versatile and cheap to implement – it has been used extensively in many live audio systems for sound, lighting and synchronization control. It is definitely worth investigating how the chapter 5 project could be extended both in its own right (to incorporate multiple sensor inputs), but also in how the core system of sensor input and MIDI input could be adapted and applied to many other system applications.

In chapter 9, the Arduino was used as a controller that takes digital (switch) or analogue (piezo, potentiometer) sensor inputs and then uses selection instructions to determine what outputs to produce. This approach (though valid) does not consider the huge range of possibilities of digital systems, where binary sensor information can be evaluated using combinations of logic gates. Some of this logic was introduced briefly with the Boolean operators AND/OR/NOT (&&/||/!) used in selection instructions, but these operators also describe the core function of dedicated digital components designed to implement digital logic. One practical use for logic gates is to replace more complex coding algorithms that analyse multiple sensor inputs (such as the push-button state counter in chapter 9) with dedicated digital components that performs the same analysis. Simply put, the chapter 9 loop() function code would have been significantly less complex if it executed processes based on a single data input defining the current state number – a value that could have been provided by a dedicated digital logic system that used gates to increment (and reset) a counter based on push-button input.

In some respects, it can be argued that the Arduino has not yet been fully embraced as a controller within audio projects – other than those examples that use the MIDI protocol (which is not audio in itself). This book has aimed to illustrate how easily digital control points can be added to an analogue audio circuit, and hopefully the use of digital potentiometers within an audio circuit has shown how other control circuits may be built in a similar way. Working from the chapter 9 project, the Arduino can be used with an analogue potentiometer to set the value of the variable that controls the level of diode distortion being added in an operational amplifier feedback loop. Similarly, amplifier gain and volume levels could be individually controlled by incrementing a push button, which opens up numerous possibilities for digital control of an analogue audio circuit.

Recommended further study

Computer programming with C/C#/C++ or another strictly typed language that enforces memory management
The wider scope of the MIDI protocol, which is both powerful and versatile (and easily implemented with the Arduino)
Audio DSP development platforms like JUCE (https://juce.com/) for digital audio (and also MIDI control)
Using digital logic components to reduce coding complexity – this is the true power of a microcontroller

10.1.3 Audio electronics

As noted, the main aim of this book was to introduce audio electronics, allowing the new learner to ground their interest in audio equipment within a practical context. Having said this, constraints of space (and scope) have led to significant omissions in the content – arguably the most notable of these being a more thorough discussion of inductance. In simple terms, inductors and capacitors respond in the opposite manner to the frequency of the input signal being applied, but capacitors can be used in many ways (e.g. DC blocking, power decoupling, rectification smoothing) and this is also true of the inductor. A dynamic loudspeaker (which describes most commercially available examples) is primarily an inductive component and thus will present a more complex load impedance to the signals required to drive it as an actuator. This loudspeaker impedance will also combine other capacitive and resistive elements, and so power amplification can become a very complex process.

Most electronics textbooks (and courses) will use a resistor/inductor/capacitor (RLC) circuit to teach AC signal theory, as such circuits help to illustrate the complex phase relationships between inductive and capacitive components. In addition, they better model real-world conditions for components like loudspeakers and other components such as signal and power cables, which may also contain parasitic capacitances and inductances depending on the conditions they are being used in – these mainly manifest themselves as noise added to the system. Although this book has avoided the more detailed analysis involved in an RLC circuit, it is recommended that this be studied to develop a more comprehensive understanding of the interplay between common audio circuit components.

The chapter 7 project was based around a minimal LM386 audio amplifier, which can be used as the basis of numerous other audio projects (there are many online examples of LM386 circuits). This is a good point to extend your practical learning from, but it does not cover the significant distinctions between circuits for preamplification, unity gain amplification for buffering (commonly used in audio processing) and dedicated power amplification systems that conform to several different classes of signal amplification (A/B/AB/C/D). All of these areas have very different design criteria, which goes some way to explaining the difficulties encountered when designing (and implementing) a professional multistage audio system. Even the distinction between microphone and line-level signal inputs (a feature that is now commonly provided in most audio recording equipment) requires a lot of design and planning to be performed effectively – notably in the avoidance of noise and/or crosstalk between the separate input stages. For this reason, this introductory text has simply noted these distinctions, but further study of each area will take significant time and effort.

Similarly, the chapter 8 project builds upon the minimal LM386 audio amplifier circuit to provide 2-band equalization, but this circuit is a first-order passive design intended to show how at the simplest level a filter is effectively a frequency-dependent potential divider. In practice, first-order filters are most commonly encountered in guitar tone control circuits and cheaper amplifiers that are primarily focussed on reducing component counts. Higher-order filters can produce much more musical results (given the steeper slope they achieve) but the design calculations are much more complex and in practice introduce significant attenuation of the signal due to the additional components involved – this is also why a bandpass filter was not implemented in the chapter 8 project. Higher-order filters typically require active (amplified) designs in order to be effective, where some form of amplification stage is built into the filter design (rather than simply amplifying the result as in the chapter 8 and 9 projects). In addition, controlling filter parameters such as bandwidth, centre frequency, Q and slope (usually provided in more expensive equipment) requires further design to determine how best to implement these variances within the filter stage. The Baxandall filter design is a very good example of how filtering can be extended beyond the basics discussed in this book – it is considered to be a very powerful design for implementing practical audio filtering and control.

The projects in this book have deliberately been designed for breadboard implementation, both to reduce costs and also allow you to learn as you build. Having said this, the final results of these projects are nowhere near the level of commercial audio equipment, and thus translating circuits onto either Veroboard or printed circuit board (PCB) layouts is a more advanced skill than developing the simpler row/column layouts used in this book. Working with real circuit designs can become very complex, and so more powerful tools are required. There are many PCB design tools available, including ORCAD (www.orcad.com/), which offers a free version of the application to registered students (much like the Autodesk product suite). The chapter 9 project split the breadboard between control (top columns) and signal (bottom columns) both for design clarity and also to reduce noise as much as possible – notably in providing a dedicated earth path between input and output signal connector blocks. This separation of power and signal paths can be significantly extended on Veroboard and PCB by connecting each of them to a separate side of the board – a useful principle to remember.

Recommended further study

Inductance and RLC (resistor, capacitor, inductor) circuits to better understand loudspeaker behaviour
Expanding audio system design into preamplifier, processing and power amplification stages
Higher-order filter design and active filter circuits – notably the Baxandall filter design
Designing and building audio circuits on Veroboard or PCB

10.2 Final notes

In completing the material of this book, you have covered a significant amount of ground in audio electronics in a short space of time. The aim of this introductory text is to get you working with circuits quickly, without skipping the core theoretical concepts that underpin them. Audio electronics is a wonderfully complex area where many disciplines (e.g. acoustics, music, programming) overlap – this can be a very creative space to work in. Having said this, the perceived 'niche' aspect of audio can also lead to a misunderstanding of both its complexity and value – it is not uncommon to find linear solutions being proposed to musical problems!

Whatever direction your learning now takes, it is hoped that this book will provide you with some of the basic information and techniques that you will need to progress further – knowledge is always your most valuable resource. It is not possible to write a single text on all elements of audio, electronics or Arduino control, but this book has tried to bridge the gap between tutorials and textbooks. It is hoped that you will now continue to learn about audio electronics and Arduino control. If you persist, you may eventually develop your own ideas and solutions.

To quote the great Les Paul: '*No one has come up with a set of rules for originality – there aren't any*'.

APPENDIX 1

Self-study questions

The answers to all self-study questions in the book are provided here, indexed by chapter.

Chapter 1

Scales

***Q1**: What is the total current I_{tot}, as a result of adding 100mA and 3mA?*

***Q2**: What is the total voltage V_{out} as a result of adding 10μV and 7mV?*

$$I_1 = 100mA = 100 \times 10^{-3}V \qquad I_2 = 3mA = 3 \times 10^{-3}V$$

$$\text{Total Current, } I_{tot} = I_1 + I_2$$

$$I_{tot} = 100 \times 10^{-3} + 3 \times 10^{-3} = (100+3) \times 10^{-3} = \mathbf{103mA}$$

$$V_1 = 10\mu V = 10 \times 10^{-6}V \qquad V_2 = 7mV = 7 \times 10^{-3}V$$

$$\text{Total Voltage, } V_{tot} = V_1 + V_2$$

$$V_{tot} = 10 \times 10^{-6} + 7 \times 10^{-3} = 0.00001 + 0.007$$

$$= 0.00701V \approx \mathbf{7mV}$$

Q1 answer: The total current I_{tot} is 103mA.

Q2 answer: The total voltage V_{tot} is *approximately* 7mV.

Notes:

For Q1, the common unit of mA means both quantities can be added in those units – (100+3)mA. Q2 combines mV (10^{-3}) and μV(10^{-6}), so the units must be converted into a common factor (volts) to avoid introducing scaling errors. Notice that the 10μV value has little impact on the 7mV quantity – this becomes more apparent when working with resistance and reactance in chapter 6, where voltage and current may have different proportions across components.

Equations

Q1: *Add* $\frac{2}{3}+\frac{1}{4}$, *what is this fraction (as a voltage) of an Arduino 5V power supply?*

Q2: *What is the reciprocal of* $\frac{1}{3}$ *and what is the reciprocal of* $\frac{3}{4}$ *?*

$\frac{2}{3}+\frac{1}{4}=\frac{8}{12}+\frac{3}{12}=\frac{11}{12}$ *so* $\frac{11}{12}$ ***of a 5V** supply* $=\frac{11}{12}\times 5=\frac{11}{12}\times\frac{5}{1}=\frac{55}{12}=4.58\mathbf{V}$

The reciprocal of a value is $\frac{1}{answer}$ so $\frac{2}{3}$ is $\frac{1}{\frac{2}{3}}=\frac{3}{2}$ (swap the numerator and denominator)

Similarly, the reciprocal of $\frac{3}{4}=\frac{4}{3}$

Q1 answer: $\frac{2}{3}+\frac{1}{4}=\frac{11}{12}$, $\frac{11}{12}$ of a 5V supply is **4.58V**.

Q2 answer: The reciprocal of $\frac{2}{3}$ is $\frac{3}{2}$, the reciprocal of $\frac{3}{4}$ is $\frac{4}{3}$.

Notes:

These equations are designed to highlight the importance of finding a common denominator when working with fractions. The reciprocal of a fraction is simply the inversion of that fraction – as chapter 3 will show, the difficulty is often in remembering to calculate it.

Electrical charge

Q1: *What is current?*

Q2: *What is potential difference?*

Q3: *What is resistivity?*

Current is the flow of electrical charge over time – *measured in amperes (A)*
Potential difference is the difference in electric potential between two points – *measured in volts (V)*
Resistivity is the resistance of a material to the flow of charge - *measured in ohms (Ω)*

Notes:

These three statements define the fundamentals of electricity and electronics, and so it is important to understand them rather than simply memorize the phrase. In the case of current, the important thing to remember is that this is the flow of charge **over time** - charge is the amount of electrons involved. The term voltage is more commonly used for potential difference, but the units of the quantity do not adequately describe the difference between two net charges. Resistivity is perhaps the easiest of the three to remember, as it describes the resistance of a material to the flow of current. Resistors are used throughout electronics to control current, with a useful rule of thumb being that the absence of any resistance creates an open circuit - an uncontrolled path for current to flow down (which is never desirable).

Chapter 2

***Q1**: Fill in the two missing labels on this open loop system.*

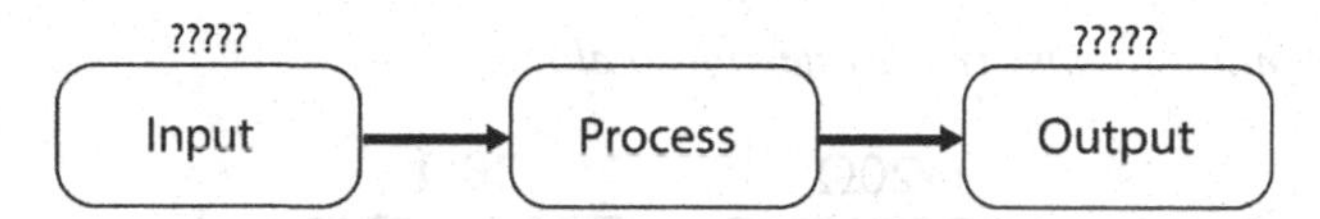

Answer: The missing labels are **sensor** (left label) and **actuator** (right label).

***Q2**: (a) What type of system is this? (b) What is the missing label on the lower right of the diagram?*

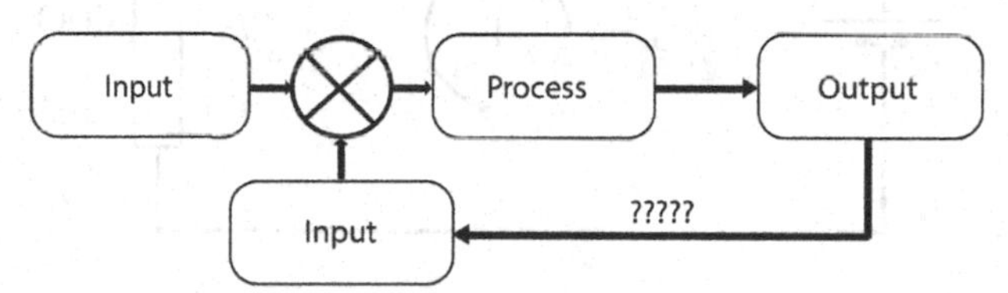

Answer: The missing label is feedback – this is a fundamental element in any system, and is crucial to the use of operational amplifiers in chapter 7.

***Q3**: Fill in the two missing labels on this open loop system.*

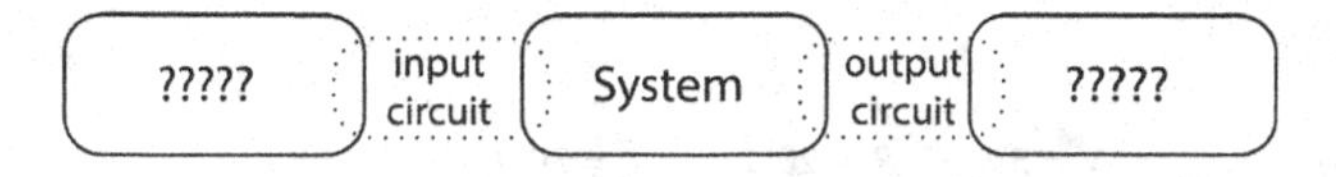

Answer: The missing labels are **source** (left label) and **load** (right label).

Notes:

For the first question, an open-loop system is shown where sensors convert physical quantities (e.g. sound, temperature, movement) into electrical signals, whilst actuators convert electrical signals into physical quantities. The second question shows a closed-loop system, where feedback is used to control the processes that the system carries out. The third question shows an open-loop system in terms of its source and load impedances, which define how different parts of a system 'see' each other in electronics terms. If you have an instrument input to an amplifier, the impedance (AC form of resistance - see chapter 6) of the instrument is the source the amplifier must be able to handle as an input. Similarly, the same amplifier may have to drive an actuator like a loudspeaker as a load - this explains why loudspeakers are usually rated by their impedance in ohms (Ω).

Chapter 3

Series circuits

Use the following schematic to answer the questions below:

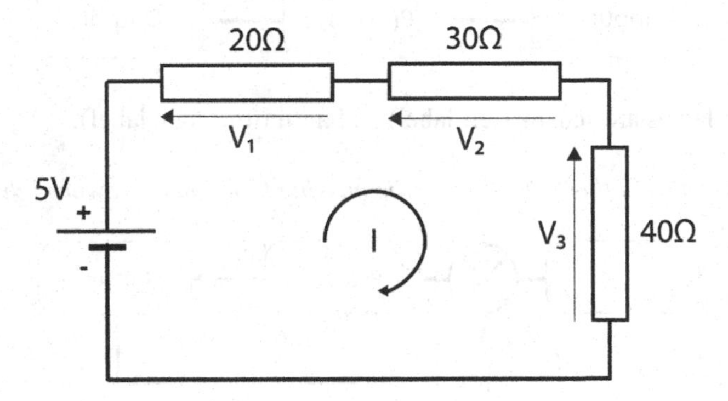

***Q1**: What is the total resistance in the circuit?*

***Q2:** What is the total current I_{tot}, in the circuit?*

***Q3**: What are the voltage drops across each of the three resistors?*

Answer:

$$V = 5V, \quad R_1 = 20\Omega, \quad R_2 = 30\Omega, \quad R_3 = 40\Omega$$

$$\textbf{Total Series Resistance, } R_{tot} = R_1 + R_2 + R_3 = 20 + 30 + 40 = \mathbf{90\Omega}$$

$$V = 5V, \quad R_{tot} = 90\Omega$$

$$\textbf{Total Current, } I = \frac{V}{R_{tot}} = \frac{5}{90} = 0.055 = \mathbf{55mA}$$

$$I = 55\text{mA}, \quad R_1 = 20\Omega, \quad R_2 = 30\Omega, \quad R_3 = 40\Omega$$

$$\mathbf{V_1} = IR_1 = 0.055 \times 20 = \mathbf{1.11V}$$

$$\mathbf{V_2} = IR_2 = 0.055 \times 30 = \mathbf{1.65V}$$

$$\mathbf{V_3} = IR_3 = 0.055 \times 40 = \mathbf{2.2V}$$

Note that the total voltage across all three resistors is ~4.96V – this is less than the 5V supply due to rounding errors.

Parallel circuits

Use the following schematic to answer the questions below:

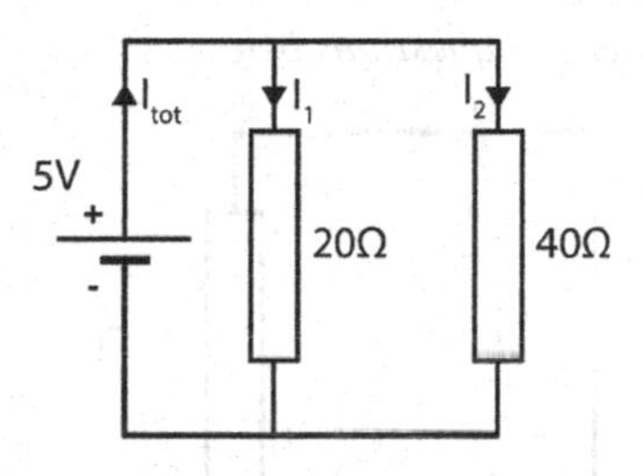

***Q4**: What is the total resistance in the circuit?*

***Q5:** What is the current I_{tot}, and the current in each branch of the circuit?*

Answer:

$$R_1 = 20\Omega, \quad R_2 = 40\Omega$$

$$\textit{Total Parallel Resistance}, \; \frac{1}{R_{tot}} = \frac{1}{R_1} + \frac{1}{R_2}$$

$$= \frac{1}{20} + \frac{1}{40} = \frac{2}{40} + \frac{1}{40}$$

$$\frac{1}{R_{tot}} = \frac{3}{40}$$

$$\mathbf{R_{tot}} = \frac{40}{3} = \mathbf{13.33\Omega}$$

$$V = 5V, R_{tot} = 13.33\Omega$$

$$\textbf{Total Current, I} = \frac{V}{R_{tot}} = \frac{5}{13.33} = 0.375 = \mathbf{37.5mA}$$

$$I_1 = \frac{V_S}{R_1} = \frac{5}{20} = \mathbf{0.25A}\ ,\ I_2 = \frac{V_S}{R_2} = \frac{5}{40} = \mathbf{0.125A}$$

As a check, the total current should be the sum of the two branch currents:

$$\text{Total Current, } I = 0.25 + 0.125 = 0.375 = 37.5mA$$

Voltage dividers

Use the following schematic to answer the questions below:

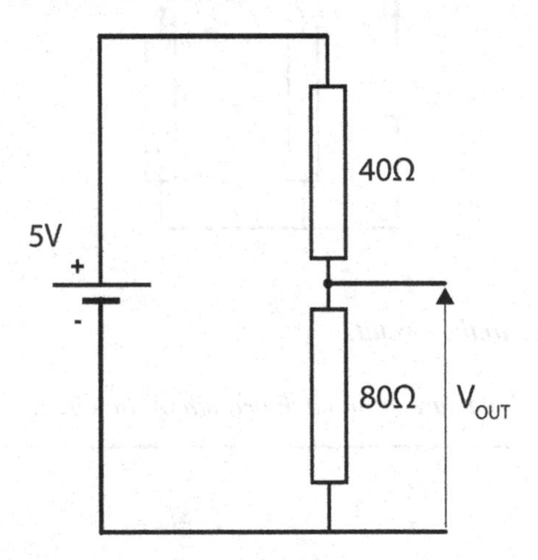

***Q6**: What is the voltage V_{out}?*

***Q7**: What is the voltage across the first resistor, V_{R1}?*

Answer:

$$V_S = 5V,\quad R_1 = 40\Omega,\quad R_2 = 80\Omega$$

$$V_{OUT} = V_S\left(\frac{R_2}{R_1 + R_2}\right) = 5\left(\frac{80}{40+80}\right) = 5 \times \frac{2}{3} = \frac{10}{3}$$

$$V_{OUT} = \mathbf{3.3\dot{3}V}$$

Knowing from KVL that all voltages in a loop sum to zero, the voltage across R1 must be:

$$V_s = V_{R1} + V_{OUT}$$

$$\therefore V_{R1} = V_S - V_{OUT} = \mathbf{5 - 3.33 \approx 1.67V}$$

Chapter 4

Programming concepts

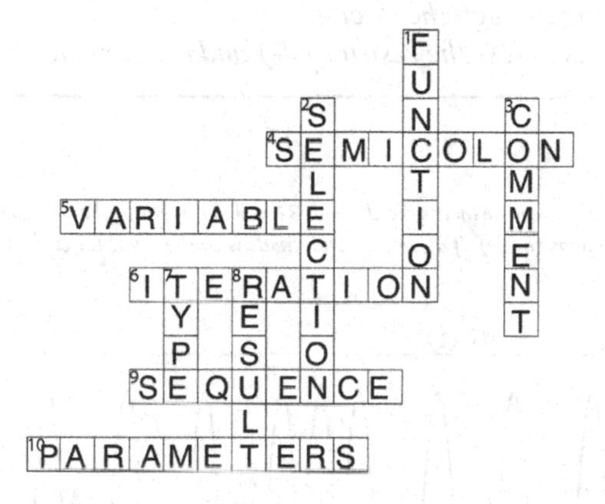

Across

(4) Every sequence instruction must be terminated with a SEMICOLON
(5) A named storage location for data VARIABLE
(6) An instruction where you repeat something more than once ITERATION
(9) An instruction that does something once SEQUENCE
(10) A function can receive input PARAMETERS

Down

(1) We can encapsulate programming instructions inside a FUNCTION
(2) An instruction where you make a choice based on a condition SELECTION
(3) Two forward slashes (//) indicate a COMMENT in your code
(7) When declaring a function, you must specify the return TYPE
(8) A function can return a RESULT

Chapter 6

LTspice simulation

***Q1**: Create an LTspice circuit to simulate the following schematic:*

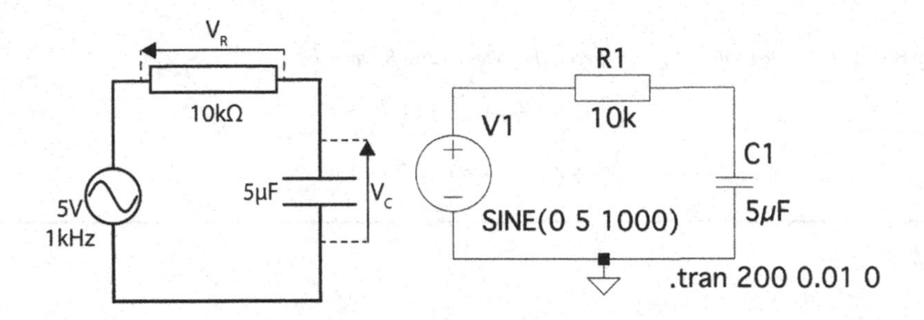

(a) *What is the current flowing through the circuit?*
(b) *What are the voltage drops across the resistor (V_R) and the capacitor (V_C)?*

Answer:

The current flowing through the circuit should be measured through R1 with a current probe in LTspice, to give the current in phase with the source (as resistors have no phase angle). If a comparison is made with the capacitor current (light grey trace below) then the difference in phase can also be seen:

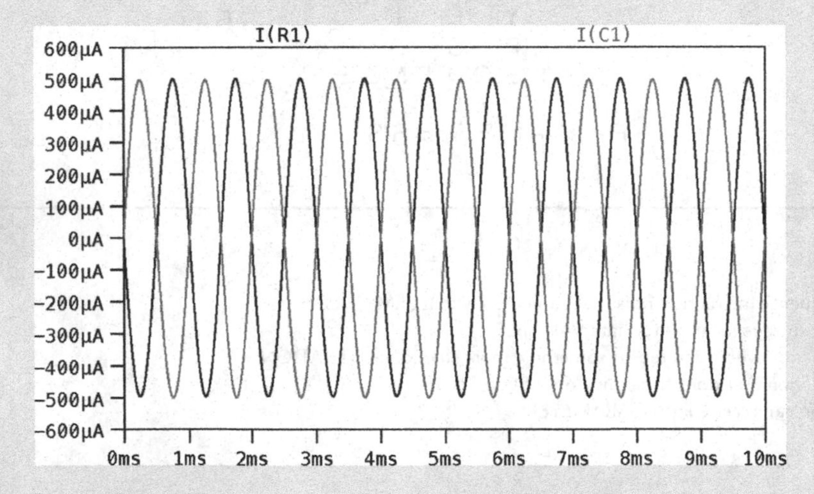

This book avoids extensive discussion of phase, but it is important to note the impact of the capacitor on the phase of the current in the circuit. Adding a cursor to the LTspice current trace will give a maximum simulated value of 497.37μA through R1 at the first peak in the waveform at a time of 256.94μs (microseconds) – this is close to the value of 0.5mA calculated in worked example 6.4.1.
The simulated voltage drops can also be measured, but a simultaneous trace shows the discrepancy in the magnitude of the capacitor voltage at this frequency:

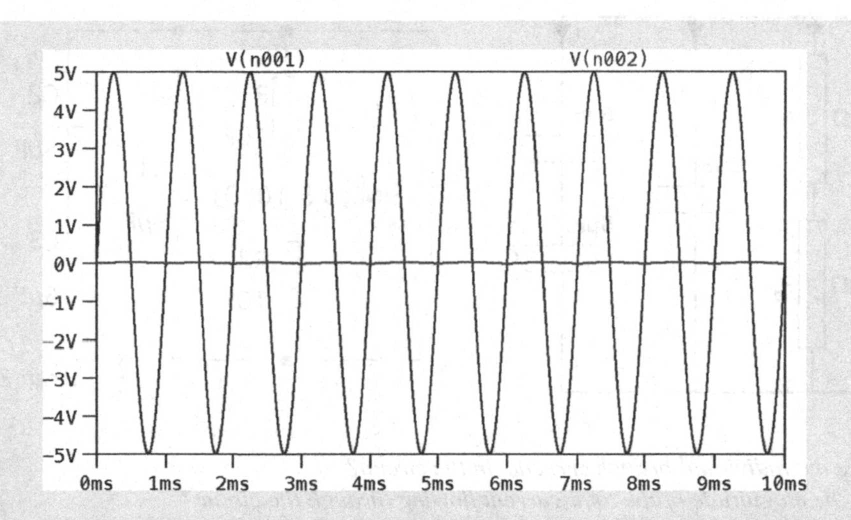

Adding a cursor to each trace gives a peak voltage of 4.99V through R1 at a time of 256.94µs, whilst the peak voltage through C1 is 31.63mV at a time of 491.32µs. This suggests a problem, as the calculated value from example 6.4.1 gives a capacitor voltage of 16mV. The problem here is the short analysis window of 0.01 seconds (used to show the frequency cycles of the input source), which does not give the simulation enough time to calculate the capacitor voltage once the circuit has stabilised at around 100ms. If the Spice Directive is changed to increase the analysis window to 100ms (.tran 200 0.1 0), the output trace shows how the capacitor voltage quickly reduces from 31mV to ~15.37mV:

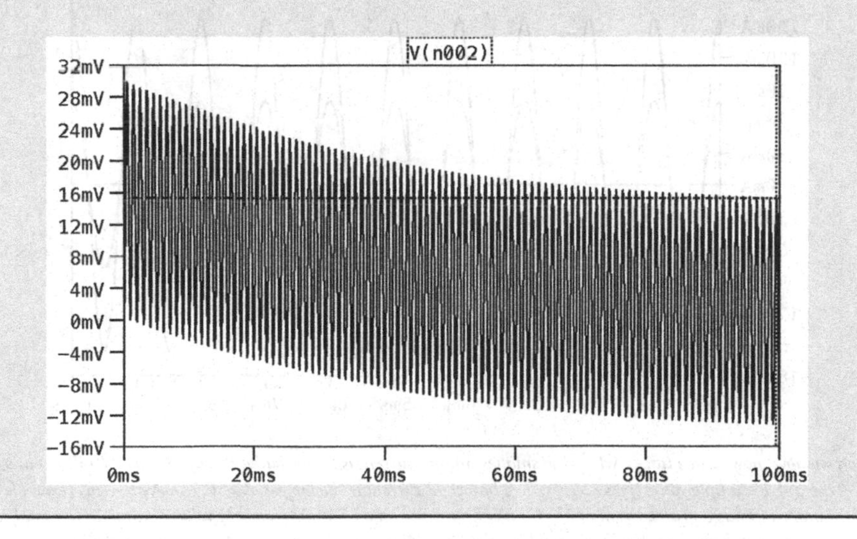

Q2: Create an LTspice circuit to simulate the following schematic:

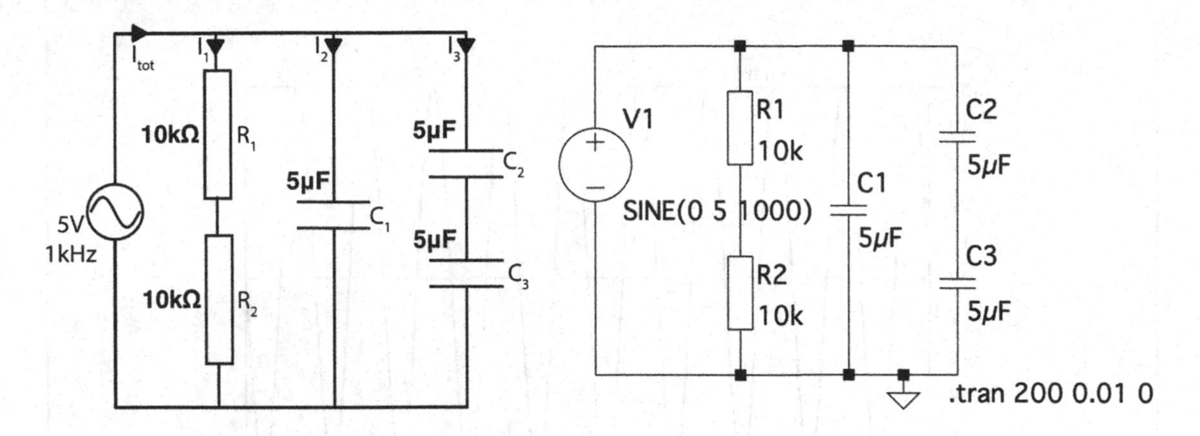

(a) What are the individual branch currents in the circuit?
(b) What is the magnitude of the total current flowing through the circuit?

Answer:

Measuring each branch in turn, LTspice will calculate the current through a component, so either R1 or R2 can be measured for branch 1 – and either C2 or C3 for branch 3:

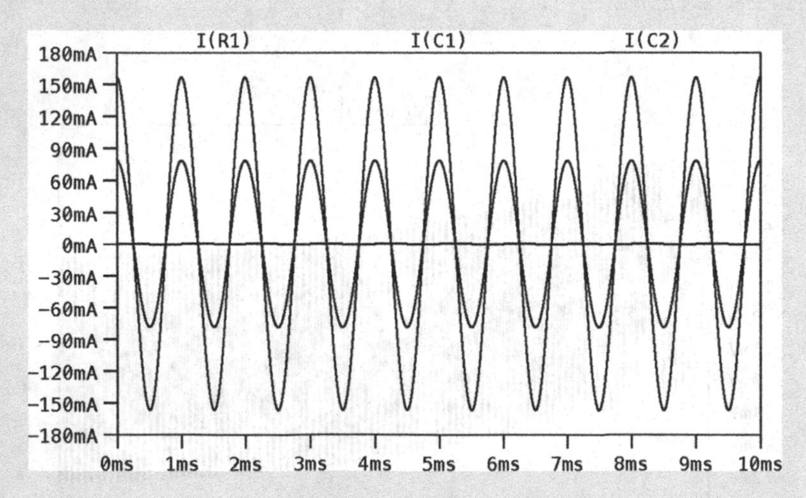

The simulation window shows three traces, where the smallest branch current is I1 measured through R1 as 249.73μA at a time of 247.17μs – the 6.4.1 example value gives 250μA. The branch I2 current is the largest at 156.73mA at ~1ms (example value 157mA), whilst branch I3 has a peak current of 78.37mA at the same time of ~1ms (example value 78.5mA). The total current in the circuit can be measured from the voltage source, which gives 235.47mA at a time of ~501μs – this is very close to the example value calculation of 235.5mA.

Chapter 7

Amplification

Q1: *Calculate the power gain A_p and decibel power gain a_p (dB) of an audio system with:*
(a) A DC input power of 2W and an output power of 4W.
(b) A DC input power of 1W and an output power of 5W.

Answer:

For an input power of 2W and output power of 4W:

$$A_{Pin} = 2W, \quad A_{Pout} = 4W$$

$$Power\ Gain, A_p = \frac{A_{Pout}}{A_{Pin}} = \frac{4}{2} = \mathbf{2}$$

$$Decibel\ Power\ Gain, a_p(dB) = 10log_{10}\left(\frac{A_{Pout}}{A_{Pin}}\right) = 10log_{10}(2) = 10 \times 0.301 \approx \mathbf{3dB}$$

For an input power of 1W and output power of 5W:

$$A_{Pin} = 1W, \quad A_{Pout} = 5W$$

$$Power\ Gain, A_p = \frac{5}{1} = \mathbf{5}$$

$$Decibel\ Power\ Gain, a_p(dB) = 10log_{10}(5) = 10 \times 0.6989 \approx \mathbf{7dB}$$

Q2: *Calculate the voltage gain A_v and decibel voltage gain a_v (dB) of an audio amplifier with:*

(a) An input signal of 100mV and an output signal of 1V.
(b) An input signal of 100mV and an output signal of 0.5V.

Answer:

For an input signal of 100mV and output signal of 1V:

$$A_{Vin} = 100mV, \quad A_{Vout} = 1V$$

$$Voltage\ Gain, A_V = \frac{A_{Vout}}{A_{Vin}} = \frac{1}{0.1} = \mathbf{10}$$

$$Decibel\ Voltage\ Gain,\ a_V(dB) = 20log_{10}\left(\frac{A_{Vout}}{A_{Vin}}\right) = 20log_{10}(10) = \mathbf{20dB}$$

For an input signal of 100mV and output signal of 0.5V:

$$A_{Vin} = 100mV, \qquad A_{Vout} = 0.5V$$

$$Voltage\ Gain, A_V = \frac{0.5}{0.1} = \mathbf{5}$$

$$Decibel\ Voltage\ Gain,\ a_V(dB) = 20log_{10}(5) = 20 \times 0.6989 \approx \mathbf{14dB}$$

Operational amplifiers

Use the following schematic for an inverting operational amplifier to answer the questions below:

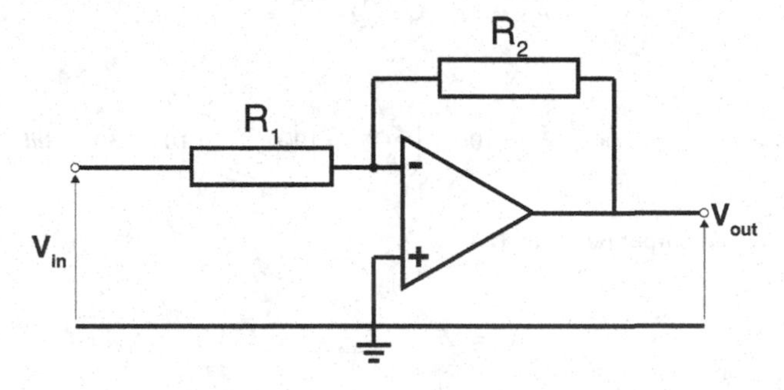

Q3: If $V_{in} = 100mV$ and $V_{out} = -0.5V$, what are the values of R_1 and R_2?

Q4: If $V_{in} = 100mV$ and $V_{out} = -2V$, what are the values of R_1 and R_2?

Answer:

$$\textbf{Inverting Amplifier Gain} = -\frac{R_2}{R_1} = \frac{V_{out}}{V_{in}} = \frac{-0.5}{0.1} = 5$$

For a gain of 5, R_2 should be five times larger than R_1.

Taking an arbitrary value of $\mathbf{R_1 = 10k\Omega}$, $\mathbf{R_2} = 5 \times R_1 = \mathbf{50k\Omega}$

$$\textbf{Inverting Amplifier Gain} = -\frac{R_2}{R_1} = \frac{V_{out}}{V_{in}} = \frac{-2}{0.1} = 20$$

Now a gain of 20 is needed, so $\mathbf{R_1 = 10k\Omega}$, $\mathbf{R_2} = 20 \times R_1 = \mathbf{200k\Omega}$

Chapter 8

First-order filters

Q1: An RC circuit is configured as a first-order low-pass filter, with a 1kΩ resistor and a 1μF capacitor. What is the cutoff frequency of the filter?

Answer:

$$\text{Cutoff Frequency, } fc = \frac{1}{2\pi RC} = \frac{1}{2\pi \times (1\times10^{3}) \times (1\times10^{-6})}$$

$$= \frac{1}{6.28\times1000\times0.000001} = \frac{1}{0.00628} \approx \mathbf{159Hz}$$

Q2: A CR circuit is configured as a first-order high-pass filter, with a 10kΩ resistor and a 10nF capacitor. What is the cutoff frequency of the filter?

Answer:

$$\text{Cutoff Frequency, } fc = \frac{1}{2\pi RC} = \frac{1}{2\pi \times (10\times10^{3}) \times (10\times10^{-9})}$$

$$= \frac{1}{6.28\times10{,}000\times0.00000001} = \frac{1}{0.000628} \approx \mathbf{1592Hz}$$

Q3: Design a low-pass filter with a cutoff of 200Hz – how close can you get to this value with a resistor of 470Ω and a single capacitor?

From appendix 3, the standard values available for a capacitor would range from 1μF to 4.7μF for E3–E12 series components. Starting with the E3 series will give an idea of the range involved:

1μF capacitor:

$$\text{Cutoff Frequency, } fc = \frac{1}{2\pi RC} = \frac{1}{2\pi \times (470) \times (1\times10^{-6})}$$

$$= \frac{1}{6.28\times470\times0.000001} = \frac{1}{0.00295} \approx \mathbf{339Hz}$$

2.2μF capacitor:

$$\text{Cutoff Frequency, } fc = \frac{1}{2p \times (470) \times (2.2\times10^{-6})}$$

$$= \frac{1}{6.28\times470\times0.0000022} = \frac{1}{0.00649} \approx \mathbf{154Hz}$$

4.7μF capacitor:

$$\textbf{Cutoff Frequency, } fc = \frac{1}{2\pi\times(470)\times(4.7\times10^{-6})}$$

$$= \frac{1}{6.28\times470\times0.0000047} = \frac{1}{0.01387} \approx \mathbf{72Hz}$$

The E3 values suggest that somewhere between 1μF and 2.2μF will be closer, so moving up to E6 the middle value is 1.5μF:

$$\textbf{Cutoff Frequency, } fc = \frac{1}{2\pi\times(470)\times(1.5\times10^{-6})}$$

$$= \frac{1}{6.28\times470\times0.0000015} = \frac{1}{0.00442} \approx \mathbf{226Hz}$$

This would be a reasonably close cutoff frequency for most purposes, but if greater accuracy is needed then the E48 series includes a value of 1.69μF:

$$\textbf{Cutoff Frequency, } fc = \frac{1}{2\pi\times(470)\times(1.69\times10^{-6})}$$

$$= \frac{1}{6.28\times470\times0.00000169} = \frac{1}{0.00498} \approx \mathbf{200Hz}$$

***Q4**: Design a high-pass filter with a cutoff of 3kHz – how close can you get to this value with a single resistor and capacitor of 0.1μF?*

The previous example included a cutoff frequency value of 339Hz for a combination of a 1μF capacitor with a 470Ω resistor, so it would be prudent to begin calculations by factoring in the 0.1μF value in this question:

1μF capacitor:

$$\textbf{Cutoff Frequency, } fc = \frac{1}{2\pi RC} = \frac{1}{2\pi\times(470)\times(0.1\times10^{-6})}$$

$$= \frac{1}{6.28\times470\times0.0000001} = \frac{1}{0.00029} \approx \mathbf{3388Hz}$$

Looking at the structure of the equation, the reciprocal relationship means that increasing the 2πRC value will decrease the cutoff frequency. Again beginning with the E3 series, the next value in the range will be 1kΩ:

$$\textbf{Cutoff Frequency, } fc = \frac{1}{2\pi\times(1000)\times(0.1\times10^{-6})}$$

$$= \frac{1}{6.28\times1000\times0.0000001} = \frac{1}{0.00063} \approx \mathbf{1592Hz}$$

This is too low a frequency, so go to the next series (E6) to find a resistor between 470Ω and 1kΩ – this gives a value of 680Ω:

$$\textbf{Cutoff Frequency, } fc = \frac{1}{2\pi\times(680)\times(0.1\times10^{-6})}$$

$$= \frac{1}{6.28\times680\times0.0000001} = \frac{1}{0.00043} \approx \mathbf{2342Hz}$$

This is still too low, so moving to E12 gives a value of 560Ω:

$$\textbf{Cutoff Frequency, } fc = \frac{1}{2\pi\times(560)\times(0.1\times10^{-6})}$$

$$= \frac{1}{6.28\times560\times0.0000001} = \frac{1}{0.00035} \approx \mathbf{2843Hz}$$

This value is closer than the first attempt of 3388Hz, but given that 3kHz is in an area of human hearing that is particularly sensitive (see section 8.1 for equal loudness curves) the calculation should perhaps aim for closer. This means moving to E24 and a value of 510Ω:

$$\textbf{Cutoff Frequency, } fc = \frac{1}{2\pi\times(510)\times(0.1\times10^{-6})}$$

$$= \frac{1}{6.28\times510\times0.0000001} = \frac{1}{0.00032} \approx \mathbf{3122Hz}$$

This is closer again, but instead of 157Hz less the value is now 122Hz more! The E48 series includes a value of 536Ω:

$$\textbf{Cutoff Frequency, } fc = \frac{1}{2\pi\times(536)\times(0.1\times10^{-6})}$$

$$= \frac{1}{6.28\times536\times0.0000001} = \frac{1}{0.00033} \approx \mathbf{2971Hz}$$

This would seem to be a close enough value to 3kHz for most practical purposes – to get any closer, the E192 series could be used, which includes a value of 530Ω (3003Hz)!

Notes:

The values chosen in questions Q1 and Q2 are intended to show the influence of scaling factors on calculations like cutoff frequency. Chapter 1 showed how scaling factors can become confusing, and it is important to note that while the resistor in Q2 has now increased by a factor of 10 from 1kΩ to 10kΩ, the change in the capacitor from 1μF to 10nF represents a factor of 100 decrease. The reciprocal relationship in the equation means that decreasing the $2\pi RC$ value will increase the cutoff frequency, so although the resistor increases by a factor of 10, the capacitor decreases by a factor of 100 – thus the cutoff frequency increases by a factor of 10 between Q1 and Q2.
The calculations performed in Q3 and Q4 may seem tedious (as they are!) and there are online filter calculators available for many filter types (including higher-order examples) that will quickly perform these calculations for you (a spreadsheet can also be used to quickly 'plug in' numerous values). The questions are intended to show how a pragmatic approach to choosing component values can get close enough to the required value in many instances. Cost is always a major factor in electronics, and when you study commercial schematics for well-known audio products it can be fascinating to see where compromises have (and have not) been made. Guitar effects pedal schematics (notably by Boss) contain some very impressive solutions that are both cost-effective and musically valid – studying the filter designs used in such circuits would be a valuable extension of the basic concepts covered in this book.

A final observation on the practicalities of audio filter design when considered relative to the purely theoretical exercise of calculating first-order RC cutoff values: the simple (and cheaper) option for the filter above would seem to be adding a second series resistor (such as a 68Ω resistor from the E6 series) to the first 470Ω value to bring the overall resistance value to ~538Ω and achieve a cutoff frequency of 2958Hz. This would be mathematically valid, but every component added to a circuit will both introduce noise and attenuate the signal. In addition, the cost of the second resistor must now be added to the build and in the highly competitive audio electronics market, the price of a piece of equipment must (usually) be kept to a minimum if it is to have any chance of market success. These factors are not important when learning about audio electronics, but when moving into the practical world of equipment sales they become critical to the success of a new product.

APPENDIX 2

AC equation derivations (chapter 6)

Deriving charge over time

Chapter 1 showed that current is the flow of charge at a specific point for a specific length of time – this can now be defined mathematically as the **rate of change of charge over time**:

$$\textit{Current}, I = \frac{dQ}{dt} \tag{6.1}$$

I is the current in amperes (A)

dQ is the rate of change of charge in coulombs (C)

dt is the rate of change of time in seconds (s)

This equation may initially seem a little unusual when compared to other DC theory terms (like Ohm's Law) that use more simple scalar relationships between quantities. With sine waves for AC analysis, the use of differentiation and integration is part of the standard mathematical mechanisms for manipulating them. AC circuits are **time-varying circuits**, and so changes in voltage, current and impedance must be analysed relative to how they change over time. The concept of rate of change is not discussed in detail in this book, but the substitution of the rate of change of charge over time ($\frac{dQ}{dt}$) for current (I) can be used to help derive some of the formulae relating to capacitors.

Deriving an RC time constant

Knowing the equation for capacitance and also knowing how a capacitor's voltage and current changes over time, a time constant (τ) can be defined that describes the rate of change in a resistor/capacitor (RC) circuit (Appendix 2 Figure 1).

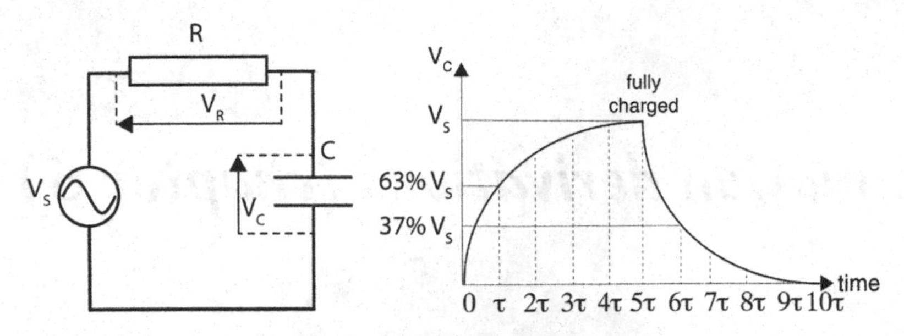

Appendix 2 Figure 1 Time constant in an RC circuit – from chapter 6, Figure 6.22. *The diagram shows the charge/discharge graph for the series combination of a resistor and capacitor. The values of both components dictate how long it takes for the capacitor to reach ~63% of its final charge level, which is known as the time constant (τ).*

The time constant (**τ**) denotes how long it takes for the capacitor to reach ~63% of the supply voltage, which is derived from the fact that the charging curve is exponential in shape – where $\frac{1}{e}$ approximates to 0.37, which is the reduction from maximum percentage in each time constant (100 − 37). It will take a total of 5 time constants to reach full charge, and a further 5 for the capacitor to discharge again – 6**τ** defines when the discharge cycle reaches ~37% (100 − 63). The mathematical proof for this exponential rise and fall involves solving the sum of the voltages (V_R and V_C) in the circuit as a differential equation, which is a more advanced mathematical process than needed for an introductory text. Nevertheless, this relationship can be stated to show how the time constant **τ** can be derived by first returning to Kirchoff's Voltage Law from chapter 3:

Kirchoff's Voltage Law

The sum of all voltages around any closed loop in a circuit is equal to zero

The current equation from 6.1 can be rearranged with the equation for capacitor charge ($Q = CV$) because a capacitor varies over time. The differential of charge is current, so **current must also be the differential of our capacitor voltage** term in the equation (capacitance C is constant):

$$\text{Charge, } Q = CV_C, \quad \text{Current, } I = \frac{dQ}{dt} = C\frac{dV_C}{dt}$$

Current is constant in a series circuit, so **Ohm's Law can be used to factor in the resistor.** This can be expressed in terms of the capacitor voltage to build an equation relating R and C:

$$\text{Current, } I = C\frac{dV_C}{dt}, \quad \text{Resistor Voltage, } V_R = IR = RC\frac{dV_C}{dt}$$

Now use **Kirchoff's Voltage Law (KVL)** to define all the voltages in the circuit (V_S, V_R *and VC*):

$$\text{Supply Voltage, } V_S = V_R + V_C = RC\frac{dV_C}{dt} + V_C$$

V_C has not been changed to allow a **first-order differential equation** to be built, which can be solved by **rearranging the terms**:

$$V_S - V_C = RC\frac{dV_C}{dt} \quad \rightarrow (V_S - V_C)dt = RCdV_C \quad \rightarrow dt = \frac{RCdV_C}{(V_S - V_C)}$$

$$\therefore \frac{dt}{RC} = \frac{dV_C}{V_S - V_C}$$

With differential terms on both sides, **integrate the entire equation** to remove them:

$$-ln(V_S - V_C) = \frac{t}{RC} + K$$

Where the term K is the constant of integration. Now **remove the natural logarithm** by changing the equation to exponential form (express the right-hand side as a power of e). Also, **split the integration constant term** out (e^K) to make it easier to solve:

$$V_S - V_C = e^{-\frac{t}{RC}+K} = e^{-\frac{t}{RC}} \times e^K$$

All the terms of the equation have been derived, apart from **the value of the integration constant (K)**. To do this, take a known time (time $t = 0$) and use the fact that the capacitor will not have begun to charge to define the value for capacitor voltage ($V_C = 0$) at that point:

$$V_S - 0 = e^{-\frac{0}{RC}} \times e^K = 1 \times e^K \quad \therefore e^K = V_S$$

Substitute this into the full equation to state the full relationship between V_C and V_S:

$$V_S - V_C = e^{-\frac{t}{RC}} \times e^K = V_S e^{-\frac{t}{RC}} \quad \rightarrow \quad V_C = V_S - V_S e^{-\frac{t}{RC}}$$

$$\therefore V_C = V_S\left(1 - e^{-\frac{t}{RC}}\right)$$

The RC term dictates how much the exponent increases over time, and thus it is known as the **time constant (τ)**:

$$V_C = V_S\left(1 - e^{-\frac{t}{\tau}}\right)$$

Deriving capacitor voltage

Capacitor voltage varies over time, based on the time constant. Equation 6.1 can be used to show that current is equivalent to the **derivative** (rate of change) of charge over time, and then the terms of this equation can be rearranged to define charge (Q) as the **integral** of current over time:

From equation 6.7 for current as the derivative of charge over time, **integrate** current to define charge:

$$\textit{if Current}, I = \frac{dQ}{dt}, \quad \textit{then } dQ = Idt, \quad \therefore \boldsymbol{Q = \int Idt}$$

Then **rearranging** equation 6.7 for capacitance:

$$\textit{Capacitance}, C = \frac{Q}{V} \therefore \boldsymbol{V = \frac{Q}{C}}$$

Substituting with the previous term for charge:

$$\textit{Voltage}, V = \frac{Q}{C} = \frac{\int Idt}{C} = \boldsymbol{\frac{1}{C}\int Idt}$$

Using the value of **AC current** from equation 6.3:

$$\textit{Instantaneous Current}, I = \boldsymbol{I_p \sin\omega t}$$

Substituting with the previous term for voltage:

$$\textit{Voltage}, V = \frac{1}{C}\int \boldsymbol{I_p \sin(\omega t)}\,dt$$

Now use **integral substitution** to solve $\int I_p \sin(\omega t)\,dt$:

$$\textit{let } u = \omega t, \quad du = \omega dt, \quad \therefore \boldsymbol{dt} = \frac{du}{\omega} = \boldsymbol{\frac{1}{\omega}du}$$

$$\int I_p \sin(\omega t)\,dt = \int I_p \sin(\boldsymbol{u}) \times \frac{1}{\omega}\boldsymbol{du}$$

Split the terms ($\frac{1}{\omega}$, I_p) out from the integration:

$$\int I_p \sin(\omega t)\,dt = I_p \times \frac{1}{\omega} \times \int \sin(u)\,du = \frac{\boldsymbol{I_p}}{\omega}\int \sin(u)\,du$$

The **integral of sin(u) is −cos(u)** and the **du term is removed** by integration – then **substitute** ωt back in for u:

$$\frac{I_p}{\omega}\int \sin(u)\,du = \frac{I_p}{\omega} \times -\cos(u)$$

$$\int I_p \sin(\omega t)\,dt = -\frac{I_p}{\omega}\cos(\omega t)$$

Now **substitute** this expression for AC current into the voltage equation:

$$V = \frac{1}{C}\left(-\frac{I_p}{\omega}\cos(\omega t)\right)$$

Now state the equation for capacitor voltage in terms of current and capacitance:

$$\textit{Capacitor Voltage}, V_C = -\frac{I_p}{\omega C}\cos\omega t \quad (6.2)$$

I_p is the instantaneous current in amperes (A)

C is the capacitance in farads (F)

ω is the angle of the sine wave signal in radians (rad)

t is the time in seconds (s)

Deriving capacitive reactance

A capacitor will react at different frequencies, which is defined as its **reactance**. The capacitance value must be fixed (the capacitor cannot expand or contract in size!) and the frequency of the input signal does not change (so $\omega = 2\pi f$ is fixed) – thus the only terms in the equation that can vary are voltage, current and time. Ignoring the effect of time for now, an increase in the peak current will also increase the peak voltage across the capacitor (the peak value is not changed by time). This is no different from Ohm's Law for a fixed resistance, where the same linear relationship between voltage and current exists:

$$\textit{Voltage}, V = IR \quad (6.3)$$

I is the current in amperes (A)

R is the resistance in ohms (Ω)

In chapter 3 (equation 3.2), this equation was rearranged to define resistance in terms of voltage and current ($R = \frac{V}{I}$). Ignoring the impact of time, the same approach can be taken to determine the **reactance of a capacitor as the ratio of peak voltage over current**:

From equations 6.3 and 6.9:

$$\textit{Current}, I = I_p \sin\omega t, \quad \textit{Voltage}, V_C = -\frac{I_p}{\omega C}\cos\omega t$$

To ignore time, use the peak values for voltage and current (i.e. removing the sine/cosine component):

$$\textit{Peak Current}, I_p, \quad \textit{Peak Voltage}, V_p = \frac{I_p}{\omega C}$$

$$\therefore \text{Capacitive Reactance}, X_C = \frac{\text{Peak Voltage}, V_P}{\text{Peak Current}, I_P} = \frac{\frac{I_P}{\omega C}}{I_P}$$

The peak current terms cancel out to state the general equation for capacitive reactance (X_C):

$$\textit{Capacitive Reactance}, X_C = \frac{1}{\omega C} \tag{6.4}$$

X_C is the reactance of the capacitor in ohms (Ω)

C is the capacitance in farads (F)

ω is the angle of the sine wave signal in radians (rad)

Looking at this equation, the reactance of a capacitor is effectively defined by its capacitance and also frequency (recall that $\omega = 2\pi f$). This means that **capacitors vary their reactance with frequency**, which is a crucial point for audio electronics.

Deriving series impedance magnitude

The previous section showed that a capacitor has a reactance in ohms (Ω) that varies with the frequency of the input signal voltage applied to it. If frequency is ignored then reactance is the ratio of voltage over current ($X_C = \frac{V}{I}$), in much the same way as it is for resistance ($R = \frac{V}{I}$). This relationship can be used to determine the total **impedance** (symbol **Z**) for a series circuit that contains both a resistor and capacitor (known as an RC circuit). To do this, define each component voltage and its associated phase angle using a **phasor diagram** (Appendix 2 Figure 2).

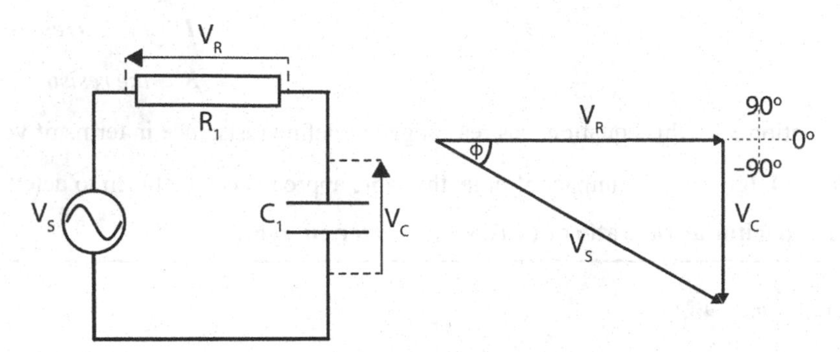

Appendix 2 Figure 2 RC series circuit phasor diagram – from chapter 6, Figure 6.28. *The left-hand diagram shows a resistor and capacitor in series, connected to an AC voltage supply (V_S). The right-hand diagram shows the resulting phasor diagram for this circuit, where the voltages in the circuit are combined to include the phase angle of each component. As it does not vary with frequency the resistor has phase angle 0°, whilst the capacitor has a phase angle of −90°.*

The phasor diagram shown in this figure is a common way of analysing an AC circuit by combining component voltages (or currents) with their associated AC phase angles. Phasors can be used to represent the phase component of a capacitor to accurately **model either the voltage levels or current flow** within an RC circuit. Thus, a phasor diagram can show how to derive the **overall impedance (Z)** of the circuit. The resistor voltage (V_R) has a phase angle of 0°, as resistive components do not vary with frequency. The previous section derived the voltage across a capacitor as $V_C = -\frac{I_p}{\omega C}\cos\omega t$ and the negative cosine component of this equation explains why the capacitor voltage in the diagram has a phase angle of −90°. The overall voltage (V_S) of the circuit forms the hypotenuse of this triangle, defining the **magnitude of this voltage** using trigonometry as follows:

From trigonometry, the **square of the hypotenuse is equal to the sum of squares** of the other two sides, so: $V_S^2 = V_R^2 + V_C^2 \therefore V_S = \sqrt{V_R^2 + V_C^2}$

(also determine the phase angle of this voltage, $\phi = \tan^{-1}\frac{V_C}{V_R}$)

Knowing the total voltage in the circuit, use Ohm's Law to substitute for current and **impedance (Z)**:

$$V_S = IZ = \sqrt{V_R^2 + V_C^2}$$

$$IZ = \sqrt{IR^2 + IX_C^2} = I\sqrt{R^2 + X_C^2}$$

Divide throughout by current (I) to state the equation for the magnitude of the total circuit impedance (Z) in an RC series circuit:

$$\textit{Impedance Magnitude, } Z = \sqrt{R^2 + X_C^2} \quad (6.5)$$

***R** is the resistance in ohms (Ω)*

X_C is the reactance of the capacitor in ohms (Ω)

This equation is very useful, allowing a series AC circuit to be analysed in much the same way as DC series circuits were in chapter 3.

Deriving series capacitance

Returning to Kirchoff's Voltage Law from chapter 3:

Kirchoff's Voltage Law

The sum of all voltages around any closed loop in a circuit is equal to zero

From KVL, the total voltage must equal the sum of the voltage drops across all the resistors. In chapter 3 (equation 3), this law proved that the total resistance ($R_{tot} = \frac{Vtot}{I}$) is the sum of all the resistances in a series circuit:

$$V_{tot} = V_1 + V_2 = IR_1 + IR_2 \quad \rightarrow \frac{V_{tot}}{I} = R_1 + R_2 \quad \therefore R_{tot} = \frac{V_{tot}}{I} = R_1 + R_2$$

In this case, the voltage across each component is added to equal the total voltage in the circuit – which is the same thing that happens when capacitors are connected together in series (Appendix 2 Figure 3).

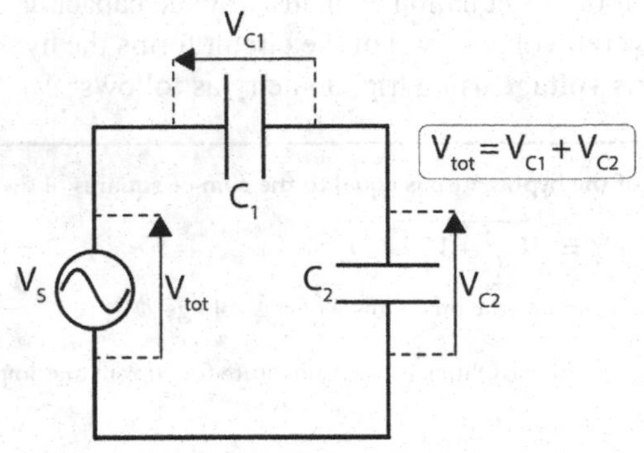

Appendix 2 Figure 3 Series capacitor circuit. *The diagram shows two capacitors connected in series, where the sum of the voltages across each capacitor (V_{C1} and V_{C2}) is equivalent to the total voltage (V_{tot}) for the circuit. An AC supply is shown in this diagram, but time is not included in magnitude calculations.*

This series circuit shows the same sum of voltages used to derive total series resistance in chapter 3, where time is omitted from the calculations to determine the magnitude only (effectively at the positive peak of the input signal). For capacitors in series, equation 6.7 can be used to define the total capacitance for the circuit in terms of the overall charge and voltage in the circuit ($C_{tot} = \frac{Q_{tot}}{V_{tot}}$).

Rearrange the terms of this equation to define each capacitor in terms of its voltage to perform the same sum of the voltage drops calculation for a capacitor series circuit:

$$V_{tot} = \frac{Q_{tot}}{C_{tot}}, \quad V_{C1} = \frac{Q_1}{C_1}, \quad V_{C2} = \frac{Q_2}{C_2}$$

$$V_{tot} = V_{C1} + V_{C2}$$

$$\therefore \frac{Q_{tot}}{C_{tot}} = \frac{Q_1}{C_1} + \frac{Q_2}{C_2}$$

Cancel out the charge (Q) terms, to derive the equation for total series circuit capacitance:

$$\textit{Total Series Capacitance}, \frac{1}{C_{tot}} = \frac{1}{C_1} + \frac{1}{C_2} + \ldots \frac{1}{C_n} \qquad (6.6)$$

C_{tot} is the total capacitance in ohms (Ω)

n *is the number of capacitors in series*

Deriving parallel capacitance

For parallel capacitors, use Kirchoff's Current Law from chapter 3 to calculate the total parallel capacitance:

Kirchoff's Current Law

The charge entering a junction is equal to the charge leaving that junction

Kirchoff's Current Law (KCL) states that the total charge (or current) entering any junction (or node) in a circuit will be the same as the current leaving that junction. Recall from chapter 3 that there will always be the same level of charge present in a circuit – free electrons cannot enter or leave other than through a voltage source. This principle can be used to derive the equation for total capacitance when capacitors are connected together in parallel (Appendix 2 Figure 4).

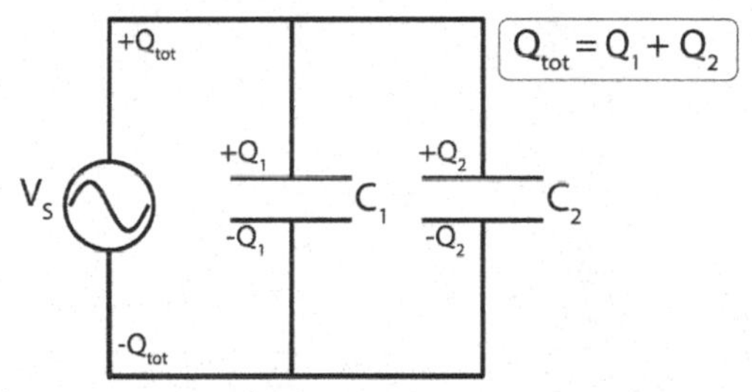

Appendix 2 Figure 4 Parallel capacitor circuit. *The diagram shows two capacitors connected in parallel, where the total charge (Q_{tot}) is the sum of all branch charges (Q_1 and Q_2) in the circuit. Again, we show an AC supply but we do not include time in our magnitude calculations.*

This parallel circuit has two capacitors that draw current from the AC supply (V_S), where the sum of all branch charges is equal to the total charge in the circuit (Q_{tot}). Although this is an AC supply, if time is again ignored to focus on the magnitude of the capacitance, the total charge in the circuit is equal to the sum of the charge across each capacitor:

Rearrange equation 6.7 in terms of charge: $Q_{tot} = C_{tot}V_{tot}$
From KCL, state that the total charge in the circuit is equal to the sum of all other charges:

$$Q_{tot} = Q_1 + Q_2$$

Now replace each charge with its (capacitance × voltage) equivalent:

$$Q_{tot} = C_{tot}V_{tot} = C_1V_1 + C_2V_2$$

Cancel out the voltage (V) terms to derive the equation for total parallel circuit capacitance:

$$\textit{Total Parallel Capacitance, } C_{tot} = C_1 + C_2 + \ldots C_n \tag{6.7}$$

C_{tot} is the total capacitance in ohms (Ω)

***n** is the number of capacitors in series*

APPENDIX 3

Standard component values

Although in theory any resistive or capacitive value can be obtained, in practice manufacturing costs require the use of standard values to maintain consistency in production. In 1952, the International Electrotechnical Commission (IEC) began a process of defining standard values for components known as the E series that is currently updated to IEC 60063:2015. The E series specifies how many values within each linear decade (1–10, 10–100, 100–1000 and so on) are available, and is based on a (broadly) logarithmic scale centred on a magnitude of 3. We recall from chapter 6 that a logarithmic scale is based on orders of magnitude, and for the E series values of 3, 6, 12, 24, 48, 96 and 192 are used to define the number of values available within each linear decade.

For example, the E3 series divides each decade into three values of 1.0, 2.2 and 4.7, whilst the E6 series divides it further into six values of 1.0, 1.5, 2.2, 3.3, 4.7, 6.8. This leads to the following table of common component values (Appendix 3 Table 1):

Appendix 3 Table 1 IEC E series component values. ***The table shows the available values in each linear decade for the series E3 to E24. Higher E series values have greater accuracy (E48 2%, E96 1%), but are more expensive.***

Series	Value																								Tolerance
3	1								2.2								4.7								>20%
6	1				1.5				2.2				3.3				4.7				6.8				20%
12	1		1.2		1.5		1.8		2.2		2.7		3.3		3.9		4.7		5.6		6.8		8.2		10%
24	1	1.1	1.2	1.3	1.5	1.6	1.8	2	2.2	2.4	2.7	3	3.3	3.6	3.9	4.3	4.7	5.1	5.6	6.2	6.8	7.5	8.2	9.1	5%

Although there are far more values available in the E48 and E96 series, these are highly specific (and hence more expensive). A better approach (particularly when beginning in electronics) is to start with the lowest series possible and determine whether a resistor is available that can be used, working upwards if none can be found. This can save a lot of time when performing gain and filter cutoff calculations, where an absolute value may not be needed in practice. An additional factor is the accuracy of the component relative to the value specified, where the table above shows how different series are linked to specific accuracy levels (known as tolerance) for the component. The table shows that while the E6 series gives a tolerance of (+/−20%) the accuracy increases for each series – E24 has a tolerance of (+/−5%), and beyond the listing are E48 (+/−2%) and E96 (+/−1%). As an example, any series could be used for a 2.2kΩ resistor but the actual value could be anywhere between 1760Ω (2200 − 440) to 2640Ω (2200 + 440) for an E6 component. In some cases (particularly digital electronics) the tolerance of the component is not critical to the effective operation of the circuit, so a pull-up resistor that serves to provide a reference voltage does not need to be of an absolute value to allow the microcontroller to distinguish between LOW and HIGH. Nor does a limiting resistor (such as the 1MΩ resistor in parallel with the piezo in the chapter 5 drum trigger project) need to be a of a specific value – it is there to protect other circuit components from damage due to voltage spikes.

Having said this, the need for more accurate component values becomes more obvious in analogue circuits such as amplifiers (to set gain) and filters (to define the cutoff frequency). In cases such as these, an accurate value is crucial to the effective operation of the circuit and so higher-tolerance components are used (which increases costs). In addition, the example questions at the end of chapter 8 show the limitations of working with standard components when a 470Ω resistor is combined with a 1.5μF capacitor in an RC filter to give a cutoff frequency of 226Hz. For audio circuits, the difference of 26Hz from the required value of 200Hz would not normally be important enough to merit a more expensive component unless the specific frequency of 200Hz was being measured (as would happen at much higher frequencies in communications circuits). For this reason, this introductory text works with lower series standard value components when building project circuits to keep costs down. It is important to note that a margin of error has been introduced as a result – a margin that is acceptable for our purposes, but may not be sufficient for more audiophile applications.

Index

Printed in the United States
by Baker & Taylor Publisher Services

Printed in the United States
by Baker & Taylor Publisher Services